AF598476

Gmelin Handbook of Inorganic Chemistry

8th Edition

The following Gmelin Formula Index volumes have been published up to now:

Formula Index

Volume 1	Ac–Au
Volume 2	B–Br_2
Volume 3	Br_3–C_3
Volume 4	C_4–C_7
Volume 5	C_8–C_{12}
Volume 6	C_{13}–C_{23}
Volume 7	C_{24}–Ca
Volume 8	Cb–Cl
Volume 9	Cm–Fr
Volume 10	Ga–I
Volume 11	In–Ns
Volume 12	O–Zr
	Elements 104–132

Formula Index 1st Supplement

Volume 1	Ac–Au (present volume)

Gmelin Handbook of Inorganic Chemistry

8th Edition

Gmelin Handbuch der Anorganischen Chemie

Achte, völlig neu bearbeitete Auflage

Prepared and issued by

Gmelin-Institut für Anorganische Chemie
der Max-Planck-Gesellschaft
zur Förderung der Wissenschaften

Director: Ekkehard Fluck

Founded by	Leopold Gmelin
8th Edition	8th Edition begun under the auspices of the Deutsche Chemische Gesellschaft by R. J. Meyer
Continued by	E.H.E. Pietsch and A. Kotowski, and by Margot Becke-Goehring

Springer-Verlag Berlin · Heidelberg · New York · Tokyo 1983

Gmelin-Institut für Anorganische Chemie
der Max-Planck-Gesellschaft zur Förderung der Wissenschaften

Gmelin Handbook of Inorganic Chemistry

8th Edition

INDEX

Formula Index

1st Supplement Volume 1

Ac–Au

AUTHORS Marie-Louise Gerwien, Helga Hartwig, Uwe Nohl, Hans-Jürgen Richter-Ditten, Paul Velić, Rudolf Warncke

CHIEF EDITOR Rudolf Warncke

Springer Verlag Berlin · Heidelberg · New York · Tokyo 1983

The volumes of the Gmelin Handbook are evaluated from 1974 up to the end of 1979

Library of Congress Catalog Card Number: Agr 25-1383

ISBN 3-540-93481-2 Springer-Verlag, Berlin · Heidelberg · New York · Tokyo
ISBN 0-387-93481-2 Springer-Verlag, New York · Heidelberg · Berlin · Tokyo

Typesetting, printing and bookbinding: Universitätsdruckerei H. Stürtz AG, Würzburg

Foreword

The Gmelin Formula Index published between 1975 and 1980 covered all volumes of the Eighth Edition of the Gmelin Handbook that had appeared up to the end of 1974 in the case of Main Volumes and up to the end of 1973 in the case of Supplement Volumes. The Gmelin Formula Index, First Supplement, continues from there and covers the handbook volumes published up to the end of 1979.

This First Supplement will consist of eight volumes, which will appear at intervals of four to six months. The basic structure of the Formula Index has been fully retained in the First Supplement: The index lists all elements, compounds, ions, and systems having definite composition that are described in the handbook text. The first column gives the empirical formula, while the second gives the conventional formula. The third column lists the pertinent pages. (The details are available in "Instructions for the Formula Index", on the next pages.)

This First Supplement was prepared and printed with extensive use of computers. In the future this will allow publication of cumulative indexes. The procedures were worked out together with the Technical Section of the *Gesellschaft für Information und Dokumentation mbH (GID)*, Frankfurt, and I take this occasion to thank them for their generous help. I would also like to thank our printers, *Universitätsdruckerei H. Stürtz AG*, Würzburg, for their advice and cooperation.

Frankfurt am Main
September 1983

Rudolf Warncke

Instructions for the Formula Index

The formula index consists of three columns. The first gives the empirical formula, the second gives the conventional formula, as well as any supplementary information or subdivisions, and the third gives the pertinent volume and page numbers.

First Column (Empirical Formula)

In the empirical formula the symbols of the elements are arranged alphabetically; C and H are not placed first. The list of empirical formulas is arranged alphabetically and by the magnitude of the subscripts. Any indefinite subscripts are placed last. Ions are always placed after the neutral species, positive ions preceding the negative.

The unsubscripted symbol is used for the element unless a specific diatomic or polyatomic species is meant (e.g., Br_2, Br_3). Transuranium elements that do not yet have an internationally recognized symbol are listed under their atomic number and placed at the end of the index. Special superscripted symbols are not used for isotopes.

H_2O is included in the empirical formula only if it is an integral part of a complex as written in the second column. Polymers of the type $(AB)_n$ are listed under AB. Multicomponent systems (mixed crystals, melts, etc.) are found under the empirical formulas of their components. However, solutions are found only under the solute.

Second Column (Conventional Formula)

The formula is written as it appears in the handbook text. However, in many cases another form is shown if there is adequate space and if it presents additional structural details. If this is not possible for isomers, then they are numbered consecutively. For elements the name is given in the second column.

Entries having the same empirical formula are arranged as follows: compounds, isotopic species, polymers, hydrates, multicomponent systems. Elements are treated in the same way.

For multicomponent systems the components are arranged in the sequence "inorganic components — organic components — water." The inorganic components are arranged alphabetically; the organic components are arranged by number of carbon atoms. If an element is a component, it is always represented by its unsubscripted symbol. Isotopic species are listed immediately after the normal species.

The concept *system* is used in its restricted sense in this index: it represents equilibrium mixtures described in phase diagrams. Ionic systems are included under their parent compounds.

The location of solubility data for compounds mentioned only briefly in the text is included under the main empirical-formula entry.

Entries for elements and compounds treated extensively in the handbook are subdivided by topics, e.g., geochemistry, preparation, or toxicity.

The concepts *solubility, solutions,* and *systems* partially overlap, and in these cases the user should always look at all three places. That is also true for the concepts *diffusion* and *systems* and for the concepts *sorption* and *system*.

In referrals to another entry in the index both the empirical and conventional formula are given. For example, "see $Al_2Na_2O_4$... $Na_2O \cdot Al_2O_3$." For referrals within the topics of a particular compound, then only the topic is given. For example "see Deposits."

Third Column (Volume and Page Numbers)

The first symbol is that of the element to which the volume belongs. Next is the abbreviated form of the type of volume followed usually by the Part or Section. The page numbers are given after a hyphen. The following abbreviations are used for type of volume:

MVol.	Main Volume (Hauptband)
SVol.	Supplement Volume (Ergänzungsband)
Org.Comp.	Organic Compounds
Org.Verb.	Organische Verbindungen
PerFHalOrg.	Perfluorohalogenoorganic Compounds of Main Group Elements
SVol.GD	Gmelin-Durrer, Metallurgy of Iron
TrU.	Transuranium Elements
Water Desalt.	Water Desalting

For example, the entry "Ag: MVol.B7-237/9" indicates that the information is to be found on pages 237 to 239 of the Silver Main Volume B 7. The entry "Fe: Org.Comp.C3-89" indicates that the desired information is to be found on page 89 of "Organic Compounds C 3" for the element iron (Fe).

Comments on System Numbers and Element Symbols

In the Formula Index itself the volume and page number citation was based on the traditional System Number (Main Volume Series) or the New Supplement Series Volume number. Today the volumes are usually arranged by the symbols of the elements. However, the symbols can be deduced easily from the system numbers. Most citations have the element symbol immediately after the system number. For example, 61 (Ag) refers to the silver volumes. The exceptions are

1 (EG)	is now	He
8 (J)	is now	I
39 (SE)	is now	Sc
69 (Ma)	is now	Tc

The old abbreviation for MVol was Hb (Hauptband), and the old abbreviation for SVol was Eb (Ergänzungsband). The volumes of the New Supplement Series are associated with the symbols of the elements as follows:

Erg.W. 1	He
Erg.W. 2	V
Erg.W. 3	Cr
Erg.W. 4, 7a, 7b, 8	Np
Erg.W. 5, 6	Co
Erg.W. 9, 12	F
Erg.W. 10	Zr
Erg.W. 11	Hf

The New Supplement Series Volumes 2 and 3 and the Volumes 10 and 11 are bound together as double volumes.

Ac–Au

Formula	Compound/System	Reference
Ac	Actinium systems	
	Ac-Pu	Np: TrU.B2-133
Ac_2O_3	Ac_2O_3 systems	
	Ac_2O_3-Pa_2O_5	Pa: SVol.2-30
Ag	Silver systems	
	Ag-Ir	Ir: SVol.1-129
	Ag-Ir-Pd	Ir: SVol.1-141
	Ag-Pu	Np: TrU.B2-19, 22, 41
		Np: TrU.B3-98/105
Ag^+	$[Ag(H_2O)_2]^+$	Ag: MVol.B6-6
–	$[Ag(H_2O)_x]^+$	Ag: MVol.B6-6/7, 15
Ag^{3+}	Ag^{3+}	Ag: MVol.B6-15
$AgAlC_6Cl_4H_6$	$AgAlCl_4 \cdot C_6H_6$	Ag: MVol.B5-94
–	$AgAlCl_4 \cdot C_6D_6$	Ag: MVol.B5-94
$AgAsCF_3HO_3$	$CF_3As(O)(OH)OAg$	F: PerFHalOrg.3-197, 198
$AgAsC_2F_6O_2$	$(CF_3)_2As(O)OAg$	F: PerFHalOrg.3-197, 198
$AgAsC_2F_{10}$	$Ag[(CF_3)_2AsF_4]$	F: PerFHalOrg.3-213
$AgAsC_3H_9NO_3$	$[AgNO_3((CH_3)_3As)]$	Ag: MVol.B7-255
$AgAsC_6H_{15}I$	$[AgI((C_2H_5)_3As)]_4$	Ag: MVol.B7-255/6
$AgAsC_9H_{21}I$	$[AgI((C_3H_7)_3As)]_4$	Ag: MVol.B7-256
$AgAsC_{10}ClH_8O_4^-$	$[Ag(ClC_6H_4As(CH_2COO)_2)]^-$	Ag: MVol.B7-261/2
$AgAsC_{10}ClH_9O_4$	$[Ag(ClC_6H_4As(CH_2COO)_2H)]$	Ag: MVol.B7-261/2
$AgAsC_{10}ClH_{10}O_4^+$	$[Ag(ClC_6H_4As(CH_2COOH)_2)]^+$	Ag: MVol.B7-261/2
$AgAsC_{10}F_6H_{10}N_2$	$[Ag(C_5H_5N)_2]AsF_6$	Ag: MVol.B6-82
$AgAsC_{10}H_9O_4^-$	$[Ag(C_6H_5As(CH_2COO)_2)]^-$	Ag: MVol.B7-258/9
$AgAsC_{10}H_{10}O_4$	$[Ag(C_6H_5As(CH_2COO)_2H)]$	Ag: MVol.B7-258/9
$AgAsC_{10}H_{11}O_4^+$	$[Ag(C_6H_5As(CH_2COOH)_2)]^+$	Ag: MVol.B7-258/9
$AgAsC_{11}F_3H_8O_4^-$	$[Ag(CF_3C_6H_4As(CH_2COO)_2)]^-$	Ag: MVol.B7-261
$AgAsC_{11}F_3H_9O_4$	$[Ag(CF_3C_6H_4As(CH_2COO)_2H)]$	Ag: MVol.B7-261
$AgAsC_{11}F_3H_{10}O_4^+$	$[Ag(CF_3C_6H_4As(CH_2COOH)_2)]^+$	Ag: MVol.B7-261
$AgAsC_{11}H_8O_6^{2-}$	$[Ag(OOCC_6H_4As(CH_2COO)_2)]^{2-}$	Ag: MVol.B7-261
$AgAsC_{11}H_9O_6^-$	$[Ag(HOOCC_6H_4As(CH_2COO)_2)]^-$	Ag: MVol.B7-261
$AgAsC_{11}H_{10}O_6$	$[Ag(OOCC_6H_4As(CH_2COOH)_2)]$	Ag: MVol.B7-261
$AgAsC_{11}H_{11}O_4^-$	$[Ag(CH_3C_6H_4As(CH_2COO)_2)]^-$	Ag: MVol.B7-260/1
$AgAsC_{11}H_{11}O_4S^-$	$[Ag(CH_3SC_6H_4As(CH_2COO)_2)]^-$	Ag: MVol.B7-262
$AgAsC_{11}H_{11}O_5^-$	$[Ag(CH_3OC_6H_4As(CH_2COO)_2)]^-$	Ag: MVol.B7-261
$AgAsC_{11}H_{11}O_6^+$	$[Ag(HOOCC_6H_4As(CH_2COOH)_2)]^+$	Ag: MVol.B7-261

$AgAsC_{11}H_{12}O_4$	$[Ag(CH_3C_6H_4As(CH_2COO)_2H)]$	Ag:	MVol.B7-260/1
$AgAsC_{11}H_{12}O_4S$	$[Ag(CH_3SC_6H_4As(CH_2COO)_2H)]$	Ag:	MVol.B7-262
$AgAsC_{11}H_{12}O_5$	$[Ag(CH_3OC_6H_4As(CH_2COO)_2H)]$	Ag:	MVol.B7-261
$AgAsC_{11}H_{13}O_4^+$	$[Ag(CH_3C_6H_4As(CH_2COOH)_2)]^+$	Ag:	MVol.B7-260/1
$AgAsC_{11}H_{13}O_4S^+$	$[Ag(CH_3SC_6H_4As(CH_2COOH)_2)]^+$	Ag:	MVol.B7-262
$AgAsC_{11}H_{13}O_5^+$	$[Ag(CH_3OC_6H_4As(CH_2COOH)_2)]^+$	Ag:	MVol.B7-261
$AgAsC_{11}H_{15}NO_3$	$[AgNO_3(CH_2CHCH_2C_6H_4As(CH_3)_2)]$	Ag:	MVol.B7-259/60
$AgAsC_{12}ClH_{12}O_4^-$	$[Ag(ClC_6H_4As(CH_2CH_2COO)_2)]^-$	Ag:	MVol.B7-262
$AgAsC_{12}ClH_{13}O_4$	$[Ag(ClC_6H_4As(C_2H_4COO)_2H)]$	Ag:	MVol.B7-262
$AgAsC_{12}F_6H_{12}$	$AgAsF_6 \cdot 2\ C_6H_6$	Ag:	MVol.B5-94
$AgAsC_{12}H_{27}I$	$[AgI((C_4H_9)_3As)]_4$	Ag:	MVol.B7-256
$AgAsC_{14}F_6H_{16}$	$AgAsF_6 \cdot 2\ C_6H_5CH_3$	Ag:	MVol.B5-97
$AgAsC_{14}H_{20}$	$AgCCC_6H_5 \cdot As(C_2H_5)_3$	Ag:	MVol.B5-18
$AgAsC_{15}H_{14}O_2$	$[Ag((C_6H_5)_2AsCH_2CH_2COO)]$	Ag:	MVol.B7-259
$AgAsC_{15}H_{15}O_2^+$	$[Ag((C_6H_5)_2AsCH_2CH_2COOH)]^+$	Ag:	MVol.B7-259
$AgAsC_{16}F_6H_{20}$	$AgAsF_6 \cdot 2\ C_6H_4(CH_3)_2$	Ag:	MVol.B5-99
$AgAsC_{18}H_{12}O_9S_3^{2-}$	$[Ag(As(C_6H_4SO_3)_3)]^{2-}$	Ag:	MVol.B7-233, 259
$AgAsC_{18}H_{15}^+$	$[Ag((C_6H_5)_3As)]^+$	Ag:	MVol.B7-256
$AgAsC_{18}H_{16}N_2O_3$	$[AgNO_3(NC_5H_4CH_2As(C_6H_5)_2)]$	Ag:	MVol.B7-258
$AgAsO_6U$	$AgUO_2AsO_4 \cdot n\ H_2O$	U:	SVol.C3-174/5
$AgAs_2BC_{46}H_{44}N_2$	$[Ag((CH_3)_2AsC_9H_6N)_2]B(C_6H_5)_4$	Ag:	MVol.B7-262/3
$AgAs_2BrC_{12}H_{12}N_2O_2$	$AgBr \cdot As_2(C_6H_3(OH)NH_2)_2$	Ag:	MVol.B7-263/4
$AgAs_2BrC_{16}H_{22}$	$[Ag((CH_3)_2AsC_6H_5)_2]Br$	Ag:	MVol.B7-258
$AgAs_2BrC_{18}H_{26}S_2$	$[AgBr((CH_3)_2AsC_6H_4SCH_3)_2]$	Ag:	MVol.B7-260
$AgAs_2BrC_{26}H_{24}$	$[AgBr(C_6H_5)_2AsC_2H_4As(C_6H_5)_2]_n$	Ag:	MVol.B7-264/5
$AgAs_2BrC_{26}H_{26}$	$[Ag((C_6H_5)_2AsCH_3)_2]Br$	Ag:	MVol.B7-257/8
$AgAs_2BrC_{28}H_{28}$	$[AgBr((C_6H_5)_2As(CH_2)_4As(C_6H_5)_2)]_n$	Ag:	MVol.B7-265
$AgAs_2C_{12}ClH_{12}N_2O_2$	$AgCl \cdot As_2(C_6H_3(OH)NH_2)_2$	Ag:	MVol.B7-263/4
$AgAs_2C_{12}ClH_{14}N_2O_6S$	$AgCl \cdot As_2(C_6H_3(OH)NH_2)_2 \cdot H_2SO_4$	Ag:	MVol.B7-264
$AgAs_2C_{12}ClH_{14}N_4O_8$	$AgCl \cdot As_2(C_6H_3(OH)NH_2)_2 \cdot 2\ HNO_3$	Ag:	MVol.B7-264
$AgAs_2C_{12}Cl_2H_{14}N_3O_5$	$AgNO_3 \cdot As_2(C_6H_3(OH)NH_2)_2 \cdot 2\ HCl$	Ag:	MVol.B7-263
$AgAs_2C_{12}Cl_3H_{14}N_2O_2$	$AgCl \cdot As_2(C_6H_3(OH)NH_2)_2 \cdot 2\ HCl$	Ag:	MVol.B7-264
$AgAs_2C_{12}H_{12}N_3O_5$	$AgNO_3 \cdot As_2(C_6H_3(OH)NH_2)_2 \cdot 2\ HCl$	Ag:	MVol.B7-263
$AgAs_2C_{12}H_{13}N_2O_3$	$AgOH \cdot As_2(C_6H_3(OH)NH_2)_2$	Ag:	MVol.B7-263/4
$AgAs_2C_{16}ClH_{22}$	$[Ag((CH_3)_2AsC_6H_5)_2]Cl$	Ag:	MVol.B7-258
$AgAs_2C_{18}ClH_{26}S_2$	$[AgCl((CH_3)_2AsC_6H_4SCH_3)_2]$	Ag:	MVol.B7-260
$AgAs_2C_{18}H_{26}IS_2$	$[AgI((CH_3)_2AsC_6H_4SCH_3)_2]$	Ag:	MVol.B7-260
$AgAs_2C_{20}H_{32}IN_2$	$AgI \cdot 2\ (CH_3)_2AsC_6H_4N(CH_3)_2$	Ag:	MVol.B7-260
$AgAs_2C_{22}ClH_{24}N_2O_4$	$[Ag((CH_3)_2AsC_9H_6N)_2]ClO_4$	Ag:	MVol.B7-262/3
$AgAs_2C_{22}F_6H_{24}N_2P$	$[Ag((CH_3)_2AsC_9H_6N)_2]PF_6$	Ag:	MVol.B7-262/3
$AgAs_2C_{22}H_{26}O_8S_2^+$	$[Ag(CH_3SC_6H_4As(CH_2COOH)_2)_2]^+$	Ag:	MVol.B7-262
$AgAs_2C_{26}ClH_{26}$	$[Ag((C_6H_5)_2AsCH_3)_2]Cl$	Ag:	MVol.B7-257/8
$AgAs_2C_{26}H_{24}NO_3$	$[AgNO_3(C_2H_4(As(C_6H_5)_2)_2)] \cdot H_2O$	Ag:	MVol.B7-264
$AgAs_2C_{26}H_{26}I$	$[Ag((C_6H_5)_2AsCH_3)_2]I$	Ag:	MVol.B7-257/8
$AgAs_2C_{26}H_{26}NO_3$	$[Ag((C_6H_5)_2AsCH_3)_2]NO_3$	Ag:	MVol.B7-257/8
$AgAs_2C_{28}H_{28}I$	$[AgI((C_6H_5)_2As(CH_2)_4As(C_6H_5)_2)]_n$	Ag:	MVol.B7-265
$AgAs_2C_{28}H_{28}NO_3$	$[AgNO_3(C_4H_8(As(C_6H_5)_2)_2)] \cdot 2\ H_2O$	Ag:	MVol.B7-265
$AgAs_2C_{28}H_{48}IP_2$	$[Ag((C_2H_5)_2PC_6H_4As(C_2H_5)_2)_2]I$	Ag:	MVol.B7-268
$AgAs_2C_{36}ClH_{30}O_4S_2$	$[Ag((C_6H_5)_3AsS)_2]ClO_4$	Ag:	MVol.B7-263
$AgAs_2C_{37}H_{30}NS$	$[AgSCN((C_6H_5)_3As)_2]$	Ag:	MVol.B7-257

Formula	Compound	Reference
$AgAs_3BBr_2C_{17}H_{23}$	$[Ag(BBr_2)(((CH_3)_2AsC_6H_4)_2AsCH_3)]$	Ag: MVol.B7-266
		B: B-Verb.3-161
$AgAs_3BrC_{17}H_{23}$	$[AgBr(((CH_3)_2AsC_6H_4)_2AsCH_3)]$	Ag: MVol.B7-266
$AgAs_3C_{21}CoH_{23}O_4$	$[AgCo(CO)_4(((CH_3)_2AsC_6H_4)_2AsCH_3)]$	Ag: MVol.B7-267
$AgAs_3C_{22}H_{23}MnO_5$	$[AgMn(CO)_5(((CH_3)_2AsC_6H_4)_2AsCH_3)]$	Ag: MVol.B7-266/7
$AgAs_3C_{55}H_{45}NO$	$[AgNCO((C_6H_5)_3As)_3]$	Ag: MVol.B7-257
$AgAs_3C_{55}H_{45}NS$	$[AgSCN((C_6H_5)_3As)_3]$	Ag: MVol.B7-257
$AgAs_4BC_{72}F_4H_{60}$	$[Ag((C_6H_5)_3As)_4]BF_4$	Ag: MVol.B7-257
$AgAs_4BrC_{72}H_{60}O_3$	$[Ag((C_6H_5)_3As)_4]BrO_3$	Ag: MVol.B7-257
$AgAs_4C_{20}ClH_{32}O_4$	$[Ag((CH_3)_2AsC_6H_4As(CH_3)_2)_2]ClO_4$	Ag: MVol.B7-265
$AgAs_4C_{20}H_{32}{}^+$	$[Ag((CH_3)_2AsC_6H_4As(CH_3)_2)_2]^+$	Ag: MVol.B7-266
$AgAs_4C_{20}H_{32}NO_3$	$[Ag((CH_3)_2AsC_6H_4As(CH_3)_2)_2]NO_3$	Ag: MVol.B7-265
$AgAs_4C_{52}H_{48}{}^+$	$[Ag((C_6H_5)_2AsC_2H_4As(C_6H_5)_2)_2]^+$	Ag: MVol.B7-264
$AgAs_4C_{72}ClH_{60}O_4$	$[Ag((C_6H_5)_3As)_4]ClO_4$	Ag: MVol.B7-257
$AgAs_4C_{72}H_{60}NO_3$	$[Ag((C_6H_5)_3As)_4]NO_3$	Ag: MVol.B7-257
$AgBCF_4H_4N_2S$	$Ag(NH_2CSNH_2)BF_4$	Ag: MVol.B7-137
$AgBC_2F_4H_3N$	$AgBF_4 \cdot CH_3NC \cdot (C_6H_5)_2PCH_3$	Ag: MVol.B5-25
–	$AgBF_4 \cdot CH_3NC \cdot (C_6H_5)_3P$	Ag: MVol.B5-25
$AgBC_2F_4H_4$	$AgBF_4 \cdot C_2H_4$	Ag: MVol.B5-29
$AgBC_2F_4H_6OS$	$Ag((CH_3)_2SO)BF_4$	Ag: MVol.B7-3
$AgBC_2F_4H_8N_4S_2$	$[Ag(NH_2CSNH_2)_2]BF_4$	Ag: MVol.B7-142
$AgBC_3F_4H_6$	$AgBF_4 \cdot 1.5\ C_2H_4$	Ag: MVol.B5-29
$AgBC_3F_4H_6S_3$	$Ag(S(CH_2S)_2CH_2)BF_4 \cdot H_2O$	Ag: MVol.B7-91/2
$AgBC_3F_4H_7NO$	$AgBF_4 \cdot OCHN(CH_3)_2$	Ag: MVol.B6-323
$AgBC_3F_4H_{12}N_6S_3$	$[Ag(NH_2CSNH_2)_3]BF_4$	Ag: MVol.B7-146
$AgBC_4Cl_2F_4H_6$	$AgBF_4 \cdot 2\ CH_2CHCl$	Ag: MVol.B5-46/7
$AgBC_4F_4H_4$	$AgBF_4 \cdot C_4H_4$	Ag: MVol.B5-83
$AgBC_4F_4H_6$	$AgBF_4 \cdot C_4H_6$	Ag: MVol.B5-36
$AgBC_4F_4H_6N_2$	$AgBF_4 \cdot 2\ CH_3CN$	Ag: MVol.B6-348
–	$AgBF_4 \cdot 2\ CH_3NC$	Ag: MVol.B5-25
$AgBC_4F_4H_8$	$AgBF_4 \cdot CH_3CHCHCH_3$	Ag: MVol.B5-33
–	$AgBF_4 \cdot 2\ C_2H_4$	Ag: MVol.B5-29
$AgBC_4F_6H_6$	$AgBF_4 \cdot 2\ CH_2CHF$	Ag: MVol.B5-46/7
$AgBC_4N_4$	$Ag[B(CN)_4]$	B: B-Verb.10-218
$AgBC_5F_4H_{10}$	$AgBF_4 \cdot CH_2CHC_3H_7$	Ag: MVol.B5-37
$AgBC_5H_{11}N_3$	$(CH_3)_3N \cdot BH_2(NC) \cdot AgCN$	B: B-Verb.15-18
$AgBC_6F_4H_4N_4$	$[Ag(NCCH_2CN)_2]BF_4$	Ag: MVol.B6-350
$AgBC_6F_4H_6$	$AgBF_4 \cdot C_6H_6$	Ag: MVol.B5-93
$AgBC_6F_4H_8$	$AgBF_4 \cdot C_6H_8$	Ag: MVol.B5-60
$AgBC_6F_4H_8N_4$	$[Ag(C_3H_4N_2)_2]BF_4$	Ag: MVol.B6-131
$AgBC_6F_4H_{10}$	$AgBF_4 \cdot C_6H_{10}$	Ag: MVol.B5-56
$AgBC_6F_4H_{10}N_2$	$AgBF_4 \cdot 2\ C_2H_5CN$	Ag: MVol.B6-349
$AgBC_6F_4H_{12}$	$AgBF_4 \cdot 2\ CH_2CHCH_3$	Ag: MVol.B5-31
–	$AgBF_4 \cdot 3\ C_2H_4$	Ag: MVol.B5-29
$AgBC_6F_4H_{18}N_6S_3$	$[Ag(CH_3NHCSNH_2)_3]BF_4$	Ag: MVol.B7-151/2
$AgBC_7F_4H_8$	$AgBF_4 \cdot C_7H_8$	Ag: MVol.B5-78/9
–	$AgBF_4 \cdot C_7H_8 \cdot H_2O$	Ag: MVol.B5-61
–	$AgBF_4 \cdot C_7H_8 \cdot x\ H_2O$	Ag: MVol.B5-78/9
$AgBC_8ClF_4H_{14}NO$	$Ag(ClC_2H_4C_6H_{10}NO)BF_4$	Ag: MVol.B6-202
$AgBC_8F_4H_8$	$AgBF_4 \cdot C_8H_8$	Ag: MVol.B5-65

Formula	Compound	Reference
$AgBC_8F_4H_8N_4$	$[Ag(C_2H_4(CN)_2)_2]BF_4$	Ag: MVol.B6-350, 351
$AgBC_8F_4H_{12}$	$AgBF_4 \cdot C_8H_{12}$	Ag: MVol.B5-65
$AgBC_8F_4H_{12}N_4$	$AgBF_4 \cdot 4\ CH_3CN = [Ag(CH_3CN)_4]BF_4$	Ag: MVol.B6-348
–	$AgBF_4 \cdot 4\ CH_3NC$	Ag: MVol.B5-25
$AgBC_8F_4H_{16}$	$AgBF_4 \cdot 2\ C_4H_8$	Ag: MVol.B5-34
$AgBC_8F_4H_{16}O_2$	$AgBF_4 \cdot 2\ CH_3COC_2H_5$	Ag: MVol.B6-209
$AgBC_9F_4H_{12}$	$AgBF_4 \cdot C_6H_3(CH_3)_3$	Ag: MVol.B5-99
$AgBC_9F_4H_{18}$	$AgBF_4 \cdot 3\ CH_2CHCH_3$	Ag: MVol.B5-31
$AgBC_9F_4H_{18}O_3$	$AgBF_4 \cdot 3\ (CH_3)_2CO$	Ag: MVol.B6-209
$AgBC_9F_4H_{24}N_6S_3$	$[Ag(CH_3NHCSNHCH_3)_3]BF_4$	Ag: MVol.B7-153
–	$[Ag(C_2H_5NHCSNH_2)_3]BF_4$	Ag: MVol.B7-154/5
$AgBC_9F_4H_{27}O_9P_3S_3$	$[Ag((CH_3O)_3PS)_3]BF_4$	Ag: MVol.B7-252
$AgBC_9H_{10}N_6$	$Ag[HB(C_3H_3N_2)_3]$	B: B-Verb.5-14
$AgBC_{10}F_4H_{10}$	$AgBF_4 \cdot C_{10}H_{10} \cdot H_2O$	Ag: MVol.B5-84/5
$AgBC_{10}F_4H_{10}N_2$	$AgBF_4 \cdot CH_3C_6H_4NC \cdot CH_3NC$	Ag: MVol.B5-25
–	$[Ag(C_5H_5N)_2]BF_4$	Ag: MVol.B6-80
$AgBC_{10}F_4H_{11}NO$	$AgBF_4 \cdot NC_9H_8OCH_3$	Ag: MVol.B5-90
$AgBC_{10}F_4H_{12}$	$AgBF_4 \cdot C_{10}H_{12}$	Ag: MVol.B5-83
$AgBC_{10}F_4H_{12}N_4$	$[Ag(C_3H_6(CN)_2)_2]BF_4$	Ag: MVol.B6-352
–	$Ag(NH_2C_5H_4N)_2BF_4$	Ag: MVol.B6-94, 95
$AgBC_{10}F_4H_{16}$	$AgBF_4 \cdot 2\ C_5H_8$	Ag: MVol.B5-54
$AgBC_{10}F_4H_{20}$	$AgBF_4 \cdot 2\ CH_2CHC_3H_7$	Ag: MVol.B5-37
$AgBC_{10}F_4H_{20}O_2$	$AgBF_4 \cdot 2\ (C_2H_5)_2CO$	Ag: MVol.B6-209
$AgBC_{10}F_4H_{22}N_2O_2$	$Ag(C_4H_9CH_2NO)_2BF_4$	Ag: MVol.B6-202
$AgBC_{12}F_4H_8$	$AgBF_4 \cdot C_{12}H_8$	Ag: MVol.B5-113
$AgBC_{12}F_4H_{10}$	$AgBF_4 \cdot (C_6H_5)_2$	Ag: MVol.B5-103
–	$AgBF_4 \cdot C_{12}H_{10}$	Ag: MVol.B5-112
$AgBC_{12}F_4H_{10}S$	$AgBF_4 \cdot S(C_6H_5)_2$	Ag: MVol.B7-13
$AgBC_{12}F_4H_{12}$	$AgBF_4 \cdot 2\ C_6H_6$	Ag: MVol.B5-93
$AgBC_{12}F_4H_{20}$	$AgBF_4 \cdot 2\ C_6H_{10}$	Ag: MVol.B5-56
$AgBC_{12}F_4H_{20}O_2$	$AgBF_4 \cdot 2\ (CH_2)_5CO$	Ag: MVol.B6-210
$AgBC_{12}F_4H_{24}$	$AgBF_4 \cdot 2\ C_6H_{12}$	Ag: MVol.B5-40
–	$AgBF_4 \cdot 3\ C_4H_8$	Ag: MVol.B5-34
$AgBC_{12}F_4H_{24}O_6$	$Ag(C_4H_8O_2)_3BF_4$	Ag: MVol.B6-221
$AgBC_{12}F_4H_{30}N_6S_3$	$[Ag(CH_3NHCSNHC_2H_5)_3]BF_4$	Ag: MVol.B7-155/6
$AgBC_{12}F_4H_{30}O_3$	$Ag(C_2H_5OC_2H_5)_3BF_4$	Ag: MVol.B6-218
$AgBC_{12}F_4H_{30}O_6$	$Ag(CH_3OC_2H_4OCH_3)_3BF_4$	Ag: MVol.B6-218
$AgBC_{14}F_4FeH_{14}O_2$	$AgBF_4 \cdot C_{10}H_8Fe(COCH_3)_2$	Ag: MVol.B6-217
$AgBC_{14}F_4H_{10}$	$AgBF_4 \cdot C_6H_5CCC_6H_5$	Ag: MVol.B5-104
$AgBC_{14}F_4H_{10}N_2$	$[Ag(C_6H_5CN)_2]BF_4$	Ag: MVol.B6-352
$AgBC_{14}F_4H_{12}$	$AgBF_4 \cdot C_6H_5CHCHC_6H_5$	Ag: MVol.B5-103
$AgBC_{14}F_4H_{14}$	$AgBF_4 \cdot C_6H_5CH_2CH_2C_6H_5$	Ag: MVol.B5-103
$AgBC_{14}F_4H_{16}$	$AgBF_4 \cdot 2\ C_7H_8$	Ag: MVol.B5-61
$AgBC_{14}F_4H_{24}$	$AgBF_4 \cdot 2\ C_7H_{12}$	Ag: MVol.B5-60
$AgBC_{14}F_4H_{28}$	$AgBF_4 \cdot 2\ CH_2CHC_5H_{11}$	Ag: MVol.B5-44
–	$AgBF_4 \cdot 2\ CH_2CHCH_2C(CH_3)_3$	Ag: MVol.B5-44
$AgBC_{14}H_8O_6$	$Ag[B(OC_6H_4COO)_2] \cdot H_2O$	B: B-Verb.8-138
$AgBC_{15}F_4H_{10}$	$AgBF_4 \cdot C_{15}H_{10}$	Ag: MVol.B5-115
$AgBC_{15}F_4H_{16}NP$	$AgBF_4 \cdot CH_3NC \cdot (C_6H_5)_2PCH_3$	Ag: MVol.B5-25
$AgBC_{15}F_4H_{24}$	$AgBF_4 \cdot 3\ C_5H_8$	Ag: MVol.B5-54

Formula	Compound	Reference
$AgBC_{15}F_4H_{30}$	$AgBF_4 \cdot 3\ CH_2CHC_3H_7$	Ag: MVol.B5-37
$AgBC_{15}F_4H_{30}N_6S_3$	$[Ag(C_2H_5NHCSNHC_2H_5)_3]BF_4$	Ag: MVol.B7-155
$AgBC_{16}F_4H_8N_4$	$[Ag(C_6H_4(CN)_2)_2]BF_4$	Ag: MVol.B6-353
$AgBC_{16}F_4H_{14}$	$AgBF_4 \cdot C_6H_5(CH)_4C_6H_5$	Ag: MVol.B5-104
$AgBC_{16}F_4H_{20}$	$AgBF_4 \cdot 2\ C_6H_4(CH_3)_2$	Ag: MVol.B5-99
–	$AgBF_4 \cdot 2\ C_8H_{10}$	Ag: MVol.B5-65
$AgBC_{16}F_4H_{24}$	$AgBF_4 \cdot 2\ C_8H_{12}$	Ag: MVol.B5-65
$AgBC_{16}F_4H_{28}$	$AgBF_4 \cdot 2\ C_6H_{11}CHCH_2$	Ag: MVol.B5-60
–	$AgBF_4 \cdot 2\ C_7H_{11}CH_3$	Ag: MVol.B5-60
$AgBC_{16}F_4H_{30}N_2O_2$	$Ag(C_2H_5C_6H_{10}NO)_2BF_4$	Ag: MVol.B6-202
$AgBC_{16}F_4H_{32}$	$AgBF_4 \cdot 2\ CH_2C(CH_3)CH_2C(CH_3)_3$	Ag: MVol.B5-45
–	$AgBF_4 \cdot 2\ CH_2CHC_6H_{13}$	Ag: MVol.B5-44
$AgBC_{16}H_{16}N_4$	$Ag[B(N(CHCH)_2)_4]$	B: B-Verb.8-154
$AgBC_{18}F_4H_{14}$	$AgBF_4 \cdot C_5H_4C(C_6H_5)_2$	Ag: MVol.B5-106
$AgBC_{18}F_4H_{15}N$	$[Ag(C_6H_5)_3N]BF_4$	Ag: MVol.B6-53
$AgBC_{18}F_4H_{16}N_4$	$Ag(NH_2C_9H_6N)_2BF_4$	Ag: MVol.B6-113/4
$AgBC_{18}F_4H_{20}O_2$	$AgBF_4 \cdot 2\ CH_3COC_6H_4CH_3$	Ag: MVol.B6-210
$AgBC_{18}F_4H_{30}O_3$	$AgBF_4 \cdot 3\ (CH_2)_5CO$	Ag: MVol.B6-210
$AgBC_{18}F_4H_{32}$	$AgBF_4 \cdot 2\ C_6H_{11}CH_2CHCH_2$	Ag: MVol.B5-60
$AgBC_{18}F_4H_{36}$	$AgBF_4 \cdot 2\ CH_2CHCH(CH_3)CH_2C(CH_3)_3$	Ag: MVol.B5-45
$AgBC_{18}H_{42}O_{20}$	$AgBO_2 \cdot 3\ HOCH_2(CHOH)_4CH_2OH$	Ag: MVol.B6-206
$AgBC_{19}H_{15}N$	$Ag[(C_6H_5)_3BCN]$	B: B-Verb.8-97/8
$AgBC_{20}F_4H_{18}NP$	$AgBF_4 \cdot CH_3NC \cdot (C_6H_5)_3P$	Ag: MVol.B5-25
$AgBC_{20}F_4H_{20}$	$AgBF_4 \cdot 2\ C_{10}H_{10}$	Ag: MVol.B5-84
$AgBC_{20}F_4H_{20}N_4$	$Ag(C_5H_5N)_4BF_4$	Ag: MVol.B6-80
$AgBC_{20}F_4H_{40}$	$AgBF_4 \cdot 2\ CH_2CHC_8H_{17}$	Ag: MVol.B5-45
$AgBC_{22}H_{19}N$	$Ag[(CHCH)_2NB(C_6H_5)_3]$	B: B-Verb.8-157
$AgBC_{24}Cl_3F_4H_{21}O_3$	$AgBF_4 \cdot 3\ CH_3COC_6H_4Cl$	Ag: MVol.B6-210
$AgBC_{24}F_4H_{16}$	$AgBF_4 \cdot 2\ C_{12}H_8$	Ag: MVol.B5-113
$AgBC_{24}F_4H_{20}$	$AgBF_4 \cdot 2\ C_{12}H_{10}$	Ag: MVol.B5-112
$AgBC_{24}F_4H_{24}O_3$	$AgBF_4 \cdot 3\ CH_3COC_6H_5$	Ag: MVol.B6-210
$AgBC_{24}F_4H_{30}N_2O_2$	$AgBF_4 \cdot 2\ C_6H_5C_6H_{10}NO$	Ag: MVol.B6-130
$AgBC_{24}F_4H_{48}$	$AgBF_4 \cdot 2\ CH_2CHC_{10}H_{21}$	Ag: MVol.B5-45
$AgBC_{24}H_{20}$	$Ag[B(C_6H_5)_4]$	B: B-Verb.8-176/8, 184, 188, 192, 198
$AgBC_{26}F_4H_{24}$	$AgBF_4 \cdot 2\ (C_6H_5)_2CH_2$	Ag: MVol.B5-103
$AgBC_{26}F_4H_{52}$	$AgBF_4 \cdot 2\ CH_2CHC_{11}H_{23}$	Ag: MVol.B5-45
$AgBC_{27}F_4H_{30}O_3$	$AgBF_4 \cdot 3\ CH_3C_6H_4COCH_3$	Ag: MVol.B6-210
$AgBC_{28}F_4H_{56}$	$AgBF_4 \cdot 2\ CH_2CHC_{12}H_{25}$	Ag: MVol.B5-45
$AgBC_{29}F_4H_{37}NO_5$	$AgBF_4 \cdot C_{29}H_{37}NO_5 \cdot 2\ (CH_3)_2CHOH$	Ag: MVol.B5-119 Ag: MVol.B6-71
$AgBC_{30}F_4H_{30}$	$AgBF_4 \cdot 3\ C_{10}H_{10}$	Ag: MVol.B5-83/4
$AgBC_{30}F_4H_{32}N_4$	$AgBF_4 \cdot 2\ CH_3C_6H_4NCHNHC_6H_4CH_3$	Ag: MVol.B6-340, 342
$AgBC_{32}F_4H_{20}$	$AgBF_4 \cdot 2\ C_{16}H_{10}$	Ag: MVol.B5-116
$AgBC_{32}F_4H_{28}N_4$	$[Ag(C_6H_5CH_2CN)_4]BF_4$	Ag: MVol.B6-350
$AgBC_{32}F_4H_{64}$	$AgBF_4 \cdot 2\ CH_2CHC_{14}H_{29}$	Ag: MVol.B5-45
$AgBC_{33}F_4H_{36}N_6O_3$	$[Ag(C_6H_5C_3HN_2O(CH_3)_2)_3]BF_4$	Ag: MVol.B6-134
$AgBC_{35}F_4H_{53}NO_7$	$AgBF_4 \cdot C_{29}H_{37}NO_5 \cdot 2\ (CH_3)_2CHOH$	Ag: MVol.B5-119 Ag: MVol.B6-71
$AgBC_{36}F_4H_{24}$	$AgBF_4 \cdot 3\ C_{12}H_8$	Ag: MVol.B5-113

$AgBC_{36}F_4H_{30}P_2$	$[Ag(P(C_6H_5)_3)_2]BF_4$	Ag: MVol.B7-211
$AgBC_{36}F_4H_{72}$	$AgBF_4 \cdot 2\ CH_2CHC_{16}H_{33}$	Ag: MVol.B5-45
$AgBC_{36}H_{34}P_2$	$[AgBH_4(P(C_6H_5)_3)_2]$	Ag: MVol.B7-210
$AgBC_{36}H_{56}O_{12}P_4$ ...	$[Ag(P(OCH_3)_3)_4]B(C_6H_5)_4$	Ag: MVol.B7-244/5
$AgBC_{40}F_4H_{36}N_2P_2$ ..	$AgBF_4 \cdot 2\ CH_3NC \cdot 2\ P(C_6H_5)_3$	Ag: MVol.B5-25
$AgBC_{42}F_4H_{42}P_2$	$[Ag(P(C_6H_4CH_3)_3)_2]BF_4$	Ag: MVol.B7-224
$AgBC_{48}F_4H_{30}$	$AgBF_4 \cdot 3\ C_{16}H_{10}$	Ag: MVol.B5-116
$AgBC_{52}F_4H_{40}P_4$	$[Ag((C_6H_5)_2PCCP(C_6H_5)_2)_2]_n(BF_4)_n$	Ag: MVol.B7-235
$AgBC_{55}H_{48}NP_3$	$Ag[H_3BCN][(C_6H_5)_3P]_3$	B: B-Verb.8-48/9
–	$[AgNCBH_3(P(C_6H_5)_3)_3] \cdot CHCl_3$	Ag: MVol.B7-216
$AgBC_{56}Cl_3H_{49}NP_3$ ..	$[AgNCBH_3(P(C_6H_5)_3)_3] \cdot CHCl_3$	Ag: MVol.B7-216
$AgBC_{56}H_{64}O_8P_4$	$[Ag(C_6H_5P(OCH_3)_2)_4][B(C_6H_5)_4]$	Ag: MVol.B7-241
$AgBC_{57}H_{53}O_2P_3$	$[(C_6H_5)_3P]_3Ag[H_3BCOOC_2H_5]$	B: B-Verb.8-218
$AgBC_{63}F_4H_{63}P_3$	$[Ag(P(C_6H_4CH_3)_3)_3]BF_4$	Ag: MVol.B7-225
$AgBC_{64}F_4H_{40}$	$AgBF_4 \cdot 4\ C_{16}H_{10}$	Ag: MVol.B5-116
$AgBC_{64}H_{80}O_8P_4$	$[Ag(C_6H_5P(OC_2H_5)_2)_4][B(C_6H_5)_4]$	Ag: MVol.B7-241
$AgBC_{72}F_4H_{60}Sb_4$ ...	$[Ag((C_6H_5)_3Sb)_4]BF_4$	Ag: MVol.B7-268/9
$AgBC_{76}H_{72}O_4P_4$	$[Ag((C_6H_5)_2POCH_3)_4][B(C_6H_5)_4]$	Ag: MVol.B7-240
$AgBC_{80}H_{80}O_4P_4$	$[Ag((C_6H_5)_2POC_2H_5)_4][B(C_6H_5)_4]$	Ag: MVol.B7-240
$AgBC_{84}F_4H_{84}P_4$	$[Ag(P(C_6H_4CH_3)_3)_4]BF_4$	Ag: MVol.B7-226
$AgBC_{85}H_{87}NP_4$	$[Ag(P(C_6H_4CH_3)_3)_4]BH_3CN$	Ag: MVol.B7-226/7
$AgBC_{90}F_4H_{75}O_5P_5$ ..	$AgBF_4 \cdot 5\ (C_6H_5)_3PO$	Ag: MVol.B7-244
$AgBC_{5x}F_4H_{6x}$	$AgBF_4 \cdot n\ C_5H_6$	Ag: MVol.B5-54
$AgBF_3HO$	$Ag[F_3B(OH)]$	B: B-Verb.8-74, 82
$AgBF_4$	$Ag[BF_4]$	B: B-Verb.10-199, 201, 206
–	$AgBF_4$ systems	
	$AgBF_4$-$CH_2CHC_3H_7$	Ag: MVol.B5-37
$AgBF_4H_6N_2$	$[Ag(NH_3)_2]BF_4$	Ag: MVol.B6-28
$AgBH_4$	$Ag[BH_4]$	B: B-Verb.8-35/6
$AgBH_4O_4$	$Ag[B(OH)_4] = B_2O_3 \cdot Ag_2O \cdot 4\ H_2O$	B: B-Verb.7-149
$AgBH_{10}N_2$	$[Ag(NH_3)_2]BH_4$	Ag: MVol.B6-28
$AgBO_2$	$AgBO_2 = B_2O_3 \cdot Ag_2O$	B: B-Verb.7-74
$AgBO_8S_2$	$Ag[B(SO_4)_2]$	B: B-Verb.8-103
AgBS	AgBS	B: B-Verb.3-15/6
AgBSe	AgBSe	B: B-Verb.3-75, 81
$AgB_2C_4H_6O_5$	$Ag(OCOCH_3) \cdot B_2O(OCOCH_3)$	B: B-Verb.8-109
$AgB_2C_8H_{14}N_7O_3$	$AgNO_3 \cdot B_2(CN)_4 \cdot 2\ NH(CH_3)_2$	Ag: MVol.B7-269
$AgB_2C_{10}H_{15}O_{11}$	$Ag[(CH_3COO)_5B_2O]$	B: B-Verb.13-113
–	$AgCH_3COO \cdot B_2O(CH_3COO)_4 \cdot (CH_3CO)_2O$.	Ag: MVol.B5-128
$AgB_2C_{14}H_{21}O_{14}$	$AgCH_3COO \cdot B_2O(CH_3COO)_4 \cdot (CH_3CO)_2O$.	Ag: MVol.B5-128
AgB_2F_7	$Ag[B_2F_7]$	B: B-Verb.19-340
$AgB_3C_{36}H_{38}P_2$	$[AgB_3H_8(P(C_6H_5)_3)_2]$	Ag: MVol.B7-210
		B: B-Verb.3-188
$AgB_3C_{42}H_{50}P_2$	$[AgB_3H_8(P(C_6H_4CH_3)_3)_2]$	Ag: MVol.B7-227/8
$AgB_3C_{63}H_{71}P_3$	$[AgB_3H_8(P(C_6H_4CH_3)_3)_3]$	Ag: MVol.B7-229
$AgB_3C_{84}H_{92}P_4$	$[Ag(P(C_6H_4CH_3)_3)_4]B_3H_8$	Ag: MVol.B7-225/6
AgB_3F_{10}	$Ag[B_3F_{10}]$	B: B-Verb.19-341
$AgB_9C_{36}H_{42}P_2S$	$[AgB_9H_{12}S(P(C_6H_5)_3)_2]$	Ag: MVol.B7-210/1
$AgB_9C_{42}H_{54}P_2S$	$[AgB_9H_{12}S(P(C_6H_4CH_3)_3)_2]$	Ag: MVol.B7-227/8

Formula	Compound	Reference
$AgB_9C_{54}H_{57}P_3S$	$[AgB_9H_{12}S(P(C_6H_5)_3)_3]$	Ag: MVol.B7-215/6
$AgB_9C_{54}H_{59}P_3$	$[AgB_9H_{14}(P(C_6H_5)_3)_3]$	Ag: MVol.B7-215
		B: B-Verb.3-196
$AgB_9C_{63}H_{75}P_3S$	$[Ag(P(C_6H_4CH_3)_3)_3]B_9H_{12}S$	Ag: MVol.B7-224/5
$AgB_9C_{84}H_{96}P_4S$	$[Ag(P(C_6H_4CH_3)_3)_4]B_9H_{12}S$	Ag: MVol.B7-226
$AgB_9C_{84}H_{98}P_4$	$[Ag(P(C_6H_4CH_3)_3)_4]B_9H_{14}$	Ag: MVol.B7-226
$AgB_{10}C_{54}H_{58}P_3$	$[AgB_{10}H_{13}(P(C_6H_5)_3)_3] \cdot 0{,}5\ CH_2Cl_2$	Ag: MVol.B7-215
$AgB_{10}C_{54.5}ClH_{59}P_3$	$[AgB_{10}H_{13}(P(C_6H_5)_3)_3] \cdot 0.5\ CH_2Cl_2$	Ag: MVol.B7-215
$AgB_{10}C_{84}H_{97}P_4$	$[Ag(P(C_6H_4CH_3)_3)_4]B_{10}H_{13}$	Ag: MVol.B7-226
$AgB_{11}C_{84}H_{98}P_4$	$[Ag(P(C_6H_4CH_3)_3)_4]B_{11}H_{14}$	Ag: MVol.B7-226
$AgB_{18}C_4H_{22}^{3-}$	$[Ag(C_2B_9H_{11})_2]^{3-}$	Ag: MVol.B7-269
		B: B-Verb.6-62
$AgB_{18}C_{28}H_{82}N_3$	$[N(C_2H_5)_4]_3[Ag(B_9H_9(CH)_2)_2]$	Ag: MVol.B7-269
$AgBeF_4H_6N_2$	$[Ag(NH_3)_2]BeF_4$	Ag: MVol.B6-31
–	$[Ag(NH_3)_2]BeF_4 \cdot 2\ H_2O$	Ag: MVol.B6-31
$AgBiC_{18}ClH_{15}O_4$	$AgClO_4 \cdot (C_6H_5)_3Bi$	Ag: MVol.B7-269
		Bi: Org.Verb.-152
AgBr	AgBr systems	
	$AgBr-MnBr_2$	Mn: MVol.C5-305
	$AgBr-NH_3$	Ag: MVol.B6-22
$AgBrCH_4N_2S$	$Ag(NH_2CSNH_2)Br$	Ag: MVol.B7-136/7
$AgBrCH_5N$	$AgBr \cdot CH_3NH_2$	Ag: MVol.B6-42
$AgBrCH_5N_3S$	$Ag(NH_2CSNHNH_2)Br$	Ag: MVol.B7-167
$AgBrC_2H_8N_4S_2$	$[Ag(NH_2CSNH_2)_2]Br$	Ag: MVol.B7-141/2
$AgBrC_3F_6$	$AgCBr(CF_3)_2$	Ag: MVol.B5-5
$AgBrC_3H_2N_2$	$[Ag(BrC_3H_2N_2)]$	Ag: MVol.B6-131
$AgBrC_3H_6S_3$	$Ag(S(CH_2S)_2CH_2)Br$	Ag: MVol.B7-90/1
$AgBrC_3H_{12}N_6S_3$	$[AgBr(NH_2CSNH_2)_3]$	Ag: MVol.B7-134
$AgBrC_4H_2N_2O_2$	$AgC_4H_2BrN_2O_2$	Ag: MVol.B6-152
$AgBrC_4H_4N_2O$	$AgC(N_2)COCHBrCH_3$	Ag: MVol.B5-22
$AgBrC_4H_8N_2S$	$Ag(CH_2CHCH_2NHCSNH_2)Br$	Ag: MVol.B7-157
$AgBrC_5H_5N$	$AgBr \cdot C_5H_5N$	Ag: MVol.B6-80
$AgBrC_5H_6$	AgC_5H_6Br	Ag: MVol.B5-8
$AgBrC_5H_{12}N_2S$	$Ag((C_2H_5)_2NCSNH_2)Br$	Ag: MVol.B7-155
$AgBrC_6H_3N_3$	$[Ag(C_6H_3BrN_3)]$	Ag: MVol.B6-181
$AgBrC_6H_4N_2O$	$[Ag(BrC_6H_4NNO)]$	Ag: MVol.B6-293
$AgBrC_6H_4N_2O_3S$	$Ag(BrC_6H_4NNSO_3)$	Ag: MVol.B6-294
$AgBrC_6H_5^+$	$[AgC_6H_5Br]^+$	Ag: MVol.B5-117
$AgBrC_6H_8N_2$	$AgBr \cdot C_6H_5NHNH_2$	Ag: MVol.B6-36
$AgBrC_6H_{12}N_4S_2$	$[Ag(C_3H_6N_2S)_2]Br$	Ag: MVol.B7-50
$AgBrC_6H_{14}N_2S_2$	$Ag(HCSN(CH_3)_2)_2Br$	Ag: MVol.B7-122
$AgBrC_6H_{15}O_3P$	$AgBr \cdot P(OC_2H_5)_3$	Ag: MVol.B7-246
$AgBrC_7H_4S_2$	$[Ag(BrC_6H_4CSS)]$	Ag: MVol.B7-95
$AgBrC_7N_4$	$AgC(CN)_2CBrC(CN)_2$	Ag: MVol.B6-355
$AgBrC_8H_4$	$AgCCC_6H_4Br$	Ag: MVol.B5-19
$AgBrC_8H_4O$	$AgCCOC_6H_4Br$	Ag: MVol.B7-246
$AgBrC_8H_6O_2S$	$[Ag(BrC_6H_4SCH_2COO)]$	Ag: MVol.B7-27/8
$AgBrC_8H_6O_2Se$	$[Ag(BrC_6H_4SeCH_2COO)]$	Ag: MVol.B7-191/2, 194, 196
$AgBrC_8H_7^+$	$[Ag(C_6H_5CHCHBr)]^+$	Ag: MVol.B5-117

$AgBrC_8H_7O_2S^+$ .... $[Ag(BrC_6H_4SCH_2COOH)]^+$ .... Ag: MVol.B7-27/8
$AgBrC_8H_7O_2Se^+$ ... $[Ag(BrC_6H_4SeCH_2COOH)]^+$ .... Ag: MVol.B7-194
$AgBrC_9H_{10}N_2O_5$ .... $[Ag(C_9H_{10}BrN_2O_5)]$ .... Ag: MVol.B6-157
$AgBrC_9H_{21}O_3P$ ..... $[AgBr(i-C_3H_7O)_3P]_3$ .... Ag: MVol.B7-248
$AgBrC_{10}FeH_8$ ...... $AgC_5H_4FeC_5H_4Br$ .... Ag: MVol.B5-24
$AgBrC_{10}H_{24}N_4S_2$ ... $Ag((C_2H_5)_2NCSNH_2)_2Br$ .... Ag: MVol.B7-155
$AgBrC_{12}H_7N_2^+$ ..... $[Ag(BrC_{12}H_7N_2)]^+$ .... Ag: MVol.B6-128
$AgBrC_{12}H_8N_4O_2$ .... $[Ag(BrC_6H_4N_3C_6H_4NO_2)]$ .... Ag: MVol.B6-300
$AgBrC_{12}H_9N_3$ ...... $[Ag(C_6H_5N_3C_6H_4Br)]$ .... Ag: MVol.B6-300
$AgBrC_{12}H_{30}O_6P_2$ ... $[AgBr(P(OC_2H_5)_3)_2]$ .... Ag: MVol.B7-246
$AgBrC_{13}H_{10}N_2$ ..... $Ag(CH_3C_{12}H_7N_2)Br$ .... Ag: MVol.B6-126
$AgBrC_{13}H_{10}N_4S$ .... $[Ag(BrC_6H_4NNCSNNHC_6H_5)]$ .... Ag: MVol.B7-183/4
$AgBrC_{14}H_{10}N_2OS$ ... $Ag(C_6H_5C_8H_5BrN_2OS) \cdot HNO_3$ .... Ag: MVol.B7-65/6
$AgBrC_{14}H_{11}N_3O_4S$ .. $Ag(C_6H_5C_8H_5BrN_2OS) \cdot HNO_3$ .... Ag: MVol.B7-65/6
$AgBrC_{14}H_{12}N_2$ ..... $Ag((CH_3)_2C_{12}H_6N_2)Br$ .... Ag: MVol.B6-127
$AgBrC_{14}H_{12}N_2OS$ ... $Ag(C_6H_5C_8H_6N_2OS) \cdot HBr$ .... Ag: MVol.B7-65/6
$AgBrC_{14}H_{13}P$ ...... $[AgBr(C_6H_4(CH_2)_2PC_6H_5)]_n$ .... Ag: MVol.B7-221/2
$AgBrC_{14}H_{16}N_2$ ..... $Ag((CH_3)_4C_{10}H_4N_2)Br$ .... Ag: MVol.B6-121
$AgBrC_{14}H_{28}N_4O_2S_2$. $[Ag((CH_3)_3C_4H_5N_2OS)_2]Br$ .... Ag: MVol.B7-64/5
$AgBrC_{18}H_{15}P$ ...... $[AgBr(P(C_6H_5)_3)]$ .... Ag: MVol.B7-208
$AgBrC_{18}H_{38}N_4O_3S_2$. $[Ag((CH_3)_3C_4H_5N_2OS)_2]Br \cdot (C_2H_5)_2O$ ...... Ag: MVol.B7-65
$AgBrC_{18}H_{45}O_9P_3$ ... $[AgBr(P(OC_2H_5)_3)_3]$ .... Ag: MVol.B7-247
$AgBrC_{19}H_{22}N_2O_4S$ .. $Ag(NH_2C_9H_6N)C_{10}H_{14}BrO_4S$ .... Ag: MVol.B6-114
$AgBrC_{21}H_{36}N_6S_3$ ... $[Ag((CH_3)_3C_4H_3N_2S)_3]Br$ .... Ag: MVol.B7-64
$AgBrC_{21}H_{42}N_6O_3S_3$. $[Ag((CH_3)_3C_4H_5N_2OS)_3]Br$ .... Ag: MVol.B7-64/5
$AgBrC_{24}H_{60}O_{12}P_4$ ... $[Ag(P(OC_2H_5)_3)_4]Br$ .... Ag: MVol.B7-247
$AgBrC_{25}H_{52}N_6O_4S_3$. $[Ag((CH_3)_3C_4H_5N_2OS)_3]Br \cdot (C_2H_5)_2O$ ...... Ag: MVol.B7-65
$AgBrC_{26}H_{20}P_2$ ..... $[AgBr((C_6H_5)_2PCCP(C_6H_5)_2)]$ .... Ag: MVol.B7-235/6
$AgBrC_{26}H_{24}P_2$ ..... $[AgBr((C_6H_5)_2PC_2H_4P(C_6H_5)_2)]_n$ .... Ag: MVol.B7-235
$AgBrC_{28}H_{26}P_2$ ..... $[AgBr(C_6H_4(CH_2)_2PC_6H_5)_2]$ .... Ag: MVol.B7-222
$AgBrC_{28}H_{28}N_2O_6S$ .. $Ag(C_9H_7NO)_2C_{10}H_{14}BrO_4S$ .... Ag: MVol.B6-112
$AgBrC_{28}H_{28}P_2$ ..... $[AgBr((C_6H_5)_2P(CH_2)_4P(C_6H_5)_2)]_n$ .... Ag: MVol.B7-237
$AgBrC_{28}H_{28}P_2S$ .... $[AgBr[(C_6H_5)_2PCH_2CH_2]_2S]_2$ .... Ag: MVol.B7-237/9
$AgBrC_{30}H_{32}N_2O_4S_3$. $[Ag(CH_3SC_9H_6N)_2]C_{10}H_{14}BrO_4S$ .... Ag: MVol.B7-61/2
$AgBrC_{30}H_{34}P_2S^+$ .... $[AgBr((C_6H_5)_2(CH_3)PCH_2CH_2)_2S]^+$ .... Ag: MVol.B7-238/9
$AgBrC_{36}H_{30}P_2$ ..... $[AgBr(P(C_6H_5)_3)_2]$ .... Ag: MVol.B7-210
$AgBrC_{42}H_{38}P_2$ ..... $[AgBr(CH_2CHCH_2C_6H_4P(C_6H_5)_2)_2]$ .... Ag: MVol.B7-231
$AgBrC_{42}H_{42}P_2$ ..... $[AgBr(P(C_6H_4CH_3)_3)_2]$ .... Ag: MVol.B7-227/8
$AgBrC_{63}H_{63}P_3$ ..... $[AgBr(P(C_6H_4CH_3)_3)_3]$ .... Ag: MVol.B7-229
$AgBrC_{72}H_{60}O_3P_4$ ... $[Ag(P(C_6H_5)_3)_4]BrO_3$ .... Ag: MVol.B7-216/7
$AgBrC_{84}H_{84}P_4$ ..... $[Ag(P(C_6H_4CH_3)_3)_4]Br$ .... Ag: MVol.B7-225/6
$AgBrH_3N$ .... $AgBr \cdot NH_3$ .... Ag: MVol.B6-22
– .... $[AgBr(NH_3)]$ .... Ag: MVol.B6-8
$AgBrH_4N_2$ .... $AgBr \cdot N_2H_4$ .... Ag: MVol.B6-36
$AgBrH_{4.5}N_{1.5}$ ...... $AgBr \cdot 1.5\ NH_3$ .... Ag: MVol.B6-22
$AgBrH_6N_2$ .... $[AgBr(NH_3)_2]$ .... Ag: MVol.B6-8, 10
– .... $[Ag(NH_3)_2]Br$ .... Ag: MVol.B6-22
$AgBrH_6N_2O_3$ ...... $AgBrO_3 \cdot 2\ NH_3 = [Ag(NH_3)_2]BrO_3$ ...... Ag: MVol.B6-23
$AgBrH_9N_3$ .... $AgBr \cdot 3\ NH_3$ .... Ag: MVol.B6-22/3
$AgBrH_9N_3O_3$ ...... $AgBrO_3 \cdot 3\ NH_3$ .... Ag: MVol.B6-23

Formula	Compound	Reference
$AgBrH_{10}N_2O_2$	$[Ag(NH_3)_2(H_2O)_2]Br$	Ag: MVol.B6-22
$AgBr_2C_2H_2^+$	$[AgCHBrCHBr]^+$	Ag: MVol.B5-46
$AgBr_2C_4HN_2O_2$	$AgC_4HBr_2N_2O_2$	Ag: MVol.B6-148
$AgBr_2C_6H_3N_2O_3S$	$Ag(Br_2C_6H_3NNSO_3)$	Ag: MVol.B6-294
$AgBr_2C_6H_4^+$	$[AgC_6H_4Br_2]^+$	Ag: MVol.B5-117
$AgBr_2C_6H_7N_6O_6$	$Ag(BrC_3H_3N_2)_2NO_3 \cdot HNO_3$	Ag: MVol.B6-140
$AgBr_2C_8H_2NO_2$	$[Ag(C_8H_2Br_2NO_2)]$	Ag: MVol.B6-71
$AgBr_2C_{10}ClH_8N_2O_4$	$[Ag(BrC_5H_4N)_2]ClO_4$	Ag: MVol.B6-99
$AgBr_2C_{10}H_{12}Li$	$Li[Ag(C_5H_6Br)_2]$	Ag: MVol.B5-8
$AgBr_2C_{12}H_{12}N_2^+$	$[Ag(BrC_6H_4NH_2)_2]^+$	Ag: MVol.B6-52
$AgBr_2C_{13}H_9N_4S$	$[Ag(BrC_6H_4NNCSNNHC_6H_4Br)]$	Ag: MVol.B7-183/4
$AgBr_2C_{15}H_{13}N_4S$	$[Ag(CS(NNC_6H_3BrCH_3)_2H)]$	Ag: MVol.B7-185
$AgBr_2C_{16}H_{12}O_4S_2^-$	$[Ag(BrC_6H_4SCH_2COO)_2]^-$	Ag: MVol.B7-27/8
$AgBr_2C_{16}H_{12}O_4Se_2^-$	$[Ag(BrC_6H_4SeCH_2COO)_2]^-$	Ag: MVol.B7-191/2, 194
$AgBr_2C_{24}H_{14}N_4^+$	$[Ag(BrC_{12}H_7N_2)_2]^+$	Ag: MVol.B6-128
$AgBr_2C_{36}H_{82}O_{12}P_3$	$HAgBr_2 \cdot 3\,(C_4H_9O)_3PO \cdot x\,H_2O$	Ag: MVol.B7-250
$AgBr_2C_{43}H_{24}N_4O_8S$	$[Ag(C_{12}H_8N_2)_2]C_{19}H_8Br_2O_8S$	Ag: MVol.B6-208
$AgBr_2H_3N^-$	$[AgBr_2(NH_3)]^-$	Ag: MVol.B6-8, 10
$AgBr_2H_6N_2^-$	$[AgBr_2(NH_3)_2]^-$	Ag: MVol.B6-8, 10
$AgBr_2H_6N_2O_6^-$	$[Ag(BrO_3)_2(NH_3)_2]^-$	Ag: MVol.B6-10
$AgBr_3C_{12}Cl_3H_4N_3$	$[Ag(Br_3C_6H_2N_3C_6H_2Cl_3)]$	Ag: MVol.B6-300
$AgBr_3H_3N^{2-}$	$[AgBr_3(NH_3)]^{2-}$	Ag: MVol.B6-8, 10
$AgC_{0.5}H_{2.5}IN_{0.5}$	$AgI \cdot 0.5\,CH_3NH_2$	Ag: MVol.B6-42
$AgC_{0.9}H_3IN_{0.6}$	$AgI \cdot 0.3\,C_3H_6(NH_2)_2$	Ag: MVol.B6-62
$AgCClH_4N_2O_4S$	$Ag(NH_2CSNH_2)ClO_4$	Ag: MVol.B7-136
$AgCClH_5N$	$AgCl \cdot CH_3NH_2$	Ag: MVol.B6-41
$AgCClH_5N_3S$	$Ag(NH_2CSNHNH_2)Cl$	Ag: MVol.B7-165/7
$AgCF_3S$	$Ag(SCF_3)$	Ag: MVol.B7-4/5
$AgCF_4NO_4S_2$	$Ag[CF_3SO_2NSO_2F]$	S: S-N-Verb.1-50
$AgCF_6HNP$	$AgPF_6 \cdot HCN = [Ag(HCN)]PF_6$	Ag: MVol.B6-346
$AgCHN^+$	$[Ag(HCN)]^+$	Ag: MVol.B6-346
$AgCHN_2O_2$	$[Ag(CH(NO)NO)]$	Ag: MVol.B6-320
$AgCHN_4$	$[AgCHN_4]$	Ag: MVol.B6-191/2
$AgCHN_4O$	$[Ag(CHN_4O)]$	Ag: MVol.B6-194
$AgCHN_4O_3S$	$[Ag(CN_4SO_3H)]$	Ag: MVol.B6-193
$AgCHO_2$	$AgHCOO$	Ag: MVol.B5-120/1
$AgCHS_2$	$[Ag(HCSS)]$	Ag: MVol.B7-94
$AgCH_2NO$	$Ag(OCHNH)$	Ag: MVol.B6-322
$AgCH_2NO_2$	$AgNH_2COO$	Ag: MVol.B6-29
$AgCH_2N_3O_2$	$[Ag(NH_2C(NO)NO)]$	Ag: MVol.B6-320
$AgCH_2N_3O_3$	$Ag(NHCONHNO_2)$	Ag: MVol.B6-335
$AgCH_3$	$AgCH_3$	Ag: MVol.B5-3/4
$AgCH_3N_2$	$AgCN \cdot NH_3$	Ag: MVol.B6-30
$AgCH_3N_2O$	$AgNCO \cdot NH_3$	Ag: MVol.B6-30
$AgCH_3N_2S$	$AgSC(NH_2)NH$	Ag: MVol.B7-133
–	$AgSCN \cdot NH_3$	Ag: MVol.B6-30
–	$[AgSCN(NH_3)]$	Ag: MVol.B6-10
$AgCH_3N_2S_2$	$[Ag(NH_2NHCSS)]$	Ag: MVol.B7-119
$AgCH_3N_4O$	$[Ag(ONNHC(NH)NH)]$	Ag: MVol.B6-337
$AgCH_3N_4O_2$	$[Ag(NO_2NHC(NH)NH)]$	Ag: MVol.B6-337

$AgCH_3S$	$Ag(SCH_3)$	Ag: MVol.B7-4
$AgCH_4IN_2O_3S$	$Ag(NH_2CSNH_2)IO_3$	Ag: MVol.B7-137
$AgCH_4IN_2S$	$Ag(NH_2CSNH_2)I$	Ag: MVol.B7-137
$AgCH_4I_3N_2S^{2-}$	$[AgI_3(NH_2CSNH_2)]^{2-}$	Ag: MVol.B7-134
$AgCH_4N_2S^+$	$[Ag(NH_2CSNH_2)]^+$	Ag: MVol.B7-128/33
$AgCH_4N_3$	$AgCN \cdot N_2H_4$	Ag: MVol.B6-36
–	$[Ag(NHC(NH)NH_2)]$	Ag: MVol.B6-337
$AgCH_4N_3O_3S$	$Ag(NH_2CSNH_2)NO_3$	Ag: MVol.B7-136
$AgCH_4N_3O_3Se$	$AgNO_3 \cdot NH_2CSeNH_2$	Ag: MVol.B7-196/7
$AgCH_4N_3O_4$	$AgNO_3 \cdot CO(NH_2)_2$	Ag: MVol.B6-333/4
$AgCH_4N_3S$	$AgSCN \cdot N_2H_4$	Ag: MVol.B6-36
$AgCH_4O^+$	$[Ag(CH_3OH)]^+$	Ag: MVol.B6-206
$AgCH_5IN$	$AgI \cdot CH_3NH_2$	Ag: MVol.B6-42
$AgCH_5N^+$	$[Ag(CH_3NH_2)]^+$	Ag: MVol.B6-38/41
$AgCH_5N_3O^+$	$[Ag(NH_2CONHNH_2)]^+$	Ag: MVol.B6-336
$AgCH_5N_3S^+$	$[Ag(NH_2CSNHNH_2)]^+$	Ag: MVol.B7-165
$AgCH_5N_4O_3$	$[Ag(NH_2C(NH)NH_2)]NO_3$	Ag: MVol.B6-337
$AgCH_5N_4O_3S$	$Ag(NH_2CSNHNH_2)NO_3$	Ag: MVol.B7-165
$AgCH_6N_3O$	$AgNCO \cdot 2\ NH_3$	Ag: MVol.B6-30
$AgCH_6N_3S$	$[Ag(NHCSNH_2)(NH_3)]$	Ag: MVol.B7-144
–	$AgSCN \cdot 2\ NH_3$	Ag: MVol.B6-30
$AgCH_7N_2O_2$	$[Ag(NH_3)_2]HCOO$	Ag: MVol.B6-28/9
$AgCH_8N_3O_2$	$[Ag(NH_3)_2]NH_2COO$	Ag: MVol.B6-29
$AgCNSe$	$Ag(SeCN)$	C: MVol.D6-229
$AgCN_3S_2$	$[Ag(CN_3S_2)]$	Ag: MVol.B7-82
AgC_2ClF_4	$AgCFClCF_3$	Ag: MVol.B5-5
AgC_2ClH_2	$AgCHCHCl$	Ag: MVol.B5-6
$AgC_2ClH_6O_5S$	$Ag((CH_3)_2SO)ClO_4$	Ag: MVol.B7-3
$AgC_2ClH_8N_2O_4$	$[Ag(C_2H_8N_2)]ClO_4$	Ag: MVol.B6-58
$AgC_2ClH_8N_4O_4S_2$	$[Ag(NH_2CSNH_2)_2]ClO_4$	Ag: MVol.B7-141
$AgC_2ClH_8N_4O_4Se_2$	$AgClO_4 \cdot 2\ NH_2CSeNH_2$	Ag: MVol.B7-196/7
$AgC_2ClH_8N_4S_2$	$[Ag(NH_2CSNH_2)_2]Cl$	Ag: MVol.B7-139/41
$AgC_2ClH_8N_4Se_2$	$AgCl \cdot 2\ NH_2CSeNH_2$	Ag: MVol.B7-196
$AgC_2ClH_{10}N_2O_4$	$AgClO_4 \cdot 2\ CH_3NH_2$	Ag: MVol.B6-41
$AgC_2Cl_2H_2^+$	$[AgCHClCHCl]^+$	Ag: MVol.B5-46
$AgC_2Cl_3H_8N_2^{2-}$	$[AgCl_3(C_2H_8N_2)]^{2-}$	Ag: MVol.B6-55/6
$AgC_2CoH_{14}N_{10}O_8S_2$	$[Ag(NH_2CSNH_2)_2][Co(NO_2)_4(NH_3)_2]$	Ag: MVol.B7-143
$AgC_2F_2H_2^+$	$[AgCHFCHF]^+$	Ag: MVol.B5-46
$AgC_2F_2H_2N_3O_2$	$[Ag(NO_2CF_2C(NH)NH)]$	Ag: MVol.B6-341
		F: PerFHalOrg.7-3/4, 30
$AgC_2F_3H_2N_2$	$[Ag(CF_3C(NH)NH)]$	Ag: MVol.B6-341
		F: PerFHalOrg.7-3, 30
$AgC_2F_3S_2$	$[Ag(CF_3CSS)]$	Ag: MVol.B7-94
AgC_2F_5S	$AgCCSF_5$	Ag: MVol.B5-20
$AgC_2F_6H_2N_2P$	$[Ag(HCN)_2]PF_6$	Ag: MVol.B6-346
$AgC_2F_6O_2P$	$(CF_3)_2P(O)OAg$	F: PerFHalOrg.3-43, 50
$AgC_2F_{10}P$	$Ag[(CF_3)_2PF_4]$	F: PerFHalOrg.3-134, 137
AgC_2H	$AgCCH$	Ag: MVol.B7-246
$AgC_2HN_4O_3$	$[Ag(C_2N_3(OH)NO_2)]$	Ag: MVol.B6-183
$AgC_2H_2^+$	$[AgC_2H_2]^+$	Ag: MVol.B5-30

Formula	Compound	Reference
$AgC_2H_2I_2^+$	$[AgCHICHI]^+$	Ag: MVol.B5-46
AgC_2H_2NO	$[Ag(C_2H_2NO)]$	Ag: MVol.B6-267
$AgC_2H_2N_2^+$	$[Ag(HCN)_2]^+$	Ag: MVol.B6-346
$AgC_2H_2N_3$	$[Ag(C_2H_2N_3)]$	Ag: MVol.B6-182, 183
$AgC_2H_2N_3O_2$	$[Ag(C_2H_2N_3O_2)]$	Ag: MVol.B6-186
$AgC_2H_2N_3O_3S$	$AgNO_3 \cdot C_2H_2N_2S$	Ag: MVol.B7-79/80
$AgC_2H_2N_3S_2$	$[Ag(NHC_2HN_2S_2)]$	Ag: MVol.B7-81
$AgC_2H_2NaO_2S$	$Na[Ag(SCH_2COO)]$	Ag: MVol.B7-22
$AgC_2H_2O_2S^-$	$[Ag(SCH_2COO)]^-$	Ag: MVol.B7-21
$AgC_2H_3N^+$	$[Ag(CH_3CN)]^+$	Ag: MVol.B6-346/7
$AgC_2H_3N_2O_2$	$[Ag(CH_3C(NO)NO)]$	Ag: MVol.B6-320
$AgC_2H_3N_2O_3$	$AgNO_3 \cdot CH_3CN$	Ag: MVol.B6-347
–	$AgNO_3 \cdot CD_3CN$	Ag: MVol.B6-347
–	$Ag(ONNCOOCH_3)$	Ag: MVol.B6-330
$AgC_2H_3N_2O_4$	$Ag(NO_2NCOOCH_3)$	Ag: MVol.B6-330
$AgC_2H_3N_4$	$[Ag(CH_3CN_4)]$	Ag: MVol.B6-192/3
–	$[Ag(NHC(NH)NHCN)]$	Ag: MVol.B6-338
$AgC_2H_3N_4O$	$Ag[CH(NNH)_2CO]$	Ag: MVol.B6-196
–	$Ag[N(CH_3)NNNC(O)]$	Ag: MVol.B6-193
–	$Ag[N(O)NNNC(CH_3)]$	Ag: MVol.B6-194
$AgC_2H_3N_4O_2$	$AgH(CO(NHN)_2CO)$	Ag: MVol.B6-336
–	$[Ag(OC_2N_3(OH)NH_2)]$	Ag: MVol.B6-186
$AgC_2H_3N_4O_2S$	$[Ag(CH_3SO_2CN_4)]$	Ag: MVol.B6-193
$AgC_2H_3N_4O_3$	$AgNO_3 \cdot C_2H_3N_3$	Ag: MVol.B6-185
$AgC_2H_3N_4O_3S$	$[Ag(CH_3CN_4SO_3)]$	Ag: MVol.B6-193/4
–	$AgNO_3 \cdot NH_2C_2HN_2S$	Ag: MVol.B7-79/80
$AgC_2H_3N_4O_4$	$Ag(NO_2NHCONHCONH)$	Ag: MVol.B6-336
$AgC_2H_3N_4S$	$[Ag(CH_3SCN_4)]$	Ag: MVol.B6-193
–	$[Ag(NH_2C_2HN_3S)]$	Ag: MVol.B7-55
$AgC_2H_3OS_2$	$[Ag(CH_3OCSS)]$	Ag: MVol.B7-96
		C: MVol.D4-257
$AgC_2H_3O_2$	$AgCH_3COO$	
	Chemical reactions	Ag: MVol.B5-125/8
	Formation and preparation	Ag: MVol.B5-121/3
	Physical properties	Ag: MVol.B5-123/5
	Solubility	Ag: MVol.B5-128/35
	Special forms	Ag: MVol.B5-122
–	$AgCH_3{}^{14}COO$	Ag: MVol.B5-122, 127
–	$AgCH_2DCOO$	Ag: MVol.B5-122
–	$AgCD_3COO$	Ag: MVol.B5-122, 126
–	$AgCH_3COO \cdot H_2O$	Ag: MVol.B5-124
–	$AgCH_3COO$ solutions	
	$AgCH_3COO$-H_2O	Ag: MVol.B5-135/42
	$AgCH_3COO$-NH_3	Ag: MVol.B5-142
	$AgCH_3COO$-organic solvents	Ag: MVol.B5-142/6
	$AgCH_3COO$-SO_2	Ag: MVol.B5-142
$AgC_2H_4^+$	$[AgC_2H_4]^+$	Ag: MVol.B5-27/8
AgC_2H_4NO	$Ag(CH_3CONH)$	Ag: MVol.B6-324
AgC_2H_4NOS	$Ag(HCSN(CH_3)O) \cdot H_2O$	Ag: MVol.B7-123

Formula	Compound	Reference
$AgC_2H_4NO_2$	$[Ag(NH_2CH_2COO)]$	Ag: MVol.B6-234/9, 250
–	$Ag(NH_2CH_2COO) \cdot 0.5\ H_2O$	Ag: MVol.B6-240
$AgC_2H_4NO_3$	$AgNO_3 \cdot C_2H_4$	Ag: MVol.B5-28
AgC_2H_4NS	$[Ag(CH_3CSNH)]$	Ag: MVol.B7-124
$AgC_2H_4NS_2$	$[Ag(SCSNHCH_3)]$	Ag: MVol.B7-102
$AgC_2H_4N_3S$	$Ag(NH_2CSNH_2)CN$	Ag: MVol.B7-137
$AgC_2H_4N_3S_2$	$[Ag(NH_2CSNCSNH_2)]$	Ag: MVol.B7-164
$AgC_2H_4N_5$	$[Ag(CH_3NHCN_4)]$	Ag: MVol.B6-192/3
$AgC_2H_4N_5O_3$	$[Ag(NO_2NHCONHC(NH)NH)]$	Ag: MVol.B6-339
–	$AgNO_3 \cdot NH_2C(NH)NHCN$	Ag: MVol.B6-338
AgC_2H_5	AgC_2H_5	Ag: MVol.B5-4
AgC_2H_5IN	$Ag(C_4H_{10}N_2)_{0.5}I$	Ag: MVol.B6-164/5
$AgC_2H_5N^+$	$[Ag(C_2H_4NH)]^+$	Ag: MVol.B6-67
$AgC_2H_5NO_2{}^+$	$[Ag(NH_2CH_2COOH)]^+$	Ag: MVol.B6-234, 250
$AgC_2H_5NO_3{}^-$	$[Ag(OH)(NH_2CH_2COO)]^-$	Ag: MVol.B6-234/6
$AgC_2H_5NS^+$	$[Ag(CH_3CSNH_2)]^+$	Ag: MVol.B7-123
$AgC_2H_5N_2O_3S$	$[Ag(CH_3CSNH_2)]NO_3$	Ag: MVol.B7-124
$AgC_2H_5N_2O_4$	$AgNO_3 \cdot CH_3CONH_2$	Ag: MVol.B6-323/4
$AgC_2H_5N_2O_5$	$[Ag(NO_3)(NH_2CH_2COOH)]_2$	Ag: MVol.B6-236/8
$AgC_2H_5N_4$	$AgC_2H_2N_3 \cdot NH_3$	Ag: MVol.B6-183
$AgC_2H_5N_4S$	$[Ag(NH_2C(NH)NHCSNH)]$	Ag: MVol.B7-162
AgC_2H_5OS	$[Ag(SCH_2CH_2OH)]$	Ag: MVol.B7-6
$AgC_2H_5O_2S_2$	$[Ag(C_2H_5SO_2S)]$	Ag: MVol.B7-96
AgC_2H_6NO	$CH_3CH(OH)NHAg$	Ag: MVol.B6-18
$AgC_2H_6NO_2S$	$NH_4[Ag(SCH_2COO)]$	Ag: MVol.B7-22
$AgC_2H_6NO_3S$	$AgNHSO_2OC_2H_5$	S: S-N-Verb.1-97
–	$[Ag(NH_2C_2H_4SO_3)]$	Ag: MVol.B6-262
–	$AgNO_3 \cdot S(CH_3)_2$	Ag: MVol.B7-10/1
$AgC_2H_6NO_3S_2$	$AgNO_3 \cdot CH_3SSCH_3$	Ag: MVol.B7-15
$AgC_2H_6NO_4S$	$Ag((CH_3)_2SO)NO_3$	Ag: MVol.B7-2/3
–	$[Ag(NH_2C_2H_4SO_4)]$	Ag: MVol.B6-262
$AgC_2H_6NO_4S_2$	$[Ag(CH_3CSNH_2)]HSO_4$	Ag: MVol.B7-124
AgC_2H_6NS	$[Ag(SCH_2CH_2NH_2)]_n$	Ag: MVol.B7-16
$AgC_2H_6N_2S^+$	$[Ag(CH_3NHCSNH_2)]^+$	Ag: MVol.B7-150
$AgC_2H_6N_3O_3S$	$Ag(CH_3NHCSNH_2)NO_3$	Ag: MVol.B7-150
$AgC_2H_6N_4S^+$	$[Ag(NH_2C(NH)NHCSNH_2)]^+$	Ag: MVol.B7-162
$AgC_2H_6OS^+$	$[Ag((CH_3)_2SO)]^+$	Ag: MVol.B7-2
$AgC_2H_7N^+$	$[Ag(C_2H_5NH_2)]^+$	Ag: MVol.B6-43
–	$[Ag(NH(CH_3)_2)]^+$	Ag: MVol.B6-47
$AgC_2H_7NO^+$	$[Ag(NH_2C_2H_4OH)]^+$	Ag: MVol.B6-226/7
$AgC_2H_7NO_4P$	$[Ag(NH_2C_2H_4OPO_3H)]$	Ag: MVol.B7-250
$AgC_2H_7N_2O_2$	$Ag(NH_2CH_2COO) \cdot NH_3$	Ag: MVol.B6-29, 240
$AgC_2H_8IN_4S_2$	$[AgI(NH_2CSNH_2)_2]$	Ag: MVol.B7-134
$AgC_2H_8I_2N_4S_2{}^-$	$[AgI_2(NH_2CSNH_2)_2]^-$	Ag: MVol.B7-134
$AgC_2H_8IrN_2O_6S_2$	$Ag[Ir(SO_3)_2(C_2H_4(NH_2)_2)]$	Ir: SVol.2-159
$AgC_2H_8N_2{}^+$	$[Ag(C_2H_8N_2)]^+$	Ag: MVol.B6-55/7
$AgC_2H_8N_4S_2{}^+$	$[Ag(NH_2CSNH_2)_2]^+$	Ag: MVol.B7-128/33
$AgC_2H_8N_5O_3S_2$	$[Ag(NH_2CSNH_2)_2]NO_3$	Ag: MVol.B7-139
$AgC_2H_8N_5O_3Se_2$	$AgNO_3 \cdot 2\ NH_2CSeNH_2$	Ag: MVol.B7-196/7
$AgC_2H_8N_5O_5$	$AgNO_3 \cdot 2\ CO(NH_2)_2$	Ag: MVol.B6-333/4

Formula	Compound	Reference
$AgC_2H_8N_5O_5$	$AgNO_3 \cdot 2\ CO(NH_2)_2 \cdot 4\ H_2O$	Ag: MVol.B6-333/4
–	$AgNO_3 \cdot 2\ CO(NH_2)_2 \cdot 6\ H_2O$	Ag: MVol.B6-333/4
$AgC_2H_8O_2^+$	$[Ag(CH_3OH)_2]^+$	Ag: MVol.B6-206
$AgC_2H_9N_2^{2+}$	$[AgH(C_2H_8N_2)]^{2+}$	Ag: MVol.B6-55/7
$AgC_2H_9N_2O_2$	$[Ag(NH_3)_2]CH_3COO$	Ag: MVol.B6-29
$AgC_2H_{10}IN_2$	$AgI \cdot 2\ CH_3NH_2$	Ag: MVol.B6-42
$AgC_2H_{10}N_2^+$	$[Ag(CH_3NH_2)_2]^+$	Ag: MVol.B6-38/41
$AgC_2H_{10}N_3O$	$Ag(CH_3CONH) \cdot 2\ NH_3$	Ag: MVol.B6-324
$AgC_2H_{10}N_3O_2$	$Ag(NH_2CH_2COO) \cdot 2\ NH_3$	Ag: MVol.B6-29, 240
$AgC_2H_{10}N_6O_2^+$	$[Ag(NH_2CONHNH_2)_2]^+$	Ag: MVol.B6-336
$AgC_2H_{10}N_6S_2^+$	$[Ag(NH_2CSNHNH_2)_2]^+$	Ag: MVol.B7-165
$AgC_2H_{10}N_6Se_2^+$	$[Ag(NH_2CSeNHNH_2)_2]^+$	Ag: MVol.B7-197
$AgC_2NO_6^{2-}$	$[AgC_2O_4(NO_2)]^{2-}$	Ag: MVol.B5-186/7
$AgC_2N_2O_8^{3-}$	$[AgC_2O_4(NO_2)_2]^{3-}$	Ag: MVol.B5-186/7
$AgC_2N_3O_4$	$AgC(NO_2)_2CN$	Ag: MVol.B6-354
$AgC_2O_4^-$	$[AgC_2O_4]^-$	Ag: MVol.B5-186/7
AgC_3ClF_6	$AgCCl(CF_3)_2$	Ag: MVol.B5-5
$AgC_3ClH_2N^+$	$[AgCH_2CClCN]^+$	Ag: MVol.B5-50
$AgC_3ClH_2N_2$	$[Ag(ClC_3H_2N_2)]$	Ag: MVol.B6-131
$AgC_3ClH_3N_3$	$[Ag(ClC_2N_3CH_3)]$	Ag: MVol.B6-185
$AgC_3ClH_6O_4S_3$	$Ag(S(CH_2S)_2CH_2)ClO_4 \cdot H_2O$	Ag: MVol.B7-91
$AgC_3ClH_6S_3$	$Ag(S(CH_2S)_2CH_2)Cl$	Ag: MVol.B7-90/1
$AgC_3ClH_7NO_5$	$AgClO_4 \cdot OCHN(CH_3)_2$	Ag: MVol.B6-323
$AgC_3ClH_9O_3P$	$[AgCl(P(OCH_3)_3)]_4$	Ag: MVol.B7-244
AgC_3ClH_9P	$[AgCl(P(CH_3)_3)]_4$	Ag: MVol.B7-200/1
$AgC_3ClH_{10}N_2O_4$	$[Ag(NH_2C_3H_6NH_2)]ClO_4$	Ag: MVol.B6-62
$AgC_3ClH_{12}N_6O_4S_3$	$[Ag(NH_2CSNH_2)_3]ClO_4$	Ag: MVol.B7-144/6
–	$[Ag(NH_2CSNH_2)_3]ClO_4 \cdot 0.5\ H_2O$	Ag: MVol.B7-146
$AgC_3ClH_{12}N_6S_3$	$[AgCl(NH_2CSNH_2)_3]$	Ag: MVol.B7-134
–	$[Ag(NH_2CSNH_2)_3]Cl$	Ag: MVol.B7-144
$AgC_3ClH_{15}N_9S_3$	$Ag(NH_2CSNHNH_2)_3Cl$	Ag: MVol.B7-169
$AgC_3Cl_2HN_2$	$[Ag(Cl_2C_3HN_2)]$	Ag: MVol.B6-131
$AgC_3Cl_3H_2OS_2$	$[Ag(CCl_3CH_2OCSS)]$	Ag: MVol.B7-97
$AgC_3FH_{12}N_6O_3S_4$	$[Ag(NH_2CSNH_2)_3]SO_3F \cdot 0.5\ H_2O$	Ag: MVol.B7-146
AgC_3F_3	$AgCCCF_3$	Ag: MVol.B5-15
$AgC_3F_3H_4N_2O_2S$	$Ag(NH_2CSNH_2)CF_3COO$	Ag: MVol.B7-137
$AgC_3F_3H_6IP$	$[AgI(CF_3P(CH_3)_2)]_4$	Ag: MVol.B7-202
$AgC_3F_4NO_3$	$ONCF_2CF_2C(O)OAg$	F: PerFHalOrg.7-177
AgC_3F_5	$AgC(CF_3)CF_2$	Ag: MVol.B5-7
$AgC_3F_5H_2N_2$	$[Ag(C_2F_5C(NH)NH)]$	Ag: MVol.B6-341
		F: PerFHalOrg.7-3/4, 30
AgC_3F_6H	$AgCH(CF_3)_2$	Ag: MVol.B5-5
AgC_3F_7	$AgCF(CF_3)_2$	Ag: MVol.B5-5
$AgC_3F_9H_3O_3Sb$	$Ag[Sb(CF_3)_3(OH)_3]$	F: PerFHalOrg.3-226, 230
$AgC_3H_2IN_2$	$[Ag(IC_3H_2N_2)]$	Ag: MVol.B6-131
$AgC_3H_2NOS_2$	$[Ag(C_3H_2NOS_2)]$	Ag: MVol.B7-73
$AgC_3H_2NO_2S$	$[Ag(C_3H_2NO_2S)]$	Ag: MVol.B7-70
$AgC_3H_2N_3O_2$	$[Ag(C_3H_2N_3O_2)]$	Ag: MVol.B6-308/12
–	$Ag(C_3H_2N_3O_2) \cdot H_2O$	Ag: MVol.B6-308/12
–	$[Ag(NO_2C_3H_2N_2)]$	Ag: MVol.B6-131

$AgC_3H_2N_3O_3$	$[Ag(C_3N_3(OH)_2O)]$	Ag: MVol.B6-189
$AgC_3H_2N_5$	[Ag(NCNHC(NH)NCN)]	Ag: MVol.B6-338
AgC_3H_3	$AgCCCH_3$	Ag: MVol.B5-13/4
$AgC_3H_3N^+$	$[AgCH_2CHCN]^+$	Ag: MVol.B5-50
$AgC_3H_3N_2$	[Ag(NCHNCHCH)]	Ag: MVol.B6-137
–	[Ag(NNCHCHCH)]	Ag: MVol.B6-131
$AgC_3H_3N_2O$	$AgCNNCOCH_3$	Ag: MVol.B5-22
$AgC_3H_3N_2O_2$	$[Ag(C_3H_3N_2O_2)]$	Ag: MVol.B6-147
$AgC_3H_3N_2S$	$[AgC_3H_3N_2S]$	Ag: MVol.B7-49
$AgC_3H_3N_4O_2$	$AgNHC_3N_3(OH)_2$	Ag: MVol.B6-189
–	$AgNHC_3N_3(OH)_2 \cdot 0.5\ H_2O$	Ag: MVol.B6-189
–	$AgNO_2 \cdot C_3H_3N_3$	Ag: MVol.B6-187
$AgC_3H_3N_4O_6$	$[Ag(CH_3CN)][C(NO_2)_3]$	Ag: MVol.B6-348
$AgC_3H_3N_4S_2$	$[Ag(C_3H_3N_4S_2)]$	Ag: MVol.B7-68
$AgC_3H_4^+$	$[Ag(CH_2CCH_2)]^+$	Ag: MVol.B5-32
$AgC_3H_4NO_2$	$[AgC_3H_4NO_2]$	Ag: MVol.B6-203
$AgC_3H_4NO_2S_2$	$[Ag(CH_2(COOH)NHCSS)]$	Ag: MVol.B7-103
$AgC_3H_4NO_3$	$Ag(OOCC(NOH)CH_3)$	Ag: MVol.B6-306
$AgC_3H_4N_2^+$	$[AgC_3H_4N_2]^+$	Ag: MVol.B6-136/7
$AgC_3H_4N_3O$	$[Ag(CH_3CONNHCN)]$	Ag: MVol.B6-332
–	$[Ag(NHC_3H_3N_2O)]$	Ag: MVol.B6-147
$AgC_3H_4N_3O_3S$	$AgNO_3 \cdot CH_3C_2HN_2S$	Ag: MVol.B7-80
$AgC_3H_4N_3O_3S_2$	$AgNO_3 \cdot CH_3SC_2HN_2S$	Ag: MVol.B7-80
$AgC_3H_4N_3O_4S$	$AgNO_3 \cdot C_3H_4N_2OS$	Ag: MVol.B7-71
$AgC_3H_4N_3S$	$Ag(CH_3C_2HN_3S) \cdot n\ AgCl$	Ag: MVol.B7-55/6
–	$Ag(CH_3C_2HN_3S) \cdot n\ AgNO_3$	Ag: MVol.B7-55/6
$AgC_3H_4N_3S_2$	$[Ag(CH_3SC_2N_2SNH)]$	Ag: MVol.B7-81
$AgC_3H_4N_5S$	$[Ag(C_3H_4N_5S)]$	Ag: MVol.B7-68 C: MVol.D6-44
$AgC_3H_4O_2S^-$	$[Ag(CH_3CH(S)COO)]^-$	Ag: MVol.B7-29
AgC_3H_5	$AgCHCHCH_3$	Ag: MVol.B5-7
$AgC_3H_5NNaO_2S$	$AgSCH_2CH(NH_2)COONa$	Ag: MVol.B6-256
$AgC_3H_5N_2O_2$	$[Ag(NH_2C_3H_3NO_2)]$	Ag: MVol.B6-203
$AgC_3H_5N_2O_3$	$[Ag(NH_2CONHCH_2COO)]$	Ag: MVol.B6-240
–	$Ag(ONNCOOC_2H_5)$	Ag: MVol.B6-330
$AgC_3H_5N_2O_3S$	$[Ag(C_3H_5N_2O_3S)]$	Ag: MVol.B7-76
$AgC_3H_5N_2O_4$	$Ag(NO_2NCOOC_2H_5)$	Ag: MVol.B6-330
$AgC_3H_5N_4$	$[Ag(C_2H_5CN_4)]$	Ag: MVol.B6-192/3
$AgC_3H_5N_4O$	$[Ag(C_2H_5CN_4O)]$	Ag: MVol.B6-194
$AgC_3H_5N_4O_2$	$Ag(NHC(NH_2)NHNCHCOO) \cdot H_2O$	Ag: MVol.B6-345
$AgC_3H_5N_4O_3S$	$AgNO_3 \cdot CH_3C_2N_2SNH_2$	Ag: MVol.B7-80
$AgC_3H_5N_4O_4$	$AgNO_3 \cdot CH_3CONHNHCN$	Ag: MVol.B6-332
$AgC_3H_5N_4S$	$[Ag(CH_3SC_2N_3NH_2)]$	Ag: MVol.B7-55
$AgC_3H_5OS_2$	$Ag[S_2COC_2H_5]$	Ag: MVol.B7-97 C: MVol.D4-257
$AgC_3H_5O_2S$	$[Ag(CH_3CH(SH)COO)]$	Ag: MVol.B7-29
–	$[Ag(CH_3CH(SH)COO)] \cdot H_2O$	Ag: MVol.B7-30
–	$[Ag(CH_3CH_2OCOS)]$	Ag: MVol.B7-97 C: MVol.D4-237
–	$[Ag(CH_3SCH_2COO)]$	Ag: MVol.B7-25/6

Formula	Compound	Reference
$AgC_3H_5O_2S$	$[Ag(HSC_2H_4COO)] \cdot H_2O$	Ag: MVol.B7-30
–	$[Ag(HSC_2H_4COO)] \cdot 2\,H_2O$	Ag: MVol.B7-30
$AgC_3H_5O_3S_3^{2-}$	$[Ag(SCH_2CH(S)CH_2SO_3)]^{2-}$	Ag: MVol.B7-46
$AgC_3H_5O_6S_3$	$[Ag(CH(SO_2CH_2)_2SO_2)]$	Ag: MVol.B7-92
AgC_3H_5S	$Ag(SCH_2CHCH_2)$	Ag: MVol.B7-7
$AgC_3H_6^+$	$[Ag(CH_2CHCH_3)]^+$	Ag: MVol.B5-30
$AgC_3H_6IN_2S$	$[Ag(C_3H_6N_2S)]I$	Ag: MVol.B7-50
$AgC_3H_6IS_3$	$Ag(S(CH_2S)_2CH_2)I$	Ag: MVol.B7-90/1
AgC_3H_6NOS	$[Ag((CH_3)_2NCOS)]$	Ag: MVol.B7-99/100
$AgC_3H_6NO_2$	$[Ag(CH_3CH(NH_2)COO)]$	Ag: MVol.B6-241/2
–	$[Ag(CH_3NHCH_2COO)]$	Ag: MVol.B6-240
–	$Ag(NHCOOC_2H_5)$	Ag: MVol.B6-330
–	$[Ag(NH_2CH_2CH_2COO)]$	Ag: MVol.B6-243
$AgC_3H_6NO_2S$	$AgSCH_2CH(NH_2)COOH$	Ag: MVol.B6-256
$AgC_3H_6NO_3$	$[Ag(HOCH_2CH(NH_2)COO)]$	Ag: MVol.B6-245
–	$AgNO_3 \cdot CH_2CHCH_3$	Ag: MVol.B5-31
$AgC_3H_6NO_3S_3$	$Ag(S(CH_2S)_2CH_2)NO_3$	Ag: MVol.B7-88/9
–	$Ag(S(CH_2S)_2CH_2)NO_3 \cdot H_2O$	Ag: MVol.B7-89
$AgC_3H_6NS_2$	$[Ag((CH_3)_2NCSS)]$	Ag: MVol.B7-105/6
–	$[Ag(SCSNHC_2H_5)]$	Ag: MVol.B7-102
$AgC_3H_6NSe_2$	$Ag(SeSeCN(CH_3)_2)$	Ag: MVol.B7-197
$AgC_3H_6N_2O_3P$	$AgC(N_2)PO(OCH_3)_2$	Ag: MVol.B5-23
$AgC_3H_6N_2S^+$	$[Ag(C_3H_6N_2S)]^+$	Ag: MVol.B7-49
$AgC_3H_6N_3O_3$	$Ag(NO_2NHCONC_2H_5)$	Ag: MVol.B6-335
$AgC_3H_6N_3O_3S$	$[Ag(C_3H_6N_2S)]NO_3$	Ag: MVol.B7-50
$AgC_3H_6N_3O_4S$	$Ag(CH_3CONHCSNH_2)NO_3$	Ag: MVol.B7-161
$AgC_3H_6N_3O_5S_2$	$Ag[CH_3NSO_2][NSO_2][CH_3NCO]$	S: S-N-Verb.1-24
$AgC_3H_6N_5$	$[Ag(C_2H_5NHCN_4)]$	Ag: MVol.B6-192/3
–	$[Ag(N(CH_3)_2CN_4)]$	Ag: MVol.B6-192/3
$AgC_3H_6N_5O_4$	$[Ag(NO_2(NHNO_2)C_3H_5N_2)]$	Ag: MVol.B6-147
$AgC_3H_6N_7O_3$	$AgNO_3 \cdot C_3N_3(NH_2)_3$	Ag: MVol.B6-188
$AgC_3H_6O^+$	$[AgCH_2CHCH_2OH]^+$	Ag: MVol.B5-47
–	$[Ag(CH_3COCH_3)]^+$	Ag: MVol.B6-209
$AgC_3H_6OS_2^-$	$[Ag(SCH(CH_2OH)CH_2S)]^-$	Ag: MVol.B7-8/9
$AgC_3H_6O_2S^+$	$[Ag(CH_3SCH_2COOH)]^+$	Ag: MVol.B7-25/6
$AgC_3H_6S_3^+$	$[Ag(S(CH_2S)_2CH_2)]^+$	Ag: MVol.B7-89
AgC_3H_7	AgC_3H_7	Ag: MVol.B5-4
AgC_3H_7INS	$Ag(HCSN(CH_3)_2)I$	Ag: MVol.B7-122/3
$AgC_3H_7NO^+$	$[Ag(OCHN(CH_3)_2)]^+$	Ag: MVol.B6-322/3
$AgC_3H_7N_4OS$	$[Ag(CH_3CO(NH)_2CSNNH_2)]$	Ag: MVol.B7-172
$AgC_3H_7O_2S$	$[Ag(SC_3H_5(OH)_2)]$	Ag: MVol.B7-6/7
AgC_3H_7S	$Ag(SC_3H_7)$	Ag: MVol.B7-5
$AgC_3H_8N^{2+}$	$[AgCH_2CHCH_2NH_3]^{2+}$	Ag: MVol.B5-49/50
–	$[Ag(H_2O)_x(CH_2CHCH_2NH_3)]^{2+}$	Ag: MVol.B6-45
$AgC_3H_8NO_3S$	$[Ag(NH_2C_3H_6SO_3)]$	Ag: MVol.B6-262
–	$AgNO_3 \cdot CH_3SC_2H_5$	Ag: MVol.B7-11
$AgC_3H_8N_2S^+$	$[Ag(CH_3NHCSNHCH_3)]^+$	Ag: MVol.B7-152
–	$[Ag(C_2H_5NHCSNH_2)]^+$	Ag: MVol.B7-154
$AgC_3H_8N_3O_2$	$[Ag(NH_3)_2]CH_2(CN)COO$	Ag: MVol.B6-29
$AgC_3H_8N_3O_3S$	$Ag(CH_3NHCSNHCH_3)NO_3$	Ag: MVol.B7-152

Formula	Compound	Reference
$AgC_3H_8N_3O_3S$	$AgNO_3 \cdot (CH_3)_2NCSNH_2$	Ag: MVol.B7-152
–	$Ag(C_2H_5NHCSNH_2)NO_3$	Ag: MVol.B7-154
$AgC_3H_8N_3O_4$	$AgNO_3 \cdot CO(NHCH_3)_2$	Ag: MVol.B6-335
$AgC_3H_8N_5O_5$	$AgNO_3 \cdot CH_2(CONHNH_2)_2 \cdot H_2O$	Ag: MVol.B6-331/2
$AgC_3H_8N_5S_2$	$[Ag(NH_2CSNH_2)_2]CN$	Ag: MVol.B7-142
$AgC_3H_8N_5S_3$	$Ag(NH_2CSNH_2)_2SCN$	Ag: MVol.B7-142/3
$AgC_3H_8OS_3^+$	$[Ag(H_2O)(S(CH_2S)_2CH_2)]_n^{n+}$	Ag: MVol.B7-91
AgC_3H_9IP	$[AgI(P(CH_3)_3)]_4$	Ag: MVol.B7-201
$AgC_3H_9N^+$	$[Ag(C_3H_7NH_2)]^+$	Ag: MVol.B6-44
–	$[Ag(N(CH_3)_3)]^+$	Ag: MVol.B6-49
$AgC_3H_9NO^+$	$[Ag(NH_2C_2H_4OCH_3)]^+$	Ag: MVol.B6-232
–	$[Ag(NH_2C_3H_6OH)]^+$	Ag: MVol.B6-228
$AgC_3H_9NO_2^+$	$[Ag(NH_2CH_2CHOHCH_2OH)]^+$	Ag: MVol.B6-230/1
$AgC_3H_9NO_3P$	$[AgNO_3(P(CH_3)_3)]$	Ag: MVol.B7-199/200
$AgC_3H_9N_2O_2$	$AgCH_3CH(NH_2)COO \cdot NH_3$	Ag: MVol.B6-29, 242
–	$Ag(NH_2CH_2CH_2COO) \cdot NH_3$	Ag: MVol.B6-243
$AgC_3H_9N_4O_2S_2$	$[Ag(NH_2CSNH_2)_2]HCOO$	Ag: MVol.B7-142
$AgC_3H_{10}N_2^+$	$[Ag(C_3H_6(NH_2)_2)]^+$	Ag: MVol.B6-61/2
$AgC_3H_{10}N_2O^+$	$[Ag(HOCH(CH_2NH_2)_2)]^+$	Ag: MVol.B6-230
$AgC_3H_{10}N_2O^{2+}$	$[Ag((NH_2CH_2)_2CHOH)]^{2+}$	Ag: MVol.B7-272
$AgC_3H_{10}N_3O_4$	$AgNO_3 \cdot HOCH(CH_2NH_2)_2 \cdot 0.5\ H_2O$	Ag: MVol.B6-230
$AgC_3H_{10}N_7S_3$	$[AgNCS(NH_2CSNHNH_2)_2]$	Ag: MVol.B7-169
$AgC_3H_{11}N_2^{2+}$	$[Ag(NH_3C_3H_6NH_2)]^{2+}$	Ag: MVol.B6-61/2
$AgC_3H_{11}N_3^+$	$[Ag(NH_2CH(CH_2NH_2)_2)]^+$	Ag: MVol.B6-65
$AgC_3H_{12}IN_6S_3$	$[AgI(NH_2CSNH_2)_3]$	Ag: MVol.B7-134
$AgC_3H_{12}N_3^{2+}$	$[Ag(NH_3CH(CH_2NH_2)_2)]^{2+}$	Ag: MVol.B6-65
$AgC_3H_{12}N_3O_2$	$AgCH_3CH(NH_2)COO \cdot 2\ NH_3$	Ag: MVol.B6-29, 242
–	$Ag(NH_2CH_2CH_2COO) \cdot 2\ NH_3$	Ag: MVol.B6-243
$AgC_3H_{12}N_6S_3^+$	$[Ag(NH_2CSNH_2)_3]^+$	Ag: MVol.B7-128/33
$AgC_3H_{12}N_7O_3S_3$	$[Ag(NH_2CSNH_2)_3]NO_3$	Ag: MVol.B7-144
$AgC_3H_{12}N_7O_3Se_3$	$AgNO_3 \cdot 3\ NH_2CSeNH_2$	Ag: MVol.B7-196/7
$AgC_3H_{14}N_6O_4PS_3$	$[Ag(NH_2CSNH_2)_3]H_2PO_4$	Ag: MVol.B7-147
$AgC_3H_{15}N_9S_3^+$	$[Ag(NH_2CSNHNH_2)_3]^+$	Ag: MVol.B7-165
$AgC_3H_{15}N_9Se_3^+$	$[Ag(NH_2CSeNHNH_2)_3]^+$	Ag: MVol.B7-197
$AgC_3H_{15}N_{10}O_3$	$[Ag(NH_2C(NH)NH_2)_3]NO_3$	Ag: MVol.B6-337
$AgC_3H_{2x+8}NO_x^{2+}$	$Ag(H_2O)_xCH_2CHCH_2NH_3]^{2+}$	Ag: MVol.B6-45
AgC_3N_3O	$[Ag((NC)_2CNO)]$	Ag: MVol.B6-306
$AgC_3N_3O_2$	$AgC(NO_2)(CN)_2$	Ag: MVol.B6-354
$AgC_4ClH_3N_2^+$	$[Ag(C_4H_3ClN_2)]^+$	Ag: MVol.B6-159
$AgC_4ClH_4N_2O_4S_2$	$AgClO_4 \cdot C_2H_4(SCN)_2$	Ag: MVol.B7-15
$AgC_4ClH_4O_4$	$AgClO_4 \cdot C_4H_4$	Ag: MVol.B5-83
$AgC_4ClH_5N_3O_3$	$Ag(Cl(CH_3)C_3H_2N_2)NO_3$	Ag: MVol.B6-131
$AgC_4ClH_7O^+$	$[AgCHClCHOC_2H_5]^+$	Ag: MVol.B5-48
$AgC_4ClH_8N_2S$	$Ag(CH_2CHCH_2NHCSNH_2)Cl$	Ag: MVol.B7-156/7
$AgC_4ClH_8N_4O_4S_4$	$[Ag(NH_2CSCSNH_2)_2]ClO_4$	Ag: MVol.B7-186/7
$AgC_4ClH_8O_4S_2$	$AgClO_4 \cdot S(CH_2CH_2)_2S$	Ag: MVol.B7-87
$AgC_4ClH_8O_6$	$Ag(C_4H_8O_2)ClO_4$	Ag: MVol.B6-220/1
$AgC_4ClH_9NO_5$	$AgClO_4 \cdot CH_3CON(CH_3)_2 \cdot H_2O$	Ag: MVol.B6-325
$AgC_4ClH_{11}NO_4S$	$[Ag(SCH_2N(CH_3)_3)]ClO_4$	Ag: MVol.B7-16
$AgC_4ClH_{11}NS$	$[Ag(SCH_2N(CH_3)_3)]Cl$	Ag: MVol.B7-16

Formula	Compound	Reference
$AgC_4ClH_{12}N_2O_4$	$[Ag(NH_2C_4H_8NH_2)]ClO_4$	Ag: MVol.B6-63
$AgC_4ClH_{12}N_4O_4S_2$	$[Ag(CH_3NHCSNH_2)_2]ClO_4$	Ag: MVol.B7-151
$AgC_4ClH_{12}O_6S_2$	$[Ag((CH_3)_2SO)_2]ClO_4$	Ag: MVol.B7-3
$AgC_4ClH_{12}O_6Se_2$	$[Ag((CH_3)_2SeO)_2]ClO_4$	Ag: MVol.B7-191
$AgC_4ClH_{16}N_4$	$[AgCl(C_2H_8N_2)_2]$	Ag: MVol.B6-55/6
$AgC_4Cl_2H_{16}N_4^-$	$[AgCl_2(C_2H_8N_2)_2]^-$	Ag: MVol.B6-55/6
$AgC_4Cl_4H_6N_2Tl$	$AgCl \cdot TlCl_3 \cdot 2\ CH_3CN$ = $[Ag(CH_3CN)_2]TlCl_4$	Ag: MVol.B6-349
$AgC_4FH_2N_2O_2$	$[Ag(C_4H_2FN_2O_2)]$	Ag: MVol.B6-152
$AgC_4FH_{16}N_8O_3S_5$	$Ag(NH_2CSNH_2)_4SO_3F$	Ag: MVol.B7-148
AgC_4F_2H	$AgCCCHCF_2$	Ag: MVol.B5-15
$AgC_4F_3H_8N_4O_2S_2$	$[Ag(NH_2CSNH_2)_2]CF_3COO$	Ag: MVol.B7-142
$AgC_4F_3N_2$	$AgC(CN)_2CF_3$	Ag: MVol.B6-355
AgC_4F_5	$AgCCC_2F_5$	Ag: MVol.B5-15
$AgC_4F_6H_6N_2P$	$AgPF_6 \cdot 2\ CH_3CN$	Ag: MVol.B6-348
$AgC_4F_6H_6N_2W$	$[Ag(CH_3CN)_2]WF_6$	Ag: MVol.B6-349
$AgC_4F_6NO_3$	$ONCF_2CF_2CF_2C(O)OAg$	F: PerFHalOrg.7-178/9, 185
$AgC_4F_6NO_4S$	$Ag[NSO_2][OC(CF_3)_2CO]$	F: PerFHalOrg.5-57/8, 80, 103 S: S-N-Verb.1-40
AgC_4F_7	$AgC(CF_3)CFCF_3$	Ag: MVol.B5-7/8
–	$Ag(CFCF_2CF_2CF_2)$	Ag: MVol.B5-6
AgC_4F_7HNO	$Ag[C_3F_7C(O)NH]$	F: PerFHalOrg.7-18, 22, 46
$AgC_4F_7H_2N_2$	$[Ag(C_3F_7C(NH)NH)]$	Ag: MVol.B6-340, 341 F: PerFHalOrg.7-3/4, 31
AgC_4F_9	$AgC(CF_3)_3$	Ag: MVol.B5-5
AgC_4F_9S	$AgSC(CF_3)_3$	Ag: MVol.B5-5
$AgC_4H_2IN_2O_2$	$AgC_4H_2IN_2O_2$	Ag: MVol.B6-152
–	$[Ag(IC_3H_2N_2COO)]$	Ag: MVol.B6-132
$AgC_4H_2NO_2$	$AgN(COCH)_2$	Ag: MVol.B6-69
$AgC_4H_2N_3O_4$	$[Ag(C_3HN_3O_2COOH)]$	Ag: MVol.B6-309/12
–	$[Ag(ONC_4H_2N_2O_3)]$	Ag: MVol.B6-314
–	$[Ag(ONC_4H_2N_2O_3)] \cdot 0.5\ H_2O$	Ag: MVol.B6-314
–	$[Ag(ONC_4H_2N_2O_3)] \cdot H_2O$	Ag: MVol.B6-314
–	$[Ag(ONC_4H_2N_2O_3)] \cdot 3\ H_2O$	Ag: MVol.B6-314
$AgC_4H_2N_3O_5$	$Ag(NO_2C_4H_2N_2O_3)$	Ag: MVol.B6-155
–	$Ag(NO_2C_4H_2N_2O_3) \cdot H_2O$	Ag: MVol.B6-155
–	$Ag(NO_2C_4H_2N_2O_3) \cdot 4\ H_2O$	Ag: MVol.B6-155
$AgC_4H_3N_2O_2$	$AgC_4H_3N_2O_2$	Ag: MVol.B6-148, 151
–	$AgC_4H_3N_2O_2 \cdot 0.5\ H_2O$	Ag: MVol.B6-162
$AgC_4H_3N_2O_3$	$[Ag(ONC_3NO_2CH_3)]$	Ag: MVol.B6-312
$AgC_4H_3N_6O$	$[Ag(C_4H_3N_6O)]$	Ag: MVol.B6-182
$AgC_4H_3O_4S^{2-}$	$[Ag(SC_2H_3(COO)_2)]^{2-}$	Ag: MVol.B7-31/2
$AgC_4H_3S_2$	$[Ag(C_4H_3S_2)]$	Ag: MVol.B7-69
$AgC_4H_4NO_2$	$[Ag(N(COCH_2)_2)]$	Ag: MVol.B6-327
–	$AgN(COCH_2)_2 \cdot 0.5\ H_2O$	Ag: MVol.B6-327/8
$AgC_4H_4NO_2S$	$[Ag(C_4H_4NO_2S)]$	Ag: MVol.B7-86
$AgC_4H_4NO_3$	$AgNO_3 \cdot C_4H_4$	Ag: MVol.B5-83

$AgC_4H_4N_2^+$	$[Ag(NCC_2H_4CN)]^+$	Ag: MVol.B6-350
–	$[Ag(N(CHCH)_2N)]^+$	Ag: MVol.B6-157/8
$AgC_4H_4N_2S_2^+$	$[Ag(C_2H_4(SCN)_2)]^+$	Ag: MVol.B7-15
$AgC_4H_4N_3O$	$Ag(NHC_4H_3N_2O)$	Ag: MVol.B6-150
$AgC_4H_4N_3O_2$	$Ag(CH_3C_3HN_3O_2) \cdot 0.5\ H_2O$	Ag: MVol.B6-308/12
$AgC_4H_4N_3O_3$	$[Ag(NCC_2H_4CN)]NO_3$ $= AgNO_3 \cdot NCC_2H_4CN$	Ag: MVol.B6-350, 351
–	$[Ag(N(CHCH)_2N)]NO_3$	Ag: MVol.B6-158/9
$AgC_4H_4N_7$	$[Ag(NCNHC_3N_3(NH)NH_2)]$	Ag: MVol.B6-188
$AgC_4H_4NaO_4S$	$Na[Ag(S(CH_2COO)_2)]$	Ag: MVol.B7-38
$AgC_4H_4O_4S^-$	$[Ag(S(CH_2COO)_2)]^-$	Ag: MVol.B7-38
AgC_4H_5	$AgCCC_2H_5$	Ag: MVol.B5-13
$AgC_4H_5N^+$	$[AgCH(CH_3)CHCN]^+$	Ag: MVol.B5-50
–	$[AgCH_2C(CH_3)CN]^+$	Ag: MVol.B5-50
$AgC_4H_5NNaO_4$	$Na[Ag(NH(CH_2COO)_2)]$	Ag: MVol.B6-275
$AgC_4H_5NO_2S_2^-$	$[Ag(CH_3CH(COO)NHCSS)]^-$	Ag: MVol.B7-103
$AgC_4H_5N_2$	$[Ag(NC(CH_3)NCHCH)]$	Ag: MVol.B6-142
–	$[Ag(NNCHCHC(CH_3))]$	Ag: MVol.B6-131
$AgC_4H_5N_2O$	$[Ag(CH_3C_3H_2N_2O)]$	Ag: MVol.B6-133
$AgC_4H_5N_2OS$	$[Ag(CH_3C_3H_2N_2OS)]$	Ag: MVol.B7-51
$AgC_4H_5N_2O_2$	$[Ag(CH_3C_3H_2N_2O_2)]$	Ag: MVol.B6-147
–	$[Ag(C(N_2)COOC_2H_5)]$	Ag: MVol.B5-23 Ag: MVol.B6-292
$AgC_4H_5N_2O_3$	$Ag(C_3H_5N_2OCOO)$	Ag: MVol.B6-147
$AgC_4H_5N_2O_4$	$[Ag(NH_2COCONHCH_2COO)]$	Ag: MVol.B6-240
$AgC_4H_5N_2S$	$[Ag(CH_3C_3H_2N_2S)]$	Ag: MVol.B7-49
$AgC_4H_5N_3^+$	$[Ag(NH_2C_4H_3N_2)]^+$	Ag: MVol.B6-159
$AgC_4H_5N_4O_3$	$AgC_3H_2N_2O_2(NHCONH_2)$	Ag: MVol.B6-148
AgC_4H_5O	$AgCCCH(CH_3)OH \cdot AgC_3H_7COO$	Ag: MVol.B5-15
$AgC_4H_5OS_2$	$[Ag(C_3H_5OCSS)]$	Ag: MVol.B7-98
$AgC_4H_5O_4S$	$[Ag(CH_2(COO)CH(SH)COOH)]$	Ag: MVol.B7-32
$AgC_4H_6^+$	$[AgC_4H_6]^+$	Ag: MVol.B5-35
$AgC_4H_6IO_4$	$AgI(CH_3COO)_2$	Ag: MVol.B5-126
$AgC_4H_6KO_4$	$KAg(CH_3COO)_2 \cdot n\ CH_3COOH$	Ag: MVol.B5-135
$AgC_4H_6LiO_4$	$Li[Ag(CH_3COO)_2]$	Ag: MVol.B5-147
$AgC_4H_6NO_2S$	$Ag(HOCH_2C_3H_3NOS)$	Ag: MVol.B7-53
$AgC_4H_6NO_3$	$[Ag(CH_3CONHCH_2COO)]$	Ag: MVol.B6-240
–	$Ag(NH_2COC_2H_4COO)$	Ag: MVol.B6-327
–	$AgNO_3 \cdot CHCC_2H_5$	Ag: MVol.B5-36
–	$AgNO_3 \cdot CH_2CHCHCH_2$	Ag: MVol.B5-35
$AgC_4H_6NO_3S$	$AgSCH_2CH(NHCHO)COOH$	Ag: MVol.B6-261
$AgC_4H_6NO_4$	$[Ag(NH_2C_2H_3(COOH)COO)]$	Ag: MVol.B6-252
$AgC_4H_6N_2^+$	$[Ag(CH_3CN)_2]^+$	Ag: MVol.B6-346/7
–	$[Ag(CH_3C_3H_3N_2)]^+$	Ag: MVol.B6-138/9
$AgC_4H_6N_3O_3$	$AgNO_3 \cdot 2\ CH_3CN$	Ag: MVol.B6-347, 348
–	$AgNO_3 \cdot 2\ CD_3CN$	Ag: MVol.B6-348
–	$AgNO_3 \cdot 2\ CH_3NC$	Ag: MVol.B5-25
$AgC_4H_6N_3O_3S_2$	$AgNO_3 \cdot C_2H_5SC_2HN_2S$	Ag: MVol.B7-80
$AgC_4H_6N_3O_5$	$[Ag(NHC(O)CH_2NHCH_2CO)NO_3]$	Ag: MVol.B6-165, 266
$AgC_4H_6N_3S$	$[Ag(C_2H_5C_2HN_3S)]$	Ag: MVol.B7-55

$AgC_4H_6N_3S_2$	$[Ag(C_2H_5NC_2HN_2S_2)$	Ag: MVol.B7-81
$AgC_4H_6O^+$	$[AgCH_3CHCHCHO]^+$	Ag: MVol.B5-49
–	$[AgCH_3COCHCH_2]^+$	Ag: MVol.B5-49
$AgC_4H_6O_2^+$	$[AgCH_2CHCOOCH_3]^+$	Ag: MVol.B5-49
–	$[AgCH_2CHOOCCH_3]^+$	Ag: MVol.B5-48
–	$[AgCH_3CHCHCOOH]^+$	Ag: MVol.B5-49
$AgC_4H_6O_4$	$Ag(CH_3COO)_2$	Ag: MVol.B5-148
$AgC_4H_6O_4^-$	$[Ag(CH_3COO)_2]^-$	Ag: MVol.B5-139, 146/7
AgC_4H_7	$AgC(CH_3)CHCH_3$	Ag: MVol.B5-7
–	$AgCHC(CH_3)_2$	Ag: MVol.B5-7
$AgC_4H_7N_2$	$AgC_4H_4N \cdot NH_3$	Ag: MVol.B6-68
$AgC_4H_7N_2O_2$	$AgN(COCH_2)_2 \cdot NH_3$	Ag: MVol.B6-30, 328
$AgC_4H_7N_2O_3$	$[Ag(NH_2CH_2CONHCH_2COO)]$	Ag: MVol.B6-265/6
–	$[Ag(NH_2COCH_2CH(NH_2)COO)]$	Ag: MVol.B6-255
$AgC_4H_7N_2O_4S$	$AgNO_3 \cdot C_4H_7NOS$	Ag: MVol.B7-85
$AgC_4H_7N_2O_5S$	$AgNO_3 \cdot HOCH_2C_3H_4NOS$ $= Ag(HOCH_2C_3H_3NOS) \cdot HNO_3$	Ag: MVol.B7-53
$AgC_4H_7N_4$	$[Ag(C_3H_7CN_4)]$	Ag: MVol.B6-192/3
$AgC_4H_7N_4O_3S$	$AgNO_3 \cdot C_2H_5C_2N_2SNH_2$	Ag: MVol.B7-80
$AgC_4H_7N_4O_7$	$[Ag((CH_3)_2CNOH)][C(NO_2)_3]$	Ag: MVol.B6-305
$AgC_4H_7N_6S_2$	$[Ag(NH_2CSNHNC_2H_2NNCSNH_2)]$	Ag: MVol.B6-282
AgC_4H_7OS	$Ag(SC(CH_3)_2CHO)$	Ag: MVol.B7-7
$AgC_4H_7OS_2$	$[Ag((CH_3)_2CHOCSS)]$	Ag: MVol.B7-97
–	$[Ag(C_3H_7OCSS)]$	Ag: MVol.B7-97 C: MVol.D4-257
$AgC_4H_7O_2S$	$[Ag(C_2H_5SCH_2COO)]$	Ag: MVol.B7-25/6
–	$[Ag(SCH_2COOC_2H_5)]$	Ag: MVol.B7-22
$AgC_4H_7O_3S$	$Ag(HOC_2H_4SCH_2COO)$	Ag: MVol.B7-26
$AgC_4H_8^+$	$[AgC_4H_8]^+$	Ag: MVol.B5-32/3
$AgC_4H_8IN_2S$	$Ag(CH_2CHCH_2NHCSNH_2)I$	Ag: MVol.B7-157
$AgC_4H_8NO_2$	$[Ag((CH_3)_2NCH_2COO)]$	Ag: MVol.B6-240
–	$[Ag(NH_2(CH_2)_3COO)]$	Ag: MVol.B6-248
$AgC_4H_8NO_2S$	$[Ag(CH_3SCH_2CH(NH_2)COO)]$	Ag: MVol.B6-257
–	$[Ag(SCH_2CH(NH_2)COOCH_3)]$	Ag: MVol.B6-257
–	$[Ag(SCH_2CH_2CH(NH_2)COOH)]$	Ag: MVol.B6-261
$AgC_4H_8NO_3S_2$	$AgNO_3 \cdot C_4H_8S_2$	Ag: MVol.B7-86, 87
$AgC_4H_8NO_3Se_2$	$AgNO_3 \cdot Se(CH_2CH_2)_2Se$	Ag: MVol.B7-198
$AgC_4H_8NO_4S$	$AgNO_3 \cdot S(CH_2CH_2)_2O$	Ag: MVol.B7-85
$AgC_4H_8NS_2$	$[Ag(SCSNHC_3H_7)]$	Ag: MVol.B7-102
$AgC_4H_8N_2O_2^-$	$[Ag(CH_3CONH)_2]^-$	Ag: MVol.B6-324
$AgC_4H_8N_2O_4$	$[Ag(NH_2CH_2COO)_2]$	Ag: MVol.B7-272
$AgC_4H_8N_2O_4^-$	$[Ag(C_2O_4)(C_2H_8N_2)]^-$	Ag: MVol.B6-55/6
–	$[Ag(NH_2CH_2COO)_2]^-$	Ag: MVol.B6-234/6, 250
$AgC_4H_8N_3O_3S$	$Ag(CH_2CHCH_2NHCSNH_2)NO_3$	Ag: MVol.B7-156
–	$AgNO_3 \cdot C_4H_8N_2S$	Ag: MVol.B7-64
$AgC_4H_8N_3O_6$	$[Ag(NH_2CH_2CONHCH_2COOH)]NO_3$	Ag: MVol.B6-266
$AgC_4H_8N_3S$	$[Ag((CH_3)_2CNNCSNH_2)]$	Ag: MVol.B6-283
$AgC_4H_8N_5O_3$	$AgNO_3 \cdot NH_2C_2N_3(CH_3)_2$	Ag: MVol.B6-186
$AgC_4H_8N_5O_5$	$AgNO_3 \cdot C_2N_3(CH_2OH)_2NH_2$	Ag: MVol.B6-186
$AgC_4H_8N_6S_2^-$	$[Ag(CN)_2(NH_2CSNH_2)_2]^-$	Ag: MVol.B7-134

Formula	Compound	Reference
$AgC_4H_8N_9O_3$	$AgNO_3 \cdot 2\ NH_2C(NH)NHCN$	Ag: MVol.B6-338
–	$AgNO_3 \cdot 2\ NH_2C_2H_2N_3$	Ag: MVol.B6-186
$AgC_4H_8O^+$	$[AgCH_2C(CH_3)CH_2OH]^+$	Ag: MVol.B5-47
–	$[AgCH_2C(CH_3)OCH_3]^+$	Ag: MVol.B5-48
–	$[AgCH_2CHCH(OH)CH_3]^+$	Ag: MVol.B5-47
–	$[AgCH_2CHCH_2CH_2OH]^+$	Ag: MVol.B5-47
–	$[AgCH_2CHOC_2H_5]^+$	Ag: MVol.B5-48
–	$[AgCH_3CHCHCH_2OH]^+$	Ag: MVol.B5-47
$AgC_4H_8O_2S^+$	$[Ag(C_2H_5SCH_2COOH)]^+$	Ag: MVol.B7-25/6
AgC_4H_9INO	$AgI \cdot NH(CH_2CH_2)_2O$	Ag: MVol.B6-204
$AgC_4H_9NO^+$	$[Ag(NH(CH_2CH_2)_2O)]^+$	Ag: MVol.B6-204
AgC_4H_9NPS	$[AgSCN(P(CH_3)_3)]$	Ag: MVol.B7-201
$AgC_4H_9N_3S^+$	$[Ag((CH_3)_2CNNHCSNH_2)]^+$	Ag: MVol.B6-283
$AgC_4H_9N_4O_3$	$Ag(ONC_3NO_2CH_3) \cdot 2\ NH_3$	Ag: MVol.B6-313
AgC_4H_9S	$(AgSC_4H_9)_n$	Ag: MVol.B7-5/6
$AgC_4H_{10}N^{2+}$	$[AgCH_3CHCHCH_2NH_3]^{2+}$	Ag: MVol.B5-49
–	$[Ag(H_2O)_x(CH_3CHCHCH_2NH_3)]^{2+}$	Ag: MVol.B6-46
$AgC_4H_{10}NO_3S$	$[Ag(NH_2C_4H_8SO_3)]$	Ag: MVol.B6-262
–	$AgNO_3 \cdot S(C_2H_5)_2$	Ag: MVol.B7-11
$AgC_4H_{10}NO_3S_2$	$AgNO_3 \cdot C_2H_4(SCH_3)_2$	Ag: MVol.B7-14
–	$AgNO_3 \cdot C_2H_5SSC_2H_5$	Ag: MVol.B7-15
$AgC_4H_{10}NS$	$[Ag(SCH_2CH_2N(CH_3)_2)]$	Ag: MVol.B7-18
$AgC_4H_{10}N_2^+$	$[Ag(C_2H_4NH)_2]^+$	Ag: MVol.B6-67
–	$[Ag(NH(CH_2CH_2)_2NH)]^+$	Ag: MVol.B6-162/4
$AgC_4H_{10}N_2O^+$	$[Ag(HOC(CH_3)_2C(NH)NH_2)]^+$	Ag: MVol.B6-339/40
$AgC_4H_{10}N_2O_4^+$	$[Ag(NH_2CH_2COOH)_2]^+$	Ag: MVol.B6-234
$AgC_4H_{10}N_2S^+$	$[Ag((CH_3)_2NCSNHCH_3)]^+$	Ag: MVol.B7-153
$AgC_4H_{10}N_2S_2^+$	$[Ag(CH_3CSNH_2)_2]^+$	Ag: MVol.B7-123
$AgC_4H_{10}N_3O_3$	$Ag(NH_2COC_2H_3(NH_2)COO) \cdot NH_3$	Ag: MVol.B6-30, 255
$AgC_4H_{10}N_3O_3S$	$Ag((CH_3)_2NCSNHCH_3)NO_3$	Ag: MVol.B7-153
$AgC_4H_{10}N_3O_3S_2$	$[Ag(CH_3CSNH_2)_2]NO_3$	Ag: MVol.B7-124
$AgC_4H_{10}N_3O_4$	$AgNO_3 \cdot HOC(CH_3)_2C(NH)NH_2$	Ag: MVol.B6-340
$AgC_4H_{10}N_5O_5$	$AgNO_3 \cdot CH_3CH(CONHNH_2)_2 \cdot 2\ H_2O$	Ag: MVol.B6-332
$AgC_4H_{10}N_9O_3$	$AgNO_3 \cdot C_2H_2(NNHC(NH)NH_2)_2$	Ag: MVol.B6-345
$AgC_4H_{10}O_2S_2^-$	$[Ag(SCH_2CH_2OH)_2]^-$	Ag: MVol.B7-6
$AgC_4H_{10}O_4S_4^-$	$[Ag(C_2H_5SO_2S)_2]^-$	Ag: MVol.B7-96
$AgC_4H_{10}S_2^+$	$[Ag(C_2H_4(SCH_3)_2)]^+$	Ag: MVol.B7-14
$AgC_4H_{11}N^+$	$[Ag(C_4H_9NH_2)]^+$	Ag: MVol.B6-45, 48
–	$[Ag(NH(C_2H_5)_2)]^+$	Ag: MVol.B6-47/8
$AgC_4H_{11}NO^+$	$[Ag(NH_2C_4H_8OH)]^+$	Ag: MVol.B6-228
$AgC_4H_{11}NOS^+$	$[Ag(NH_2C_2H_4SC_2H_4OH)]^+$	Ag: MVol.B7-18
$AgC_4H_{11}NO_2^+$	$[Ag(CH_3C(CH_2OH)_2NH_2)]^+$	Ag: MVol.B6-230/1
–	$[Ag(NH(C_2H_4OH)_2)]^+$	Ag: MVol.B6-228/9
$AgC_4H_{11}NO_3^+$	$[Ag(NH_2C(CH_2OH)_3)]^+$	Ag: MVol.B6-230/1
$AgC_4H_{11}N_2^{2+}$	$[Ag(NH(CH_2CH_2)_2NH)H]^{2+}$	Ag: MVol.B6-162/3
$AgC_4H_{11}N_2O_4S_3$	$[Ag(CH_3CSNH_2)_2]HSO_4$	Ag: MVol.B7-124
$AgC_4H_{11}N_4O_2S_2$	$[Ag(NH_2CSNH_2)_2]CH_3COO$	Ag: MVol.B7-142
$AgC_4H_{12}ITe_2$	$AgI \cdot 2\ (CH_3)_2Te$	Ag: MVol.B7-198
$AgC_4H_{12}N_2^+$	$[Ag(NH_2C_4H_8NH_2)]^+$	Ag: MVol.B6-62/3
$AgC_4H_{12}N_2NiS_2^+$	$[AgNi(SC_2H_4NH_2)_2]^+$	Ag: MVol.B7-17

Formula	Compound	Reference
$AgC_4H_{12}N_2O^+$	$[Ag(O(C_2H_4NH_2)_2)]^+$	Ag: MVol.B6-232
$AgC_4H_{12}N_2O_6S_2^-$	$[Ag(NH_2C_2H_4SO_3)_2]^-$	Ag: MVol.B6-262
$AgC_4H_{12}N_2O_8S_2^-$	$[Ag(NH_2C_2H_4SO_4)_2]^-$	Ag: MVol.B6-262
$AgC_4H_{12}N_2S^+$	$[Ag(S(CH_2CH_2NH_2)_2)]^+$	Ag: MVol.B7-18
$AgC_4H_{12}N_3O_3$	$Ag(CH_3CONHCH_2COO) \cdot 2\ NH_3$	Ag: MVol.B6-240
–	$[Ag(NH_2C_4H_8NH_2)]NO_3$	Ag: MVol.B6-63
$AgC_4H_{12}N_4S_2^+$	$[Ag(CH_3NHCSNH_2)_2]^+$	Ag: MVol.B7-150
$AgC_4H_{12}N_5O_3S_2$	$[Ag(CH_3NHCSNH_2)_2]NO_3$	Ag: MVol.B7-150
$AgC_4H_{12}N_7S_3$	$[AgCN(NH_2CSNH_2)_3]$	Ag: MVol.B7-134
$AgC_4H_{12}N_8S_2^+$	$[Ag(NH_2C(NH)NHCSNH_2)_2]^+$	Ag: MVol.B7-162
$AgC_4H_{12}O_2S_2^+$	$[Ag((CH_3)_2SO)_2]^+$	Ag: MVol.B7-2
$AgC_4H_{13}N_2^{2+}$	$[Ag(NH_3C_4H_8NH_2)]^{2+}$	Ag: MVol.B6-63
$AgC_4H_{13}N_3^+$	$[Ag(CH(CH_2NH_2)_3)]^+$	Ag: MVol.B6-65
–	$[Ag(NH(C_2H_4NH_2)_2)]^+$	Ag: MVol.B6-65
$AgC_4H_{13}N_4O_3$	$Ag(NH_2COC_2H_3(NH_2)COO) \cdot 2\ NH_3$	Ag: MVol.B6-30, 255
$AgC_4H_{13}N_6O_2S_3$	$[Ag(NH_2CSNH_2)_3]HCOO \cdot 2\ H_2O$	Ag: MVol.B7-146
$AgC_4H_{14}N_2^+$	$[Ag(C_2H_5NH_2)_2]^+$	Ag: MVol.B6-43
–	$[Ag(NH(CH_3)_2)_2]^+$	Ag: MVol.B6-47
$AgC_4H_{14}N_2O_2^+$	$[Ag(NH_2C_2H_4OH)_2]^+$	Ag: MVol.B6-226/7
$AgC_4H_{14}N_2O_8P_2^-$	$[Ag(NH_2C_2H_4OPO_3H)_2]^-$	Ag: MVol.B7-250
$AgC_4H_{14}N_3^{2+}$	$[AgH(CH(CH_2NH_2)_3)]^{2+}$	Ag: MVol.B6-65
–	$[AgH(NH(C_2H_4NH_2)_2)]^{2+}$	Ag: MVol.B6-65
$AgC_4H_{14}N_{10}^{3+}$	$[Ag(HNC(NH_2)NHC(NH_2)NH)_2]^{3+}$	Ag: MVol.B7-321
$AgC_4H_{14}N_{11}O_7S$	$[Ag(NH(C(NH_2)NH)_2)_2](NO_3)(SO_4) \cdot 6\ H_2O$	Ag: MVol.B7-320
$AgC_4H_{14}N_{13}O_9$	$[Ag(NH(C(NH_2)NH)_2)_2](NO_3)_3$	Ag: MVol.B7-320
$AgC_4H_{15}N_4$	$C_2H_8N_2AgC_2H_7N_2$	Ag: MVol.B6-57
$AgC_4H_{15}N_{10}O_5S$	$[Ag(NH(C(NH_2)NH)_2)_2](OH)(SO_4) \cdot 8\ H_2O$	Ag: MVol.B7-320
$AgC_4H_{16}N_4^+$	$[Ag(C_2H_8N_2)_2]^+$	Ag: MVol.B6-55/7
$AgC_4H_{16}N_5O_2$	$AgNO_2 \cdot 2\ C_2H_8N_2$	Ag: MVol.B6-58
$AgC_4H_{16}N_8S_4^+$	$[Ag(NH_2CSNH_2)_4]^+$	Ag: MVol.B7-128/33
$AgC_4H_{16}N_9O_3S_4$	$Ag(NH_2CSNH_2)_4NO_3$	Ag: MVol.B7-148
$AgC_4H_{17}N_4^{2+}$	$[(C_2H_8N_2)AgH(C_2H_8N_2)]^{2+}$	Ag: MVol.B6-55
$AgC_4H_{17}N_4O$	$[Ag(C_2H_8N_2)_2]OH$	Ag: MVol.B6-57
$AgC_4H_{17}N_{10}O_3$	$[Ag(NH(C(NH_2)NH)_2)_2](OH)_3 \cdot 4.5\ H_2O$	Ag: MVol.B7-320
$AgC_4H_{18}N_5O_4$	$Ag(NH_2C_2H_3(COOH)COO) \cdot 4\ NH_3$	Ag: MVol.B6-252
$AgC_4H_{2x+10}NO_x^{2+}$	$[Ag(H_2O)_xC_4H_7NH_3]^{2+}$	Ag: MVol.B6-46
$AgC_4KN_6O_8$	$AgC(NO_2)_2CN \cdot KC(NO_2)_2CN$	Ag: MVol.B6-354
AgC_4N_3	$AgC(CN)_3$	Ag: MVol.B6-353/4
AgC_4N_5	$[Ag(C_2N_3(CN)_2)]$	Ag: MVol.B6-181
$AgC_5ClF_6H_3N$	$AgCCl(CF_3)_2 \cdot CH_3CN$	Ag: MVol.B5-5
$AgC_5ClH_3NO_2S$	$[Ag(CH_3C_3ClNSCOO)]$	Ag: MVol.B7-70
AgC_5ClH_5N	$AgCl \cdot C_5H_5N$	Ag: MVol.B6-78/9
$AgC_5ClH_5N_3O_3$	$AgNO_3 \cdot CH_3C_4H_2ClN_2$	Ag: MVol.B6-160
$AgC_5ClH_5N_3O_4$	$Ag(CH_3C_4H_2ClN_2O)NO_3$	Ag: MVol.B6-162
$AgC_5ClH_6O_4$	$AgClO_4 \cdot C_5H_6$	Ag: MVol.B5-54
$AgC_5ClH_{14}N_2O_4$	$[Ag(NH_2(CH_2)_5NH_2)]ClO_4$	Ag: MVol.B6-63
–	$[Ag(NH_2C_3H_6N(CH_3)_2)]ClO_4$	Ag: MVol.B6-62
$AgC_5ClH_{20}N_{10}S_5$	$Ag(NH_2CSNH_2)_5Cl$	Ag: MVol.B7-148
$AgC_5Cl_4H_{14}N_4Rh$	$[Ag(NH_3)_2][RhCl_4(NH_3)(C_5H_5N)]$	Ag: MVol.B6-32
$AgC_5F_3H_{12}N_6O_2S_3$	$[Ag(NH_2CSNH_2)_3]CF_3COO$	Ag: MVol.B7-147

$AgC_5F_7H_3N$	$AgCF(CF_3)_2 \cdot CH_3CN$	Ag: MVol.B5-5
$AgC_5H_2N_3O_3S$	$[Ag(C_5H_2N_3O_3S)]$	Ag: MVol.B7-65
AgC_5H_3	$AgCCCCCH_3$	Ag: MVol.B5-14
$AgC_5H_3N_2O_2$	$[Ag(C_4H_3N_2(COO))]$	Ag: MVol.B6-160
$AgC_5H_3N_2O_4$	$Ag(C_4HNO_3(CONH_2))$	Ag: MVol.B6-69
–	$Ag(NHC_4H_2(NO_2)COO)$	Ag: MVol.B6-68
$AgC_5H_3N_4O_2$	$AgC_5H_3N_4O_2$	Ag: MVol.B6-166/7
$AgC_5H_3N_4S$	$[Ag(C_5H_3N_4S)]$	Ag: MVol.B7-66/7
AgC_5H_4NO	$AgOC_5H_4N$	Ag: MVol.B6-91
AgC_5H_4NOS	$[Ag(C_5H_4NOS)]$	Ag: MVol.B7-59
$AgC_5H_4NO_2S_2$	$[Ag((CH_2CO)_2NCSS)]$	Ag: MVol.B7-117
$AgC_5H_4NO_4S_2{}^{2-}$	$[Ag(N(CH_2COO)_2CSS)]^{2-}$	Ag: MVol.B7-115
$AgC_5H_4N_3O_4$	$[Ag(ONC_4HN_2O_3CH_3)]$	Ag: MVol.B6-315
AgC_5H_5IN	$AgI \cdot C_5H_5N$	Ag: MVol.B6-80
$AgC_5H_5N^+$	$[Ag(C_5H_5N)]^+$	Ag: MVol.B6-74/6
$AgC_5H_5N_2O$	$[Ag(CH_3C(O)C_3H_2N_2)]$	Ag: MVol.B6-131
$AgC_5H_5N_2O_4$	$Ag(C_5H_5NO)NO_3$	Ag: MVol.B6-104
$AgC_5H_6{}^+$	$[AgC_5H_6]^+$	Ag: MVol.B5-52/3
$AgC_5H_6KN_2OSSe$	$KAg(SCN)(SeCN) \cdot (CH_3)_2CO$	Ag: MVol.B6-209
$AgC_5H_6NO_3$	$[Ag(C_3NO_3(CH_3)_2)]$	Ag: MVol.B6-203
–	$Ag(NHC_4H_5O(COO))$	Ag: MVol.B6-69
$AgC_5H_6N_2{}^+$	$[Ag(CH_3C_4H_3N_2)]^+$	Ag: MVol.B6-159
–	$[Ag(NH_2C_5H_4N)]^+$	Ag: MVol.B6-91/2
$AgC_5H_6N_3O$	$Ag(NHC_4H_2N_2O(CH_3))$	Ag: MVol.B6-152
$AgC_5H_6N_3O_3$	$Ag(CH_3CONHC_3H_2N_2O_2) \cdot H_2O$	Ag: MVol.B6-148
–	$Ag(C_3N_3O_3(CH_3)_2) \cdot 0.5\ H_2O$	Ag: MVol.B6-190
–	$Ag(NH_2C_5H_4N)NO_3$	Ag: MVol.B6-94
–	$Ag(ND_2C_5D_4N)NO_3$	Ag: MVol.B6-94
–	$AgNO_3 \cdot CH_3C_4H_3N_2$	Ag: MVol.B6-160
$AgC_5H_6N_3O_4$	$Ag(CH_3C_4H_3N_2O)NO_3$	Ag: MVol.B6-162
–	$Ag(C_3H_2N_2O_2(CH_3)NHCOO)$	Ag: MVol.B6-148
$AgC_5H_6N_3S_2$	$[Ag(C_3H_5NC_2HN_2S_2)]$	Ag: MVol.B7-81
$AgC_5H_6N_5O_2$	$AgC_5H_3N_4O_2 \cdot NH_3$	Ag: MVol.B6-166/7
$AgC_5H_6O_4S_2{}^-$	$[Ag(CH_2(SCH_2COO)_2)]^-$	Ag: MVol.B7-39/40
AgC_5H_7	$AgCCC_3H_7$	Ag: MVol.B5-13
$AgC_5H_7NO^+$	$[Ag(NH_2CH_2C_4H_3O)]^+$	Ag: MVol.B6-222
$AgC_5H_7N_2$	$Ag[NC(C_2H_5)NCHCH]$	Ag: MVol.B6-142
–	$Ag[NCHNC(CH_3)C(CH_3)]$	Ag: MVol.B6-142
–	$Ag[NNCHC(C_2H_5)CH]$	Ag: MVol.B6-131
$AgC_5H_7N_2O_2$	$Ag((CH_3)_2C_3HN_2O_2)$	Ag: MVol.B6-147/8
–	$Ag(C_4H_5NO(CONH_2))$	Ag: MVol.B6-69
$AgC_5H_7N_2O_6$	$Ag(NO_2NCOOCH_2COOC_2H_5)$	Ag: MVol.B6-331
–	$AgNO_3 \cdot (CH_3)_2C_3HNO_3$	Ag: MVol.B6-203
AgC_5H_7O	$AgCCC(CH_3)_2OH \cdot AgC_3H_7COO$	Ag: MVol.B5-15
$AgC_5H_7O_2$	$[Ag(CH_3COCHCOCH_3)]$	Ag: MVol.B6-211/3
$AgC_5H_7O_2S$	$[AgCH_2CHCH_2SCH_2COO]$	Ag: MVol.B5-50/1
$AgC_5H_8{}^+$	$[Ag(CHCC_3H_7)]^+$	Ag: MVol.B5-38
–	$[Ag(CH_2C(CH_3)CHCH_2)]^+$	Ag: MVol.B5-38
–	$[AgC_4H_5CH_3]^+$	Ag: MVol.B5-52
–	$[AgC_4H_6CH_2]^+$	Ag: MVol.B5-52

Formula	Compound	Reference
$AgC_5H_8^+$	$[AgC_5H_8]^+$	Ag: MVol.B5-52/3
$AgC_5H_8NOS_2$	$[Ag(O(CH_2CH_2)_2NCSS)]$	Ag: MVol.B7-118
$AgC_5H_8NO_2$	$[Ag(C_3H_5CH(NH_2)COO)]$	Ag: MVol.B5-50
		Ag: MVol.B6-250
–	$[Ag(C_3H_2NO_2(CH_3)_2)]$	Ag: MVol.B6-203
–	$[Ag(NC_4H_8COO)]$	Ag: MVol.B6-254
$AgC_5H_8NO_3S$	$AgSCH_2CH_2CH(NHCHO)COOH$	Ag: MVol.B6-261
$AgC_5H_8NO_4$	$[Ag(NH_2C_3H_5(COOH)COO)]$	Ag: MVol.B6-253
$AgC_5H_8NS_2$	$[Ag(C_4H_8NCSS)]$	Ag: MVol.B7-117
$AgC_5H_8NS_3$	$[Ag(S(CH_2CH_2)_2NCSS)]$	Ag: MVol.B7-118
$AgC_5H_8N_2^+$	$[Ag((CH_3)_2C_3H_2N_2)]^+$	Ag: MVol.B6-138/9
–	$[Ag(C_2H_5C_3H_3N_2)]^+$	Ag: MVol.B6-138/9
$AgC_5H_8N_3O$	$[Ag(C_3H_7CONNHCN)]$	Ag: MVol.B6-332
$AgC_5H_8N_3O_5S$	$Ag(CH_3CONHCSNHCOCH_3)NO_3$	Ag: MVol.B7-161
$AgC_5H_8N_3O_6S$	$AgC(NO_2)_3 \cdot S(CH_2CH_2)_2$	Ag: MVol.B7-70
$AgC_5H_8N_3O_8$	$Ag(C_4H_8O_2)C(NO_2)_3$	Ag: MVol.B6-222
$AgC_5H_8N_3S$	$Ag(C_2N_3S(CH_3)C_2H_5) \cdot n\ AgCl$	Ag: MVol.B7-55/6
–	$Ag(C_2N_3S(CH_3)C_2H_5) \cdot n\ AgNO_3$	Ag: MVol.B7-55/6
$AgC_5H_8N_5$	$[Ag(C_4H_8NCN_4)]$	Ag: MVol.B6-192/3
$AgC_5H_8N_5O$	$[Ag(C_4H_8ONCN_4)]$	Ag: MVol.B6-192/3
$AgC_5H_8N_5O_5$	$AgNO_3 \cdot NH_2C_3N_3(OCH_3)_2$	Ag: MVol.B6-189
$AgC_5H_8N_7O_5$	$AgNO_3 \cdot C_3H_5N_6CH_2COOH \cdot H_2O$	Ag: MVol.B6-188
$AgC_5H_8O_2^+$	$[AgCH_2CHCH_2OOCCH_3]^+$	Ag: MVol.B5-48
$AgC_5H_8O_2S^+$	$[AgCH_2CHCH_2SCH_2COOH]^+$	Ag: MVol.B5-50/1
$AgC_5H_9NO_2^+$	$[Ag(C_3H_5CH(NH_2)COOH)]^+$	Ag: MVol.B6-250
$AgC_5H_9N_2O_2$	$AgC_4H_9N_2(COO)$	Ag: MVol.B6-163
$AgC_5H_9N_2O_4$	$Ag(CH_3CH(NNO_2)COOC_2H_5)$	Ag: MVol.B6-243
$AgC_5H_9N_3^+$	$[Ag(NH_2C_2H_4C_3H_3N_2)]^+$	Ag: MVol.B6-139
$AgC_5H_9N_4$	$[Ag(C_3H_7C_2N_3NH_2)]$	Ag: MVol.B6-183
–	$[Ag(C_4H_9CN_4)]$	Ag: MVol.B6-192/3
$AgC_5H_9N_4O_4$	$AgNO_3 \cdot C_3H_7CONHNHCN$	Ag: MVol.B6-332
$AgC_5H_9N_4S_2$	$[Ag(C_3H_5NCSNHNHCSNH_2)]$	Ag: MVol.B7-169
$AgC_5H_9OS_2$	$[Ag(C_4H_9OCSS)]$	Ag: MVol.B7-98
$AgC_5H_9O_2S$	$[Ag(C_3H_7SCH_2COO)]$	Ag: MVol.B7-25/6
$AgC_5H_9O_3S_4^{2-}$	$[Ag(SCH_2CH(S)CH_2SC_2H_4SO_3)]^{2-}$	Ag: MVol.B7-46
$AgC_5H_9O_4S_3^{2-}$	$[Ag(SCH_2CH(S)CH_2OC_2H_4SO_3)]^{2-}$	Ag: MVol.B7-45
$AgC_5H_9O_5S_4^{2-}$	$[Ag(SCH_2CH(S)CH_2SO_2C_2H_4SO_3)]^{2-}$	Ag: MVol.B7-46
AgC_5H_9Si	$AgCCSi(CH_3)_3$	Ag: MVol.B7-246
$AgC_5H_{10}^+$	$[Ag(C_5H_{10})]^+$	Ag: MVol.B5-36/7
$AgC_5H_{10}NOS$	$[Ag((C_2H_5)_2NCOS)]$	Ag: MVol.B7-99/100
$AgC_5H_{10}NO_2$	$[Ag((CH_3)_2CHCH(NH_2)COO)]$	Ag: MVol.B6-248
–	$[Ag(C_3H_7CH(NH_2)COO)]$	Ag: MVol.B6-248, 250
–	$[Ag(NH_2(CH_2)_4COO)]$	Ag: MVol.B6-248
$AgC_5H_{10}NO_2S$	$[Ag(CH_3SC_2H_4CH(NH_2)COO)]$	Ag: MVol.B6-259
–	$[Ag(SC(CH_3)_2CH(NH_2)COOH)]$	Ag: MVol.B6-258
$AgC_5H_{10}NO_2S_2$	$[Ag(N(C_2H_4OH)_2CSS)]$	Ag: MVol.B7-116
$AgC_5H_{10}NO_3S$	$AgNO_3 \cdot S(CH_2)_5$	Ag: MVol.B7-84
$AgC_5H_{10}NSSe$	$Ag(SSeCN(C_2H_5)_2)$	Ag: MVol.B7-197
$AgC_5H_{10}NS_2$	$[Ag((C_2H_5)_2NCSS)]$	Ag: MVol.B7-105/10
–	$[Ag(C_4H_9NHCSS)]$	Ag: MVol.B7-102

Formula	Compound	Reference
$AgC_5H_{10}N_2O_3P$	$AgC(N_2)PO(OC_2H_5)_2$	Ag: MVol.B5-23
$AgC_5H_{10}N_5$	$[Ag(N(C_2H_5)_2CN_4)]$	Ag: MVol.B6-192/3
$AgC_5H_{10}O^+$	$[AgCH_2CHCH(OH)C_2H_5]^+$	Ag: MVol.B5-47
–	$[AgCH_2CH(CH_2)_2CH_2OH]^+$	Ag: MVol.B5-47
–	$[AgCH_3CHC(CH_3)CH_2OH]^+$	Ag: MVol.B5-47
–	$[AgCH_3CHCHOC_2H_5]^+$	Ag: MVol.B5-48
–	$[Ag(CH_3)_2CCHCH_2OH]^+$	Ag: MVol.B5-47
$AgC_5H_{10}O_2S^+$	$[Ag(C_3H_7SCH_2COOH)]^+$	Ag: MVol.B7-25/6
$AgC_5H_{11}IN$	$AgI \cdot NH(CH_2)_5$	Ag: MVol.B6-72
$AgC_5H_{11}N^+$	$[Ag(NH(CH_2)_5)]^+$	Ag: MVol.B6-72/3
$AgC_5H_{11}N_2O_6S_2$	$AgNO_3 \cdot [CH_3C_4H_8S_2]NO_3$	Ag: MVol.B7-87
$AgC_5H_{11}O_3S$	$[Ag(SCH_2C(CH_2OH)_3)]$	Ag: MVol.B7-7
$AgC_5H_{11}S$	$(AgSC_5H_{11})_n$	Ag: MVol.B7-5/6
$AgC_5H_{12}NO_3S$	$[Ag(NH_2(CH_2)_5SO_3)]$	Ag: MVol.B6-262
$AgC_5H_{12}NO_3S_2$	$AgNO_3 \cdot CH_2(SCH_2CH_3)_2$	Ag: MVol.B7-14
$AgC_5H_{12}NS$	$[Ag(SC(CH_3)_2CH_2NHCH_3)]$	Ag: MVol.B7-17
$AgC_5H_{12}N_2{}^+$	$[Ag(CH_3C_4H_9N_2)]^+$	Ag: MVol.B6-163/4
$AgC_5H_{12}N_2O^+$	$[Ag(C_2H_5C(CH_3)(OH)C(NH)NH_2)]^+$	Ag: MVol.B6-339/40
$AgC_5H_{12}N_2S^+$	$[Ag((CH_3)_2NCSN(CH_3)_2)]^+$	Ag: MVol.B7-153
–	$[Ag(C_2H_5NHCSNHC_2H_5)]^+$	Ag: MVol.B7-155
$AgC_5H_{12}N_3O_3S$	$Ag(C_2H_5NHCSNHC_2H_5)NO_3$	Ag: MVol.B7-155
–	$Ag((C_2H_5)_2NCSNH_2)NO_3$	Ag: MVol.B7-155
$AgC_5H_{12}N_3O_4$	$AgNO_3 \cdot CO(N(CH_3)_2)_2$	Ag: MVol.B6-335
–	$AgNO_3 \cdot C_2H_5C(CH_3)(OH)C(NH)NH_2$	Ag: MVol.B6-340
$AgC_5H_{13}N^+$	$[Ag(C_5H_{11}NH_2)]^+$	Ag: MVol.B6-46
$AgC_5H_{13}NO^+$	$[Ag(NH_2C_5H_{10}OH)]^+$	Ag: MVol.B6-228
$AgC_5H_{13}NO_2{}^+$	$[Ag(CH_3N(C_2H_4OH)_2)]^+$	Ag: MVol.B6-229
–	$[Ag(NH(C_2H_4OH)C_3H_6OH)]^+$	Ag: MVol.B6-229
$AgC_5H_{13}N_2O_2$	$Ag(C_3H_7CH(NH_2)COO) \cdot NH_3$	Ag: MVol.B6-248
$AgC_5H_{14}IN_4OS_2$	$AgI \cdot 2\ NH_4SCN \cdot (CH_3)_2CO$	Ag: MVol.B6-209
$AgC_5H_{14}N_2{}^+$	$[Ag(NH_2(CH_2)_5NH_2)]^+$	Ag: MVol.B6-63
$AgC_5H_{14}N_2{}^{2+}$	$[Ag((NH_2CH_2)_2C(CH_3)_2)]^{2+}$	Ag: MVol.B7-272
$AgC_5H_{16}N_3O_2$	$Ag(NH_2(CH_2)_4COO) \cdot 2\ NH_3$	Ag: MVol.B6-249
$AgC_5H_{16}N_3O_2S$	$Ag(CH_3SC_2H_4CH(NH_2)COO) \cdot 2\ NH_3$	Ag: MVol.B6-259
$AgC_5H_{16}N_5$	$[AgCN(C_2H_8N_2)_2]$	Ag: MVol.B6-55/6
$AgC_5H_{19}N_4O_2$	$Ag(NH_2(CH_2)_4COO) \cdot 3\ NH_3$	Ag: MVol.B6-249
$AgC_5H_{19}N_4O_2S$	$Ag(CH_3SC_2H_4CH(NH_2)COO) \cdot 3\ NH_3$	Ag: MVol.B6-259
$AgC_5H_{20}N_5O_4$	$Ag(NH_2C_3H_5(COOH)COO) \cdot 4\ NH_3$	Ag: MVol.B6-253
$AgC_5H_{20}N_{10}S_5{}^+$	$[Ag(NH_2CSNH_2)_5]^+$	Ag: MVol.B7-132
AgC_5N_3O	$Ag(C(CN)_2C(CN)O)$	Ag: MVol.B6-355
$AgC_6ClH_2N_2O_4S$	$Ag(ClC_6H_2N_2OSO_3) \cdot 2\ H_2O$	Ag: MVol.B6-293
$AgC_6ClH_3N_3O_6$	$Ag(NO_2(CN)C_5H_3N)ClO_4$	Ag: MVol.B6-101
$AgC_6ClH_5{}^+$	$[AgC_6H_5Cl]^+$	Ag: MVol.B5-117
$AgC_6ClH_6N_2O_4S$	$[AgC_6H_6N_2S]ClO_4 \cdot 2\ H_2O$	Ag: MVol.B7-127
$AgC_6ClH_6O_4$	$AgClO_4 \cdot C_6H_6$	Ag: MVol.B5-91/3
$AgC_6ClH_7NO_4$	$AgClO_4 \cdot C_6H_5NH_2$	Ag: MVol.B6-51
$AgC_6ClH_7NO_4S$	$[Ag(C_6H_7NS)]ClO_4$	Ag: MVol.B7-59
$AgC_6ClH_7N_3O_3$	$AgNO_3 \cdot (CH_3)_2C_4HClN_2$	Ag: MVol.B6-160
$AgC_6ClH_8N_2$	$AgCl \cdot C_6H_5NHNH_2$	Ag: MVol.B6-36
$AgC_6ClH_8O_4$	$AgClO_4 \cdot C_6H_8$	Ag: MVol.B5-59

Formula	Compound	Reference
$AgC_6ClH_{10}O_4$	$AgClO_4 \cdot C_5H_7CH_3$	Ag: MVol.B5-54
$AgC_6ClH_{11}NO_5$	$[AgC_6H_{11}NO]ClO_4$	Ag: MVol.B6-129/30
$AgC_6ClH_{12}N_4O_3$	$AgClO_3 \cdot (CH_2)_6N_4$	Ag: MVol.B6-198
–	$AgClO_3 \cdot (CH_2)_6N_4 \cdot H_2O$	Ag: MVol.B6-198
$AgC_6ClH_{12}N_4O_4$	$AgClO_4 \cdot (CH_2)_6N_4$	Ag: MVol.B6-198
$AgC_6ClH_{12}N_8O_4$	$[Ag((CH_3)_2CN_4)_2]ClO_4$	Ag: MVol.B6-193
$AgC_6ClH_{12}O_4$	$AgClO_4 \cdot 2\ CH_2CHCH_3$	Ag: MVol.B5-31/2
$AgC_6ClH_{14}N_2S_2$	$Ag(HCSN(CH_3)_2)_2Cl$	Ag: MVol.B7-121/2
$AgC_6ClH_{15}O_3P$	$AgCl \cdot P(OC_2H_5)_3$	Ag: MVol.B7-246
$AgC_6ClH_{15}P$	$[AgCl(P(C_2H_5)_3)]$	Ag: MVol.B7-202
$AgC_6ClH_{16}N_2O_4$	$[Ag(NH_2(CH_2)_6NH_2)]ClO_4$	Ag: MVol.B6-63
$AgC_6ClH_{16}N_4O_4S_2$	$[Ag(C_2H_5NHCSNH_2)_2]ClO_4$	Ag: MVol.B7-154
$AgC_6ClH_{18}N_3O_5P$	$[Ag(OP(N(CH_3)_2)_3)]ClO_4$	Ag: MVol.B7-251
$AgC_6ClH_{18}N_6O_4S_3$	$[Ag(CH_3NHCSNH_2)_3]ClO_4$	Ag: MVol.B7-151
$AgC_6ClH_{18}N_6S_3$	$[Ag(CH_3NHCSNH_2)_3]Cl$	Ag: MVol.B7-150/1
$AgC_6ClH_{18}P_2$	$[AgCl(P(CH_3)_3)_2]_2$	Ag: MVol.B7-200/1
$AgC_6Cl_2H_3N_2O_4$	$Ag(Cl(CN)C_5H_3N)ClO_4$	Ag: MVol.B6-101
$AgC_6Cl_2H_4^+$	$[AgC_6H_4Cl_2]^+$	Ag: MVol.B5-117
$AgC_6Cl_3H_{16}N_{10}O_{12}$	$[Ag(C_2H_4(C_2H_6N_5)_2)](ClO_4)_3 \cdot 1.5\ H_2O$	Ag: MVol.B7-321/4
AgC_6Cl_5S	$Ag(SC_6Cl_5)$	Ag: MVol.B7-8
AgC_6FH_4	AgC_6H_4F	Ag: MVol.B5-9
$AgC_6FH_5^+$	$[AgC_6H_5F]^+$	Ag: MVol.B5-117
$AgC_6FH_{12}N_4$	$AgF \cdot (CH_2)_6N_4 \cdot 3\ H_2O$	Ag: MVol.B6-198
$AgC_6F_2H_6NO_4S_2$	$AgN(SO_2F)_2 \cdot C_6H_6$	Ag: MVol.B5-93
		S: S-N-Verb.1-50
AgC_6F_5	AgC_6F_5	Ag: MVol.B5-9/10
AgC_6F_5S	$Ag(SC_6F_5)$	Ag: MVol.B7-8
$AgC_6F_7H_2N_2$	$[Ag(C_3F_7C_3H_2N_2)]$	Ag: MVol.B6-133
AgC_6F_{13}	$AgC(CF_3)_2C_3F_7$	Ag: MVol.B5-6
$AgC_6F_{14}O_2P$	$(C_3F_7)_2P(O)OAg$	F: PerFHalOrg.3-50
$AgC_6H_3N_2OS$	$AgC(N_2)COSC_4H_3$	Ag: MVol.B5-22
$AgC_6H_3N_2O_3S_3$	$AgNO_3 \cdot C_6H_3NS_3$	Ag: MVol.B7-82
$AgC_6H_3O_3S_3$	$[Ag(C_5H_3OS_3COO)]$	Ag: MVol.B7-83/4
$AgC_6H_4I_2^+$	$[AgC_6H_4I_2]^+$	Ag: MVol.B5-117
$AgC_6H_4NO_2$	$Ag(NC_5H_4COO)$	Ag: MVol.B6-102
–	$Ag(OC_6H_4NO)$	Ag: MVol.B6-306
$AgC_6H_4NO_3$	$Ag(NHC_4H_3COCOO)$	Ag: MVol.B6-68
$AgC_6H_4N_2O_2S_4^-$	$[Ag(C_3H_2NOS_2)_2]^-$	Ag: MVol.B7-73
$AgC_6H_4N_3$	$[Ag(C_6H_4N_3)]$	Ag: MVol.B6-181
$AgC_6H_4N_3O_3$	$[Ag(NO_2C_6H_4NNO)]$	Ag: MVol.B6-293
$AgC_6H_4N_3O_3S$	$AgNO_3 \cdot C_6H_4N_2S$	Ag: MVol.B7-79
$AgC_6H_4N_3O_3Se$	$AgNO_3 \cdot C_6H_4N_2Se$	Ag: MVol.B7-198
$AgC_6H_4N_3O_5$	$[Ag(OC_6H_2(NO_2)_2NH_2)]$	Ag: MVol.B6-231/2
$AgC_6H_4N_3O_5S$	$Ag(NO_2C_6H_4NNSO_3)$	Ag: MVol.B6-294
AgC_6H_5	AgC_6H_5	Ag: MVol.B5-8/9
$AgC_6H_5I^+$	$[AgC_6H_5I]^+$	Ag: MVol.B5-117
$AgC_6H_5NNaO_3S_2$	$Na[Ag(SC_6H_3(NH_2)SO_3)]$	Ag: MVol.B7-43
$AgC_6H_5NO_2^+$	$[AgC_6H_5NO_2]^+$	Ag: MVol.B5-118
–	$[Ag(NC_5H_4COOH)]^+$	Ag: MVol.B6-102
$AgC_6H_5N_2O$	$[Ag(C_6H_5NNO)]$	Ag: MVol.B6-292/3

Formula	Compound	Reference
$AgC_6H_5N_2O$	$AgCNO \cdot C_5H_5N$	Ag: MVol.B6-82
–	$AgNCO \cdot C_5H_5N$	Ag: MVol.B6-81/2
$AgC_6H_5N_2O_2$	$[Ag(C_6H_5N(O)NO)]$	Ag: MVol.B6-321
$AgC_6H_5N_2O_3S$	$Ag(C_6H_5NNSO_3)$	Ag: MVol.B6-294
$AgC_6H_5N_2O_4S$	$Ag(C_6H_5SO_2NNO_2) \cdot H_2O$	Ag: MVol.B6-342
$AgC_6H_5N_2O_5$	$Ag(NC_5H_4COOH)NO_3$	Ag: MVol.B6-102
$AgC_6H_5N_2S$	$AgSCN \cdot C_5H_5N$	Ag: MVol.B6-82
$AgC_6H_5N_4O$	$[Ag(CH_3C_5H_2N_4O)]$	Ag: MVol.B6-184
$AgC_6H_5N_4O_6$	$[AgC_5H_5N]C(NO_2)_3$	Ag: MVol.B6-82
$AgC_6H_5N_4O_7$	$Ag(C_5H_5NO)C(NO_2)_3$	Ag: MVol.B6-104
$AgC_6H_5OS_3$	$[Ag(CH_3C_5H_2OS_3)]$	Ag: MVol.B7-83/4
AgC_6H_5Se	$Ag(SeC_6H_5)$	Ag: MVol.B7-191
$AgC_6H_6^+$	$[AgC_6H_6]^+$	Ag: MVol.B5-91
AgC_6H_6K	$K[Ag(CCCH_3)_2]$	Ag: MVol.B5-13, 14
$AgC_6H_6NNa_2O_6$	$Na_2[Ag(N(CH_2COO)_3)]$	Ag: MVol.B6-276
AgC_6H_6NO	$AgNC_4H_3COCH_3$	Ag: MVol.B6-68
AgC_6H_6NOS	$[Ag(CH_3C_5H_3NOS)]$	Ag: MVol.B7-59
$AgC_6H_6NO_2S$	$[Ag((CH_3)_2C_3NSCOO)]$	Ag: MVol.B7-70
–	$Ag(C_6H_5SO_2NH)$	Ag: MVol.B6-342
$AgC_6H_6NO_3S$	$[Ag(NH_2C_6H_4SO_3)]$	Ag: MVol.B6-262
$AgC_6H_6NO_4S_2^{2-}$	$[Ag(C_3H_5(COO)_2NHCSS)]^{2-}$	Ag: MVol.B7-103
$AgC_6H_6NO_6^{2-}$	$[Ag(N(CH_2COO)_3)]^{2-}$	Ag: MVol.B6-276
$AgC_6H_6N_3O_2$	$Ag(NH(CH_3CO)C_4H_2N_2O)$	Ag: MVol.B6-150
$AgC_6H_6N_3O_4$	$[Ag(ONC_4N_2O_3(CH_3)_2)]$	Ag: MVol.B6-315
$AgC_6H_6N_4O_4^-$	$[Ag(C_3H_3N_2O_2)_2]^-$	Ag: MVol.B6-147
$AgC_6H_6N_7O_3$	$AgNO_3 \cdot 2\ C_3H_3N_3$	Ag: MVol.B6-187
$AgC_6H_6N_7O_7$	$[Ag(NO_2C_3H_3N_2)_2]NO_3$	Ag: MVol.B6-131
$AgC_6H_6O^+$	$[Ag(C_6H_5OH)]^+$	Ag: MVol.B5-118
		Ag: MVol.B6-207
$AgC_6H_7N^+$	$[Ag(CH_3C_5H_4N)]^+$	Ag: MVol.B6-84/5
–	$[Ag(C_6H_5NH_2)]^+$	Ag: MVol.B6-50/1
$AgC_6H_7NO^+$	$[Ag(CH_3OC_5H_4N)]^+$	Ag: MVol.B6-91
–	$[Ag(HOCH_2C_5H_4N)]^+$	Ag: MVol.B6-91
$AgC_6H_7NO_2^+$	$[Ag(CH_3CN)(C_4H_4O_2)]^+$	Ag: MVol.B6-222
$AgC_6H_7NO_2S_2^-$	$[Ag(NC_4H_7(COO)CSS)]^-$	Ag: MVol.B7-117
$AgC_6H_7N_2O$	$Ag((CH_3)_2C_4HN_2O)$	Ag: MVol.B6-150
$AgC_6H_7N_2OS$	$[Ag(CH_3CONC_3HNSCH_3)]$	Ag: MVol.B7-71
–	$[Ag(C_2H_5SC_4H_2N_2O)]$	Ag: MVol.B7-63
–	$[Ag(C_4HN_2O(CH_3)SCH_3)]$	Ag: MVol.B7-63
$AgC_6H_7N_2O_2S$	$[Ag(NHC_6H_4SO_2NH_2)]$	Ag: MVol.B6-263
$AgC_6H_7N_2O_3$	$Ag((CH_3)_2C_3HN_2OCOO)$	Ag: MVol.B6-147
$AgC_6H_7N_4O_3$	$[Ag((CH_3)_2C_4HN_4O_3)]$	Ag: MVol.B6-317
–	$Ag(NH_2(NHCOCH_3)C_4HN_2O_2)$	Ag: MVol.B6-152
$AgC_6H_7N_8O_4$	$AgC_3H_3N_4O_2 \cdot C_3H_4N_4O_2 \cdot 1.5\ H_2O$	Ag: MVol.B6-189
$AgC_6H_8^+$	$[Ag(CH_2CHCHCHCHCH_2)]^+$	Ag: MVol.B5-40/1
–	$[Ag(C_6H_8)]^+$	Ag: MVol.B5-57/8
$AgC_6H_8NO_3$	$Ag(C_2H_3NHCOC_2H_4COO)$	Ag: MVol.B6-327
$AgC_6H_8N_2^+$	$[Ag(CH_3(CH_2CH)C_3H_2N_2)]^+$	Ag: MVol.B6-138/9
–	$[Ag((CH_3)_2C_4H_2N_2)]^+$	Ag: MVol.B6-159
–	$[Ag(NH_2CH_2C_5H_4N)]^+$	Ag: MVol.B6-96

Formula	Compound	Reference
$AgC_6H_8N_2^+$	$[Ag(NH_2(CH_3)C_5H_3N)]^+$	Ag: MVol.B6-95
$AgC_6H_8N_3O_2$	$[Ag(C_3H_3N_2CH_2CH(NH_2)COO)]$	Ag: MVol.B6-246
$AgC_6H_8N_3O_3$	$Ag(NH_2CH_2C_5H_4N)NO_3$	Ag: MVol.B6-96
–	$AgNO_3 \cdot (CH_3)_2C_4H_2N_2$	Ag: MVol.B6-160
$AgC_6H_8N_3O_4$	$Ag((CH_3)_2C_4H_2N_2O)NO_3$	Ag: MVol.B6-162
$AgC_6H_8N_3O_5$	$Ag((CH_3)_2C_4H_2N_2O_2)NO_3$	Ag: MVol.B6-162
$AgC_6H_8N_3O_5S$	$AgNO_3 \cdot NH_2C_6H_4SO_2NH_2$	Ag: MVol.B6-263
$AgC_6H_8N_4^+$	$[Ag(C_3H_4N_2)_2]^+$	Ag: MVol.B6-136/7
$AgC_6H_8N_5O_3$	$[Ag(NHCHNCHCH)_2]NO_3$	Ag: MVol.B6-137/8
–	$[Ag(NHNCHCHCH)_2]NO_3$	Ag: MVol.B6-131
$AgC_6H_8N_5O_7$	$[Ag(NH_3)_2]OC_6H_2(NO_2)_3$	Ag: MVol.B6-30
$AgC_6H_8O_4S^-$	$[Ag(S(C_2H_4COO)_2)]^-$	Ag: MVol.B7-38
$AgC_6H_8O_4S_2^-$	$[Ag(C_2H_4(SCH_2COO)_2)]^-$	Ag: MVol.B7-39/40
$AgC_6H_8O_6^+$	$[Ag(C_4HO_2(OH)_2CH(OH)CH_2OH)]^+$	Ag: MVol.B6-220
AgC_6H_9	$AgCCC(CH_3)_3$	Ag: MVol.B5-13
–	$AgCCC_4H_9$	Ag: MVol.B5-13, 14
$AgC_6H_9N_2$	$Ag(C_2H_5(CH_3)C_3HN_2)$	Ag: MVol.B6-142
–	$Ag(C_3H_7C_3H_2N_2)$	Ag: MVol.B6-142
$AgC_6H_9N_2^{2+}$	$[AgH(NH_2CH_2C_5H_4N)]^{2+}$	Ag: MVol.B6-96
$AgC_6H_9N_2O_2$	$Ag(CH_3(C_2H_5)C_3HN_2O_2)$	Ag: MVol.B6-148
$AgC_6H_9N_2O_2S$	$Ag(C_6H_5SO_2NH) \cdot NH_3$	Ag: MVol.B6-342
$AgC_6H_9N_2O_4$	$Ag(C_4H_8N_2(COOH)COO)$	Ag: MVol.B6-163
$AgC_6H_9N_2O_6$	$Ag(N(NO_2)COOCH(CH_3)COOC_2H_5) \cdot H_2O$	Ag: MVol.B6-331
$AgC_6H_9N_3^+$	$[Ag(CH_3CN)_3]^+$	Ag: MVol.B6-347
AgC_6H_9O	$2\ AgCCC(C_2H_5)(CH_3)OH \cdot AgCH_3COO$	Ag: MVol.B5-15
–	$2\ AgCCCH(C_3H_7)OH \cdot AgCH_3COO$	Ag: MVol.B5-15
$AgC_6H_9O_2S$	$[AgCH_2CH(CH_2)_2SCH_2COO]$	Ag: MVol.B5-50/1
$AgC_6H_9O_2Se$	$[Ag(CH_2CH(CH_2)_2SeCH_2COO)]$	Ag: MVol.B5-50/1
		Ag: MVol.B7-191/3, 196
$AgC_6H_9O_4S_2$	$[Ag(C_2H_4(SCH_2COO)_2H)]$	Ag: MVol.B7-40/1
$AgC_6H_9O_6^{2-}$	$[Ag(CH_3COO)_3]^{2-}$	Ag: MVol.B5-146/7
$AgC_6H_{10}^+$	$[Ag(CH_3CCC_3H_7)]^+$	Ag: MVol.B5-40/1
–	$[Ag(C_2H_5CCC_2H_5)]^+$	Ag: MVol.B5-40/1
–	$[AgC_5H_7CH_3]^+$	Ag: MVol.B5-52/3
–	$[AgC_5H_8CH_2]^+$	Ag: MVol.B5-52/3
–	$[AgC_6H_{10}]^+$	Ag: MVol.B5-54/5
$AgC_6H_{10}NO_2$	$[Ag(C_4H_7CH(NH_2)COO)]$	Ag: MVol.B6-250
$AgC_6H_{10}NO_3$	$AgNO_3 \cdot C_2H_5CCC_2H_5$	Ag: MVol.B5-41
–	$AgNO_3 \cdot C_6H_{10}$	Ag: MVol.B5-56
$AgC_6H_{10}NO_5$	$Ag(C_2H_5OCOONCOOC_2H_5)$	Ag: MVol.B6-331
$AgC_6H_{10}NO_5S_2$	$[Ag(C_6H_{10}O_2S_2)]NO_3$	Ag: MVol.B7-87
$AgC_6H_{10}NS_2$	$[Ag(C_5H_{10}NCSS)]$	Ag: MVol.B7-117
$AgC_6H_{10}N_2^+$	$[Ag(CH_3(C_2H_5)C_3H_2N_2)]^+$	Ag: MVol.B6-138/9
–	$[Ag(C_2H_5CN)_2]^+$	Ag: MVol.B6-349
$AgC_6H_{10}N_3$	$[Ag((C_2H_5)_2C_2N_3)]$	Ag: MVol.B6-185
$AgC_6H_{10}N_3O_3$	$Ag((CH_3)_2C_4H_4N_2)NO_3$	Ag: MVol.B6-160
$AgC_6H_{10}N_3O_3S_2$	$AgNO_3 \cdot (CH_3)_3CSC_2HN_2S$	Ag: MVol.B7-80
$AgC_6H_{10}N_3O_4$	$Ag(NH_2(CH_2CONH)_2CH_2COO) \cdot 2\ NH_3$	Ag: MVol.B6-269
$AgC_6H_{10}N_5$	$[Ag(C_5H_{10}NCN_4)]$	Ag: MVol.B6-192/3
$AgC_6H_{10}N_5O_3$	$[Ag((CH_2)_5CN_4)]NO_3$	Ag: MVol.B6-195

Formula	Compound	Reference
$AgC_6H_{10}O_2^+$	$[AgCH_2CH(CH_2)_2OOCCH_3]^+$	Ag: MVol.B5-48
$AgC_6H_{10}O_2S^+$	$[AgCH_2CH(CH_2)_2SCH_2COOH]^+$	Ag: MVol.B5-50/1
$AgC_6H_{10}O_2Se^+$	$[Ag(CH_2CH(CH_2)_2SeCH_2COOH)]^+$	Ag: MVol.B5-50/1
		Ag: MVol.B7-193
$AgC_6H_{10}O_4S_2^+$	$[Ag(C_2H_4(SCH_2COOH)_2)]^+$	Ag: MVol.B7-40
$AgC_6H_{10}O_4S_2^-$	$[Ag(CH_3SCH_2COO)_2]^-$	Ag: MVol.B7-25/6
$AgC_6H_{11}NO_2^+$	$[Ag(C_4H_7CH(NH_2)COOH)]^+$	Ag: MVol.B6-250
$AgC_6H_{11}N_2O_2S_2$	$[Ag(HOC_2H_4NHCSCSNC_2H_4OH)]$	Ag: MVol.B7-187/8
$AgC_6H_{11}N_4$	$[Ag(C_5H_{11}CN_4)]$	Ag: MVol.B6-192/3
$AgC_6H_{11}N_4O_3$	$AgNO_3 \cdot C_4H_9C_2H_2N_3$	Ag: MVol.B6-185
$AgC_6H_{11}N_4O_4S$	$Ag(C_6H_5SO_2NNO_2) \cdot 2\ NH_3$	Ag: MVol.B6-342
$AgC_6H_{11}OS_2$	$[Ag(C_5H_{11}OCSS)]$	Ag: MVol.B7-98
		C: MVol.D4-257
$AgC_6H_{11}O_2S$	$[Ag(C_4H_9SCH_2COO)]$	Ag: MVol.B7-25/6
$AgC_6H_{11}O_2Se$	$[Ag(CH_3(CH_2)_3SeCH_2COO)]$	Ag: MVol.B7-191/3, 196
$AgC_6H_{11}O_6S_3$	$Ag(CH_3C(SO_2CHCH_3)_2SO_2) \cdot H_2O$	Ag: MVol.B7-92
$AgC_6H_{11}S$	$(AgSC_6H_{11})_n$	Ag: MVol.B7-5/6
$AgC_6H_{12}^+$	$[Ag(C_6H_{12})]^+$	Ag: MVol.B5-39/40
$AgC_6H_{12}NO_2$	$[Ag((CH_3)_2CHCH_2CH(NH_2)COO)]$	Ag: MVol.B6-249
–	$[Ag(C_2H_5CH(CH_3)CH(NH_2)COO)]$	Ag: MVol.B6-249
–	$[Ag(C_4H_9CH(NH_2)COO)]$	Ag: MVol.B6-249, 250
–	$[Ag(NH_2(CH_2)_5COO)]$	Ag: MVol.B6-249/50
$AgC_6H_{12}NO_2S$	$[Ag(C_2H_5SC_2H_4CH(NH_2)COO)]$	Ag: MVol.B6-259
$AgC_6H_{12}NO_3S_3$	$AgNO_3 \cdot S(CH(CH_3)S)_2CHCH_3$	Ag: MVol.B7-92
$AgC_6H_{12}NO_3S_6$	$Ag(S(CH_2S)_2CH_2)_2NO_3$	Ag: MVol.B7-89/90
$AgC_6H_{12}NS_2$	$[Ag(SCSNHC_5H_{11})]$	Ag: MVol.B7-102
$AgC_6H_{12}N_2^+$	$[Ag(N(CH_2CH_2)_3N)]^+$	Ag: MVol.B6-130
$AgC_6H_{12}N_2O_2S_2^+$	$[Ag(HOC_2H_4NHCSCSNHC_2H_4OH)]^+$	Ag: MVol.B7-187/8
$AgC_6H_{12}N_2O_4^-$	$[Ag(CH_3CH(NH_2)COO)_2]^-$	Ag: MVol.B6-241
–	$[Ag(CH_3NHCH_2COO)_2]^-$	Ag: MVol.B6-240
–	$[Ag(NH_2CH_2CH_2COO)_2]^-$	Ag: MVol.B6-243
$AgC_6H_{12}N_2O_4S_2^-$	$[Ag(SCH_2CH(NH_2)COOH)_2]^-$	Ag: MVol.B6-256
$AgC_6H_{12}N_2O_6^-$	$[Ag(HOCH_2CH(NH_2)COO)_2]^-$	Ag: MVol.B6-245
$AgC_6H_{12}N_2S_4$	$[Ag(SCSN(CH_3)_2)_2]$	Ag: MVol.B7-317/8
$AgC_6H_{12}N_2S_4^+$	$[Ag(SS(CSN(CH_3)_2)_2)]^+$	Ag: MVol.B7-120
$AgC_6H_{12}N_3O_2S$	$Ag(C_6H_5SO_2NH) \cdot 2\ NH_3$	Ag: MVol.B6-342
$AgC_6H_{12}N_3O_3$	$AgNO_3 \cdot N(CH_2CH_2)_3N$	Ag: MVol.B6-130
$AgC_6H_{12}N_3O_3S$	$Ag((CH_3)_2NCSNHCH_2CHCH_2)NO_3$	Ag: MVol.B7-157
$AgC_6H_{12}N_3S$	$[Ag(C_4H_9CHNNCSNH_2)]$	Ag: MVol.B6-282
$AgC_6H_{12}N_4^+$	$[Ag((CH_2)_6N_4)]^+$	Ag: MVol.B6-197
$AgC_6H_{12}N_4S_2^+$	$[Ag(C_3H_6N_2S)_2]^+$	Ag: MVol.B7-49
$AgC_6H_{12}N_5O_3$	$AgNO_3 \cdot (CH_2)_6N_4$	Ag: MVol.B6-197/8
$AgC_6H_{12}N_5O_5S_2$	$[Ag(CH_3CONHCSNH_2)_2]NO_3$	Ag: MVol.B7-161
$AgC_6H_{12}N_{13}O_3$	$AgNO_3 \cdot 3\ NH_2C(NH)NHCN$	Ag: MVol.B6-338
$AgC_6H_{12}O^+$	$[AgCH_2CH(CH_2)_2CHOHCH_3]^+$	Ag: MVol.B5-47
–	$[AgCH_2CH(CH_2)_3CH_2OH]^+$	Ag: MVol.B5-47
–	$[AgCH_2CH(CH_2)_3OCH_3]^+$	Ag: MVol.B5-48
–	$[AgCH_2CHOCH_2CH(CH_3)_2]^+$	Ag: MVol.B5-48
–	$[AgCH_2CHOC_4H_9]^+$	Ag: MVol.B5-48
–	$[Ag(CH_3)_2CCHOC_2H_5]^+$	Ag: MVol.B5-48

Formula	Compound	Reference
$AgC_6H_{12}O^+$	$[AgC_2H_5CHCHOC_2H_5]^+$	Ag: MVol.B5-48
$AgC_6H_{12}O_2^+$	$[Ag(CH_2CHCH_2OH)_2]^+$	Ag: MVol.B5-47
$AgC_6H_{12}O_2S^+$	$[Ag(C_4H_9SCH_2COOH)]^+$	Ag: MVol.B7-25/6
$AgC_6H_{12}O_2Se^+$	$[Ag(C_4H_9SeCH_2COOH)]^+$	Ag: MVol.B7-193
$AgC_6H_{12}S_6^+$	$[Ag(S(CH_2S)_2CH_2)_2]_n^{n+}$	Ag: MVol.B7-90
$AgC_6H_{13}N^+$	$[Ag(CH_3N(CH_2)_5)]^+$	Ag: MVol.B6-73
–	$[Ag(NHC_5H_9CH_3)]^+$	Ag: MVol.B6-73
$AgC_6H_{13}N_4O_2$	$[Ag(NH_2C(NH)NHC_3H_6CH(NH_2)COO)]$	Ag: MVol.B6-251
$AgC_6H_{13}S$	$(AgSC_6H_{13})_n$	Ag: MVol.B7-5/6
$AgC_6H_{14}NO_3S$	$AgNO_3 \cdot S(CH(CH_3)_2)_2$	Ag: MVol.B7-11
–	$AgNO_3 \cdot S(C_3H_7)_2$	Ag: MVol.B7-11
$AgC_6H_{14}N_2^+$	$[Ag(CH_2CHCH_2NH_2)_2]^+$	Ag: MVol.B6-45
$AgC_6H_{14}N_2O_2^+$	$[Ag(OCHN(CH_3)_2)_2]^+$	Ag: MVol.B6-322/3
$AgC_6H_{14}N_2S_2^+$	$[Ag(HCSN(CH_3)_2)_2]^+$	Ag: MVol.B7-121
$AgC_6H_{14}N_3O_5$	$AgNO_3 \cdot 2\ CH_3CONHCH_3$	Ag: MVol.B6-325
–	$AgNO_3 \cdot 2\ (CH_3)_2NCHO$	Ag: MVol.B6-323
–	$AgNO_3 \cdot (NH_2)_2C_5H_9COOH \cdot HNO_3$	Ag: MVol.B6-251/2
$AgC_6H_{14}N_5O_5$	$AgNO_3 \cdot C_6H_{14}N_4O_2 \cdot 0.5\ H_2O$	Ag: MVol.B6-251
$AgC_6H_{14}O_3PS$	$Ag(OSP(OC_3H_7)_2)$	Ag: MVol.B7-253
$AgC_6H_{15}IO_3P$	$AgI \cdot P(OC_2H_5)_3$	Ag: MVol.B7-246
$AgC_6H_{15}IP$	$[AgI(P(C_2H_5)_3)]_4$	Ag: MVol.B7-202
$AgC_6H_{15}N^+$	$[Ag(C_6H_{13}NH_2)]^+$	Ag: MVol.B6-46
–	$[Ag(N(C_2H_5)_3)]^+$	Ag: MVol.B6-49/50
$AgC_6H_{15}NO^+$	$[Ag(NH_2C_6H_{12}OH)]^+$	Ag: MVol.B6-228
$AgC_6H_{15}NO_2^+$	$[Ag(NH(C_3H_6OH)_2)]^+$	Ag: MVol.B6-229
$AgC_6H_{15}NO_3^+$	$[Ag(N(C_2H_4OH)_3)]^+$	Ag: MVol.B6-229/30
$AgC_6H_{15}NO_6PS$	$[Ag((C_2H_5O)_3PS)]NO_3$	Ag: MVol.B7-253
$AgC_6H_{15}NO_7P$	$[AgNO_3((C_2H_5O)_3PO)]$	Ag: MVol.B7-249/50
$AgC_6H_{15}N_2O_2$	$Ag(C_4H_9CH(NH_2)COO) \cdot NH_3$	Ag: MVol.B6-249
$AgC_6H_{15}N_2O_3$	$AgNO_3 \cdot (C_2H_5)_3N$	Ag: MVol.B6-50
$AgC_6H_{15}N_3S_3^+$	$[Ag(CH_3CSNH_2)_3]^+$	Ag: MVol.B7-123/4
$AgC_6H_{15}N_4O_3S_3$	$[Ag(CH_3CSNH_2)_3]NO_3$	Ag: MVol.B7-124
$AgC_6H_{15}N_4O_8$	$AgNO_3 \cdot (NH_2)_2C_5H_9COOH \cdot HNO_3$	Ag: MVol.B6-251/2
$AgC_6H_{15}N_6O_8$	$AgNO_3 \cdot C_6H_{14}N_4O_2 \cdot HNO_3$	Ag: MVol.B6-251
$AgC_6H_{15}OP^+$	$[Ag((C_2H_5)_2PCH_2CH_2OH)]^+$	Ag: MVol.B7-203
$AgC_6H_{15}O_6S_6^{2-}$	$[Ag(C_2H_5SO_2S)_3]^{2-}$	Ag: MVol.B7-96
$AgC_6H_{16}N_2^+$	$[Ag(C_2H_4(N(CH_3)_2)_2)]^+$	Ag: MVol.B6-60
$AgC_6H_{16}N_2OS^+$	$[Ag(NH_2C_2H_4OC_2H_4SC_2H_4NH_2)]^+$	Ag: MVol.B7-19
$AgC_6H_{16}N_2O_2^+$	$[Ag(C_2H_4(OC_2H_4NH_2)_2)]^+$	Ag: MVol.B6-232
$AgC_6H_{16}N_2O_6S_2^-$	$[Ag(NH_2C_3H_6SO_3)_2]^-$	Ag: MVol.B6-262
$AgC_6H_{16}N_2S_2^+$	$[Ag(C_2H_4(SC_2H_4NH_2)_2)]^+$	Ag: MVol.B7-18
$AgC_6H_{16}N_4S_2^+$	$[Ag(CH_3NHCSNHCH_3)_2]^+$	Ag: MVol.B7-152
–	$[Ag(C_2H_5NHCSNH_2)_2]^+$	Ag: MVol.B7-154
$AgC_6H_{16}N_5O_3S_2$	$Ag(CH_3NHCSNHCH_3)_2NO_3$	Ag: MVol.B7-152
–	$[Ag(C_2H_5NHCSNH_2)_2]NO_3$	Ag: MVol.B7-154
–	$AgNO_3 \cdot 2\ (CH_3)_2NCSNH_2$	Ag: MVol.B7-152
$AgC_6H_{16}N_5O_4$	$Ag(NH_2(CH_2CONH)_2CH_2COO) \cdot 2\ NH_3$	Ag: MVol.B6-269
$AgC_6H_{16}N_6^-$	$[Ag(CN)_2(C_2H_8N_2)_2]^-$	Ag: MVol.B6-55/6
$AgC_6H_{16}N_8O_8S_{1.5}$	$[Ag(NH_2C(NH)NC(OCH_3)NH_2)_2](SO_4)_{1.5}$	Ag: MVol.B7-325
$AgC_6H_{16}N_{10}^{3+}$	$[Ag(C_2H_4(NHC(NH)NHC(NH)NH_2)_2)]^{3+}$	Ag: MVol.B7-321

Formula	Compound	Reference
$AgC_6H_{16}N_{10}O_6S_{1.5}$	$[Ag(C_2H_4(C_2H_6N_5)_2)](SO_4)_{1.5} \cdot 3.5\,H_2O$	Ag: MVol.B7-321/4
$AgC_6H_{16}N_{13}O_9$	$[Ag(C_2H_4(C_2H_6N_5)_2)](NO_3)_3$	Ag: MVol.B7-321/4
$AgC_6H_{17}NO_4^+$	$[Ag(N(C_2H_4OH)_3)(H_2O)]^+$	Ag: MVol.B6-230
$AgC_6H_{17}N_{10}O_8S_2$	$[Ag(C_2H_4(C_2H_6N_5)_2)](SO_4)(HSO_4) \cdot H_2O$	Ag: MVol.B7-321/4
$AgC_6H_{17}N_{12}O_{10}S$	$[Ag(C_2H_4(C_2H_6N_5)_2)](HSO_4)(NO_3)_2$	Ag: MVol.B7-321/4
$AgC_6H_{18}NO_9P_2$	$[AgNO_3(P(OCH_3)_3)_2]_2$	Ag: MVol.B7-244
$AgC_6H_{18}N_2^+$	$[Ag(C_3H_7NH_2)_2]^+$	Ag: MVol.B6-44
–	$[Ag(N(CH_3)_3)_2]^+$	Ag: MVol.B6-49
$AgC_6H_{18}N_2O_2^+$	$[Ag(NH_2C_2H_4OCH_3)_2]^+$	Ag: MVol.B6-232
–	$[Ag(NH_2C_3H_6OH)_2]^+$	Ag: MVol.B6-228
$AgC_6H_{18}N_2O_4^+$	$[Ag(NH_2CH_2CHOHCH_2OH)_2]^+$	Ag: MVol.B6-230/1
$AgC_6H_{18}N_3OP^+$	$[Ag(OP(N(CH_3)_2)_3)]^+$	Ag: MVol.B7-250/1
$AgC_6H_{18}N_4^+$	$[Ag(C_2H_4(NHC_2H_4NH_2)_2)]^+$	Ag: MVol.B6-66
–	$[Ag(N(C_2H_4NH_2)_3)]^+$	Ag: MVol.B6-65
$AgC_6H_{18}N_4O_3PS$	$[AgNO_3(SP(N(CH_3)_2)_3)]$	Ag: MVol.B7-253/4
$AgC_6H_{18}N_4O_4P$	$[AgNO_3(OP(N(CH_3)_2)_3)]$	Ag: MVol.B7-251
$AgC_6H_{18}N_5O_3$	$[Ag(C_2H_4(NHC_2H_4NH_2)_2)]NO_3$	Ag: MVol.B6-66
–	$AgNO_3 \cdot C_2H_4(NHC_2H_4NH_2)_2 \cdot 2\,H_2O$	Ag: MVol.B6-66
$AgC_6H_{18}N_6S_3^+$	$[Ag(CH_3NHCSNH_2)_3]^+$	Ag: MVol.B7-150
$AgC_6H_{18}N_{12}S_3^+$	$[Ag(NH_2C(NH)NHCSNH_2)_3]^+$	Ag: MVol.B7-162
$AgC_6H_{18}N_{13}O_9$	$[Ag(CH_3C_2H_6N_5)_2](NO_3)_3$	Ag: MVol.B7-321
$AgC_6H_{18}O_3S_3^+$	$[Ag((CH_3)_2SO)_3]^+$	Ag: MVol.B7-2
$AgC_6H_{19}N_4^{2+}$	$[AgH(C_2H_4(NHC_2H_4NH_2)_2)]^{2+}$	Ag: MVol.B6-66
–	$[AgH(N(C_2H_4NH_2)_3)]^{2+}$	Ag: MVol.B6-65
$AgC_6H_{19}N_{10}O_3$	$[Ag(C_2H_4(C_2H_6N_5)_2)](OH)_3 \cdot 3\,H_2O$	Ag: MVol.B7-321/4
$AgC_6H_{20}N_4^+$	$[Ag(C_3H_6(NH_2)_2)_2]^+$	Ag: MVol.B6-61/2
$AgC_6H_{20}N_4^{3+}$	$[AgH_2(C_2H_4(NHC_2H_4NH_2)_2)]^{3+}$	Ag: MVol.B6-66
–	$[AgH_2(N(C_2H_4NH_2)_3)]^{3+}$	Ag: MVol.B6-65
$AgC_6H_{21}N_3O_3^+$	$[Ag(NH_2C_2H_4OH)_3]^+$	Ag: MVol.B6-226/7
$AgC_6H_{21}N_{10}O_3$	$[Ag(CH_3C_2H_6N_5)_2](OH)_3 \cdot 5\,H_2O$	Ag: MVol.B7-321
$AgC_6H_{22}N_4^{3+}$	$[Ag(NH_3C_3H_6NH_2)_2]^{3+}$	Ag: MVol.B6-61/2
$AgC_6H_{24}N_6^+$	$[Ag(C_2H_8N_2)_3]^+$	Ag: MVol.B6-55/6
$AgC_6H_{25}N_6O$	$[Ag(C_2H_8N_2)_3]OH$	Ag: MVol.B6-57
AgC_6N_5	$[AgC_3N_2(CN)_3]$	Ag: MVol.B6-131
$AgC_{6.5}H_{5.5}N_{2.5}O_3$	$Ag(ONC_3NO_2CH_3) \cdot 0.5\,C_5H_5N$	Ag: MVol.B6-313
$AgC_7ClH_7NO_2S$	$Ag(CH_3C_6H_4SO_2NCl)$	Ag: MVol.B6-343
$AgC_7ClH_8O_4$	$AgClO_4 \cdot C_6H_5CH_3$	Ag: MVol.B5-96
$AgC_7ClH_9NO_4S$	$[Ag(NC_5H_4CH_2SCH_3)]ClO_4$	Ag: MVol.B7-59
$AgC_7ClH_{12}N_2S$	$[Ag((CH_3)_3C_4H_3N_2S)]Cl$	Ag: MVol.B7-64
$AgC_7Cl_3H_5IrKNO_4$	$AgK[Ir(C_2O_4)(C_5H_5N)Cl_3] \cdot 2\,H_2O$	Ir: SVol.2-36
$AgC_7Cl_3H_5IrNO_4$	$Ag[Ir(C_2O_4)(C_5H_5N)Cl_3]$	Ir: SVol.2-36
$AgC_7F_{12}NO_2$	$(CF_3)_2CCONAgC(CF_3)_2O$	F: PerFHalOrg.5-55/6, 102, 110
$AgC_7H_2N_5$	$AgC(CN)_2C(NH_2)C(CN)_2$	Ag: MVol.B6-355
$AgC_7H_4I_2NO_3$	$Ag(CO(ClCH)_2NCH_2COO)$	Ag: MVol.B6-106
AgC_7H_4NOS	$[Ag(C_7H_4NOS)]$	Ag: MVol.B7-53
$AgC_7H_4NO_3S$	$Ag(C_7H_4NO_3S)$	Ag: MVol.B6-344/5
$AgC_7H_4NO_4$	$Ag(NC_5H_3(COOH)COO) \cdot H_2O$	Ag: MVol.B6-104
$AgC_7H_4NS_2$	$[Ag(C_7H_4NS_2)]$	Ag: MVol.B7-77
$AgC_7H_4NaO_2S$	$Na[Ag(SC_6H_4COO)]$	Ag: MVol.B7-34

$AgC_7H_4O_2S^-$	$[Ag(SC_6H_4COO)]^-$	Ag: MVol.B7-34, 36
$AgC_7H_5N^+$	$[Ag(C_6H_5CN)]^+$	Ag: MVol.B6-352
$AgC_7H_5NNaO_2S$	$Na[Ag(SC_6H_3(NH_2)COO)]$	Ag: MVol.B7-35
$AgC_7H_5NO_4S^-$	$[AgHNSO_2C_6H_4COO]^-$	Ag: MVol.B6-344
$AgC_7H_5N_2$	$[Ag(NC_6H_4CHN)]$	Ag: MVol.B6-132
–	$[Ag(NC_6H_4NCH)]$	Ag: MVol.B6-144/5
$AgC_7H_5N_2O$	$AgC_7H_5N_2O$	Ag: MVol.B6-147
$AgC_7H_5N_2O_2$	$Ag(C_6H_5C(NO)NO) \cdot n\ NH_3$	Ag: MVol.B6-320
$AgC_7H_5N_2O_3S_3^{2-}$	$[Ag(O_3SC_6H_4NHNCSS)]^{2-}$	Ag: MVol.B7-119
$AgC_7H_5N_2S$	$[Ag(NHCSC_6H_4N)]$	Ag: MVol.B7-77
–	$[Ag(SCNC_6H_4NH)]$	Ag: MVol.B7-52
$AgC_7H_5N_4$	$[Ag(C_6H_5CN_4)]$	Ag: MVol.B6-192/3
$AgC_7H_5N_4S$	$[Ag(C_6H_5CN_4S)]$	Ag: MVol.B7-57
AgC_7H_5OS	$[Ag(C_6H_5COS)]$	Ag: MVol.B7-94
$AgC_7H_5O_3S_3$	$[Ag(C_5H_3OS_3CH_2COO)]$	Ag: MVol.B7-83/4
$AgC_7H_5S_2$	$[Ag(C_6H_5CSS)]$	Ag: MVol.B7-94/5
$AgC_7H_6NNa_2O_5S_3$	$Na_2[Ag(O_2SCH_2NHC_6H_3(S)SO_3)]$	Ag: MVol.B7-43
AgC_7H_6NO	$Ag(C_6H_5CONH)$	Ag: MVol.B6-326
$AgC_7H_6NO_2$	$[Ag(NH_2C_6H_4COO)]$	Ag: MVol.B6-254
$AgC_7H_6NO_2S$	$[Ag(HSC_6H_3(NH_2)COO)]$	Ag: MVol.B7-35
$AgC_7H_6NO_3$	$Ag(NHC_4H_2(COCH_3)COO)$	Ag: MVol.B6-68
$AgC_7H_6NO_4S$	$Ag(NH_2COC_6H_4SO_3) \cdot H_2O$	Ag: MVol.B6-344
–	$Ag(NH_2SO_2C_6H_4COO)$	Ag: MVol.B6-343, 344
$AgC_7H_6NS_2$	$[Ag(C_6H_5NHCSS)]$	Ag: MVol.B7-103
$AgC_7H_6N_3O_3$	$[Ag(NO_2C_6H_3(CH_3)NNO)]$	Ag: MVol.B6-293
$AgC_7H_6N_3O_6S$	$Ag(CH_3C_6H_3(NO_2)SO_2NNO_2)$	Ag: MVol.B6-343
AgC_7H_7	$AgC_6H_4CH_3$	Ag: MVol.B5-10
–	$AgC_6H_4CH_3 \cdot x\ AgNO_3$	Ag: MVol.B5-12
$AgC_7H_7NO_2^+$	$[Ag(NC_5H_3(CH_3)COOH)]^+$	Ag: MVol.B6-102
$AgC_7H_7N_2$	$AgCN \cdot CH_3C_5H_4N$	Ag: MVol.B6-86
$AgC_7H_7N_2O$	$AgNCO \cdot CH_3C_5H_4N$	Ag: MVol.B6-86
$AgC_7H_7N_2O_3$	$[Ag(C_2H_3C_5H_4N)]_x(NO_3)_x$	Ag: MVol.B6-90
$AgC_7H_7N_2O_3S$	$Ag(CH_3C_6H_4NNSO_3)$	Ag: MVol.B6-294
–	$Ag(C_6H_5CSNH_2)NO_3$	Ag: MVol.B7-126
$AgC_7H_7N_2O_4$	$AgNO_3 \cdot C_6H_5CONH_2$	Ag: MVol.B6-325/6
$AgC_7H_7N_2O_4S$	$Ag(CH_3C_6H_4SO_2NNO_2)$	Ag: MVol.B6-342/3
$AgC_7H_7N_2S$	$AgSCN \cdot CH_3C_5H_4N$	Ag: MVol.B6-86
$AgC_7H_7N_2S_2$	$[Ag(C_6H_5NHNHCSS)]$	Ag: MVol.B7-119
–	$[Ag(NH_2C_6H_4NHCSS)]$	Ag: MVol.B7-104
$AgC_7H_7N_4O$	$[Ag(C_6H_5N_3CONH_2)]$	Ag: MVol.B6-301/2
$AgC_7H_7N_4O_2$	$Ag((CH_3)_2C_5HN_4O_2)$	Ag: MVol.B6-166/7
$AgC_7H_7N_4O_3$	$Ag(NH_2C_7H_5N_2)NO_3$	Ag: MVol.B6-132
$AgC_7H_7N_4O_7$	$[Ag(CH_3C_5H_4NO)]C(NO_2)_3$	Ag: MVol.B6-105
AgC_7H_7O	$AgC_6H_4OCH_3$	Ag: MVol.B5-10
$AgC_7H_7OS_3$	$[Ag((CH_3)_2C_5HOS_3)]$	Ag: MVol.B7-83/4
–	$[Ag(C_2H_5C_5H_2OS_3)]$	Ag: MVol.B7-83/4
$AgC_7H_7O_4S$	$[Ag(CH_3OC_6H_4SO_3)]$	Ag: MVol.B6-218
$AgC_7H_8^+$	$[Ag(CH_2(CH)_6)]^+$	Ag: MVol.B5-60, 61
–	$[AgC_6H_5CH_3]^+$	Ag: MVol.B5-96
–	$[AgC_7H_8]^+$	Ag: MVol.B5-75, 77

Formula	Compound	Reference
$AgC_7H_8NO_2$	$Ag(C_2H_5NC_4H_3COO)$	Ag: MVol.B6-68
$AgC_7H_8NO_3$	$AgNO_3 \cdot C_7H_8$	Ag: MVol.B5-78
$AgC_7H_8NO_3S_3$	$AgNO_3 \cdot (CH_3)_2C_5H_2S_3$	Ag: MVol.B7-93
AgC_7H_8NS	$[Ag(SCH_2CH_2C_5H_4N)]$	Ag: MVol.B7-59
$AgC_7H_8N_2OS^+$	$[Ag(HOC_6H_4NHCSNH_2)]^+$	Ag: MVol.B7-159
$AgC_7H_8N_3$	$[Ag(CH_3N_3C_6H_5)]$	Ag: MVol.B6-295/6
$AgC_7H_8N_3O_2$	$Ag(C_6H_5C(NO)NO) \cdot NH_3$	Ag: MVol.B6-320
$AgC_7H_8N_3O_3S$	$Ag(C_6H_5NHCSNH_2)NO_3$	Ag: MVol.B7-158
$AgC_7H_8N_3O_4$	$AgNO_3 \cdot NH_2C_6H_4CONH_2$	Ag: MVol.B6-255
$AgC_7H_8N_5$	$[Ag((CH_3)_2NC_5H_2N_4)]$	Ag: MVol.B6-166/7
$AgC_7H_8N_5OS$	$[Ag(NC_5H_4CO(NH)_2CSNNH_2)]$	Ag: MVol.B7-172
$AgC_7H_8O^+$	$[Ag(CH_3C_6H_4OH)]^+$	Ag: MVol.B5-118
–	$[Ag(C_6H_5OCH_3)]^+$	Ag: MVol.B5-118
$AgC_7H_9N^+$	$[Ag(CH_3C_6H_4NH_2)]^+$	Ag: MVol.B6-52
–	$[Ag((CH_3)_2C_5H_3N)]^+$	Ag: MVol.B6-87/8
–	$[Ag(C_2H_5C_5H_4N)]^+$	Ag: MVol.B6-86
–	$[Ag(C_6H_5CH_2NH_2)]^+$	Ag: MVol.B6-53/4
–	$[Ag(C_6H_5NHCH_3)]^+$	Ag: MVol.B6-53
$AgC_7H_9NO_3P$	$Ag(C_6H_5CH_2OP(O)NHOH)$	Ag: MVol.B7-251
$AgC_7H_9N_2$	$[Ag(C_7H_9N_2)]$	Ag: MVol.B6-132
$AgC_7H_9N_2O$	$[Ag(C_3N_2(CH_3)_2COCH_3)]$	Ag: MVol.B6-132
–	$AgNHCOC_6H_5 \cdot NH_3$	Ag: MVol.B6-30, 326
$AgC_7H_9N_2O_2$	$Ag(NH_2C_6H_4COO) \cdot NH_3$	Ag: MVol.B6-29, 254
$AgC_7H_9N_2O_3$	$[Ag(CH_3C_6H_4NH_2)]NO_3$	Ag: MVol.B6-52
–	$Ag((CH_3)_3C_3N_2OCOO)$	Ag: MVol.B6-147
$AgC_7H_9N_4O_4$	$AgNO_3 \cdot NH_2C_6H_4CONHNH_2$	Ag: MVol.B6-255
$AgC_7H_9N_6O_8$	$AgNO_3 \cdot HNO_3 \cdot (CH_3)_2C_5H_2N_4O_2$	Ag: MVol.B6-167
AgC_7H_9O	$AgCCC_4H_6O(CH_3)$	Ag: MVol.B5-19
$AgC_7H_9O_2$	$[Ag(CH_3COC_5H_6O)]$	Ag: MVol.B6-212
$AgC_7H_{10}{}^+$	$[AgC_6H_7CH_3]^+$	Ag: MVol.B5-57/8
–	$[AgC_7H_{10}]^+$	Ag: MVol.B5-75, 77
$AgC_7H_{10}NO_3S_3$	$AgNO_3 \cdot (CH_2)_4C_3H_2S_3$	Ag: MVol.B7-83
$AgC_7H_{10}N_2{}^+$	$[Ag(CH_3(NH_2CH_2)C_5H_3N)]^+$	Ag: MVol.B6-96/7
–	$[Ag(C_2H_5(CHCH_2)C_3H_2N_2)]^+$	Ag: MVol.B6-138/9
$AgC_7H_{10}N_2O^+$	$[Ag(HOC_2H_4(CHCH_2)C_3H_2N_2)]^+$	Ag: MVol.B6-138/9
$AgC_7H_{10}N_3O_2$	$Ag(CH_3C_3HN_2O(C_2H_5)CONH)$	Ag: MVol.B6-135
$AgC_7H_{10}N_3O_3$	$[Ag(C_3N_3O_3(C_2H_5)_2)]$	Ag: MVol.B6-190
–	$Ag(NH_2C_2H_4C_5H_4N)NO_3$	Ag: MVol.B6-97
$AgC_7H_{10}N_5$	$[Ag(N(C_3H_5)_2CN_4)]$	Ag: MVol.B6-192/3
$AgC_7H_{10}N_5O_2$	$Ag((CH_3)_2C_5HN_4O_2) \cdot NH_3$	Ag: MVol.B6-167
$AgC_7H_{10}O^+$	$[AgC_7H_9OH]^+$	Ag: MVol.B5-88/9
$AgC_7H_{10}O_4S_2{}^-$	$[Ag(C_3H_6(SCH_2COO)_2)]^-$	Ag: MVol.B7-39/40
AgC_7H_{11}	$AgC_4H_2(CH_3)_3$	Ag: MVol.B5-4/5
$AgC_7H_{11}N_2$	$Ag((C_2H_5)_2C_3HN_2)$	Ag: MVol.B6-142
–	$Ag(C_4H_9C_3H_2N_2)$	Ag: MVol.B6-142
$AgC_7H_{11}N_2{}^{2+}$	$[Ag(C_5H_5NC_2H_4NH_2)]^{2+}$	Ag: MVol.B6-97/8
–	$[AgH(CH_3(NH_2CH_2)C_5H_3N)]^{2+}$	Ag: MVol.B6-96/7
$AgC_7H_{11}N_2O_2$	$Ag(C_3H_7(CH_3)C_3HN_2O_2)$	Ag: MVol.B6-148
–	$[Ag(NH_3)_2]C_6H_5COO$	Ag: MVol.B6-29
$AgC_7H_{11}N_2S$	$[Ag((CH_3)_3C_4H_2N_2S)]$	Ag: MVol.B7-64

Formula	Compound	Reference
$AgC_7H_{11}N_4$	$[Ag(C_6H_{11}CN_4)]$	Ag: MVol.B6-192/3
$AgC_7H_{11}N_4O_2$	$Ag(C_6H_5C(NO)NO) \cdot 2\ NH_3$	Ag: MVol.B6-320
$AgC_7H_{11}O$	$2\ AgCCC(C_2H_5)_2OH \cdot AgCH_3COO$	Ag: MVol.B5-15
–	$AgCCCH(C_4H_9)OH \cdot AgNO_3$	Ag: MVol.B5-15
$AgC_7H_{11}O_2$	$AgCCCH(OC_2H_5)_2$	Ag: MVol.B5-16
$AgC_7H_{11}O_2S$	$[AgCH_2CH(CH_2)_3SCH_2COO]$	Ag: MVol.B5-50/1
$AgC_7H_{11}O_2Se$	$[Ag(CH_2CH(CH_2)_3SeCH_2COO)]$	Ag: MVol.B5-50/1
		Ag: MVol.B7-191/3, 196
$AgC_7H_{12}^+$	$[AgCH_2C(CH_3)CHC(CH_3)_2]^+$	Ag: MVol.B5-42/4
–	$[AgCH_2CH(CH_2)_2C(CH_3)CH_2]^+$	Ag: MVol.B5-42/4
–	$[AgCH_2CH(CH_2)_3CHCH_2]^+$	Ag: MVol.B5-42/4
–	$[AgCH_3CCC_4H_9]^+$	Ag: MVol.B5-44
–	$[AgC_2H_5CCCH(CH_3)_2]^+$	Ag: MVol.B5-44
–	$[AgC_2H_5CCC_3H_7]^+$	Ag: MVol.B5-44
–	$[AgC_5H_7C_2H_5]^+$	Ag: MVol.B5-52/3
–	$[AgC_5H_8CHCH_3]^+$	Ag: MVol.B5-52/3
–	$[AgC_6H_9CH_3]^+$	Ag: MVol.B5-57/8
–	$[AgC_6H_{10}CH_2]^+$	Ag: MVol.B5-57/8
–	$[AgC_7H_{12}]^+$	Ag: MVol.B5-60
$AgC_7H_{12}NO_2$	$[Ag(C_5H_9CH(NH_2)COO)]$	Ag: MVol.B6-250
$AgC_7H_{12}NS_2$	$[Ag(C_6H_{11}NHCSS)]$	Ag: MVol.B7-103
$AgC_7H_{12}N_3O_2$	$Ag(NH_2C_6H_4COO) \cdot 2\ NH_3$	Ag: MVol.B6-29, 254
$AgC_7H_{12}N_3O_3S$	$[Ag((CH_3)_3C_4H_3N_2S)]NO_3$	Ag: MVol.B7-64
$AgC_7H_{12}N_5O_6S$	$Ag(CH_3C_6H_3(NO_2)SO_2NNO_2) \cdot 2\ NH_3$	Ag: MVol.B6-343
$AgC_7H_{12}O^+$	$[AgC_5H_7C_2H_4OH]^+$	Ag: MVol.B5-88/9
–	$[AgC_6H_9(CH_2)OH]^+$	Ag: MVol.B5-88/9
$AgC_7H_{12}O_2^+$	$[AgCH_2CH(CH_2)_3OOCCH_3]^+$	Ag: MVol.B5-48
$AgC_7H_{12}O_2S^+$	$[AgCH_2CH(CH_2)_3SCH_2COOH]^+$	Ag: MVol.B5-50/1
$AgC_7H_{12}O_2Se^+$	$[Ag(CH_2CHC_3H_6SeCH_2COOH)]^+$	Ag: MVol.B5-50/1
		Ag: MVol.B7-193
$AgC_7H_{13}N^+$	$[Ag(C_7H_{13}N)]^+$	Ag: MVol.B6-73
$AgC_7H_{13}NO_2^+$	$[Ag(C_5H_9CH(NH_2)COOH)]^+$	Ag: MVol.B6-250
$AgC_7H_{13}N_2OS_2$	$[Ag(HOC_2H_4N(CH_2CH_2)_2NCSS)]$	Ag: MVol.B7-117
$AgC_7H_{13}N_4$	$[Ag(C_6H_{13}CN_4)]$	Ag: MVol.B6-192/3
$AgC_7H_{13}N_4O_3S$	$Ag(C_7H_4NO_3S) \cdot 3\ NH_3$	Ag: MVol.B6-344/5
$AgC_7H_{13}N_4O_4S$	$Ag(CH_3C_6H_4SO_2NNO_2) \cdot 2\ NH_3$	Ag: MVol.B6-342/3
$AgC_7H_{13}N_6O_2$	$Ag((CH_3)_2C_5HN_4O_2) \cdot 2\ NH_3$	Ag: MVol.B6-167
$AgC_7H_{13}OS_2$	$[Ag(C_6H_{13}OCSS)]$	Ag: MVol.B7-98
		C: MVol.D4-257
$AgC_7H_{13}O_2S$	$[Ag(C_5H_{11}SCH_2COO)]$	Ag: MVol.B7-25/6
$AgC_7H_{13}S$	$(AgSC_6H_{10}CH_3)_n$	Ag: MVol.B7-5/6
$AgC_7H_{14}^+$	$[AgCH_2C(CH_3)C(CH_3)_3]^+$	Ag: MVol.B5-42/4
–	$[AgCH_2C(CH_3)CH_2CH(CH_3)_2]^+$	Ag: MVol.B5-42/4
–	$[AgCH_2C(CH_3)C_4H_9]^+$	Ag: MVol.B5-42/4
–	$[AgCH_2CHCH(CH_3)C_3H_7]^+$	Ag: MVol.B5-42/4
–	$[AgCH_2CHCH(C_2H_5)_2]^+$	Ag: MVol.B5-42/4
–	$[AgCH_2CHCH_2CH(CH_3)C_2H_5]^+$	Ag: MVol.B5-42/4
–	$[AgCH_2CH(CH_2)_2CH(CH_3)_2]^+$	Ag: MVol.B5-42/4
–	$[AgCH_2CHC_5H_{11}]^+$	Ag: MVol.B5-42/4
–	$[AgCH_3CHCHC(CH_3)_3]^+$	Ag: MVol.B5-42/4

Formula	Compound	Reference
$AgC_7H_{14}^+$	$[AgCH_3CHCHC_4H_9]^+$	Ag: MVol.B5-42/4
–	$[Ag(CH_3)_2CHCHCHC_2H_5]^+$	Ag: MVol.B5-42/4
–	$[AgC_2H_5CHCHC_3H_7]^+$	Ag: MVol.B5-42/4
$AgC_7H_{14}NOS$	$[Ag((C_3H_7)_2NCOS)]$	Ag: MVol.B7-99/100
–	$[Ag((C_3H_7)_2NCOS)]_6$	Ag: MVol.B7-100/1
$AgC_7H_{14}NO_2$	$[Ag(NH_2(CH_2)_6COO)]$	Ag: MVol.B6-250
$AgC_7H_{14}NS_2$	$[Ag(CH_3N(C_5H_{11})CSS)]$	Ag: MVol.B7-105/6, 111/2
–	$[Ag(C_2H_5N(C_4H_9)CSS)]$	Ag: MVol.B7-105/6, 111/2
–	$[Ag((C_3H_7)_2NCSS)]$	Ag: MVol.B7-105/6, 110/1
–	$[Ag(SCSNHC_6H_{13})]$	Ag: MVol.B7-102
$AgC_7H_{14}N_3O_4S$	$[Ag((CH_3)_3C_4H_5N_2OS)]NO_3$	Ag: MVol.B7-64/5
$AgC_7H_{14}N_5$	$[Ag(N(C_3H_7)_2CN_4)]$	Ag: MVol.B6-192/3
$AgC_7H_{14}O^+$	$[AgCH_2CH(CH_2)_4CH_2OH]^+$	Ag: MVol.B5-47
–	$[AgCH_2CH(CH_2)_4OCH_3]^+$	Ag: MVol.B5-48
–	$[AgCH_3CHCHOCH_2CH(CH_3)_2]^+$	Ag: MVol.B5-48
–	$[Ag(CH_3)_2CHCHCHOC_2H_5]^+$	Ag: MVol.B5-48
$AgC_7H_{14}O_2S^+$	$[Ag(C_5H_{11}SCH_2COOH)]^+$	Ag: MVol.B7-25/6
$AgC_7H_{15}N^+$	$[Ag(C_2H_5N(CH_2)_5)]^+$	Ag: MVol.B6-73
–	$[Ag(NHC_5H_8(CH_3)_2)]^+$	Ag: MVol.B6-73
–	$[Ag(NHC_5H_9C_2H_5)]^+$	Ag: MVol.B6-73
$AgC_7H_{15}NPS$	$[AgSCN(P(C_2H_5)_3)]$	Ag: MVol.B7-202
$AgC_7H_{15}N_2S_2$	$[Ag((C_2H_5)_2NC_2H_4NHCSS)]$	Ag: MVol.B7-104
$AgC_7H_{16}NOS$	$[Ag(SCH(CH_3)CH_2NHC_3H_6OCH_3)]$	Ag: MVol.B7-17
AgC_8ClH_4	$AgCCC_6H_4Cl$	Ag: MVol.B5-19
$AgC_8ClH_6N_2$	$Ag(Cl(CH_3)C_7H_3N_2)$	Ag: MVol.B6-146
$AgC_8ClH_6O_2S$	$[Ag(ClC_6H_4SCH_2COO)]$	Ag: MVol.B7-27/8
$AgC_8ClH_6O_2Se$	$[Ag(ClC_6H_4SeCH_2COO)]$	Ag: MVol.B7-191/2, 194, 196
$AgC_8ClH_7^+$	$[Ag(ClC_6H_4CHCH_2)]^+$	Ag: MVol.B5-117
AgC_8ClH_7NOS	$[Ag(SCH_2CONHC_6H_4Cl)]$	Ag: MVol.B7-23/4
$AgC_8ClH_7O_2S^+$	$[Ag(ClC_6H_4SCH_2COOH)]^+$	Ag: MVol.B7-27/8
$AgC_8ClH_7O_2Se^+$	$[Ag(ClC_6H_4SeCH_2COOH)]^+$	Ag: MVol.B7-194
$AgC_8ClH_8N_4O_4$	$[Ag(C_2H_4(CN)_2)_2]ClO_4$	Ag: MVol.B6-350, 351
$AgC_8ClH_8O_4$	$AgClO_4 \cdot C_6H_5CHCH_2$	Ag: MVol.B5-101
AgC_8ClH_9NS	$[Ag(CH_3CSNHC_6H_5)]Cl$	Ag: MVol.B7-126
$AgC_8ClH_{10}O_4$	$AgClO_4 \cdot C_6H_4(CH_3)_2$	Ag: MVol.B5-99
$AgC_8ClH_{11}NO_4S$	$[AgClO_4(NC_5H_4C_2H_4SCH_3)]_n$	Ag: MVol.B7-59
$AgC_8ClH_{12}N_4O_4$	$[Ag(CH_3C_3H_3N_2)_2]ClO_4$	Ag: MVol.B6-140
–	$AgClO_4 \cdot 4\ CH_3CN = [Ag(CH_3CN)_4]ClO_4$	Ag: MVol.B6-348
–	$AgClO_4 \cdot N_2(C(CH_3)_2CN)_2$	Ag: MVol.B6-285
$AgC_8ClH_{12}O_4$	$AgClO_4 \cdot C_8H_{12}$	Ag: MVol.B5-65
$AgC_8ClH_{14}O_4$	$AgClO_4 \cdot CH_2CH(CH_2)_4CHCH_2$	Ag: MVol.B5-45
$AgC_8ClH_{16}N_4O_4S_4$	$[Ag(CH_3NHCSCSNHCH_3)_2]ClO_4$	Ag: MVol.B7-187
$AgC_8ClH_{16}O_8$	$Ag(C_4H_8O_2)_2ClO_4$	Ag: MVol.B6-220/1
$AgC_8ClH_{18}O_4S$	$AgClO_4 \cdot S(C_4H_9)_2$	Ag: MVol.B7-12
$AgC_8ClH_{20}N_4S_4$	$[Ag(CH_3CSNH_2)_4]Cl$	Ag: MVol.B7-124/5
$AgC_8ClH_{22}N_2O_4$	$AgClO_4 \cdot 2\ (C_2H_5)_2NH$	Ag: MVol.B6-48
$AgC_8Cl_2H_6N_4^+$	$[Ag(C_4H_3ClN_2)_2]^+$	Ag: MVol.B6-159
$AgC_8Cl_4H_{20}IrS_2$	$Ag[Ir(S(C_2H_5)_2)_2Cl_4]$	Ir: SVol.2-166
$AgC_8Cl_6H_{12}N_4Sb$	$AgCl \cdot SbCl_5 \cdot 4\ CH_3CN = [Ag(CH_3CN)_4]SbCl_6$	Ag: MVol.B6-348

Formula	Compound	Reference
$AgC_8CoH_{16}N_3O_7S_2$	$[AgCo(CH_3SC_2H_3(NH_2)COO)_2]NO_3 \cdot H_2O$	Ag: MVol.B6-258
$AgC_8CuH_{16}N_3O_7S_2$	$[AgCu(CH_3SC_2H_3(NH_2)COO)_2]NO_3$	Ag: MVol.B6-257/8
$AgC_8F_2H_4KN_4O_4$	$K[Ag(C_4H_2FN_2O_2)_2]$	Ag: MVol.B6-152
$AgC_8F_2H_4N_4O_4^-$	$[Ag(C_4H_2FN_2O_2)_2]^-$	Ag: MVol.B6-152
$AgC_8F_3H_8N_2O_5$	$Ag(C_8H_8F_3N_2O_5)$	Ag: MVol.B6-319
$AgC_8F_{14}N_3$	$NC(C_3F_7)NAgC(C_3F_7)N$	F: PerFHalOrg.5-71/2
$AgC_8H_4MnO_3$	$AgC_5H_4Mn(CO)_3$	Ag: MVol.B5-23
$AgC_8H_4NOS_2$	$[Ag(C_8H_4NOS_2)]$	Ag: MVol.B7-85
$AgC_8H_4NO_2$	$AgCCC_6H_4NO_2$	Ag: MVol.B5-19
–	$[Ag(C_8H_4NO_2)]$	Ag: MVol.B6-70/1
–	$Ag(C_8H_4NO_2) \cdot 0.5\ H_2O$	Ag: MVol.B6-328
$AgC_8H_4NO_2S$	$AgCCSC_6H_4NO_2$	Ag: MVol.B5-20
–	$[Ag(NC(O)C_6H_4SC(O))]$	Ag: MVol.B7-85
–	$[Ag(ONCC(O)C_6H_4S)]$	Ag: MVol.B7-69
$AgC_8H_4NO_2S_2$	$[Ag(C_4H_3OCHC_3NOS_2)]$	Ag: MVol.B7-76
AgC_8H_5	$AgCCC_6H_5$	Ag: MVol.B5-16/7
		Ag: MVol.B7-246
$AgC_8H_5N_2O$	$AgC(N_2)COC_6H_5$	Ag: MVol.B5-22
–	$Ag(C_6H_5CONCN)$	Ag: MVol.B6-326
$AgC_8H_5N_2O_2$	$[Ag(NC(NOH)C(O)C_6H_4)]$	Ag: MVol.B6-316
–	$[Ag(OCNNC(OH)C_6H_4)]$	Ag: MVol.B6-180
$AgC_8H_5N_2O_5$	$AgNO_3 \cdot C_8H_5NO_2$	Ag: MVol.B6-328
$AgC_8H_5N_2S_3$	$[Ag(C_6H_5C_2N_2S_3)]$	Ag: MVol.B7-81
$AgC_8H_5N_6O_8$	$[AgH(ONC_4H_2N_2O_3)_2]$	Ag: MVol.B6-314
–	$[AgH(ONC_4H_2N_2O_3)_2] \cdot 3\ H_2O$	Ag: MVol.B6-314
–	$[AgH(ONC_4H_2N_2O_3)_2] \cdot 3.8\ H_2O$	Ag: MVol.B6-314
AgC_8H_5O	$AgCCC_6H_4OH \cdot AgNO_3$	Ag: MVol.B5-18
–	$AgCCOC_6H_5$	Ag: MVol.B7-246
AgC_8H_5S	$AgCCSC_6H_5$	Ag: MVol.B5-20
AgC_8H_5Se	$AgCCSeC_6H_5$	Ag: MVol.B5-20
$AgC_8H_6^+$	$[AgC_6H_5CCH]^+$	Ag: MVol.B5-102
$AgC_8H_6NNa_2O_4S_2$	$Na_2[Ag(SC_6H_3(NHCH_2SO_2)COO)]$	Ag: MVol.B7-35
$AgC_8H_6NO_4$	$Ag(OC_6H_3(NO)COOCH_3)$	Ag: MVol.B6-307
$AgC_8H_6NO_4S$	$[Ag(NO_2C_6H_4SCH_2COO)]$	Ag: MVol.B7-27/8
$AgC_8H_6NO_4Se$	$[Ag(NO_2C_6H_4SeCH_2COO)]$	Ag: MVol.B7-191/4, 196
$AgC_8H_6NO_5$	$AgNO_3 \cdot C_6H_4(CHO)_2$	Ag: MVol.B6-209
$AgC_8H_6N_2O_2^+$	$[Ag(NO_2C_6H_4CH_2CN)]^+$	Ag: MVol.B6-350
$AgC_8H_6N_3$	$Ag(NC_5H_4C_3H_2N_2) \cdot 0.5\ H_2O$	Ag: MVol.B6-142
$AgC_8H_6N_3O$	$[Ag(C_6H_5CONNHCN)]$	Ag: MVol.B6-332
$AgC_8H_6N_3OS$	$[Ag(C_6H_5C_2HN_3OS)]$	Ag: MVol.B6-184
$AgC_8H_6N_3O_2$	$[Ag(C_6H_5C_2HN_3O_2)]$	Ag: MVol.B6-184
–	$[Ag(HOC_6H_4CONNHCN)]$	Ag: MVol.B6-332
$AgC_8H_6N_3O_3$	$AgNO_3 \cdot C_8H_6N_2$	Ag: MVol.B6-129
$AgC_8H_6N_3O_3S$	$AgNO_3 \cdot C_6H_5C_2HN_2S$	Ag: MVol.B7-79/80
$AgC_8H_6N_3O_6$	$Ag(NO_2NCOOCH_2C_6H_4NO_2)$	Ag: MVol.B6-330
$AgC_8H_6N_3S$	$Ag(C_6H_5C_2HN_3S) \cdot n\ AgCl$	Ag: MVol.B7-55/6
–	$Ag(C_6H_5C_2HN_3S) \cdot n\ AgNO_3$	Ag: MVol.B7-55/6
–	$[Ag(SCNCHN(C_6H_5)N)]$	Ag: MVol.B7-55
–	$[Ag(SCNNCHN(C_6H_5))]$	Ag: MVol.B7-56
$AgC_8H_6N_3S_2$	$[Ag(C_6H_5NC_2HN_2S_2)]$	Ag: MVol.B7-81

Formula	Compound	Reference
$AgC_8H_6N_4O_4^-$	$[Ag(C_4H_3N_2O_2)_2]^-$	Ag: MVol.B6-151
$AgC_8H_6N_5O_3$	$Ag(N_2C_4H_3C_4H_3N_2)NO_3$	Ag: MVol.B6-152
$AgC_8H_6N_7O_7$	$AgOC_6H_2(NO_2)_3 \cdot NH_2C(NH)NHCN$	Ag: MVol.B6-338
$AgC_8H_6S^+$	$[Ag(C_8H_6S)]^+$	Ag: MVol.B7-69
AgC_8H_7	$AgCHCHC_6H_5$	Ag: MVol.B5-7
$AgC_8H_7KNO_4S$	$KAg(C_6H_5SO_2NHCHCOO)$	Ag: MVol.B6-241
$AgC_8H_7N^+$	$[Ag(C_6H_5CH_2CN)]^+$	Ag: MVol.B6-350
–	$[Ag(C_8H_7N)]^+$	Ag: MVol.B6-70
$AgC_8H_7NO_2^+$	$[Ag(CH_2CHC_6H_4NO_2)]^+$	Ag: MVol.B5-118
$AgC_8H_7NO_4S^+$	$[Ag(NO_2C_6H_4SCH_2COOH)]^+$	Ag: MVol.B7-27/8
$AgC_8H_7NO_4Se^+$	$[Ag(NO_2C_6H_4SeCH_2COOH)]^+$	Ag: MVol.B7-193/4
$AgC_8H_7N_2$	$Ag(CH_3C_7H_4N_2)$	Ag: MVol.B6-145
$AgC_8H_7N_2O_3S$	$[Ag(SCH_2CONHC_6H_4NO_2)]$	Ag: MVol.B7-23/4
$AgC_8H_7N_2O_4S$	$[Ag(SCH_2CONHC_6H_3(OH)NO_2)] \cdot H_2O$	Ag: MVol.B7-24
$AgC_8H_7N_2O_4S_3$	$[Ag(NO_2C_6H_4SO_2NCSSCH_3)]$	Ag: MVol.B7-104
$AgC_8H_7N_2S$	$[Ag(C_8H_7N_2S)]$	Ag: MVol.B7-53
$AgC_8H_7N_4$	$[Ag(CH_3C_6H_4CN_4)]$	Ag: MVol.B6-192/3
–	$[Ag(C_6H_5CH_2CN_4)]$	Ag: MVol.B6-192/3
$AgC_8H_7N_4O_3S$	$AgNO_3 \cdot C_6H_5C_2N_2SNH_2$	Ag: MVol.B7-80
$AgC_8H_7N_4O_5$	$AgNO_3 \cdot HOC_6H_4CON_2H_2CN$	Ag: MVol.B6-332
$AgC_8H_7O_2S$	$[Ag(CH_3COCHCOC_4H_3S)]$	Ag: MVol.B6-212
–	$[Ag(CH_3SC_6H_4COO)]$	Ag: MVol.B7-35, 36
–	$[Ag(C_6H_5SCH_2COO)]$	Ag: MVol.B7-27/8
–	$[Ag(SC_6H_4COOCH_3)]$	Ag: MVol.B7-34/5
$AgC_8H_7O_2Se$	$[Ag(C_6H_5SeCH_2COO)]$	Ag: MVol.B7-191/3, 196
$AgC_8H_7O_3$	$[Ag(C_4H_3OCOCHCOCH_3)]$	Ag: MVol.B6-211/2
–	$Ag(C_6H_5OCH_2COO)$	Ag: MVol.B6-218
$AgC_8H_7S_2$	$[Ag(C_6H_5CH_2CSS)]$	Ag: MVol.B7-95
$AgC_8H_8^+$	$[AgC_6H_5CHCH_2]^+$	Ag: MVol.B5-101
–	$[AgC_8H_8]^+$	Ag: MVol.B5-61/2
AgC_8H_8NO	$Ag(CH_3CONC_6H_5)$	Ag: MVol.B6-325
AgC_8H_8NOS	$[Ag(SCH_2CONHC_6H_5)]$	Ag: MVol.B7-23
$AgC_8H_8NO_2$	$Ag(CH_3OC_6H_4CONH)$	Ag: MVol.B6-326
–	$AgNC_4H_2(COCH_3)_2$	Ag: MVol.B6-68
$AgC_8H_8NO_2S$	$[Ag(NH_2C_6H_4SCH_2COO)]$	Ag: MVol.B7-27/8
–	$[Ag(SCH_2CONHC_6H_4OH)]$	Ag: MVol.B7-23/4
$AgC_8H_8NO_2S_3$	$[Ag(C_6H_5SO_2NCSSCH_3)]$	Ag: MVol.B7-104
$AgC_8H_8NO_3$	$AgNC_4H_2(COCH_3)COOCH_3$	Ag: MVol.B6-68
–	$AgNO_3 \cdot C_8H_8$	Ag: MVol.B5-63/4
$AgC_8H_8NO_4S$	$Ag(C_6H_5SO_2NHCH_2COO)$	Ag: MVol.B6-241
–	$[Ag(C_6H_5SO_2NO(COCH_3))]$	Ag: MVol.B6-307
$AgC_8H_8NS_2$	$[Ag(CH_3C_6H_4NHCSS)]$	Ag: MVol.B7-103
–	$[Ag(C_6H_5CH_2NHCSS)]$	Ag: MVol.B7-103
$AgC_8H_8N_2O_4^-$	$[Ag(N(COCH_2)_2)_2]^-$	Ag: MVol.B6-327
$AgC_8H_8N_3O$	$[Ag(CH_3CON_3C_6H_5)]$	Ag: MVol.B6-296
$AgC_8H_8N_3O_2$	$Ag(C_6H_2N_3O_2(CH_3)_2)$	Ag: MVol.B6-153
–	$[Ag(C_6H_5N_3COOCH_3)]$	Ag: MVol.B6-301/2
$AgC_8H_8N_3O_3S$	$AgNO_3 \cdot NH_2C_7H_3NSCH_3$	Ag: MVol.B7-77
$AgC_8H_8N_3S_2$	$Ag(C_6H_5NHCSNH_2)SCN$	Ag: MVol.B7-158
$AgC_8H_8N_4^+$	$[Ag(C_4H_4N_2)_2]^+$	Ag: MVol.B6-157/8

Formula	Compound	Reference
$AgC_8H_8N_4^+$	$[Ag(NCC_2H_4CN)_2]^+$	Ag: MVol.B6-350
$AgC_8H_8N_4O_8S_2$	$[Ag(N(CHCH)_2N)_2]S_2O_8$	Ag: MVol.B7-297
$AgC_8H_8N_5$	$[Ag(C_6H_5CH_2NHCN_4)]$	Ag: MVol.B6-192/3
$AgC_8H_8N_5O_3$	$4\,[AgNO_3 \cdot 2\,C_2H_4(CN)_2] \cdot H_2O$	Ag: MVol.B6-350
–	$AgNO_3 \cdot 2\,C_2H_4(CN)_2 \cdot H_2O$	Ag: MVol.B6-350
$AgC_8H_8N_5O_5$	$Ag(C_4H_4N_2O)_2NO_3$	Ag: MVol.B6-162
$AgC_8H_8O^+$	$[Ag(CH_3COC_6H_5)]^+$	Ag: MVol.B5-118
		Ag: MVol.B6-210
–	$[Ag(C_6H_5OCHCH_2)]^+$	Ag: MVol.B5-118
$AgC_8H_8O_2S^+$	$[Ag(C_6H_5SCH_2COOH)]^+$	Ag: MVol.B7-27/8
$AgC_8H_8O_2Se^+$	$[Ag(C_6H_5SeCH_2COOH)]^+$	Ag: MVol.B7-193
$AgC_8H_8O_8S_2^{3-}$	$[Ag(S(CH_2COO)_2)_2]^{3-}$	Ag: MVol.B7-38
AgC_8H_9	$AgC_6H_3(CH_3)_2 \cdot x\,AgNO_3$	Ag: MVol.B5-12
$AgC_8H_9NO_2S^+$	$[Ag(NH_2C_6H_4SCH_2COOH)]^+$	Ag: MVol.B7-27/8
$AgC_8H_9N_2$	$AgC_8H_6N \cdot NH_3$	Ag: MVol.B6-70
$AgC_8H_9N_2O_3$	$AgNO_3 \cdot CH_2CHCH_2C_5H_4N$ $= [Ag(CH_2CHCH_2C_5H_4N)]NO_3$	Ag: MVol.B5-50
		Ag: MVol.B6-90/1
$AgC_8H_9N_2O_3S$	$Ag(CH_3CONHC_6H_4SO_2NH)$	Ag: MVol.B6-263
$AgC_8H_9N_2O_3S_2$	$[Ag(SCH_2CONHC_6H_4SO_2NH_2)]$	Ag: MVol.B7-23/4
$AgC_8H_9N_2O_4S$	$[Ag(NH_2C_6H_4SO_2NHCH_2COO)]$	Ag: MVol.B6-263
$AgC_8H_9N_4OS$	$[Ag(C_6H_5CO(NH)_2CSNNH_2)]$	Ag: MVol.B7-172
$AgC_8H_9N_6O_3$	$AgNO_3 \cdot C_6H_5C_2N_3(NH_2)_2$	Ag: MVol.B6-182
AgC_8H_9O	$AgC_6H_4OC_2H_5 \cdot x\,AgNO_3$	Ag: MVol.B5-12
$AgC_8H_9OS_3$	$[Ag(C_3H_7C_5H_2OS_3)]$	Ag: MVol.B7-83/4
$AgC_8H_9O_2$	$AgC_6H_3(OCH_3)_2$	Ag: MVol.B5-10
$AgC_8H_9O_3S_2$	$[Ag(C_2H_5SC_6H_4SO_3)]$	Ag: MVol.B7-44
$AgC_8H_{10}^+$	$[AgC_6H_4(CH_3)_2]^+$	Ag: MVol.B5-97/8
–	$[AgC_6H_5C_2H_5]^+$	Ag: MVol.B5-97/8
–	$[AgC_8H_{10}]^+$	Ag: MVol.B5-81/2
$AgC_8H_{10}IO_2$	$Ag((CH_3)_2C(CH_2CO)_2CI)$	Ag: MVol.B6-214
$AgC_8H_{10}NO_2S$	$Ag(C_2H_5SO_2NC_6H_5)$	Ag: MVol.B6-342
$AgC_8H_{10}NO_3$	1) $AgNO_3 \cdot C_8H_{10}$	Ag: MVol.B5-63
–	2) $AgNO_3 \cdot C_8H_{10}$	Ag: MVol.B5-82/3
$AgC_8H_{10}NO_3S$	$[Ag((CH_3)_2NC_6H_4SO_3)]$	Ag: MVol.B6-262
$AgC_8H_{10}NO_4$	$Ag((CH_3)_2C(CH_2CO)_2CNO_2)$	Ag: MVol.B6-214
$AgC_8H_{10}N_2NaO_3$	$Na[Ag((C_2H_5)_2C_4N_2O_3)]$	Ag: MVol.B6-154
–	$Na[Ag((C_2H_5)_2C_4N_2O_3)] \cdot 2\,H_2O$	Ag: MVol.B6-155/6
$AgC_8H_{10}N_2O^+$	$[Ag(HOCH_2C(NH)NHC_6H_5)]^+$	Ag: MVol.B6-339/40
$AgC_8H_{10}N_2O_3^-$	$[Ag((C_2H_5)_2C_4N_2O_3)]^-$	Ag: MVol.B6-154
$AgC_8H_{10}N_2O_4S_4^{3-}$	$[Ag(CH_3CH(COO)NHCSS)_2]^{3-}$	Ag: MVol.B7-103
$AgC_8H_{10}N_2O_8^{3-}$	$[Ag(NH_2C_2H_3(COO)_2)_2]^{3-}$	Ag: MVol.B6-252
$AgC_8H_{10}N_3$	$[Ag(CH_3N_3CH_2C_6H_5)]$	Ag: MVol.B6-296
–	$[Ag(CH_3N_3C_6H_4CH_3)]$	Ag: MVol.B6-296
–	$[Ag(C_2H_5N_3C_6H_5)]$	Ag: MVol.B6-295/6
$AgC_8H_{10}N_3O_2$	$AgN(CO)_2C_6H_4 \cdot 2\,NH_3$	Ag: MVol.B6-30, 328
$AgC_8H_{10}N_3O_3S$	$Ag((CH_3)_2NC_6H_4NNSO_3)$	Ag: MVol.B6-294
$AgC_8H_{10}N_3O_4$	$AgNO_3 \cdot HOCH_2C(NH)NHC_6H_5$	Ag: MVol.B6-340
$AgC_8H_{10}N_3O_6S$	$AgNO_3 \cdot NH_2C_6H_4SO_2NHCOCH_3 \cdot H_2O$	Ag: MVol.B6-263
$AgC_8H_{10}N_4O_4^-$	$[Ag(CH_3C_3H_2N_2O_2)_2]^-$	Ag: MVol.B6-147

$AgC_8H_{10}N_6^+$	$[Ag(NH_2C_4H_3N_2)_2]^+$	Ag: MVol.B6-159
$AgC_8H_{10}N_7O_7S_2$	$[Ag(NH_2CSNH_2)_2]OC_6H_2(NO_2)_3$	Ag: MVol.B7-143
$AgC_8H_{11}IP$	$AgI \cdot (CH_3)_2PC_6H_5$	Ag: MVol.B7-220
$AgC_8H_{11}N^+$	$[Ag(CH_3(C_2H_5)C_5H_3N)]^+$	Ag: MVol.B6-89
–	$[Ag((CH_3)_2C_6H_3NH_2)]^+$	Ag: MVol.B6-52/3
–	$[Ag((CH_3)_3C_5H_2N)]^+$	Ag: MVol.B6-89
–	$[Ag(C_3H_7C_5H_4N)]^+$	Ag: MVol.B6-87
–	$[Ag(C_6H_5N(CH_3)_2)]^+$	Ag: MVol.B6-53
–	$[Ag(C_6H_5NHC_2H_5)]^+$	Ag: MVol.B6-53
$AgC_8H_{11}N_2$	$[Ag(CH_3C_7H_8N_2)]$	Ag: MVol.B6-132
$AgC_8H_{11}N_2O$	$Ag(CH_3CONC_6H_5) \cdot NH_3$	Ag: MVol.B6-325
$AgC_8H_{11}N_2O_2S$	$[Ag(CH_3C_3N_2(SCH_3)COOC_2H_5)]$	Ag: MVol.B7-49
$AgC_8H_{11}O$	$AgCCC_6H_{10}OH$	Ag: MVol.B5-15
$AgC_8H_{11}O_2$	$[Ag(CH_3COC_6H_8O)]$	Ag: MVol.B6-211/2
$AgC_8H_{12}^+$	$[AgC_6H_9CHCH_2]^+$	Ag: MVol.B5-57/8
–	$[AgC_7H_{10}CH_2]^+$	Ag: MVol.B5-75, 77
–	1) $[AgC_8H_{12}]^+$	Ag: MVol.B5-61/2
–	2) $[AgC_8H_{12}]^+$	Ag: MVol.B5-79
$AgC_8H_{12}NO_3$	$AgNO_3 \cdot C_8H_{12}$	Ag: MVol.B5-63
$AgC_8H_{12}N_3O_3$	$AgNO_3 \cdot (CH_3)_4C_4N_2$	Ag: MVol.B6-160
$AgC_8H_{12}N_4^+$	$[Ag(CH_3CN)_4]^+$	Ag: MVol.B6-347
–	$[Ag(CH_3C_3H_3N_2)_2]^+$	Ag: MVol.B6-138/9
$AgC_8H_{12}N_5O_3$	$[Ag(N(CH_3)CHNCHCH)_2]NO_3$	Ag: MVol.B6-140
–	$[Ag(NHC(CH_3)NCHCH)_2]NO_3$	Ag: MVol.B6-140
–	$[Ag(NHNCHCHC(CH_3))]NO_3$	Ag: MVol.B6-131
$AgC_8H_{12}N_7O_3$	$AgNO_3 \cdot C_2H_5C_2HN_3C_2HN_3C_2H_5$	Ag: MVol.B6-187
$AgC_8H_{12}N_7O_5$	$AgNO_3 \cdot C_4H_8(CONHNHCN)_2$	Ag: MVol.B6-333
$AgC_8H_{12}N_8^+$	$[Ag((CH_2)_3CN_4)_2]^+$	Ag: MVol.B6-194
$AgC_8H_{12}O^+$	$[AgC_7H_9CH_2OH]^+$	Ag: MVol.B5-88/9
–	$[AgC_8H_{12}O]^+$	Ag: MVol.B5-88/9
$AgC_8H_{12}O_4S^-$	$[Ag(S(C_3H_6COO)_2)]^-$	Ag: MVol.B7-39
$AgC_8H_{12}O_4S_2^-$	$[Ag(C_4H_8(SCH_2COO)_2)]^-$	Ag: MVol.B7-39/40
$AgC_8H_{12}O_4S_3^-$	$[Ag(S(C_2H_4SCH_2COO)_2)]^-$	Ag: MVol.B7-42/3
AgC_8H_{13}	$AgCCC_6H_{13}$	Ag: MVol.B5-13, 14
$AgC_8H_{13}N_2$	$Ag(CH_3(C_2H_5)_2C_3N_2)$	Ag: MVol.B6-142
$AgC_8H_{13}N_2^{2+}$	$[Ag(C_5H_5NC_3H_6NH_2)]^{2+}$	Ag: MVol.B6-97/8
$AgC_8H_{13}N_4$	$[Ag(C_6H_{11}CH_2CN_4)]$	Ag: MVol.B6-192/3
$AgC_8H_{13}O_6S$	$[Ag((CH_3)_2C(CH_2CO)_2CSO_3H)(H_2O)]$	Ag: MVol.B6-215/6
$AgC_8H_{14}^+$	$[AgCH_2C(CH_3)(CH_2)_2C(CH_3)CH_2]^+$	Ag: MVol.B5-42/4
–	$[AgCH_2CH(CH_2)_4CHCH_2]^+$	Ag: MVol.B5-42/4
–	$[AgCH_3CCC_5H_{11}]^+$	Ag: MVol.B5-44
–	$[Ag(CH_3)_2CCHCHC(CH_3)_2]^+$	Ag: MVol.B5-42/4
–	$[AgC_2H_5CCC(CH_3)_3]^+$	Ag: MVol.B5-44
–	$[AgC_2H_5CCC_4H_9]^+$	Ag: MVol.B5-44
–	$[AgC_3H_7CCC_3H_7]^+$	Ag: MVol.B5-44
–	$[AgC_5H_9CH_2CHCH_2]^+$	Ag: MVol.B5-52/3
–	$[AgC_6H_8(CH_3)_2]^+$	Ag: MVol.B5-57/8
–	$[AgC_6H_9C_2H_5]^+$	Ag: MVol.B5-57/8
–	$[AgC_6H_{10}CHCH_3]^+$	Ag: MVol.B5-57/8
–	$[AgC_6H_{11}CHCH_2]^+$	Ag: MVol.B5-57/8

Formula	Compound	Reference
$AgC_8H_{14}^+$	$[AgC_7H_{12}CH_2]^+$	Ag: MVol.B5-60
–	$[AgC_8H_{14}]^+$	Ag: MVol.B5-61/2
$AgC_8H_{14}NO_3$	$Ag(C_4H_9CH(NHCOCH_3)COO)$	Ag: MVol.B6-249
–	$AgNO_3 \cdot C_8H_{14}$	Ag: MVol.B5-62
$AgC_8H_{14}NO_5$	$Ag(C_3H_7OCOONCOOC_3H_7)$	Ag: MVol.B6-331
$AgC_8H_{14}N_3$	$[Ag((C_3H_7)_2C_2N_3)]$	Ag: MVol.B6-185
$AgC_8H_{14}N_4O_6^-$	$[Ag(NH_2CH_2CONHCH_2COO)_2]^-$	Ag: MVol.B6-265/6
–	$[Ag(NH_2COCH_2CH(NH_2)COO)_2]^-$	Ag: MVol.B6-255
$AgC_8H_{14}O^+$	$[AgC_8H_{13}OH]^+$	Ag: MVol.B5-88/9
$AgC_8H_{14}O_2^+$	$[AgCH_2CH(CH_2)_2CH(CH_3)OOCCH_3]^+$	Ag: MVol.B5-48
–	$[AgCH_2CH(CH_2)_4OOCCH_3]^+$	Ag: MVol.B5-48
$AgC_8H_{14}O_4S_2^-$	$[Ag(C_2H_5SCH_2COO)_2]^-$	Ag: MVol.B7-25/6
$AgC_8H_{14}O_4S_3^+$	$[Ag(S(C_2H_4SCH_2COOH)_2)]^+$	Ag: MVol.B7-43
$AgC_8H_{15}N_2O_2S_2$	$[Ag(CSCS(NCH_2CH(OH)CH_3)_2H)]$	Ag: MVol.B7-189
$AgC_8H_{15}N_4$	$[Ag(C_7H_{15}CN_4)]$	Ag: MVol.B6-192/3
$AgC_8H_{15}N_6OS_2$	$AgH(C_2H_5OC_4H_5(NNCSNH_2)_2) \cdot 0.5\ HNO_3$	Ag: MVol.B6-283
$AgC_8H_{15}O_2S$	$[Ag(C_6H_{13}SCH_2COO)]$	Ag: MVol.B7-25/6
$AgC_8H_{15.5}N_{6.5}O_{2.5}S_2$	$AgH(C_2H_5OC_4H_5(NNCSNH_2)_2) \cdot 0.5\ HNO_3$	Ag: MVol.B6-283
$AgC_8H_{16}^+$	$[AgCH_2C(C_2H_5)C_4H_9]^+$	Ag: MVol.B5-42/4
–	$[AgCH_2CHC_6H_{13}]^+$	Ag: MVol.B5-42/4
–	$[AgCH_3CHCHC_5H_{11}]^+$	Ag: MVol.B5-42/4
–	$[Ag(CH_3)_2CHCHCHCH(CH_3)_2]^+$	Ag: MVol.B5-42/4
–	$[Ag(CH_3)_3CCHCHC_2H_5]^+$	Ag: MVol.B5-42/4
–	$[AgC_3H_7CHCHC_3H_7]^+$	Ag: MVol.B5-42/4
$AgC_8H_{16}I_2LiO_2$	$LiAgI_2 \cdot 2\ C_4H_8O$	Ag: MVol.B6-220
$AgC_8H_{16}NO_3S_4$	$AgNO_3 \cdot 2\ S(CH_2CH_2)_2S$	Ag: MVol.B7-87
$AgC_8H_{16}NO_6S_2$	$[Ag(HOCH_2(CHOH)_5N(CH_3)CSS)]$	Ag: MVol.B7-116
$AgC_8H_{16}NS_2$	$[Ag(SCSNHC_7H_{15})]$	Ag: MVol.B7-102
$AgC_8H_{16}N_2O_4^-$	$[Ag((CH_3)_2NCH_2COO)_2]^-$	Ag: MVol.B6-240
–	$[Ag(NH_2(CH_2)_3COO)_2]^-$	Ag: MVol.B6-248
$AgC_8H_{16}N_3S$	$[Ag(CH_3(CH_2)_5CHNNCSNH_2)]$	Ag: MVol.B6-282
$AgC_8H_{16}N_4S_2^+$	$[Ag(CH_2CHCH_2NHCSNH_2)_2]^+$	Ag: MVol.B7-156
$AgC_8H_{16}N_5O_3$	$AgNO_3 \cdot 2\ CH_3C_3H_5N_2$	Ag: MVol.B6-147
$AgC_8H_{16}N_9O_3$	$AgNO_3 \cdot 2\ C_2N_3(CH_3)_2NH_2$	Ag: MVol.B6-186
$AgC_8H_{16}O_2^+$	$[Ag(CH_3CHCHCH_2OH)_2]^+$	Ag: MVol.B5-47
$AgC_8H_{16}O_2S^+$	$[Ag(C_6H_{13}SCH_2COOH)]^+$	Ag: MVol.B7-25/6
$AgC_8H_{17}IP$	$[AgI((C_2H_5)_2PCH_2CH_2CHCH_2)]_4$	Ag: MVol.B7-206/7
$AgC_8H_{17}N^+$	$[Ag(NHC_5H_9C_3H_7)]^+$	Ag: MVol.B6-73
$AgC_8H_{17}N_2S_2$	$[Ag((C_2H_5)_2NC_2H_4N(CH_3)CSS)]$	Ag: MVol.B7-115
–	$[Ag((C_2H_5)_2NC_3H_6NHCSS)]$	Ag: MVol.B7-104
$AgC_8H_{17}S$	$(AgSC_8H_{17})_n$	Ag: MVol.B7-5/6
$AgC_8H_{18}NO_3S$	$AgNO_3 \cdot S(C_4H_9)_2$	Ag: MVol.B7-11/2
$AgC_8H_{18}NO_6S_2$	$AgNO_3 \cdot O(C_2H_4SC_2H_4OH)_2$	Ag: MVol.B7-14
$AgC_8H_{18}NS$	$[Ag(SCH_2CH_2N(C_3H_7)_2)]$	Ag: MVol.B7-18
$AgC_8H_{18}N_2O_2^+$	$[Ag(NH(CH_2CH_2)_2O)_2]^+$	Ag: MVol.B6-204
$AgC_8H_{18}N_3O_5$	$AgNO_3 \cdot 2\ CH_3CON(CH_3)_2$	Ag: MVol.B6-325
$AgC_8H_{18}N_6S_2^+$	$[Ag((CH_3)_2CNNHCSNH_2)_2]^+$	Ag: MVol.B6-283
$AgC_8H_{18}N_7O_3S_2$	$AgNO_3 \cdot 2\ (CH_3)_2CNNHCSNH_2$	Ag: MVol.B6-283
$AgC_8H_{19}N^+$	$[Ag(C_5H_{11}C(CH_3)_2NH_2)]^+$	Ag: MVol.B6-46
$AgC_8H_{19}NO_2^+$	$[Ag(NH(C_4H_8OH)_2)]^+$	Ag: MVol.B6-229

$AgC_8H_{19}NO_3^+$	$[Ag(N(C_2H_4OH)_3)(C_2H_4)]^+$	Ag: MVol.B6-229/30
$AgC_8H_{20}N_2O_6S_2^-$	$[Ag(NH_2C_4H_8SO_3)_2]^-$	Ag: MVol.B6-262
$AgC_8H_{20}N_4^+$	$[Ag(NH(CH_2CH_2)_2NH)_2]^+$	Ag: MVol.B6-162/4
$AgC_8H_{20}N_4O_2^+$	$[Ag(HOC(CH_3)_2C(NH)NH_2)_2]^+$	Ag: MVol.B6-339/40
$AgC_8H_{20}N_4S_2^+$	$[Ag((CH_3)_2NCSNHCH_3)_2]^+$	Ag: MVol.B7-153
$AgC_8H_{20}N_4S_4^+$	$[Ag(CH_3CSNH_2)_4]^+$	Ag: MVol.B7-123/4
$AgC_8H_{20}N_5O_3$	$[Ag(NH_2C_2H_4NC_2H_4)_2]NO_3$	Ag: MVol.B6-67
$AgC_8H_{20}N_5O_3S_4$	$[Ag(CH_3CSNH_2)_4]NO_3$	Ag: MVol.B7-124
$AgC_8H_{20}N_{15}O_9$	$[Ag(C_4H_8N_2(C_2H_6N_5)_2)](NO_3)_3$	Ag: MVol.B7-325
$AgC_8H_{21}N_4O_3$	$[Ag(HOC(CH_3)_2C(NH)NH_2)_2]OH$	Ag: MVol.B6-339/40
$AgC_8H_{22}N_2^+$	$[Ag(C_4H_9NH_2)_2]^+$	Ag: MVol.B6-45
–	$[Ag(NH(C_2H_5)_2)_2]^+$	Ag: MVol.B6-47/8
$AgC_8H_{22}N_2O_2^+$	$[Ag((CH_3)_2NC_2H_4OH)_2]^+$	Ag: MVol.B6-227
–	$[Ag(C_2H_5NHC_2H_4OH)_2]^+$	Ag: MVol.B6-227
–	$[Ag(NH_2C_4H_8OH)_2]^+$	Ag: MVol.B6-228
$AgC_8H_{22}N_2O_2S_2^+$	$[Ag(NH_2C_2H_4SC_2H_4OH)_2]^+$	Ag: MVol.B7-18
$AgC_8H_{22}N_2O_4^+$	$[Ag(CH_3C(CH_2OH)_2NH_2)_2]^+$	Ag: MVol.B6-230/1
–	$[Ag(NH(C_2H_4OH)_2)_2]^+$	Ag: MVol.B6-228/9
$AgC_8H_{22}N_2O_6^+$	$[Ag(NH_2C(CH_2OH)_3)_2]^+$	Ag: MVol.B6-230/1
$AgC_8H_{22}N_3O_3$	$AgNO_3 \cdot 2\ (C_2H_5)_2NH$	Ag: MVol.B6-48
$AgC_8H_{23}N_5^+$	$[Ag(NH(C_2H_4NHC_2H_4NH_2)_2)]^+$	Ag: MVol.B6-66
$AgC_8H_{23}N_{12}O_3$	$[Ag(C_4H_8N_2(C_2H_6N_5)_2)](OH)_3$	Ag: MVol.B7-325
$AgC_8H_{24}N_4^+$	$[Ag(NH_2C_2H_4N(CH_3)_2)_2]^+$	Ag: MVol.B6-60
$AgC_8H_{24}N_4Ni_2S_4^+$	$[Ag(Ni(SC_2H_4NH_2)_2)_2]^+$	Ag: MVol.B7-17
$AgC_8H_{24}N_5^{2+}$	$[AgH(NH(C_2H_4NHC_2H_4NH_2)_2)]^{2+}$	Ag: MVol.B6-66
$AgC_8H_{24}N_5Ni_2O_3S_4$	$[Ag(Ni(SC_2H_4NH_2)_2)_2]NO_3$	Ag: MVol.B7-17
$AgC_8H_{24}N_{16}S_4^+$	$[Ag(NH_2C(NH)NHCSNH_2)_4]^+$	Ag: MVol.B7-162
$AgC_8H_{24}O_4S_4^+$	$[Ag((CH_3)_2SO)_4]^+$	Ag: MVol.B7-2
$AgC_8H_{25}N_5^{3+}$	$[AgH_2(NH(C_2H_4NHC_2H_4NH_2)_2)]^{3+}$	Ag: MVol.B6-66
$AgC_8H_{26}N_5^{4+}$	$[AgH_3(NH(C_2H_4NHC_2H_4NH_2)_2)]^{4+}$	Ag: MVol.B6-66
$AgC_8H_{26}N_6^+$	$[Ag(NH(C_2H_4NH_2)_2)_2]^+$	Ag: MVol.B6-65
AgC_8N_5	$AgC(CN)_2C(CN)C(CN)_2$	Ag: MVol.B6-355
$AgC_9CaH_{12}N_7$	$CaAg(CN)_3 \cdot (CH_2)_6N_4 \cdot 3\ H_2O$	Ag: MVol.B6-200
$AgC_9ClH_8N_2$	$Ag(Cl(CH_3)_2C_7H_2N_2)$	Ag: MVol.B6-146
$AgC_9ClH_8O_4$	$AgClO_4 \cdot C_9H_8$	Ag: MVol.B5-106/7
$AgC_9ClH_{12}O_4$	$AgClO_4 \cdot C_6H_3(CH_3)_3$	Ag: MVol.B5-99
$AgC_9ClH_{18}N_6S_3$	$[Ag(C_3H_6N_2S)_3]Cl$	Ag: MVol.B7-50
$AgC_9ClH_{21}O_3P$	$[AgCl(i\text{-}C_3H_7O)_3P]_3$	Ag: MVol.B7-248
$AgC_9ClH_{24}N_6O_4S_3$	$[Ag(CH_3NHCSNHCH_3)_3]ClO_4$	Ag: MVol.B7-152
–	$[Ag(C_2H_5NHCSNH_2)_3]ClO_4$	Ag: MVol.B7-154
$AgC_9ClH_{27}P_3$	$[AgCl(P(CH_3)_3)_3]$	Ag: MVol.B7-200/1
$AgC_9Cl_3H_6N_2$	$Ag(Cl_3(CH_3)_2C_7N_2)$	Ag: MVol.B6-146
$AgC_9Cl_6H_5OS_2$	$[Ag(C_7H_3Cl_6CH_2OCSS)]$	Ag: MVol.B7-98/9
$AgC_9FH_{10}N_2O_5$	$[Ag(C_9H_{10}FN_2O_5)]$	Ag: MVol.B6-157
$AgC_9FH_{24}N_6O_3S_4$	$[Ag(CH_3NHCSNHCH_3)_3]SO_3F$	Ag: MVol.B7-153
$AgC_9F_3H_6O_2S$	$[Ag(CF_3C_6H_4SCH_2COO)]$	Ag: MVol.B7-27/8
$AgC_9F_3H_7O_2S^+$	$[Ag(CF_3C_6H_4SCH_2COOH)]^+$	Ag: MVol.B7-27/8
$AgC_9F_3H_{12}N_4O_3S$	$AgCF_3SO_3 \cdot 4\ CH_3CN$	Ag: MVol.B6-348
$AgC_9F_6H_{18}Sb$	$AgSbF_6 \cdot 3\ CH_2CHCH_3$	Ag: MVol.B5-31/2
$AgC_9F_9H_9O_3Sb$	$Ag[Sb(CF_3)_3(OH)_3] \cdot C_6H_6$	F: PerFHalOrg.3-226

Formula	Compound	Reference
AgC_9H_4NO	$AgCCC_7H_4NO$	Ag: MVol.B5-20
$AgC_9H_4NO_2S_2$	$[Ag(C_8H_4O_2NCSS)]$	Ag: MVol.B7-118
$AgC_9H_4NO_4$	$[Ag(NO_2C_9H_4O_2)]$	Ag: MVol.B6-216
$AgC_9H_5N_2O_3$	$[Ag(C_2N_2O(C_6H_5)COO)]$	Ag: MVol.B6-204
–	$[Ag(ONC_3NO_2C_6H_5)]$	Ag: MVol.B6-313
AgC_9H_6NO	$[AgC_9H_6NO]$	Ag: MVol.B6-109, 110
$AgC_9H_6NO_2S$	$[Ag(NCC_6H_4SCH_2COO)]$	Ag: MVol.B7-27/8
$AgC_9H_6NO_3S_2$	$[Ag(SC_9H_5NSO_3H)]$	Ag: MVol.B7-61
AgC_9H_6NS	$[Ag(C_9H_6NS)]$	Ag: MVol.B7-60
$AgC_9H_6NS_2$	$[Ag(C_8H_6NCSS)]$	Ag: MVol.B7-118
$AgC_9H_6N_2O_3^+$	$[Ag(C_9H_5N(OH)NO_2)]^+$	Ag: MVol.B6-112/3
$AgC_9H_6N_3O_2$	$[Ag(C_6H_5C(NO)C(NOH)CN)]$	Ag: MVol.B6-318
–	$[Ag(C_6H_5C_3HN_3O_2)]$	Ag: MVol.B6-308/12
–	$[Ag(C_6H_5C_3HN_3O_2)]_n$	Ag: MVol.B6-308/12
–	$Ag(C_6H_5C_3HN_3O_2) \cdot H_2O$	Ag: MVol.B6-308/12
–	$Ag(C_6H_5C_3HN_3O_2) \cdot 2\ H_2O$	Ag: MVol.B6-308/12
–	$Ag(C_6H_5C_3HN_3O_2) \cdot x\ H_2O$	Ag: MVol.B6-308/12
$AgC_9H_6N_3O_3$	$[Ag(C_6H_5C_3HN_3O_3)]$	Ag: MVol.B6-190
$AgC_9H_6O_4Se^-$	$[Ag(OCOC_6H_4SeCH_2COO)]^-$	Ag: MVol.B7-191/3
$AgC_9H_7N^+$	$[Ag(C_9H_7N)]^+$	Ag: MVol.B6-107/8
$AgC_9H_7NNaO_3S$	$Na[Ag(SC_6H_3(NHCOCH_3)COO)]$	Ag: MVol.B7-35
$AgC_9H_7NO_2S^+$	$[Ag(NCC_6H_4SCH_2COOH)]^+$	Ag: MVol.B7-27/8
$AgC_9H_7NO_3S^-$	$[Ag(SC_6H_3(NHCOCH_3)COO)]^-$	Ag: MVol.B7-35
$AgC_9H_7N_2$	$AgCN \cdot CH_3C_6H_4NC$	Ag: MVol.B5-25
–	$[Ag(NC(C_6H_5)NCHCH)]$	Ag: MVol.B6-142
–	$[Ag(NCHNC(C_6H_5)CH)]$	Ag: MVol.B6-142
–	$[Ag(NCHNCHC(C_6H_5))]$	Ag: MVol.B6-142
–	$[Ag(NNC(C_6H_5)CHCH)]$	Ag: MVol.B6-131
–	$[Ag(NNCHC(C_6H_5)CH)]$	Ag: MVol.B6-131
–	$[Ag(NNCHCHC(C_6H_5))]$	Ag: MVol.B6-131
$AgC_9H_7N_2O$	$Ag(CH_3C_6H_4CONCN)$	Ag: MVol.B6-326
–	$Ag(C_4H_4NCOC_4H_3N)$	Ag: MVol.B6-68
$AgC_9H_7N_2O_2$	$Ag(CH_3C_7H_4N_2COO)$	Ag: MVol.B6-146
–	$[Ag(CH_3(NO)C_8H_4NO)]$	Ag: MVol.B6-316
$AgC_9H_7N_2S$	$[Ag(C_6H_5C_3H_2N_2S)]$	Ag: MVol.B7-49
$AgC_9H_7N_4O_3$	$[Ag(CH_3C_2N_3(O)C_6H_4NO_2)]$	Ag: MVol.B6-185
$AgC_9H_7N_4O_3S$	$[Ag(NHC_3H_2NOSNC_6H_4NO_2)]$	Ag: MVol.B7-72
$AgC_9H_7N_4O_4$	$Ag(NOC_3H_3N_2OC_6H_4NO_2)$	Ag: MVol.B6-135
AgC_9H_7O	$AgCCC_6H_4OCH_3 \cdot AgNO_3$	Ag: MVol.B5-19
–	$AgCCOC_6H_4CH_3$	Ag: MVol.B7-246
AgC_9H_7OS	$AgCCSC_6H_4OCH_3$	Ag: MVol.B5-20
$AgC_9H_7O_4S$	$[Ag(C_6H_4(COOH)SCH_2COO)]$	Ag: MVol.B7-27/8
$AgC_9H_7O_5S_3$	$[Ag(C_5HOS_3(COOCH_3)_2)]$	Ag: MVol.B7-83/4
AgC_9H_7S	$AgCCSC_6H_4CH_3$	Ag: MVol.B5-20
$AgC_9H_8^+$	$[AgC_9H_8]^+$	Ag: MVol.B5-106
$AgC_9H_8NOS_2$	$[Ag(CH_3CON(C_6H_5)CSS)]$	Ag: MVol.B7-116
$AgC_9H_8NO_3$	$Ag(C_6H_5CONHCH_2COO) \cdot 0.5\ H_2O$	Ag: MVol.B6-240/1
$AgC_9H_8NO_3S$	$[Ag(HSC_6H_3(NHCOCH_3)COO)]$	Ag: MVol.B7-35
–	$[Ag(SCH_2CONHC_6H_4COOH)]$	Ag: MVol.B7-23/4
$AgC_9H_8N_3$	$Ag(C_5H_3N(CH_3)C_3H_2N_2) \cdot 1.5\ H_2O$	Ag: MVol.B6-142

Formula	Compound	Reference
$AgC_9H_8N_3O$	$[Ag(C_6H_5CH_2CONNHCN)]$	Ag: MVol.B6-332
$AgC_9H_8N_3OS$	$[Ag(C_2N_3OS(CH_3)C_6H_5)]$	Ag: MVol.B7-56
$AgC_9H_8N_3O_2$	$[Ag(C_6H_5(CH_3)C_2N_3O_2)]$	Ag: MVol.B6-184
$AgC_9H_8N_3O_2S$	$[Ag(C_7H_5O_2CHNNCSNH_2)]$	Ag: MVol.B6-284
$AgC_9H_8N_3O_2S_2$	$[Ag(NHC_6H_4SO_2NHC_3H_2NS)]$	Ag: MVol.B6-264
$AgC_9H_8N_3O_3$	$Ag(NH_2C_9H_6N)NO_3 \cdot CH_3CN$	Ag: MVol.B6-114
$AgC_9H_8N_3O_3S$	$[Ag(CH_3SO_2C_2N_3(O)C_6H_5)]$	Ag: MVol.B7-57
$AgC_9H_8N_3O_3S_2$	$AgNO_3 \cdot C_6H_5CH_2SC_2HN_2S$	Ag: MVol.B7-80
$AgC_9H_8N_3S$	$Ag(C_2N_3S(CH_3)C_6H_5) \cdot n\ AgCl$	Ag: MVol.B7-55/6
–	$Ag(C_2N_3S(CH_3)C_6H_5) \cdot n\ AgNO_3$	Ag: MVol.B7-55/6
$AgC_9H_8N_5O_3$	$[Ag(NHC_3H_3N_3OC_6H_4NO_2)]$	Ag: MVol.B6-190
$AgC_9H_8O_4S^+$	$[Ag(C_6H_4(COOH)SCH_2COOH)]^+$	Ag: MVol.B7-27/8
$AgC_9H_8O_4Se^+$	$[Ag(HOCOC_6H_4SeCH_2COOH)]^+$	Ag: MVol.B7-193
AgC_9H_9	$AgCCC_7H_9$	Ag: MVol.B5-20
$AgC_9H_9NO_5^-$	$[Ag(OC_4H_3CH_2N(CH_2COO)_2)]^-$	Ag: MVol.B6-275
$AgC_9H_9N_2$	$Ag((CH_3)_2C_7H_3N_2)$	Ag: MVol.B6-145/6
$AgC_9H_9N_2O$	$Ag(CH_3OCH_2C_7H_4N_2)$	Ag: MVol.B6-146
$AgC_9H_9N_2O_2S$	$[Ag(NO_2C_7H_3NS(CH_3)_2)]$	Ag: MVol.B7-78
$AgC_9H_9N_2O_3$	$Ag(C_6H_5NHCONHCH_2COO)$	Ag: MVol.B6-335
$AgC_9H_9N_3S_3$	$Ag(C_6H_5C_2N_2S_3) \cdot 0.5\ C_2H_8N_2$	Ag: MVol.B7-81
$AgC_9H_9N_4$	$[Ag(C_6H_5C_2H_4CN_4)]$	Ag: MVol.B6-192/3
$AgC_9H_9N_4O_2$	$Ag(CH_3C_3HN_3O_2) \cdot C_5H_5N$	Ag: MVol.B6-308/12
–	$Ag(C_6H_5C_3HN_3O_2) \cdot NH_3$	Ag: MVol.B6-308/12
$AgC_9H_9N_4O_3S$	$AgNO_3 \cdot C_6H_5CH_2C_2N_2SNH_2$	Ag: MVol.B7-80
$AgC_9H_9N_4O_4$	$AgNO_3 \cdot C_6H_5CH_2CON_2H_2CN$	Ag: MVol.B6-332
$AgC_9H_9N_4S$	$[Ag(CH_3C_6H_3(NH_2)C_2HN_3S)]$	Ag: MVol.B7-55
–	$[Ag(CH_3N_3C_7H_3NSCH_3)]$	Ag: MVol.B6-302
–	$[Ag(NH_2C_2N_3SC_6H_4CH_3)]$	Ag: MVol.B7-55
$AgC_9H_9O_2S$	$[Ag(CH_3C_6H_4SCH_2COO)]$	Ag: MVol.B7-27/8
–	$[Ag(C_2H_5SC_6H_4COO)]$	Ag: MVol.B7-36
–	$[Ag(C_6H_5CH_2SCH_2COO)]$	Ag: MVol.B7-29
$AgC_9H_9O_2SSe$	$[Ag(CH_3SC_6H_4SeCH_2COO)]$	Ag: MVol.B7-191/2, 195/6
$AgC_9H_9O_2S_2$	$[Ag(CH_3SC_6H_4SCH_2COO)]$	Ag: MVol.B7-27/8
$AgC_9H_9O_2Se$	$[Ag(CH_3C_6H_4SeCH_2COO)]$	Ag: MVol.B7-191/2, 194, 196
$AgC_9H_9O_3S$	$[Ag(CH_3OC_6H_4SCH_2COO)]$	Ag: MVol.B7-27/8
$AgC_9H_9O_3Se$	$[Ag(CH_3OC_6H_4SeCH_2COO)]$	Ag: MVol.B7-191/2, 195/6
$AgC_9H_{10}^+$	$[AgCH_3C_6H_4CHCH_2]^+$	Ag: MVol.B5-102
–	$[AgC_6H_5CHCHCH_3]^+$	Ag: MVol.B5-102
$AgC_9H_{10}IN_2O_5$	$[Ag(C_9H_{10}IN_2O_5)]$	Ag: MVol.B6-157
$AgC_9H_{10}NOS$	$[Ag(SCH_2CON(CH_3)C_6H_5)]$	Ag: MVol.B7-23
$AgC_9H_{10}NO_2$	$[Ag(C_6H_5CH_2CH(NH_2)COO)]$	Ag: MVol.B6-244
$AgC_9H_{10}NO_2S_3$	$[Ag(CH_3C_6H_4SO_2NCSSCH_3)]$	Ag: MVol.B7-104
$AgC_9H_{10}NO_3$	$[Ag(HOC_6H_4CH_2CH(NH_2)COO)]$	Ag: MVol.B6-245/6
$AgC_9H_{10}NO_4$	$AgNO_3 \cdot CH_2CHC_6H_4OCH_3$	Ag: MVol.B5-118
$AgC_9H_{10}NO_4S$	$[Ag(CH_3C_6H_4SO_2NHCH_2COO)]$	Ag: MVol.B6-263
–	$[Ag(C_6H_5SO_2NHCH(CH_3)COO)]$	Ag: MVol.B6-242
–	$[Ag(C_6H_5SO_2NHCH_2CH_2COO)]$	Ag: MVol.B6-244
$AgC_9H_{10}NS$	$[Ag((CH_3)_2C_7H_4NS)]$	Ag: MVol.B7-78
$AgC_9H_{10}NS_2$	$[Ag(C_2H_5N(C_6H_5)CSS)]$	Ag: MVol.B7-116
$AgC_9H_{10}N_3$	$[Ag(C_3H_5N_3C_6H_5)]$	Ag: MVol.B6-296

$AgC_9H_{10}N_3O_2$ ...... $[Ag(C_6H_5N_3COOC_2H_5)]$ .... Ag: MVol.B6-301/2
$AgC_9H_{10}N_5$ .... $[Ag(C_6H_5CH_2N(CH_3)CN_4)]$ .... Ag: MVol.B6-192/3
$AgC_9H_{10}O^+$ .... $[Ag(CH_2CHC_6H_4OCH_3)]^+$ .... Ag: MVol.B5-118
$AgC_9H_{10}O_2^+$ .... $[Ag(C_6H_5COOC_2H_5)]^+$ .... Ag: MVol.B5-118
$AgC_9H_{10}O_2S^+$ ...... $[Ag(CH_3C_6H_4SCH_2COOH)]^+$ .... Ag: MVol.B7-27/8
$AgC_9H_{10}O_2SSe^+$ .... $[Ag(CH_3SC_6H_4SeCH_2COOH)]^+$ .... Ag: MVol.B7-195
$AgC_9H_{10}O_2S_2^+$ ..... $[Ag(CH_3SC_6H_4SCH_2COOH)]^+$ .... Ag: MVol.B7-27/8
$AgC_9H_{10}O_2Se^+$ ..... $[Ag(CH_3C_6H_4SeCH_2COOH)]^+$ .... Ag: MVol.B7-194
$AgC_9H_{10}O_3S^+$ ...... $[Ag(CH_3OC_6H_4SCH_2COOH)]^+$ .... Ag: MVol.B7-27/8
$AgC_9H_{10}O_3Se^+$ ..... $[Ag(CH_3OC_6H_4SeCH_2COOH)]^+$ .... Ag: MVol.B7-195
$AgC_9H_{11}N_2O_5$ ...... $[Ag(C_9H_{11}N_2O_5)]$ .... Ag: MVol.B6-157
$AgC_9H_{11}N_4O_3$ ...... $Ag(ONC_3NO_2C_6H_5) \cdot 2\ NH_3$ .... Ag: MVol.B6-313
$AgC_9H_{11}OS_3$ .... $[Ag(CH_3C_5HOS_3C_3H_7)]$ .... Ag: MVol.B7-83/4
– .... $[Ag(C_4H_9C_5H_2OS_3)]$ .... Ag: MVol.B7-83/4
– .... $[Ag(C_5HOS_3(C_2H_5)_2)]$ .... Ag: MVol.B7-83/4
$AgC_9H_{11}O_3$ .... $AgC_6H_2(OCH_3)_3$ .... Ag: MVol.B5-10/1
$AgC_9H_{12}^+$ .... $[AgC_6H_3(CH_3)_3]^+$ .... Ag: MVol.B5-97/8
– .... $[AgC_6H_5C_3H_7]^+$ .... Ag: MVol.B5-97/8
$AgC_9H_{12}N$ .... $AgC_6H_4CH_2N(CH_3)_2$ .... Ag: MVol.B5-11/2
$AgC_9H_{12}NO_3$ .... $AgNO_3 \cdot C_9H_{12}$ .... Ag: MVol.B5-66
$AgC_9H_{12}N_2O^+$ ...... $[Ag(CH_3C(C_6H_5)(OH)C(NH)NH_2)]^+$ .... Ag: MVol.B6-339/40
$AgC_9H_{12}N_2O_3^-$ ...... $[Ag(C_2H_5(C_3H_7)C_4N_2O_3)]^-$ .... Ag: MVol.B6-155
$AgC_9H_{12}N_3O_4$ ...... $AgNO_3 \cdot CH_3C(C_6H_5)(OH)C(NH)NH_2$ .... Ag: MVol.B6-340
$AgC_9H_{12}N_3P^+$ ...... $[Ag(P(C_2H_4CN)_3)]^+$ .... Ag: MVol.B7-203
$AgC_9H_{12}N_7Na_2$ ..... $Na_2Ag(CN)_3 \cdot (CH_2)_6N_4 \cdot 2\ H_2O$ .... Ag: MVol.B6-199
$AgC_9H_{12}O_2^+$ .... $[AgC_7H_9COOCH_3]^+$ .... Ag: MVol.B5-88/9
– .... $[AgC_7H_9OCOCH_3]^+$ .... Ag: MVol.B5-89
$AgC_9H_{13}N^+$ .... $[Ag(CH_3C_6H_4N(CH_3)_2)]^+$ .... Ag: MVol.B6-53
$AgC_9H_{13}NO_5PS$ ..... $Ag((CH_3O)_2P(O)NSO_2C_6H_4CH_3)$ .... Ag: MVol.B7-251
$AgC_9H_{13}N_2O_2$ ...... $[Ag(NH_3)_2]C_6H_5CHCHCOO$ .... Ag: MVol.B6-29
$AgC_9H_{13}N_2O_3$ ...... $Ag(CH_3(C_2H_5)_2C_4N_2O_3)$ .... Ag: MVol.B6-156
$AgC_9H_{13}O_2$ .... $AgCCC_5H_6O(CH_3)_2OH$ .... Ag: MVol.B5-19/20
– .... $[Ag(CH_3COC_7H_{10}O)]$ .... Ag: MVol.B6-211/2
$AgC_9H_{14}^+$ .... $[AgC_7H_8(CH_3)_2]^+$ .... Ag: MVol.B5-75, 77
$AgC_9H_{14}NO_3$ .... $AgNO_3 \cdot C_9H_{14}$ .... Ag: MVol.B5-66
$AgC_9H_{14}N_3O_3$ ...... $Ag(C_6H_5CONHCH_2COO) \cdot 2\ NH_3$ .... Ag: MVol.B6-241
$AgC_9H_{14}O_4S_2^-$ ...... $[Ag(C_5H_{10}(SCH_2COO)_2)]^-$ .... Ag: MVol.B7-39/40
$AgC_9H_{15}N_2$ .... $Ag((C_3H_7)_2C_3HN_2)$ .... Ag: MVol.B6-142
$AgC_9H_{15}N_2^{2+}$ ...... $[Ag(C_5H_5NC_4H_8NH_2)]^{2+}$ .... Ag: MVol.B6-97/8
$AgC_9H_{15}N_4$ .... $[Ag(C_6H_{11}C_2H_4CN_4)]$ .... Ag: MVol.B6-192/3
$AgC_9H_{16}^+$ .... $[AgCH_2CH(CH_2)_5CHCH_2]^+$ .... Ag: MVol.B5-42/4
– .... $[Ag(CH_3)_2CHCCC(CH_3)_3]^+$ .... Ag: MVol.B5-44
– .... $[AgC_6H_9CH(CH_3)_2]^+$ .... Ag: MVol.B5-57/8
– .... $[AgC_6H_{11}CH_2CHCH_2]^+$ .... Ag: MVol.B5-57/8
– .... $[AgC_8H_{13}CH_3]^+$ .... Ag: MVol.B5-61/2
$AgC_9H_{16}NS_2$ .... $[Ag(C_2H_5N(C_6H_{11})CSS)]$ .... Ag: MVol.B7-116
$AgC_9H_{16}N_3O_2$ ...... $Ag(C_6H_5CH_2CH(NH_2)COO) \cdot 2\ NH_3$ .... Ag: MVol.B6-244
$AgC_9H_{16}O_2^+$ .... $[AgCH_2CH(CH_2)_5OOCCH_3]^+$ .... Ag: MVol.B5-48
$AgC_9H_{17}N^+$ .... $[Ag(C_9H_{17}N)]^+$ .... Ag: MVol.B6-116
$AgC_9H_{17}N_4$ .... $[Ag(C_8H_{17}CN_4)]$ .... Ag: MVol.B6-192/3

Formula	Compound	Reference
$AgC_9H_{17}N_4O_3$	$Ag(C_6H_5CONHCH_2COO) \cdot 3\ NH_3$	Ag: MVol.B6-241
$AgC_9H_{17}OS_2$	$[Ag(C_8H_{17}OCSS)]$	Ag: MVol.B7-98
$AgC_9H_{18}NOS$	$[Ag((C_4H_9)_2NCOS)]$	Ag: MVol.B7-99/100
$AgC_9H_{18}NO_2$	$[Ag(NH_2(CH_2)_8COO)]$	Ag: MVol.B6-250
$AgC_9H_{18}NS_2$	$[Ag((C_4H_9)_2NCSS)]$	Ag: MVol.B7-105/6, 112/3
–	$[Ag(SCSNHC_8H_{17})]$	Ag: MVol.B7-102
$AgC_9H_{18}NSe_2$	$Ag(SeSeCN(C_4H_9)_2)$	Ag: MVol.B7-197
$AgC_9H_{18}N_4O_3$	$AgNO_3 \cdot 1.5\ N(CH_2CH_2)_3N$	Ag: MVol.B6-130
$AgC_9H_{18}N_5$	$[Ag(N(C_4H_9)_2CN_4)]$	Ag: MVol.B6-192/3
$AgC_9H_{18}N_6S_3{}^+$	$[Ag(C_3H_6N_2S)_3]^+$	Ag: MVol.B7-49
$AgC_9H_{19}N_2S_2$	$[Ag((C_2H_5)_2NC_3H_6N(CH_3)CSS)]$	Ag: MVol.B7-115
$AgC_9H_{21}IO_3P$	$[AgI(i\text{-}C_3H_7O)_3P]_3$	Ag: MVol.B7-248
$AgC_9H_{21}IP$	$[AgI(P(CH(CH_3)_2)_3)]_4$	Ag: MVol.B7-205
–	$[AgI(P(C_3H_7)_3)]_4$	Ag: MVol.B7-203/4
$AgC_9H_{21}NO_3{}^+$	$[Ag(N(CH_2CHOHCH_3)_3)]^+$	Ag: MVol.B6-230
$AgC_9H_{21}NO_6PS$	$[Ag(PS(C_3H_7O)_3)]NO_3$	Ag: MVol.B7-253
$AgC_9H_{21}NPS_2$	$[Ag(S_2CN(CH_3)_2)P(C_2H_5)_3]$	Ag: MVol.B7-202/3
$AgC_9H_{21}N_3O_3{}^+$	$[Ag(OCHN(CH_3)_2)_3]^+$	Ag: MVol.B6-322/3
$AgC_9H_{23}N_6O_6S_3$	$[Ag(NH_2CSNH_2)_3]CH_3COO \cdot 2\ CH_3COOH$	Ag: MVol.B7-146
$AgC_9H_{24}N_6S_3{}^+$	$[Ag(CH_3NHCSNHCH_3)_3]^+$	Ag: MVol.B7-152
–	$[Ag(C_2H_5NHCSNH_2)_3]^+$	Ag: MVol.B7-154
$AgC_9H_{24}N_7O_3S_3$	$AgNO_3 \cdot 3\ (CH_3)_2NCSNH_2$	Ag: MVol.B7-152
$AgC_{10}ClCoH_{14}O_8$	$[Co(CH(COCH_3)_2)_2] \cdot AgClO_4$	Ag: MVol.B6-213
$AgC_{10}ClCuH_{14}O_8$	$[Cu(CH(COCH_3)_2)_2] \cdot AgClO_4$	Ag: MVol.B6-213
$AgC_{10}ClCuH_{20}N_2O_8S_2$	$[AgCu(CH_3SC_3H_5(NH_2)COO)_2]ClO_4$	Ag: MVol.B6-259
$AgC_{10}ClFeH_8$	$AgC_5H_3ClFeC_5H_5$	Ag: MVol.B5-24
–	$AgC_5H_4FeC_5H_4Cl$	Ag: MVol.B5-23/4
$AgC_{10}ClH_7N_3O_2$	$[Ag(CH_3C_3N_3O_2C_6H_4Cl)]$	Ag: MVol.B6-308/12
$AgC_{10}ClH_8N_2O_4$	$Ag(NC_5H_4C_5H_4N)ClO_4$	Ag: MVol.B6-120
$AgC_{10}ClH_8N_4O_{10}$	$Ag(NO_2C_5H_4NO)_2ClO_4$	Ag: MVol.B6-106
$AgC_{10}ClH_{8.5}N_{3.5}O_2$	$Ag(CH_3C_3N_3O_2C_6H_4Cl) \cdot 0.5\ NH_3$	Ag: MVol.B6-308/12
$AgC_{10}ClH_{10}N_2$	$AgCl \cdot 2\ C_5H_5N$	Ag: MVol.B6-78/9
$AgC_{10}ClH_{10}N_2O_4$	$AgClO_4 \cdot 2\ C_5H_5N = [Ag(C_5H_5N)_2]ClO_4$	Ag: MVol.B6-79
$AgC_{10}ClH_{10}N_2O_8$	$Ag(C_4H_3OCHNOH)_2ClO_4$	Ag: MVol.B6-305
$AgC_{10}ClH_{11}NS_2$	$[Ag(C_3H_7N(C_6H_4Cl)CSS)]$	Ag: MVol.B7-116
$AgC_{10}ClH_{12}N_4O_4$	$[Ag(C_3H_6(CN)_2)_2]ClO_4$	Ag: MVol.B6-352
–	$[Ag(NH_2C_5H_4N)_2]ClO_4$	Ag: MVol.B6-93
$AgC_{10}ClH_{12}O_4$	1) $AgClO_4 \cdot C_{10}H_{12}$	Ag: MVol.B5-83
–	2) $AgClO_4 \cdot C_{10}H_{12}$	Ag: MVol.B5-109
$AgC_{10}ClH_{13}N_5O_2$	$Ag(CH_3C_3N_3O_2C_6H_4Cl) \cdot 2\ NH_3$	Ag: MVol.B6-308/12
$AgC_{10}ClH_{14}NiO_8$	$[Ni(CH(COCH_3)_2)_2] \cdot AgClO_4$	Ag: MVol.B6-213
$AgC_{10}ClH_{14}O_4S$	$AgClO_4 \cdot C_6H_5SC_4H_9$	Ag: MVol.B7-14
$AgC_{10}ClH_{14}O_8Pd$	$[Pd(CH(COCH_3)_2)_2] \cdot AgClO_4$	Ag: MVol.B6-213
$AgC_{10}ClH_{16}O_4$	$AgClO_4 \cdot 2\ C_5H_8$	Ag: MVol.B5-54
$AgC_{10}ClH_{17}NO_6$	$[Ag(H_2O)(C_5H_{11}C_5H_4NO)]ClO_4$	Ag: MVol.B6-105
$AgC_{10}ClH_{18}O_4$	$AgClO_4 \cdot CH_2CH(CH_2)_6CHCH_2$	Ag: MVol.B5-45
$AgC_{10}Cl_2H_8N_3O_3$	$[AgNO_3(ClC_5H_4N)_2]$	Ag: MVol.B6-99
$AgC_{10}Cl_2H_{10}N_5O_3$	$AgNO_3 \cdot 2\ CH_3C_4H_2ClN_2$	Ag: MVol.B6-160
$AgC_{10}Cl_2H_{24}N_4O_8$	$Ag(C_{10}H_{24}N_4)(ClO_4)_2$	Ag: MVol.B7-300/1
$AgC_{10}Cl_3H_8N_2O_4$	$[Ag(ClC_5H_4N)_2]ClO_4$	Ag: MVol.B6-99

Formula	Compound	Reference
$AgC_{10}Cl_3H_8N_2O_6$	$Ag(ClC_5H_4NO)_2ClO_4$	Ag: MVol.B6-106
$AgC_{10}Cl_3H_{11}IrN_2O$	$Ag[Ir(C_5H_5N)_2Cl_3(OH)]$	Ir: SVol.2-73
$AgC_{10}Cl_3H_{24}N_4O_{12}$	$[Ag(C_{10}H_{24}N_4)](ClO_4)_3$	Ag: MVol.B7-325
$AgC_{10}Cl_4H_{10}IrN_2$	$Ag[Ir(C_5H_5N)_2Cl_4]$	Ir: SVol.2-70
$AgC_{10}Cl_6H_{10}N_2Sb$	$[Ag(C_5H_5N)_2]SbCl_6$	Ag: MVol.B6-83
$AgC_{10}FH_{10}N_2$	$AgF \cdot 2\ C_5H_5N \cdot 5\ H_2O$	Ag: MVol.B6-78
$AgC_{10}F_6H_9O_2$	$AgCH(COCF_3)_2 \cdot C_5H_8$	Ag: MVol.B5-75/6
$AgC_{10}F_6H_{10}N_2P$	$[Ag(C_5H_5N)_2]PF_6$	Ag: MVol.B6-82
$AgC_{10}F_6H_{10}N_2Sb$	$[Ag(C_5H_5N)_2]SbF_6$	Ag: MVol.B6-82
$AgC_{10}FeH_9$	$AgC_5H_4FeC_5H_5$	Ag: MVol.B5-23
$AgC_{10}FeH_{12}N_7O_3$	$(CH_3NC)_4Fe(CN)_2 \cdot AgNO_3$	Fe: Org.Verb.B4-67
$AgC_{10}H_3N_2O_4$	$Ag(C_{10}H_3N_2O_4)$	Ag: MVol.B6-328
$AgC_{10}H_5$	$AgCCCCC_6H_5$	Ag: MVol.B5-14
$AgC_{10}H_5N_4O_2$	$[Ag(C_{10}H_5N_4O_2)]$	Ag: MVol.B6-180
$AgC_{10}H_6NOS_2$	$[Ag(C_6H_5CHC_3NOS_2)]$	Ag: MVol.B7-73/4
$AgC_{10}H_6NOS_3$	$[Ag(NC_5H_4C_5H_2OS_3)]$	Ag: MVol.B7-83/4
$AgC_{10}H_6NO_2$	$Ag(NC_9H_6COO)$	Ag: MVol.B6-115
–	$[Ag(OC_{10}H_6NO)]$	Ag: MVol.B6-307
$AgC_{10}H_6N_4O_4$	$[Ag(C_4H_3N_2COO)_2]$	Ag: MVol.B7-298
$AgC_{10}H_6N_5O_6$	$[Ag(CH_3C_3N_3O_2C_6H_3(NO_2)_2)]$	Ag: MVol.B6-309/12
$AgC_{10}H_7$	$AgC_{10}H_7 \cdot x\ AgNO_3$	Ag: MVol.B5-12
$AgC_{10}H_7NO_2^+$	$[Ag(NC_9H_6COOH)]^+$	Ag: MVol.B6-115
$AgC_{10}H_7N_2$	$AgCCC_7H_4N_2CH_3$	Ag: MVol.B5-20
$AgC_{10}H_7N_2O$	$Ag(C_6H_5C_4H_2N_2O)$	Ag: MVol.B6-150
$AgC_{10}H_7N_2O_3$	$Ag(C_2N_2O(C_6H_4CH_3)COO)$	Ag: MVol.B6-204
$AgC_{10}H_7N_4O_4$	$Ag(CH_3C_3N_3O_2C_6H_4NO_2) \cdot H_2O$	Ag: MVol.B6-308/12
$AgC_{10}H_7N_8S_2$	$Ag(C_5H_3N_4S) \cdot C_5H_4N_4S$	Ag: MVol.B7-66
$AgC_{10}H_8^+$	1) $[AgC_{10}H_8]^+$	Ag: MVol.B5-79
–	2) $[AgC_{10}H_8]^+$	Ag: MVol.B5-107/8
$AgC_{10}H_8NO$	$[Ag(CH_3C_9H_5NO)]$	Ag: MVol.B6-109
$AgC_{10}H_8NO_2S$	$[Ag(C_7H_4NSC_2H_4COO)]$	Ag: MVol.B7-78
$AgC_{10}H_8NS_2$	$[Ag(CH_3SC_9H_5NS)]$	Ag: MVol.B7-61
$AgC_{10}H_8N_2^+$	$[Ag(NC_5H_4C_5H_4N)]^+$	Ag: MVol.B6-118/9
$AgC_{10}H_8N_2^{2+}$	$[Ag(NC_5H_4C_5H_4N)]^{2+}$	Ag: MVol.B7-287
$AgC_{10}H_8N_2O_8S_4^{5-}$	$[Ag(N(CH_2COO)_2CSS)_2]^{5-}$	Ag: MVol.B7-116
$AgC_{10}H_8N_3O_2$	$[Ag(CH_3C_6H_4C_2HN_3COO)]$	Ag: MVol.B6-182
–	$[Ag(C_6H_5(CH_3)C_3N_3O_2)]$	Ag: MVol.B6-308/12
–	$Ag(C_6H_5(CH_3)C_3N_3O_2) \cdot H_2O$	Ag: MVol.B6-308/12
–	$[Ag(C_6H_5CON(COCH_3)NCN)]$	Ag: MVol.B6-332
$AgC_{10}H_8N_3O_3$	$Ag(NC_5H_4C_5H_4N)NO_3$	Ag: MVol.B6-119, 121
$AgC_{10}H_8N_4O_6$	$[Ag(NO_3)_2(NC_5H_4C_5H_4N)]$	Ag: MVol.B7-288/9
$AgC_{10}H_8N_5O_8$	$Ag(NO_2C_6H_4C_3HN_2O(NO_2)CH_3)NO_3$	Ag: MVol.B6-134
$AgC_{10}H_8O_4S_2^-$	$[Ag(C_6H_4(SCH_2COO)_2)]^-$	Ag: MVol.B7-42
$AgC_{10}H_9N^+$	$[Ag(C_{10}H_7NH_2)]^+$	Ag: MVol.B6-54
$AgC_{10}H_9NO^+$	$[Ag(CH_3CN)(C_8H_6O)]^+$	Ag: MVol.B6-222
$AgC_{10}H_9NO_2^+$	$[Ag(CH_3CN)(C_8H_6O_2)]^+$	Ag: MVol.B6-222
$AgC_{10}H_9NO_4^-$	$[Ag(C_6H_5N(CH_2COO)_2)]^-$	Ag: MVol.B6-275
$AgC_{10}H_9N_2O$	$[Ag(CH_3(C_6H_5)C_3HN_2O)]$	Ag: MVol.B6-134
$AgC_{10}H_9N_2O_2$	$Ag(C_7H_4N_2CH_2COOCH_3)$	Ag: MVol.B6-146
$AgC_{10}H_9N_2O_2S$	$[Ag(CH_3SO_2NC_9H_6N)]$	Ag: MVol.B6-115

$AgC_{10}H_9N_2S$	$[Ag(CH_3C_3HN_2SC_6H_5)]$	Ag: MVol.B7-48
$AgC_{10}H_9N_4O_2S$	$[Ag(NHC_6H_4SO_2NHC_4H_3N_2)]$	Ag: MVol.B6-264
$AgC_{10}H_9N_4O_4$	$[AgC_2H_2N_3(C_4N(CH_3)_2(COO)_2H)]$	Ag: MVol.B6-186
$AgC_{10}H_9O_2$	$[Ag(C_6H_5COCHCOCH_3)]$	Ag: MVol.B6-212
$AgC_{10}H_9O_2S$	$[Ag(C_3H_5SC_6H_4COO)]$	Ag: MVol.B7-36
$AgC_{10}H_{10}MnN_2O_4$	$[Ag(C_5H_5N)_2]MnO_4$	Ag: MVol.B6-83
$AgC_{10}H_{10}N$	$AgCCC(NH_2)(CH_3)C_6H_5$	Ag: MVol.B5-16
$AgC_{10}H_{10}NO_3$	$[Ag(C_6H_5CONHCH(CH_3)COO)]$	Ag: MVol.B6-242
–	$[Ag(C_6H_5CONHCH_2CH_2COO)]$	Ag: MVol.B6-243
$AgC_{10}H_{10}N_2^+$	$[Ag(C_5H_5N)_2]^+$	Ag: MVol.B6-74/6
$AgC_{10}H_{10}N_2NaO_3$	$Na[Ag((C_3H_5)_2C_4N_2O_3)]$	Ag: MVol.B6-154
$AgC_{10}H_{10}N_2O_2^+$	$[Ag(C_5H_5NO)_2]^+$	Ag: MVol.B6-104
$AgC_{10}H_{10}N_2O_3^-$	$[Ag((C_3H_5)_2C_4N_2O_3)]^-$	Ag: MVol.B6-154/5
$AgC_{10}H_{10}N_2O_4^-$	$[Ag(NC_5H_4CH_2N(CH_2COO)_2)]^-$	Ag: MVol.B6-275
$AgC_{10}H_{10}N_3O_3$	$AgNO_3 \cdot 2\ C_5H_5N = [Ag(C_5H_5N)_2]NO_3$	Ag: MVol.B6-76/7
–	$[Ag(C_5H_5N)_2]NO_3 \cdot H_2O$	Ag: MVol.B6-77
–	$AgNO_3 \cdot 2\ C_5H_5N \cdot n\ H_2O$	Ag: MVol.B6-77/8
$AgC_{10}H_{10}N_3O_7$	$Ag(C_4H_3OCHNOH)_2NO_3$	Ag: MVol.B6-305
$AgC_{10}H_{10}N_4O_6$	$Ag(C_5H_5N)_2(NO_3)_2$	Ag: MVol.B7-272/3
$AgC_{10}H_{10}N_5O_7S$	$Ag(C_3H_5NHCSNH_2)OC_6H_2(NO_2)_3$	Ag: MVol.B7-157
$AgC_{10}H_{10}N_{11}O_7$	$AgOC_6H_2(NO_2)_3 \cdot 2\ NH_2C(NH)NHCN$	Ag: MVol.B6-338
$AgC_{10}H_{11}N_2O$	$Ag(C_2H_5OCH_2C_7H_4N_2)$	Ag: MVol.B6-146
$AgC_{10}H_{11}N_2O_2S$	$[Ag(SCH_2CONHC_6H_4NHCOCH_3)] \cdot H_2O$	Ag: MVol.B7-24
$AgC_{10}H_{11}N_2O_5S$	$[Ag(CH_3CONHC_6H_4SO_2NHCH_2COO)]$	Ag: MVol.B6-263
$AgC_{10}H_{11}N_2S$	$[Ag(CH_3C_3H_3NSNC_6H_5)]$	Ag: MVol.B7-71
$AgC_{10}H_{11}N_4$	$[Ag(C_6H_5C_3H_6CN_4)]$	Ag: MVol.B6-192/3
$AgC_{10}H_{11}N_4S$	$[Ag(C_2H_5N_3C_7H_3NSCH_3)]$	Ag: MVol.B6-302
$AgC_{10}H_{11}O_2$	$Ag(C_3H_7C_7H_4O_2)$	Ag: MVol.B6-211
$AgC_{10}H_{11}O_2S$	$[Ag(C_3H_7SC_6H_4COO)]$	Ag: MVol.B7-36
$AgC_{10}H_{11}O_3Se$	$[Ag(C_2H_5OC_6H_4SeCH_2COO)]$	Ag: MVol.B7-191/2, 195
$AgC_{10}H_{12}NO_2S_3$	$[Ag(C_6H_5SO_2NCSSC_3H_7)]$	Ag: MVol.B7-104
$AgC_{10}H_{12}NO_3$	$AgNO_3 \cdot C_{10}H_{12}$	Ag: MVol.B5-83
$AgC_{10}H_{12}N_2O_3^-$	$[Ag(C_3H_5(C_3H_7)C_4N_2O_3)]^-$	Ag: MVol.B6-155
$AgC_{10}H_{12}N_2O_8^{3-}$	$[Ag(C_2H_4(N(CH_2COO)_2)_2)]^{3-}$	Ag: MVol.B6-276, 277
$AgC_{10}H_{12}N_3O_6$	$[Ag(C_3N_3(OH)_2C(COOC_2H_5)_2)]$	Ag: MVol.B6-189
$AgC_{10}H_{12}N_4^+$	$[Ag(CH_3C_4H_3N_2)_2]^+$	Ag: MVol.B6-159
–	$[Ag(NH_2C_5H_4N)_2]^+$	Ag: MVol.B6-91/2
$AgC_{10}H_{12}N_5$	$[Ag(C_6H_5CH_2N(C_2H_5)CN_4)]$	Ag: MVol.B6-192/3
$AgC_{10}H_{12}N_5O_3$	$Ag(NH_2C_5H_4N)_2NO_3$	Ag: MVol.B6-94/5
$AgC_{10}H_{12}N_5O_3S$	$N(CH_3)_4SO_3C(CN)_2C(CN)_2Ag$	Ag: MVol.B6-352
$AgC_{10}H_{12}N_5O_5$	$[Ag(CH_3COC_3H_3N_2)_2]NO_3$	Ag: MVol.B6-140
$AgC_{10}H_{12}O_3Se^+$	$[Ag(C_2H_5OC_6H_4SeCH_2COOH)]^+$	Ag: MVol.B7-195
$AgC_{10}H_{12}O_8S_4^{3-}$	$[Ag(CH_2(SCH_2COO)_2)_2]^{3-}$	Ag: MVol.B7-39/40
$AgC_{10}H_{13}$	$AgCH_2C(CH_3)_2C_6H_5 \cdot P(C_4H_9)_3$	Ag: MVol.B5-4
$AgC_{10}H_{13}N_2O_8^{2-}$	$[Ag(C_2H_4(N(CH_2COO)_2)_2H)]^{2-}$	Ag: MVol.B6-276
$AgC_{10}H_{13}N_4O$	$[Ag(NH_2C_5H_4N)_2]OH$	Ag: MVol.B6-92
$AgC_{10}H_{13}N_4O_3$	$[Ag((CH_3)_3CN(O)NNC_6H_4NO_2)]$	Ag: MVol.B6-303/4
$AgC_{10}H_{13}N_6O$	$[Ag((CH_3)_3CN(O)NNC_6H_4N_3)]$	Ag: MVol.B6-303/4
$AgC_{10}H_{13}OS_3$	$[Ag(C_5H_{11}C_5H_2OS_3)]$	Ag: MVol.B7-83/4
$AgC_{10}H_{14}^+$	$[AgC_6H_2(CH_3)_4]^+$	Ag: MVol.B5-97/8

Formula	Compound	Reference
$AgC_{10}H_{14}^+$	$[AgC_6H_4(C_2H_5)_2]^+$	Ag: MVol.B5-98
–	$[AgC_6H_5C_4H_9]^+$	Ag: MVol.B5-97/8
–	$[AgC_{10}H_{14}]^+$	Ag: MVol.B5-81/2
$AgC_{10}H_{14}NO_3$	$AgNO_3 \cdot C_{10}H_{14}$	Ag: MVol.B5-69
$AgC_{10}H_{14}NO_4$	$[Ag((CH_2CO)_2N(CH_2)_5COO)]$	Ag: MVol.B6-250
$AgC_{10}H_{14}N_2^+$	$[Ag(C_6H_5C_4H_9N_2)]^+$	Ag: MVol.B6-163/4
$AgC_{10}H_{14}N_2O^+$	$[Ag(NC_5H_4CON(C_2H_5)_2)]^+$	Ag: MVol.B6-103
$AgC_{10}H_{14}N_2O_3^-$	$[Ag(C_2H_5(C_4H_9)C_4N_2O_3)]^-$	Ag: MVol.B6-155
$AgC_{10}H_{14}N_2O_8^{3-}$	$[Ag(NH_2C_3H_5(COO)_2)_2]^{3-}$	Ag: MVol.B6-253
$AgC_{10}H_{14}N_3$	$[Ag(C_4H_9N_3C_6H_5)]$	Ag: MVol.B6-295/6
$AgC_{10}H_{14}N_3O$	$[Ag((CH_3)_3CN(O)NNC_6H_5)]$	Ag: MVol.B6-303/4
$AgC_{10}H_{14}N_3O_3$	$AgNO_3 \cdot NC_5H_4C_4H_7NCH_3$	Ag: MVol.B6-107
$AgC_{10}H_{14}O_4S_2^-$	$[Ag(CH_2CHCH_2SCH_2COO)_2]^-$	Ag: MVol.B5-50/1
$AgC_{10}H_{15}$	$AgCCCHCHC_6H_{13}$	Ag: MVol.B5-14
$AgC_{10}H_{15}IP$	$[AgI((C_2H_5)_2PC_6H_5)]_4$	Ag: MVol.B7-221
$AgC_{10}H_{15}N^+$	$[Ag(C_6H_5N(C_2H_5)_2)]^+$	Ag: MVol.B6-53
$AgC_{10}H_{15}NO_4^-$	$[Ag(C_6H_{11}N(CH_2COO)_2)]^-$	Ag: MVol.B6-275
$AgC_{10}H_{15}NO_5^-$	$[Ag(C_5H_9OCH_2N(CH_2COO)_2)]^-$	Ag: MVol.B6-275
–	$[Ag(HOC_6H_{10}N(CH_2COO)_2)]^-$	Ag: MVol.B6-275
$AgC_{10}H_{15}N_2OS$	$[Ag(C_4HN_2OS(CH_3)C_5H_{11})]$	Ag: MVol.B7-64
$AgC_{10}H_{15}N_2O_2$	$[Ag(C_{10}H_{15}N_2O_2)]$	Ag: MVol.B6-321
$AgC_{10}H_{15}N_2O_7^{2-}$	$[Ag(C_2H_4N_2(CH_2COO)_3C_2H_4OH)]^{2-}$	Ag: MVol.B6-276
$AgC_{10}H_{15}N_2S$	$[Ag(C_4HN_2S(CH_3)_3C_3H_5)]$	Ag: MVol.B7-64
$AgC_{10}H_{15}N_3O_6S^-$	$[AgC_{10}H_{15}N_3O_6S]^-$	Ag: MVol.B6-260
$AgC_{10}H_{15}N_6O_2S$	$Ag(NHC_6H_4SO_2NHC_4H_3N_2) \cdot 2\ NH_3$	Ag: MVol.B6-264
$AgC_{10}H_{16}^+$	$[Ag(CH_3C_6H_6CH(CH_3)_2)]^+$	Ag: MVol.B5-57/8
–	$[Ag(CH_3C_6H_8C(CH_2)CH_3)]^+$	Ag: MVol.B5-57/8
–	$[AgC_7H_7(CH_3)_3]^+$	Ag: MVol.B5-77
–	$[AgC_7H_8(CH_3)_2CH_2]^+$	Ag: MVol.B5-77
$AgC_{10}H_{16}INP$	$AgI \cdot (CH_3)_2PC_6H_4N(CH_3)_2$	Ag: MVol.B7-232
$AgC_{10}H_{16}NO$	$AgCCC_5H_6N(CH_3)_3OH$	Ag: MVol.B5-19/20
$AgC_{10}H_{16}NO_3$	$AgNO_3 \cdot C_{10}H_{16}$	Ag: MVol.B5-68/9
$AgC_{10}H_{16}N_2O_4^-$	$[Ag(C_3H_5CH(NH_2)COO)_2]^-$	Ag: MVol.B5-50
		Ag: MVol.B6-250
$AgC_{10}H_{16}N_3O_6S$	$[AgC_{10}H_{16}N_3O_6S]$	Ag: MVol.B6-260
–	$AgC_{10}H_{16}N_3O_6S \cdot H_2O$	Ag: MVol.B6-260
$AgC_{10}H_{16}N_4^+$	$[Ag((CH_3)_2C_3H_2N_2)_2]^+$	Ag: MVol.B6-138/9
–	$[Ag(C_2H_5C_3H_3N_2)_2]^+$	Ag: MVol.B6-138/9
$AgC_{10}H_{16}N_8^+$	$[Ag((CH_2)_4CN_4)_2]^+$	Ag: MVol.B6-194
$AgC_{10}H_{16}O_4S_2^+$	$[Ag(CH_2CHCH_2SCH_2COOH)_2]^+$	Ag: MVol.B5-50/1
$AgC_{10}H_{16}O_4S_2^-$	$[Ag(C_6H_{12}(SCH_2COO)_2)]^-$	Ag: MVol.B7-39/40
$AgC_{10}H_{16}O_8S_4^+$	$[Ag(CH_2(SCH_2COOH)_2)_2]^+$	Ag: MVol.B7-40
$AgC_{10}H_{17}N_2$	$Ag(CH_3(C_3H_7)_2C_3N_2)$	Ag: MVol.B6-142
$AgC_{10}H_{17}N_2^{2+}$	$[Ag(C_5H_5NC_5H_{10}NH_2)]^{2+}$	Ag: MVol.B6-97/8
$AgC_{10}H_{17}N_2O_3$	$[Ag(C_{10}H_{17}N_2O_3)]$	Ag: MVol.B6-321
$AgC_{10}H_{17}N_4$	$[Ag(C_6H_{11}(CH_2)_3CN_4)]$	Ag: MVol.B6-192/3
$AgC_{10}H_{18}^+$	$[AgCH_2CH(CH_2)_6CHCH_2]^+$	Ag: MVol.B5-42/4
–	$[Ag(CH_3)_3CCCC(CH_3)_3]^+$	Ag: MVol.B5-44
$AgC_{10}H_{18}NO_5$	$Ag(C_4H_9OCOONCOOC_4H_9)$	Ag: MVol.B6-331
$AgC_{10}H_{18}N_6^+$	$[Ag(NH_2C_2H_4C_3H_3N_2)_2]^+$	Ag: MVol.B6-139, 143

Formula	Compound	Reference
$AgC_{10}H_{18}N_8O_6Tl$	$[AgTl(NH_2C_2H_4C_3H_3N_2)_2](NO_3)_2$	Ag: MVol.B6-143/4
$AgC_{10}H_{18}O_4S_2^-$	$[Ag(C_3H_7SCH_2COO)_2]^-$	Ag: MVol.B7-25/6
$AgC_{10}H_{19}OS_2$	$[Ag(C_9H_{19}OCSS)]$	Ag: MVol.B7-98
		C: MVol.D4-257
$AgC_{10}H_{20}LiN_2O_4S_2$	$LiAg(CH_3SC_2H_4CH(NH_2)COO)_2$	Ag: MVol.B6-259
$AgC_{10}H_{20}NO_3S_2$	$AgNO_3 \cdot 2\ S(CH_2)_5$	Ag: MVol.B7-84
$AgC_{10}H_{20}NS_2$	$[Ag(SCSNHC_9H_{19})]$	Ag: MVol.B7-102
$AgC_{10}H_{20}N_2O_4^-$	$[Ag(C_3H_7CH(NH_2)COO)_2]^-$	Ag: MVol.B6-248, 250
–	$[Ag(NH_2(CH_2)_4COO)_2]^-$	Ag: MVol.B6-248
$AgC_{10}H_{20}N_2O_4S_2^-$	$[Ag(CH_3SC_2H_4CH(NH_2)COO)_2]^-$	Ag: MVol.B6-259
$AgC_{10}H_{20}N_2S_4$	$[Ag(SCSN(C_2H_5)_2)_2]$	Ag: MVol.B7-317/8
$AgC_{10}H_{20}N_2Se_4$	$[Ag((C_2H_5)_2NCSeSe)_2]$	Ag: MVol.B7-319
$AgC_{10}H_{21}NPS$	$[AgSCN(P(CH(CH_3)_2)_3)]$	Ag: MVol.B7-205
–	$[AgSCN(P(C_3H_7)_3)]$	Ag: MVol.B7-204/5
$AgC_{10}H_{21}S$	$(AgSC_{10}H_{21})_n$	Ag: MVol.B7-5/6
$AgC_{10}H_{22}NO_3S$	$AgNO_3 \cdot S(C_5H_{11})_2$	Ag: MVol.B7-12
$AgC_{10}H_{22}N_2^+$	$[Ag(NH(CH_2)_5)_2]^+$	Ag: MVol.B6-72/3
$AgC_{10}H_{23}NO_2^+$	$[Ag(NH(C_5H_{10}OH)_2)]^+$	Ag: MVol.B6-229
$AgC_{10}H_{24}IP_2$	$[AgI((C_2H_5)_2PC_2H_4P(C_2H_5)_2)]$	Ag: MVol.B7-234
$AgC_{10}H_{24}N_2O_4S_2^{3+}$	$[Ag(CH_3SC_2H_4CH(NH_3)COOH)_2]^{3+}$	Ag: MVol.B6-258
$AgC_{10}H_{24}N_2O_6S_2^-$	$[Ag(NH_2(CH_2)_5SO_3)_2]^-$	Ag: MVol.B6-262
$AgC_{10}H_{24}N_4^+$	$[Ag(CH_3C_4H_9N_2)_2]^+$	Ag: MVol.B6-163/4
$AgC_{10}H_{24}N_4O_2^+$	$[Ag(C_2H_5C(CH_3)(OH)C(NH)NH_2)_2]^+$	Ag: MVol.B6-339/40
$AgC_{10}H_{24}N_4S_2^+$	$[Ag(C_2H_5NHCSNHC_2H_5)_2]^+$	Ag: MVol.B7-155
$AgC_{10}H_{24}N_5O_3S_2$	$[Ag(C_2H_5NHCSNHC_2H_5)_2]NO_3$	Ag: MVol.B7-155
–	$Ag((C_2H_5)_2NCSNH_2)_2NO_3$	Ag: MVol.B7-155
$AgC_{10}H_{24}N_8O_8S_{1.5}$	$[Ag(NH_2C(NH)NC(OC_3H_7)NH_2)_2](SO_4)_{1.5}$	Ag: MVol.B7-325
$AgC_{10}H_{25}N_4O_3$	$[Ag(C_2H_5C(CH_3)(OH)C(NH)NH_2)_2]OH$	Ag: MVol.B6-339/40
$AgC_{10}H_{26}N_2^+$	$[Ag(C_5H_{11}NH_2)_2]^+$	Ag: MVol.B6-46
$AgC_{10}H_{26}N_2O_2^+$	$[Ag(NH_2C_5H_{10}OH)_2]^+$	Ag: MVol.B6-228
$AgC_{10}H_{26}N_2O_4^+$	$[Ag(CH_3N(C_2H_4OH)_2)_2]^+$	Ag: MVol.B6-229
–	$[Ag(NH(C_2H_4OH)C_3H_6OH)_2]^+$	Ag: MVol.B6-229
$AgC_{10}H_{28}N_7O_6S$	$AgC_{10}H_{16}N_3O_6S \cdot 4\ NH_3$	Ag: MVol.B6-260
$AgC_{10}N_7$	$AgC(CN)_2C(CN)NC(CN)C(CN)_2$	Ag: MVol.B6-356
$AgC_{11}ClH_{11}NO_4S$	$[Ag(CH_3C_9H_5NSCH_3)]ClO_4$	Ag: MVol.B7-62
$AgC_{11}Cl_2H_{11}N_2O_7S$	$Ag(C_{11}H_{11}N_2O_7SCl_2)$	Ag: MVol.B6-318
$AgC_{11}Cl_5H_{13}N_2OSb$	$[Ag(C_5H_5N)_2][SbCl_5(OCH_3)]$	Ag: MVol.B6-83
$AgC_{11}F_6H_{11}O_2$	$AgCH(COCF_3)_2 \cdot C_6H_{10}$	Ag: MVol.B5-75/6
$AgC_{11}H_5O_4$	$Ag(C_6H_5C_5O_4)$	Ag: MVol.B6-217
$AgC_{11}H_7N_2$	$AgCCC_3H_2N_2C_6H_5$	Ag: MVol.B5-20
$AgC_{11}H_7N_4$	$[Ag(C_{10}H_7CN_4)]$	Ag: MVol.B6-192/3
$AgC_{11}H_7N_4O_2$	$[Ag(CH_3C_{10}H_4N_4O_2)]$	Ag: MVol.B6-180
$AgC_{11}H_7OS_3$	$[Ag(C_6H_5C_5H_2OS_3)]$	Ag: MVol.B7-83/4
$AgC_{11}H_7S_2$	$[Ag(C_{10}H_7CSS)]$	Ag: MVol.B7-95
$AgC_{11}H_8NO$	$AgNC_4H_3COC_6H_5$	Ag: MVol.B6-68
$AgC_{11}H_8NO_2$	$Ag(C_6H_5NC_4H_3COO)$	Ag: MVol.B6-68
$AgC_{11}H_8NO_2S_2$	$[Ag(CH_3OC_6H_4CHC_3NOS_2)]$	Ag: MVol.B7-76
$AgC_{11}H_8NO_4S$	$[Ag(C_6H_5C_3HNO_2SCH_2COO)]$	Ag: MVol.B7-71
$AgC_{11}H_8NO_6$	$[Ag(C_9H_2O_2(OCH_3)_2NO_2)]$	Ag: MVol.B6-216
$AgC_{11}H_8N_3O$	$[Ag(NC_5H_4NNC_6H_4O)]$	Ag: MVol.B6-287

Formula	Compound	Reference
$AgC_{11}H_8N_5O_5$	$[AgNO_3(CH_3C_{10}H_5N_4O_2)]$	Ag: MVol.B6-176
$AgC_{11}H_9NO_2^+$	$[Ag(C_9H_5N(OH)COCH_3)]^+$	Ag: MVol.B6-112/3
$AgC_{11}H_9NO_6^-$	$[Ag(HOCOC_6H_4N(CH_2COO)_2)]^-$	Ag: MVol.B6-275
$AgC_{11}H_{10}N_2^+$	$[Ag(CH_2(C_5H_4N)_2)]^+$	Ag: MVol.B6-123
$AgC_{11}H_{10}N_3$	$[Ag(C_{10}H_7N_3CH_3)]$	Ag: MVol.B6-297
$AgC_{11}H_{10}N_3O$	$Ag(CH_3NHC_6H_4C_4H_2N_2O)$	Ag: MVol.B6-150
$AgC_{11}H_{10}N_3OS_2$	$[Ag((CH_3)_2NC_6H_4NC_3NOS_2)]$	Ag: MVol.B7-75
$AgC_{11}H_{10}N_3O_2S$	$[Ag(NHC_6H_4SO_2NHC_5H_4N)]$	Ag: MVol.B6-264
$AgC_{11}H_{10}N_3O_3S_3$	$[Ag(SCH_2CONHC_6H_4SO_2NHC_3H_2NS)]$	Ag: MVol.B7-23/4
$AgC_{11}H_{10}O_4S_2^-$	$[Ag(CH_3C_6H_3(SCH_2COO)_2)]^-$	Ag: MVol.B7-42
$AgC_{11}H_{11}$	$AgCCC_6H_2(CH_3)_3$	Ag: MVol.B5-19
$AgC_{11}H_{11}NO_5^-$	$[Ag(CH_3OC_6H_4N(CH_2COO)_2)]^-$	Ag: MVol.B6-275
$AgC_{11}H_{11}N_2^{2+}$	$[AgH(CH_2(C_5H_4N)_2)]^{2+}$	Ag: MVol.B6-123
$AgC_{11}H_{11}N_2O_2$	$[Ag(C_8H_6NC_2H_3(NH_2)COO)]$	Ag: MVol.B6-247
$AgC_{11}H_{11}N_2O_2S$	$[Ag(C_{11}H_{11}N_2O_2S)]$	Ag: MVol.B7-78
$AgC_{11}H_{11}N_2O_3$	$Ag((CH_3)_2C_9H_5N)NO_3$	Ag: MVol.B6-109
$AgC_{11}H_{11}N_2O_4$	$Ag(C_2H_5OC_9H_6N)NO_3$	Ag: MVol.B6-113
–	$[Ag(C_6H_5(CONHCH_2)_2COO)]$	Ag: MVol.B6-267/8
$AgC_{11}H_{11}N_4O_3$	$Ag(NH_2C_9H_6N)NO_3 \cdot CH_3CN$	Ag: MVol.B6-114
$AgC_{11}H_{11}O_2$	$[Ag(CH_3COC(C_6H_5)COCH_3)]$	Ag: MVol.B6-212
$AgC_{11}H_{11}O_2S$	$[Ag(C_4H_7SC_6H_4COO)]$	Ag: MVol.B7-36
$AgC_{11}H_{12}NO_3$	$[Ag(C_6H_5C_2H_3(NHCOCH_3)COO)]$	Ag: MVol.B6-245
$AgC_{11}H_{12}NO_3S$	$[Ag(SCH_2CONHC_6H_4COOC_2H_5)]$	Ag: MVol.B7-23/4
–	$[Ag(SC_2H_3(NHCOC_6H_5)COOCH_3)]$	Ag: MVol.B6-257
$AgC_{11}H_{12}NS$	$[Ag(C_{11}H_{12}NS)]$	Ag: MVol.B7-78
$AgC_{11}H_{12}N_3O_3S$	$AgNO_3 \cdot C_6H_5C_3HN_2S(CH_3)_2 \cdot H_2O$	Ag: MVol.B7-49
$AgC_{11}H_{12}N_3O_4$	$[Ag(C_6H_5NH(CONHCH_2)_2COO)]$	Ag: MVol.B6-267/8
$AgC_{11}H_{12}O_2^+$	$[Ag(C_6H_5C_2H_2COOC_2H_5)]^+$	Ag: MVol.B5-118
$AgC_{11}H_{13}N_2S$	$[Ag(CH_3C_3H_3NSNC_6H_4CH_3)]$	Ag: MVol.B7-71
$AgC_{11}H_{13}N_4S$	$[Ag(C_3H_7N_3C_7H_3NSCH_3)]$	Ag: MVol.B6-302
$AgC_{11}H_{13}O_2S$	$[Ag(C_4H_9SC_6H_4COO)]$	Ag: MVol.B7-36
$AgC_{11}H_{14}N$	$AgCCC_6H_5 \cdot (CH_3)_2CHNH_2$	Ag: MVol.B5-17
$AgC_{11}H_{14}N_2O_3^-$	$[Ag(C_2H_5(CCH_3CHC_2H_5)C_4N_2O_3)]^-$	Ag: MVol.B6-155
$AgC_{11}H_{14}P$	$AgCCC_6H_5 \cdot P(CH_3)_3$	Ag: MVol.B5-17/8
$AgC_{11}H_{15}NPS$	$[AgSCN((C_2H_5)_2PC_6H_5)]$	Ag: MVol.B7-221
$AgC_{11}H_{15}N_2O_3S_3$	$[Ag(C_{11}H_{15}NS_3)]NO_3$	Ag: MVol.B7-93
$AgC_{11}H_{15}OS_3$	$[Ag(C_5HOS_3(C_3H_7)_2)]$	Ag: MVol.B7-83/4
–	$[Ag(C_6H_{13}C_5H_2OS_3)]$	Ag: MVol.B7-83/4
$AgC_{11}H_{16}^+$	$[AgC_6H(CH_3)_5]^+$	Ag: MVol.B5-98
–	$[AgC_6H_5C_5H_{11}]^+$	Ag: MVol.B5-97
$AgC_{11}H_{16}N_2O_3^-$	$[Ag(C_2H_5(C_5H_{11})C_4N_2O_3)]^-$	Ag: MVol.B6-155
$AgC_{11}H_{16}N_3$	$[Ag(C_5H_{11}N_3C_6H_5)]$	Ag: MVol.B6-296
$AgC_{11}H_{17}N_4O_2$	$Ag(C_8H_6NC_2H_3(NH_2)COO) \cdot 2\ NH_3$	Ag: MVol.B6-247
$AgC_{11}H_{18}KN_2O_3S_2$	$KAg(SCN)_2 \cdot 3\ (CH_3)_2CO$	Ag: MVol.B6-209
$AgC_{11}H_{19}N_2$	$Ag((C_4H_9)_2C_3HN_2)$	Ag: MVol.B6-142
$AgC_{11}H_{19}N_2S$	$[Ag(C_4HN_2S(CH_3)_3C_4H_9)]$	Ag: MVol.B7-64
$AgC_{11}H_{19}N_4$	$[Ag(C_6H_{11}(CH_2)_4CN_4)]$	Ag: MVol.B6-192/3
$AgC_{11}H_{20}N_2O_3^+$	$[AgC_7H_9CH_2N(CH_3)_3NO_3]^+$	Ag: MVol.B5-88/9
$AgC_{11}H_{21}OS_2$	$[Ag(C_{10}H_{21}OCSS)]$	Ag: MVol.B7-98
$AgC_{11}H_{22}NOS$	$[Ag((C_5H_{11})_2NCOS)]$	Ag: MVol.B7-99, 102

Formula	Compound	Reference
$AgC_{11}H_{22}NS_2$	$[Ag((C_5H_{11})_2NCSS)]$	Ag: MVol.B7-105/6, 113
–	$[Ag(SCSNHC_{10}H_{21})]$	Ag: MVol.B7-102
$AgC_{11}H_{22}N_5$	$[Ag(N(C_5H_{11})_2CN_4)]$	Ag: MVol.B6-192/3
$AgC_{11}H_{22}O^+$	$[AgCH_2CH(CH_2)_8CH_2OH]^+$	Ag: MVol.B5-47
$AgC_{11}H_{23}N_2S_2$	$[Ag((C_2H_5)_2NC_5H_{10}N(CH_3)CSS)]$	Ag: MVol.B7-115
$AgC_{11}H_{24}NS$	$[Ag(SCH(CH_3)CH_2N(C_4H_9)_2)]$	Ag: MVol.B7-18
$AgC_{11}H_{25}NPS_2$	$[Ag(S_2CN(C_2H_5)_2)P(C_2H_5)_3]$	Ag: MVol.B7-202/3
$AgC_{12}ClCuH_{10}N_4O_6$	$[Cu(NC_5H_4CHNO)_2Ag]ClO_4 \cdot H_2O$	Ag: MVol.B6-305
$AgC_{12}ClH_6N_6O_8$	$Ag(NO_2(CN)C_5H_3N)_2ClO_4$	Ag: MVol.B6-101
$AgC_{12}ClH_7N_2^+$	$[Ag(ClC_{12}H_7N_2)]^+$	Ag: MVol.B6-128
$AgC_{12}ClH_8N_4O_4$	$[Ag(NCC_5H_4N)_2]ClO_4$	Ag: MVol.B6-99, 100, 101
$AgC_{12}ClH_8N_4O_6$	$Ag(NCC_5H_4NO)_2ClO_4$	Ag: MVol.B6-106
$AgC_{12}ClH_8O_4$	$AgClO_4 \cdot C_{12}H_8$	Ag: MVol.B5-112/3
$AgC_{12}ClH_{10}N_2O_8$	$Ag(NC_5H_4COOH)_2ClO_4$	Ag: MVol.B6-102
$AgC_{12}ClH_{10}N_4O_4$	$[AgClO_4 \cdot N_2(CHC_5H_4N)_2]_2$	Ag: MVol.B6-295
$AgC_{12}ClH_{10}O_4$	$AgClO_4 \cdot C_{12}H_{10}$	Ag: MVol.B5-111/2
$AgC_{12}ClH_{10}O_4S$	$AgClO_4 \cdot S(C_6H_5)_2$	Ag: MVol.B7-13
$AgC_{12}ClH_{12}O_4$	$AgClO_4 \cdot 2\ C_6H_6$	Ag: MVol.B5-91, 93
–	$AgClO_4 \cdot C_{10}H_6(CH_3)_2$	Ag: MVol.B5-109
$AgC_{12}ClH_{14}N_2O_4$	$[Ag(CH_3C_5H_4N)_2]ClO_4$	Ag: MVol.B6-86
–	$AgClO_4 \cdot 2\ C_6H_5NH_2$	Ag: MVol.B6-51
$AgC_{12}ClH_{14}N_2O_6$	$[Ag(CH_3C_5H_4NO)_2]ClO_4$	Ag: MVol.B6-105
$AgC_{12}ClH_{14}N_2O_8$	$Ag(CH_3OC_5H_4NO)_2ClO_4$	Ag: MVol.B6-105
$AgC_{12}ClH_{15}NO_6S$	$[Ag(C_6H_5CH_2CSNHCH_2COOC_2H_5)]ClO_4$	Ag: MVol.B7-127
$AgC_{12}ClH_{16}N_4O_4$	$Ag(CH_3NHC_5H_4N)_2ClO_4$	Ag: MVol.B6-98
–	$Ag(NCC_4H_8CN)_2ClO_4$	Ag: MVol.B6-352
–	$Ag(NH_2(CH_3)C_5H_3N)_2ClO_4$	Ag: MVol.B6-96
$AgC_{12}ClH_{18}O_4$	$AgClO_4 \cdot 3\ CH_3CCCH_3$	Ag: MVol.B5-36
–	$AgClO_4 \cdot C_{12}H_{18}$	Ag: MVol.B5-74
$AgC_{12}ClH_{20}O_4$	$AgClO_4 \cdot 2\ C_6H_{10}$	Ag: MVol.B5-55, 56
$AgC_{12}ClH_{24}N_6S_3$	$AgCl \cdot 3\ C_4H_8N_2S$	Ag: MVol.B7-64
$AgC_{12}ClH_{24}O_7S_3$	$AgClO_4 \cdot 3\ S(CH_2CH_2)_2O$	Ag: MVol.B7-85
$AgC_{12}ClH_{24}O_8$	$AgClO_4 \cdot 4\ (CH_3)_2CO$	Ag: MVol.B6-209
$AgC_{12}ClH_{24}O_{10}$	$Ag(C_4H_8O_2)_3ClO_4$	Ag: MVol.B6-220/1
$AgC_{12}ClH_{30}N_6O_4S_3$	$[Ag(CH_3NHCSNHC_2H_5)_3]ClO_4$	Ag: MVol.B7-155/6
$AgC_{12}ClH_{30}O_6P_2$	$[AgCl(P(OC_2H_5)_3)_2]$	Ag: MVol.B7-246
$AgC_{12}ClH_{30}O_{10}P_2$	$[Ag(P(OC_2H_5)_3)_2]ClO_4$	Ag: MVol.B7-246
$AgC_{12}ClH_{32}N_4O_4$	$[Ag(NH_2C_2H_4N(C_2H_5)_2)_2]ClO_4$	Ag: MVol.B6-60
$AgC_{12}ClH_{33}PSi_3$	$[AgCl(P(CH_2Si(CH_3)_3)_3)]$	Ag: MVol.B7-255
$AgC_{12}ClH_{36}N_6O_4P_2S_2$	$[Ag(SP(N(CH_3)_2)_3)_2]ClO_4$	Ag: MVol.B7-254
$AgC_{12}ClH_{36}N_6O_6P_2$	$[Ag(OP(N(CH_3)_2)_3)_2]ClO_4$	Ag: MVol.B7-251
$AgC_{12}ClH_{36}O_{16}P_4$	$[Ag(P(OCH_3)_3)_4]ClO_4$	Ag: MVol.B7-244/5
$AgC_{12}ClH_{36}P_4$	$[Ag(P(CH_3)_3)_4]Cl$	Ag: MVol.B7-200/1
$AgC_{12}Cl_2H_7N_2O_3S$	$Ag(ClC_6H_4NNC_6H_3(Cl)SO_3)$	Ag: MVol.B6-290
$AgC_{12}Cl_2H_8N_3$	$[Ag(ClC_6H_4N_3C_6H_4Cl)]$	Ag: MVol.B6-300
$AgC_{12}Cl_2H_{11}N_2O_6S$	$Ag(C_{12}H_{11}Cl_2N_2O_6S)$	Ag: MVol.B6-319
$AgC_{12}Cl_2H_{12}N_2^+$	$[Ag(ClC_6H_4NH_2)_2]^+$	Ag: MVol.B6-52
$AgC_{12}Cl_3H_6N_4O_4$	$Ag(Cl(CN)C_5H_3N)_2ClO_4$	Ag: MVol.B6-101
$AgC_{12}Cl_4H_4O_2S^-$	$[AgS(C_6H_2(O)Cl_2)_2]^-$	Ag: MVol.B7-13
$AgC_{12}Cl_4H_{14}IrN_2$	$Ag[Ir(CH_3C_5H_4N)_2Cl_4]$	Ir: SVol.2-75

Formula	Compound	Reference
$AgC_{12}Cl_6H_4N_3$	$[Ag(Cl_3C_6H_2N_3C_6H_2Cl_3)]$	Ag: MVol.B6-300
$AgC_{12}Cl_6H_{12}NO_3Sb_2$	$AgNO_3 \cdot 2\ Sb(CHCHCl)_3$	Ag: MVol.B5-51
$AgC_{12}CuH_{10}N_4O_2^+$	$[Cu(NC_5H_4CHNO)_2Ag]^+$	Ag: MVol.B6-305
$AgC_{12}F_6H_{12}Nb$	$AgNbF_6 \cdot 2\ C_6H_6$	Ag: MVol.B5-94
$AgC_{12}F_6H_{12}P$	$AgPF_6 \cdot 2\ C_6H_6$	Ag: MVol.B5-94
$AgC_{12}F_6H_{12}Sb$	$AgSbF_6 \cdot 2\ C_6H_6$	Ag: MVol.B5-94
$AgC_{12}F_6H_{12}Ta$	$AgTaF_6 \cdot 2\ C_6H_6$	Ag: MVol.B5-94
$AgC_{12}F_6H_{13}O_2$	$AgCH(COCF_3)_2 \cdot C_7H_{12}$	Ag: MVol.B5-75/6
$AgC_{12}F_6H_{24}O_6Sb$	$Ag(C_4H_8O_2)_3SbF_6$	Ag: MVol.B6-221
$AgC_{12}F_{10}Li$	$LiAg(C_6F_5)_2$	Ag: MVol.B5-10
$AgC_{12}F_{10}S_2^-$	$[Ag(SC_6F_5)_2]^-$	Ag: MVol.B7-7
$AgC_{12}H_4N_4$	$Ag(C(CN)_2C_6H_4C(CN)_2)$	Ag: MVol.B6-355
$AgC_{12}H_4N_7O_{12}$	$AgN(C_6H_2(NO_2)_3)_2$	Ag: MVol.B6-53
$AgC_{12}H_6NO_3$	$[Ag(C_{12}H_6NO_3)]$	Ag: MVol.B6-317
$AgC_{12}H_6NO_5$	$[Ag(C_{12}H_6NO_5)]$	Ag: MVol.B6-204
$AgC_{12}H_7N_2O$	$[Ag(C_{12}H_7N_2O)]$	Ag: MVol.B6-180
$AgC_{12}H_7N_4O_5$	$[Ag(NO_2C_6H_4NNC_6H_3(O)NO_2)]$	Ag: MVol.B6-285
–	$Ag(NO_2C_{12}H_7N_2)NO_3$	Ag: MVol.B6-128
$AgC_{12}H_8N$	$AgN(C_6H_4)_2 \cdot n\ NH_3$	Ag: MVol.B6-71
$AgC_{12}H_8NOS_2$	$[Ag(C_6H_5CHCHCHC_3NOS_2)]$	Ag: MVol.B7-76
$AgC_{12}H_8N_2^+$	$[Ag(C_{12}H_8N_2)]^+$	Ag: MVol.B6-124
$AgC_{12}H_8N_2O_4$	$[Ag(C_5H_4NCOO)_2]$	Ag: MVol.B7-275/9
–	$Ag(C_5H_4NCOO)_2 \cdot H_2O$	Ag: MVol.B7-279
$AgC_{12}H_8N_3$	$Ag(C_5H_4NC_7H_4N_2)$	Ag: MVol.B6-146
$AgC_{12}H_8N_3O_3$	$Ag(C_{12}H_8N_2)NO_3$	Ag: MVol.B6-180
$AgC_{12}H_8N_4^+$	$[Ag(NCC_5H_4N)_2]^+$	Ag: MVol.B6-99, 100
$AgC_{12}H_8N_5O_3$	$[AgNO_3(NCC_5H_4N)_2]$	Ag: MVol.B6-100
$AgC_{12}H_8N_5O_4$	$[Ag(NO_2C_6H_4N_3C_6H_4NO_2)]$	Ag: MVol.B6-298/9
$AgC_{12}H_8O_6S_3$	$[Ag(S(C_6H_4SO_3)_2)]$	Ag: MVol.B7-45
$AgC_{12}H_8S^+$	$[Ag(C_{12}H_8S)]^+$	Ag: MVol.B7-69
$AgC_{12}H_8S_2^+$	$[Ag(C_{12}H_8S_2)]^+$	Ag: MVol.B7-86/7
$AgC_{12}H_9N_2O$	$[Ag(C_6H_5NNC_6H_4O)]$	Ag: MVol.B6-285
$AgC_{12}H_9N_2O_3S$	$Ag(C_6H_5NNC_6H_4SO_3)$	Ag: MVol.B6-289
$AgC_{12}H_9N_2O_4$	$Ag(NC_5H_4COO) \cdot NC_5H_4COOH$	Ag: MVol.B6-102
$AgC_{12}H_9N_2O_4S$	$[Ag(C_6H_5NN(O)C_6H_4SO_3)]$	Ag: MVol.B6-292
$AgC_{12}H_9N_2S$	$[Ag(NC_5H_4CSNC_6H_5)]$	Ag: MVol.B7-127
$AgC_{12}H_9N_4O_2$	$[Ag(NO_2C_6H_4N_3C_6H_5)]$	Ag: MVol.B6-298
$AgC_{12}H_9N_4O_3$	$[Ag(C_6H_5N(NO)C_6H_4N(O)NO)]$	Ag: MVol.B6-321
$AgC_{12}H_9N_4O_3S$	$Ag(C_6H_5NNC_6H_4NNSO_3)$	Ag: MVol.B6-294
$AgC_{12}H_9OS_3$	$[Ag(CH_3C_5HOS_3C_6H_5)]$	Ag: MVol.B7-83/4
–	$[Ag(C_6H_5CH_2C_5H_2OS_3)]$	Ag: MVol.B7-83/4
$AgC_{12}H_9O_3SSe$	$[Ag(C_6H_5SeC_6H_4SO_3)]$	Ag: MVol.B7-191
$AgC_{12}H_9O_3S_2$	$[Ag(C_6H_5SC_6H_4SO_3)]$	Ag: MVol.B7-44/5
$AgC_{12}H_9O_4$	$[Ag(C_9H_4O_2COOC_2H_5)]$	Ag: MVol.B6-216
$AgC_{12}H_{10}^+$	$[AgC_6H_5C_6H_5]^+$	Ag: MVol.B5-102
$AgC_{12}H_{10}IO_2$	$Ag(C_6H_5CH(CH_2CO)_2Cl)$	Ag: MVol.B6-214/5
$AgC_{12}H_{10}KN_2O_3$	$K[Ag(C_2H_5(C_6H_5)C_4N_2O_3)] \cdot 5\ H_2O$	Ag: MVol.B6-156
$AgC_{12}H_{10}NOS$	$[Ag(SCH_2CONHC_{10}H_7)]$	Ag: MVol.B7-24/5
$AgC_{12}H_{10}NO_2$	$C_6H_5COOAg \cdot C_5H_5N$	Ag: MVol.B6-81
$AgC_{12}H_{10}NO_3$	$AgNO_3 \cdot C_{12}H_{10}$	Ag: MVol.B5-109

Formula	Compound	Reference
$AgC_{12}H_{10}N_2NaO_3$	$Na[Ag(C_2H_5C_4N_2O_3C_6H_5)]$	Ag: MVol.B6-154/6
$AgC_{12}H_{10}N_2O_3^-$	$[Ag(C_2H_5(C_6H_5)C_4N_2O_3)]^-$	Ag: MVol.B6-155
$AgC_{12}H_{10}N_2O_4^+$	$[Ag(NC_5H_4COOH)_2]^+$	Ag: MVol.B6-102
$AgC_{12}H_{10}N_3$	$[Ag(C_6H_5N_3C_6H_5)]$	Ag: MVol.B6-296
$AgC_{12}H_{10}N_3O_3S$	$Ag(C_6H_5NHC_6H_4NNSO_3)$	Ag: MVol.B6-294
$AgC_{12}H_{10}N_3O_7$	$Ag(NC_5H_4COOH)_2NO_3$	Ag: MVol.B6-102
$AgC_{12}H_{10}N_5O_3$	$[AgNO_3 \cdot N_2(CHC_5H_4N)_2]_2$	Ag: MVol.B6-295
$AgC_{12}H_{10}N_5O_5$	$[Ag((NO_2)_2C_6H_3C_3N_2(CH_3)_2CONH)]$	Ag: MVol.B6-132
$AgC_{12}H_{10}O_3PS$	$Ag(OSP(OC_6H_5)_2)$	Ag: MVol.B7-253
$AgC_{12}H_{11}N_2$	$AgC_{12}H_8N \cdot NH_3$	Ag: MVol.B6-71
$AgC_{12}H_{11}N_2OS_2$	$[Ag((CH_3)_2NC_6H_4CHC_3NOS_2)]$	Ag: MVol.B7-74/5
$AgC_{12}H_{11}N_2O_2$	$NH_2C_6H_4COOAg \cdot C_5H_5N$	Ag: MVol.B6-81
$AgC_{12}H_{11}N_2O_2S$	$Ag(C_6H_5C_3N_2O(CH_3)_2COS)$	Ag: MVol.B6-135
$AgC_{12}H_{11}N_2O_3$	$Ag(C_2H_5(C_6H_5)C_4HN_2O_3)$	Ag: MVol.B6-155
$AgC_{12}H_{11}N_4S_2$	$[Ag(C_{10}H_7NHCSNNHCSNH_2)]$	Ag: MVol.B7-169/70
$AgC_{12}H_{11}N_6O_4$	$[Ag(NO_2C_6H_4N_3C_6H_4NO_2)(NH_3)]$	Ag: MVol.B6-299
$AgC_{12}H_{11}O_2$	$AgCCC(OCOCH_3)(CH_3)C_6H_5$	Ag: MVol.B5-16
$AgC_{12}H_{12}^+$	$[Ag(C_6H_6)_2]^+$	Ag: MVol.B5-94
$AgC_{12}H_{12}I_2N_2^+$	$[Ag(IC_6H_4NH_2)_2]^+$	Ag: MVol.B6-52
$AgC_{12}H_{12}KN_6O_8$	$AgK(ONC_4N_2O_3(CH_3)_2)_2$	Ag: MVol.B6-315
$AgC_{12}H_{12}NO_3$	$AgNO_3 \cdot C_{12}H_{12}$	Ag: MVol.B5-109
$AgC_{12}H_{12}NO_4$	$AgNO_3 \cdot C_8H_7COC_3H_5$	Ag: MVol.B5-90
$AgC_{12}H_{12}NO_5$	$AgNO_3 \cdot 2\ C_6H_5OH$	Ag: MVol.B5-118
		Ag: MVol.B6-206/7
$AgC_{12}H_{12}N_2^+$	$[Ag(C_2H_4(C_5H_4N)_2)]^+$	Ag: MVol.B6-123
$AgC_{12}H_{12}N_2O_2^+$	$[Ag(C_2H_4(NCHC_4H_3O)_2)]^+$	Ag: MVol.B6-280
$AgC_{12}H_{12}N_2O_6S_2^-$	$[Ag(NH_2C_6H_4SO_3)_2]^-$	Ag: MVol.B6-262
$AgC_{12}H_{12}N_2O_8S_4^{5-}$	$[Ag(C_3H_5(COO)_2NHCSS)_2]^{5-}$	Ag: MVol.B7-103
$AgC_{12}H_{12}N_2S_2^+$	$[Ag(C_2H_4(NCHC_4H_3S)_2)]^+$	Ag: MVol.B6-280
$AgC_{12}H_{12}N_3$	$[Ag(C_{10}H_7N_3C_2H_5)]$	Ag: MVol.B6-297
$AgC_{12}H_{12}N_3O_2$	$[Ag(CH_3C_3N_3O_2C_2H_4C_6H_5)]$	Ag: MVol.B6-309/12
$AgC_{12}H_{12}N_3O_3$	$AgNO_3 \cdot NH_2C_6H_4C_6H_4NH_2$	Ag: MVol.B6-64
$AgC_{12}H_{12}N_4O_2^+$	$[Ag(NC_5H_4CONH_2)_2]^+$	Ag: MVol.B6-103
$AgC_{12}H_{12}N_4O_4^+$	$[Ag(C_6H_4(NO_2)NH_2)_2]^+$	Ag: MVol.B6-51
$AgC_{12}H_{12}N_5O_3$	$AgNO_3 \cdot C_6H_5NNC_6H_3(NH_2)_2$	Ag: MVol.B6-286
$AgC_{12}H_{12}N_5O_5$	$Ag(NC_5H_4CONH_2)_2NO_3$	Ag: MVol.B6-103
–	$Ag(NC_5H_4CONH_2)_2NO_3 \cdot H_2O$	Ag: MVol.B6-103
$AgC_{12}H_{13}NOS^+$	$[Ag(NC_8H_6SC_4H_7O)]^+$	Ag: MVol.B7-48
$AgC_{12}H_{13}NO_5PS$	$Ag((CH_3O)_2P(O)NSO_2C_{10}H_7)$	Ag: MVol.B7-251
$AgC_{12}H_{13}N_2^{2+}$	$[AgH(C_2H_4(C_5H_4N)_2)]^{2+}$	Ag: MVol.B6-123
$AgC_{12}H_{13}N_2O_2$	$CH_3COOAg \cdot 2\ C_5H_5N$	Ag: MVol.B6-81
$AgC_{12}H_{13}N_2O_4$	$[AgC_{12}H_{13}N_2O_4]$	Ag: MVol.B6-267/8
$AgC_{12}H_{13}N_2O_4S$	$[Ag(CH_3C_7H_3NS(NO_2)CH_2COOC_2H_5)]$	Ag: MVol.B7-78
$AgC_{12}H_{13}N_2O_5$	$[AgC_{12}H_{13}N_2O_5]$	Ag: MVol.B6-267/8
$AgC_{12}H_{13}N_3^+$	$[Ag(NH(CH_2C_5H_4N)_2)]^+$	Ag: MVol.B6-98
$AgC_{12}H_{13}N_6O_8$	$[AgH(ONC_4N_2O_3(CH_3)_2)_2]$	Ag: MVol.B6-315
$AgC_{12}H_{13}O_2S$	$[Ag(C_5H_9SC_6H_4COO)]$	Ag: MVol.B7-36
$AgC_{12}H_{14}NO_2S$	$[Ag(CH_3C_7H_4NSCH_2COOC_2H_5)]$	Ag: MVol.B7-78
$AgC_{12}H_{14}NS$	$[Ag(C_{12}H_{14}NS)]$	Ag: MVol.B7-78
$AgC_{12}H_{14}N_2^+$	$[Ag(CH_3C_5H_4N)_2]^+$	Ag: MVol.B6-84/5

Formula	Compound	Reference
$AgC_{12}H_{14}N_2^+$	$[Ag(C_6H_5NH_2)_2]^+$	Ag: MVol.B6-50/1
$AgC_{12}H_{14}N_2O_2^+$	$[Ag(CH_3C_5H_4NO)_2]^+$	Ag: MVol.B6-105
–	$[Ag(CH_3OC_5H_4N)_2]^+$	Ag: MVol.B6-91
–	$[Ag(HOCH_2C_5H_4N)_2]^+$	Ag: MVol.B6-91
$AgC_{12}H_{14}N_2O_3^-$	$[Ag(C_2H_5(C_6H_9)C_4N_2O_3)]^-$	Ag: MVol.B6-155
$AgC_{12}H_{14}N_2O_6P$	$Ag(C_{12}H_{14}N_2O_6P)$	Ag: MVol.B6-319
$AgC_{12}H_{14}N_3$	$AgC_{12}H_8N \cdot 2\ NH_3$	Ag: MVol.B6-71
$AgC_{12}H_{14}N_3O_3$	$Ag(CH_3C_5H_4N)_2NO_3$	Ag: MVol.B6-85
$AgC_{12}H_{14}N_3O_9S$	$Ag(C_{12}H_{14}N_3O_9S)$	Ag: MVol.B6-318
$AgC_{12}H_{15}$	$AgCCC_7H_7(CH_3)_2CH_2$	Ag: MVol.B5-19
$AgC_{12}H_{15}N_2O_3$	$Ag(C_6H_9C_4N_2O_3(CH_3)_2)$	Ag: MVol.B6-155/6
$AgC_{12}H_{15}N_2O_7S$	$Ag(C_{12}H_{15}N_2O_7S)$	Ag: MVol.B6-318
$AgC_{12}H_{15}N_4S$	$[Ag(C_4H_9N_3C_7H_3NSCH_3)]$	Ag: MVol.B6-302
$AgC_{12}H_{15}O_2S$	$[Ag(C_5H_{11}SC_6H_4COO)]$	Ag: MVol.B7-36
$AgC_{12}H_{16}^+$	$[AgC_{12}H_{16}]^+$	Ag: MVol.B5-86
$AgC_{12}H_{16}NO_3$	$AgNO_3 \cdot C_{10}H_{10}(CH_3)_2$	Ag: MVol.B5-83
$AgC_{12}H_{16}N_4^+$	$[Ag(CH_3(CH_2CH)C_3H_2N_2)_2]^+$	Ag: MVol.B6-138/9
–	$[Ag((CH_3)_2C_4H_2N_2)_2]^+$	Ag: MVol.B6-159
–	$[Ag(NH_2CH_2C_5H_4N)_2]^+$	Ag: MVol.B6-96
–	$[Ag(NH_2(CH_3)C_5H_3N)_2]^+$	Ag: MVol.B6-95
$AgC_{12}H_{16}N_5O_2$	$[Ag(C_7H_6N_4O_2CH_2C_4H_8N)]$	Ag: MVol.B6-166, 168
$AgC_{12}H_{16}N_5O_3$	$Ag(CH_3NHC_5H_4N)_2NO_3$	Ag: MVol.B6-98
–	$Ag(NH_2(CH_3)C_5H_3N)_2NO_3$	Ag: MVol.B6-95
$AgC_{12}H_{16}N_5O_4$	$[Ag(C(CH_3)_2NOC(CH_3)_2N(O)N_2C_6H_4NO_2)]$	Ag: MVol.B6-303/4
$AgC_{12}H_{16}N_6O_4^-$	$[Ag(C_3H_3N_2C_2H_3(NH_2)COO)_2]^-$	Ag: MVol.B6-246
$AgC_{12}H_{16}Na_3O_8S_4$	$Na_3[Ag(C_2H_4(SCH_2COO)_2)_2] \cdot 10\ H_2O$	Ag: MVol.B7-41
$AgC_{12}H_{16}O_8S_2^{3-}$	$[Ag(S(C_2H_4COO)_2)_2]^{3-}$	Ag: MVol.B7-38
$AgC_{12}H_{16}O_8S_4^{3-}$	$[Ag(C_2H_4(SCH_2COO)_2)_2]^{3-}$	Ag: MVol.B7-39/40
$AgC_{12}H_{17}N_4O_2$	$[Ag(C(CH_3)_2NOC(CH_3)_2N(O)N_2C_6H_5)]$	Ag: MVol.B6-303/4
$AgC_{12}H_{17}O$	$AgCCC_7H_7(CH_3)_3OH$	Ag: MVol.B5-19
$AgC_{12}H_{18}^+$	$[AgC_6(CH_3)_6]^+$	Ag: MVol.B5-98
–	$[AgC_6H_3(C_2H_5)_3]^+$	Ag: MVol.B5-98
–	$[AgC_6H_4(C_3H_7)_2]^+$	Ag: MVol.B5-98
$AgC_{12}H_{18}NO_3$	$AgNO_3 \cdot C_{10}H_{12}(CH_3)_2 \cdot H_2O$	Ag: MVol.B5-70
–	$AgNO_3 \cdot C_{12}H_{18}$	Ag: MVol.B5-74
$AgC_{12}H_{18}N_2^+$	$[Ag(CH_3(CH_3C_6H_4)C_4H_8N_2)]^+$	Ag: MVol.B6-163/4
$AgC_{12}H_{18}N_2O^+$	$[Ag(CH_3(CH_3OC_6H_4)C_4H_8N_2)]^+$	Ag: MVol.B6-163/4
$AgC_{12}H_{18}N_3O$	$[Ag(C_2H_5N_3C_6H_4C(CH_3)(C_2H_5)OH)]$	Ag: MVol.B6-301
$AgC_{12}H_{18}N_6O_4^+$	$[Ag(C_3H_3N_2C_2H_3(NH_2)COOH)_2]^+$	Ag: MVol.B6-246
$AgC_{12}H_{18}O^+$	$[AgC_{11}H_{15}CH_2OH]^+$	Ag: MVol.B5-89
$AgC_{12}H_{18}O_4S_2^-$	$[Ag(CH_2CH(CH_2)_2SCH_2COO)_2]^-$	Ag: MVol.B5-50/1
$AgC_{12}H_{18}O_4Se_2^-$	$[Ag(CH_2CH(CH_2)_2SeCH_2COO)_2]^-$	Ag: MVol.B5-50/1
		Ag: MVol.B7-191/3, 196
$AgC_{12}H_{19}O_2$	$AgCC(CH_2)_8COOCH_3$	Ag: MVol.B5-16
$AgC_{12}H_{19}O_8S_4$	$[Ag(C_6H_9O_4S_2)(C_6H_{10}O_4S_2)]$	Ag: MVol.B7-41
$AgC_{12}H_{20}INP$	$[AgI((C_2H_5)_2PC_6H_4N(CH_3)_2)]_4$	Ag: MVol.B7-232
$AgC_{12}H_{20}NO_3$	$AgNO_3 \cdot 2\ C_6H_{10}$	Ag: MVol.B5-56
$AgC_{12}H_{20}N_2O_4^-$	$[Ag(C_4H_7CH(NH_2)COO)_2]^-$	Ag: MVol.B6-250
$AgC_{12}H_{20}N_2S_4$	$[Ag(SCSN(CH_2)_5)_2]$	Ag: MVol.B7-317
$AgC_{12}H_{20}N_4^+$	$[Ag(CH_3(C_2H_5)C_3H_2N_2)_2]^+$	Ag: MVol.B6-138/9

$AgC_{12}H_{20}N_8^+$ ...... $[Ag((CH_2)_5CN_4)_2]^+$ .... Ag: MVol.B6-194
$AgC_{12}H_{20}N_9O_3$ ...... $[AgNO_3((CH_2)_5CN_4)_2]_2$ .... Ag: MVol.B6-195
$AgC_{12}H_{20}O_4S_2^+$ ..... $[Ag(CH_2CH(CH_2)_2SCH_2COOH)_2]^+$ .... Ag: MVol.B5-50/1
$AgC_{12}H_{20}O_4Se_2^+$ .... $[Ag(CH_2CH(CH_2)_2SeCH_2COOH)_2]^+$ .... Ag: MVol.B5-50/1
Ag: MVol.B7-193
$AgC_{12}H_{20}O_8S_4^+$ ..... $[Ag(C_2H_4(SCH_2COOH)_2)_2]^+$ .... Ag: MVol.B7-40
$AgC_{12}H_{21}N_4$ .... $[Ag(C_6H_{11}(CH_2)_5CN_4)]$ .... Ag: MVol.B6-192/3
$AgC_{12}H_{22}NO_3$ ...... $AgNO_3 \cdot C_{12}H_{22}$ .... Ag: MVol.B5-73
$AgC_{12}H_{22}N_2^+$ ...... $[Ag(NH(CH_2CHCH_2)_2)_2]^+$ .... Ag: MVol.B6-48
$AgC_{12}H_{22}O_4S_2^-$ ..... $[Ag(C_4H_9SCH_2COO)_2]^-$ .... Ag: MVol.B7-25/6
$AgC_{12}H_{22}O_4Se_2^-$ .... $[Ag(CH_3(CH_2)_3SeCH_2COO)_2]^-$ .... Ag: MVol.B7-191/3, 196
$AgC_{12}H_{23}N_4O_4S_4$ .... $[Ag(C_6H_{11}N_2O_2S_2)(C_6H_{12}N_2O_2S_2)]$ .... Ag: MVol.B7-187/8
$AgC_{12}H_{24}LiN_2O_4$ .... $LiAg(C_4H_9CH(NH_2)COO)_2$ .... Ag: MVol.B6-249
$AgC_{12}H_{24}N_2O_4^-$ ..... $[Ag(C_4H_9CH(NH_2)COO)_2]^-$ .... Ag: MVol.B6-249/50
– .... $[Ag(NH_2(CH_2)_5COO)_2]^-$ .... Ag: MVol.B6-249/50
$AgC_{12}H_{24}N_4^+$ ...... $[Ag(N(CH_2CH_2)_3N)_2]^+$ .... Ag: MVol.B6-130
$AgC_{12}H_{24}N_4O_4S_4^+$ ... $[Ag(HOC_2H_4NHCSCSNHC_2H_4OH)_2]^+$ ...... Ag: MVol.B7-187/8
$AgC_{12}H_{24}N_8^+$ ...... $[Ag((CH_2)_6N_4)_2]^+$ .... Ag: MVol.B6-197/8
$AgC_{12}H_{24}N_9O_3$ ...... $AgNO_3 \cdot 2\ C_2N_3(C_2H_5)_2NH_2$ .... Ag: MVol.B6-186
$AgC_{12}H_{24}N_9O_3S_4$ .... $[Ag(C_3H_6N_2S)_4]NO_3$ .... Ag: MVol.B7-50
$AgC_{12}H_{24}N_9O_7$ ...... $AgNO_3 \cdot 2\ C_2N_3(C_2H_4OH)_2NH_2$ .... Ag: MVol.B6-186
$AgC_{12}H_{24}O_4S_2^+$ ..... $[Ag(C_4H_9SCH_2COOH)_2]^+$ .... Ag: MVol.B7-26
– .... $[Ag(S(C_2H_4OC_2H_4OC_2H_4)_2S)]^+$ .... Ag: MVol.B7-93
$AgC_{12}H_{24}O_4Se_2^+$ .... $[Ag(C_4H_9SeCH_2COOH)_2]^+$ .... Ag: MVol.B7-193
$AgC_{12}H_{24}O_6^+$ ...... $[Ag(O(C_2H_4OC_2H_4OC_2H_4)_2O)]^+$ .... Ag: MVol.B6-219
$AgC_{12}H_{25}NO_5^+$ ..... $[Ag(NH(C_2H_4OC_2H_4OC_2H_4)_2O)]^+$ .... Ag: MVol.B6-204
$AgC_{12}H_{25}N_2S_2$ ...... $[Ag((C_2H_5)_2NC_6H_{12}N(CH_3)CSS)]$ .... Ag: MVol.B7-115
$AgC_{12}H_{25}S$ .... $(AgSC_{12}H_{25})_n$ .... Ag: MVol.B7-5/6
$AgC_{12}H_{26}NS$ .... $[Ag(SC_2H_4NHC_{10}H_{21})]$ .... Ag: MVol.B7-17
$AgC_{12}H_{26}N_2^+$ ...... $[Ag(CH_3N(CH_2)_5)_2]^+$ .... Ag: MVol.B6-73
– .... $[Ag(NHC_5H_9CH_3)_2]^+$ .... Ag: MVol.B6-73
$AgC_{12}H_{26}N_2O_4^+$ .... $[Ag(NH(C_2H_4OC_2H_4OC_2H_4)_2NH)]^+$ .... Ag: MVol.B6-204
$AgC_{12}H_{26}N_8O_4^-$ ..... $[Ag(NH_2C(NH)NHC_3H_6CH(NH_2)COO)_2]^-$ .... Ag: MVol.B6-251
$AgC_{12}H_{27}IP$ .... $[AgI(P(C_4H_9)_3)]_4$ .... Ag: MVol.B7-205/6
$AgC_{12}H_{27}N^+$ .... $[Ag(N(C_4H_9)_3)]^+$ .... Ag: MVol.B6-49
$AgC_{12}H_{27}NO_6P$ ..... $AgNO_3 \cdot (C_4H_9O)_3P$ .... Ag: MVol.B7-248
$AgC_{12}H_{27}NO_6PS$ .... $[Ag(PS(OC_4H_9\text{-}i)_3)]NO_3$ .... Ag: MVol.B7-253
$AgC_{12}H_{27}N_9S_3^+$ ..... $[Ag((CH_3)_2CNNHCSNH_2)_3]^+$ .... Ag: MVol.B6-283
$AgC_{12}H_{27}O_4PS$ ..... $Ag(OSP(OC_4H_9)_3)$ .... Ag: MVol.B7-253
$AgC_{12}H_{27}P^+$ .... $[Ag(P(C_4H_9)_3)]^+$ .... Ag: MVol.B7-205
$AgC_{12}H_{28}N_4O_4^+$ .... $[Ag(OCHN(CH_3)_2)_4]^+$ .... Ag: MVol.B6-322/3
$AgC_{12}H_{28}N_8O_8S_{1.5}$ ... $[Ag(NH_2C(NH)NC(OC_4H_9)NH_2)_2](SO_4)_{1.5}$ ... Ag: MVol.B7-325
$AgC_{12}H_{30}NO_9P_2$ .... $[AgNO_3(P(OC_2H_5)_3)_2]$ .... Ag: MVol.B7-246
$AgC_{12}H_{30}N_2^+$ ...... $[Ag(C_6H_{13}NH_2)_2]^+$ .... Ag: MVol.B6-46
– .... $[Ag(N(C_2H_5)_3)_2]^+$ .... Ag: MVol.B6-49/50
– .... $[Ag(NH(C_3H_7)_2)_2]^+$ .... Ag: MVol.B6-48
$AgC_{12}H_{30}N_2O_2^+$ .... $[Ag((C_2H_5)_2NC_2H_4OH)_2]^+$ .... Ag: MVol.B6-227
– .... $[Ag(NH_2C_6H_{12}OH)_2]^+$ .... Ag: MVol.B6-228
$AgC_{12}H_{30}N_2O_4^+$ .... $[Ag(NH(C_3H_6OH)_2)_2]^+$ .... Ag: MVol.B6-229
$AgC_{12}H_{30}N_2O_6^+$ .... $[Ag(N(C_2H_4OH)_3)_2]^+$ .... Ag: MVol.B6-229/30

$AgC_{12}H_{30}N_3O_3$ ...... $AgNO_3 \cdot 2\ (C_2H_5)_3N$ .... Ag: MVol.B6-50
$AgC_{12}H_{30}N_6S_3{}^+$ ..... $[Ag((CH_3)_2NCSNHCH_3)_3]^+$ .... Ag: MVol.B7-153
$AgC_{12}H_{30}O_2P_2{}^+$ ..... $[Ag((C_2H_5)_2PCH_2CH_2OH)_2]^+$ .... Ag: MVol.B7-203
$AgC_{12}H_{30}O_6P_2{}^+$ ..... $[Ag(P(OC_2H_5)_3)_2]^+$ .... Ag: MVol.B7-246
$AgC_{12}H_{32}N_4{}^+$ ...... $[Ag(C_2H_4(N(CH_3)_2)_2)_2]^+$ .... Ag: MVol.B6-60
$AgC_{12}H_{32}N_5O_3$ ...... $[Ag(C_2H_4(N(CH_3)_2)_2)_2]NO_3$ .... Ag: MVol.B6-60/1
$AgC_{12}H_{33}N_3O_6{}^+$ .... $[Ag(NH(C_2H_4OH)_2)_3]^+$ .... Ag: MVol.B6-228/9
$AgC_{12}H_{36}NO_{15}P_4$ .... $[Ag(P(OCH_3)_3)_4]NO_3$ .... Ag: MVol.B7-244/5
$AgC_{12}H_{36}N_6O_2P_2{}^+$ ... $[Ag(OP(N(CH_3)_2)_3)_2]^+$ .... Ag: MVol.B7-250/1
$AgC_{13}ClH_8N_2O_3$ ..... $Ag(ClC_6H_4NNC_6H_3(OH)COO)$ .... Ag: MVol.B6-287
$AgC_{13}ClH_{10}NOP$ .... $AgCl \cdot (C_6H_5)_2PNCO$ .... Ag: MVol.B7-241
$AgC_{13}ClH_{10}O_4$ ...... $AgClO_4 \cdot C_{13}H_{10}$ .... Ag: MVol.B5-111
$AgC_{13}ClH_{11}N_3$ ...... $[Ag(ClC_6H_4N_3C_6H_4CH_3)]$ .... Ag: MVol.B6-300
$AgC_{13}Cl_2H_9N_4S$ ..... $[Ag(ClC_6H_4NNCSNNHC_6H_4Cl)]$ .... Ag: MVol.B7-183/4
$AgC_{13}F_3H_{12}O_2$ ...... $Ag(CF_3COCHCOCH_3) \cdot C_8H_8$ .... Ag: MVol.B5-75/6
$AgC_{13}F_3H_{16}O_2$ ...... $Ag(CF_3COCHCOCH_3) \cdot C_8H_{12}$ .... Ag: MVol.B5-75/6
$AgC_{13}F_6H_9O_2$ ...... $AgCH(COCF_3)_2 \cdot C_8H_8$ .... Ag: MVol.B5-75/6
$AgC_{13}F_6H_{13}O_2$ ...... $AgCH(COCF_3)_2 \cdot C_8H_{12}$ .... Ag: MVol.B5-75/6
$AgC_{13}F_6H_{15}O_2$ ...... $AgCH(COCF_3)_2 \cdot C_8H_{14}$ .... Ag: MVol.B5-75/6
$AgC_{13}FeH_7N_{1.5}O_4$ ... $Ag(C_{12}H_8N_2)_{0.75}HFe(CO)_4$ .... Ag: MVol.B6-125
$AgC_{13}H_7O$ .... $AgCCC_{11}H_7O$ .... Ag: MVol.B5-19
$AgC_{13}H_8NO_6S_4{}^{2-}$ .... $[Ag(N(C_6H_4SO_3)_2CSS)]^{2-}$ .... Ag: MVol.B7-116
$AgC_{13}H_8N_3OS$ ...... $[Ag(C_3H_2NSNNC_{10}H_6O)]$ .... Ag: MVol.B6-288
$AgC_{13}H_9I_2N_4S$ ...... $[Ag(IC_6H_4NNCSNNHC_6H_4I)]$ .... Ag: MVol.B7-183/4
$AgC_{13}H_9N_2$ .... $AgCCC_4HN_2(CH_3)(C_6H_5) \cdot 0.5\ H_2O$ .... Ag: MVol.B5-20
– .... $Ag(C_6H_5C_7H_4N_2)$ .... Ag: MVol.B6-146
$AgC_{13}H_9N_2O_2$ ...... $Ag(C_6H_5NNC_6H_4COO)$ .... Ag: MVol.B6-287
$AgC_{13}H_9N_2S$ .... $[Ag(C_{10}H_7C_3H_2N_2S)]$ .... Ag: MVol.B7-49
$AgC_{13}H_9N_4O_4$ ...... $[Ag(NO_2C_6H_4NCHNC_6H_4NO_2)]$ .... Ag: MVol.B6-341
$AgC_{13}H_9N_4S$ .... $[Ag(C_7H_4NSN_3C_6H_5)]$ .... Ag: MVol.B6-303
$AgC_{13}H_9O$ .... $AgCCC_{11}H_8OH$ .... Ag: MVol.B5-19
$AgC_{13}H_9O_2S$ .... $[Ag(C_6H_5SC_6H_4COO)]$ .... Ag: MVol.B7-36/7
$AgC_{13}H_9O_3$ .... $[Ag(C_6H_5COCHCOC_4H_3O)]$ .... Ag: MVol.B6-212
$AgC_{13}H_{10}NO$ .... $AgNC_4H_3COCHCHC_6H_5$ .... Ag: MVol.B6-68
$AgC_{13}H_{10}NO_2S_2$ .... $[Ag(C_6H_5SO_2NCSC_6H_5)]$ .... Ag: MVol.B7-126
$AgC_{13}H_{10}N_2{}^+$ ...... $[Ag(CH_3C_{12}H_7N_2)]^+$ .... Ag: MVol.B6-126
$AgC_{13}H_{10}N_2OP$ ..... $AgC(N_2)PO(C_6H_5)_2$ .... Ag: MVol.B5-23
$AgC_{13}H_{10}N_3O$ ...... $[Ag(C_6H_5CON_3C_6H_5)]$ .... Ag: MVol.B6-296
$AgC_{13}H_{10}N_3O_2$ ...... $[Ag(NO_2C_6H_4NCHNC_6H_5)]$ .... Ag: MVol.B6-341
$AgC_{13}H_{10}N_3O_3$ ...... $Ag(C_{13}H_{10}N_2)NO_3$ .... Ag: MVol.B6-180
$AgC_{13}H_{10}N_5$ .... $[Ag(C_6H_5NCN_4C_6H_5)]$ .... Ag: MVol.B6-196
$AgC_{13}H_{11}N_2$ .... $[Ag(C_6H_5NCHNC_6H_5)]$ .... Ag: MVol.B6-341
$AgC_{13}H_{11}N_2O_2$ ...... $AgC_{13}H_{11}N_2O_2$ .... Ag: MVol.B6-247
$AgC_{13}H_{11}N_2O_4S$ .... $Ag(C_6H_5NHCONHC_6H_4SO_3)$ .... Ag: MVol.B6-335
$AgC_{13}H_{11}N_2O_5S$ .... $AgNO_3 \cdot C_{13}H_{11}NO_2S$ .... Ag: MVol.B7-69
$AgC_{13}H_{11}N_4O$ ...... $[Ag(C_6H_5N_3CONHC_6H_5)]$ .... Ag: MVol.B6-301/2
$AgC_{13}H_{11}N_4O_5S$ .... $AgNO_3 \cdot NO_2C_6H_4SNC(C_6H_5)NH_2$ .... Ag: MVol.B6-340, 341
$AgC_{13}H_{11}N_4S$ ...... $[Ag(C_6H_5NNCSNNHC_6H_5)]$ .... Ag: MVol.B7-173, 175/81
$AgC_{13}H_{12}{}^+$ .... $[AgCH_2(C_6H_5)_2]^+$ .... Ag: MVol.B5-102
$AgC_{13}H_{12}N_3$ .... $[Ag(C_6H_5N_3C_6H_4CH_3)]$ .... Ag: MVol.B6-296

Formula	Compound	Reference
$AgC_{13}H_{12}N_3$	$[Ag(C_{10}H_7N_3C_3H_5)]$	Ag: MVol.B6-297
$AgC_{13}H_{12}N_3O$	$[Ag(CH_3OC_6H_4N_3C_6H_5)]$	Ag: MVol.B6-296
$AgC_{13}H_{12}N_3O_3S$	$Ag(C_6H_5NHCSNHC_6H_5)NO_3$	Ag: MVol.B7-159
$AgC_{13}H_{12}N_3O_3S_2$	$[Ag(SCH_2CONHC_6H_4SO_2NHC_5H_4N)]$	Ag: MVol.B7-23/4
$AgC_{13}H_{13}N_2O_3$	$Ag(C_2H_5(C_6H_5)C_4N_2O_3CH_3)$	Ag: MVol.B6-156
–	$Ag_2(C_2H_5(C_6H_5)C_4N_2O_3CH_3)_2$	Ag: MVol.B6-155
$AgC_{13}H_{13}N_3^+$	$[Ag(C_6H_5NHC(NH)NHC_6H_5)]^+$	Ag: MVol.B6-337
$AgC_{13}H_{14}N_2^+$	$[Ag(C_3H_6(C_5H_4N)_2)]^+$	Ag: MVol.B6-123
$AgC_{13}H_{14}N_3$	$[Ag(C_{10}H_7N_3C_3H_7)]$	Ag: MVol.B6-297
$AgC_{13}H_{14}N_3O_2S$	$[Ag(C_4HN_2S(CH_3)_3C_6H_4NO_2)]$	Ag: MVol.B7-64
$AgC_{13}H_{14}N_3O_3$	$AgNO_3 \cdot CH_2(C_6H_4NH_2)_2$	Ag: MVol.B6-64
–	$[Ag(OC_2N_3(C_6H_5)CHCH_3COOC_2H_5)]$	Ag: MVol.B6-183
$AgC_{13}H_{15}NOS^+$	$[Ag(NC_8H_6SC_5H_9O)]^+$	Ag: MVol.B7-48
$AgC_{13}H_{15}N_2OS$	$[Ag(C_8H_5N_2S(C_3H_5)OC_2H_5)]$	Ag: MVol.B7-65/6
$AgC_{13}H_{15}N_2O_4$	$[Ag(C_6H_5CONHCH_2CONHC_3H_6COO)]$	Ag: MVol.B6-267/8
$AgC_{13}H_{15}N_2O_6S$	$Ag(C_{13}H_{15}N_2O_6S)$	Ag: MVol.B6-319
$AgC_{13}H_{15}N_2S$	$Ag(C_4HN_2S(CH_3)_3C_6H_5)]$	Ag: MVol.B7-64
$AgC_{13}H_{16}NO_3$	$Ag(C_4H_9CH(NHCOC_6H_5)COO)$	Ag: MVol.B6-249
$AgC_{13}H_{17}N_4S$	$[Ag(C_5H_{11}N_3C_7H_3NSCH_3)]$	Ag: MVol.B6-302
$AgC_{13}H_{17}O_2S$	$[Ag(C_6H_{13}SC_6H_4COO)]$	Ag: MVol.B7-36
$AgC_{13}H_{18}NO_2S_3$	$[Ag(C_6H_5SO_2NCSSC_6H_{13})]$	Ag: MVol.B7-104
$AgC_{13}H_{18}NS$	$[Ag((C_3H_7)_2C_7H_4NS)]$	Ag: MVol.B7-78
$AgC_{13}H_{18}O_2^+$	$[AgC_{11}H_{15}COOCH_3]^+$	Ag: MVol.B5-89
$AgC_{13}H_{22}N_5O_4S_2$	$Ag(NHC_6H_4SO_2NHC_3H_2NS) \cdot 2\ HOC_2H_4NH_2$	Ag: MVol.B6-265
$AgC_{13}H_{24}O_2^+$	$[AgCH_2CH(CH_2)_9OOCCH_3]^+$	Ag: MVol.B5-48
$AgC_{13}H_{27}NPS$	$[AgSCN(P(C_4H_9)_3)]$	Ag: MVol.B7-206
$AgC_{13}H_{29}NPS_2$	$[Ag(S_2CN(C_3H_7)_2)P(C_2H_5)_3]$	Ag: MVol.B7-202/3
$AgC_{13.5}H_{9.5}N_{0.5}S_2$	$[Ag(C_{10}H_7CSS)] \cdot 0.5\ C_5H_5N$	Ag: MVol.B7-95
$AgC_{14}ClH_9NOS$	$[Ag(CH_3OC_{13}H_6(Cl)NS)]$	Ag: MVol.B7-60
$AgC_{14}ClH_{10}$	$AgCClC(C_6H_5)_2$	Ag: MVol.B5-8
$AgC_{14}ClH_{10}O_4$	$AgClO_4 \cdot C_6H_5CCC_6H_5$	Ag: MVol.B5-104
–	$AgClO_4 \cdot C_{14}H_{10}$	Ag: MVol.B5-110
$AgC_{14}ClH_{12}N_2$	$Ag((CH_3)_2C_{12}H_6N_2)Cl$	Ag: MVol.B6-127
$AgC_{14}ClH_{12}N_2OS$	$Ag(C_6H_5C_8H_6N_2OS) \cdot HCl$	Ag: MVol.B7-65/6
$AgC_{14}ClH_{12}N_2O_5S$	$Ag(C_6H_5C_8H_6N_2OS) \cdot HClO_4$	Ag: MVol.B7-65/6
$AgC_{14}ClH_{12}N_4O_4$	$[Ag(CH_3(CN)C_5H_3N)_2]ClO_4$	Ag: MVol.B6-101
$AgC_{14}ClH_{12}O_4$	$AgClO_4 \cdot C_6H_5CHCHC_6H_5$	Ag: MVol.B5-103
$AgC_{14}ClH_{13}P$	$[AgCl(C_6H_4(CH_2)_2PC_6H_5)]_n$	Ag: MVol.B7-221/2
$AgC_{14}ClH_{14}N_2O_4$	$[Ag(C_2H_3C_5H_4N)_2]ClO_4$	Ag: MVol.B6-90
$AgC_{14}ClH_{14}N_4O_4$	$[AgClO_4 \cdot N_2(CCH_3C_5H_4N)_2]_2$	Ag: MVol.B6-295
–	$[Ag(C_2H_4(NCHC_5H_4N)_2)]ClO_4$	Ag: MVol.B6-280
$AgC_{14}ClH_{14}O_4$	$AgClO_4 \cdot C_6H_5CH_2CH_2C_6H_5$	Ag: MVol.B5-103
$AgC_{14}ClH_{16}N_2$	$Ag((CH_3)_4C_{10}H_4N_2)Cl$	Ag: MVol.B6-121
$AgC_{14}ClH_{16}N_4O_4S_2$	$[Ag(C_6H_5NHCSNH_2)_2]ClO_4$	Ag: MVol.B7-158
$AgC_{14}ClH_{16}O_2S_2$	$[AgCl((CH_3)_2C_5H_2OS)_2]$	Ag: MVol.B7-58
$AgC_{14}ClH_{16}O_4$	$AgClO_4 \cdot 2\ C_7H_8$	Ag: MVol.B5-61
$AgC_{14}ClH_{18}N_2O_4$	$[Ag((CH_3)_2C_5H_3N)_2]ClO_4$	Ag: MVol.B6-88/9
–	$[Ag(C_2H_5C_5H_4N)_2]ClO_4$	Ag: MVol.B6-86
$AgC_{14}ClH_{18}N_2O_6$	$[Ag((CH_3)_2C_5H_3NO)_2]ClO_4$	Ag: MVol.B6-105
–	$[Ag(C_2H_5C_5H_4NO)_2]ClO_4$	Ag: MVol.B6-105

Formula	Compound	Reference
$AgC_{14}ClH_{24}N_4S_2$	$[Ag((CH_3)_3C_4H_3N_2S)_2]Cl$	Ag: MVol.B7-64
$AgC_{14}ClH_{24}O_4$	$AgClO_4 \cdot 2\ C_6H_9CH_3$	Ag: MVol.B5-59
–	$AgClO_4 \cdot 2\ C_6H_{10}CH_2$	Ag: MVol.B5-59
–	$AgClO_4 \cdot 2\ C_7H_{12}$	Ag: MVol.B5-61
$AgC_{14}ClH_{28}N_4O_2S_2$	$[Ag((CH_3)_3C_4H_5N_2OS)_2]Cl$	Ag: MVol.B7-64/5
$AgC_{14}ClH_{38}P_2Si_2$	$[Ag(CH(Si(CH_3)_3)P(CH_3)_3)_2]Cl$	Ag: MVol.B5-24
$AgC_{14}Cl_2H_9N_8S_2$	$Ag(ClC_6H_4CN_4S) \cdot ClC_6H_4CHN_4S \cdot 2\ H_2O$	Ag: MVol.B7-57/8
$AgC_{14}Cl_2H_{32}N_4O_8$	$[Ag((CH_3)_4C_{10}H_{20}N_4)](ClO_4)_2$	Ag: MVol.B7-300/1
$AgC_{14}Cl_3H_8$	$AgCClC(C_6H_4Cl)_2$	Ag: MVol.B5-8
$AgC_{14}CoH_8N_2O_4$	$[Ag(NC_5H_4C_5H_4N)][Co(CO)_4]$	Ag: MVol.B6-120/1
$AgC_{14}FH_{12}N_2$	$Ag((CH_3)_2C_{12}H_6N_2)F$ $= [Ag((CH_3)_2C_{12}H_6N_2)_2][AgF_2]$	Ag: MVol.B6-127
$AgC_{14}F_6H_{16}P$	$AgPF_6 \cdot 2\ C_6H_5CH_3$	Ag: MVol.B5-97
$AgC_{14}F_6H_{16}Sb$	$AgSbF_6 \cdot 2\ C_6H_5CH_3$	Ag: MVol.B5-97
$AgC_{14}H_8NO_3$	$[Ag(C_{13}H_7N(OH)COO)]$	Ag: MVol.B6-116
$AgC_{14}H_8NO_4$	$Ag(C_{13}H_6N(OH)_2COO) \cdot AgOH$	Ag: MVol.B6-116
$AgC_{14}H_8N_2O_8$	$[Ag(C_5H_3N(COO)COOH)_2]$	Ag: MVol.B7-281/2, 285
–	$[Ag(C_5H_3N(COO)COOH)_2] \cdot H_2O$	Ag: MVol.B7-283/4, 285
–	$[Ag(C_5H_3N(COO)COOH)_2] \cdot 2\ H_2O$	Ag: MVol.B7-280/1, 282/3
–	$Ag(C_5H_3N(COO)COOH)_2 \cdot 4\ H_2O$	Ag: MVol.B7-282
–	$Ag(C_5H_3N(COO)COOH)_2$ solutions $Ag(C_5H_3N(COO)COOH)_2$-H_2O	Ag: MVol.B7-284
$AgC_{14}H_9NNaO_4$	$NaAg(OC_6H_4CHNC_6H_3(OH)COO)$	Ag: MVol.B6-284
$AgC_{14}H_9N_2O_2S$	$[Ag(C_4H_3SC_3HN_2(C_6H_5)COO)]$	Ag: MVol.B7-70
$AgC_{14}H_9O_6S$	$[Ag(S(C_6H_3(OH)COO)_2H)] \cdot 2\ H_2O$	Ag: MVol.B7-39
$AgC_{14}H_{10}^+$	$[AgC_{14}H_{10}]^+$	Ag: MVol.B5-109
$AgC_{14}H_{10}NOS$	$[Ag(CH_3OC_{13}H_7NS)]$	Ag: MVol.B7-59
$AgC_{14}H_{10}N_2^+$	$[Ag(C_6H_5CN)_2]^+$	Ag: MVol.B6-352
$AgC_{14}H_{10}N_3$	$[Ag(C_2N_3(C_6H_5)_2)]$	Ag: MVol.B6-186
–	$[AgNC_4H_2(C_5H_4N)_2]$	Ag: MVol.B6-107
$AgC_{14}H_{10}N_3OS$	$[Ag((C_6H_5)_2C_2N_3OS)]$	Ag: MVol.B7-56
$AgC_{14}H_{10}N_3O_3$	$AgNO_3 \cdot 2\ C_6H_5CN$	Ag: MVol.B6-352
$AgC_{14}H_{10}N_3O_3S$	$AgNO_3 \cdot (C_6H_5)_2C_2N_2S$	Ag: MVol.B7-81
$AgC_{14}H_{10}N_3O_4$	$AgNO_3 \cdot (C_6H_5)_2C_2N_2O$	Ag: MVol.B6-203
$AgC_{14}H_{11}NO^+$	$[Ag(CH_3CN)(C_{12}H_8O)]^+$	Ag: MVol.B6-222
$AgC_{14}H_{11}NO_2^+$	$[Ag(CH_3CN)(C_{12}H_8O_2)]^+$	Ag: MVol.B6-222
$AgC_{14}H_{11}N_2O$	$Ag(C_6H_5OCH_2C_7H_4N_2)$	Ag: MVol.B6-146
$AgC_{14}H_{11}N_2OS$	$Ag(C_6H_5C_8H_6N_2OS) \cdot HX$ (X = Cl, ClO_4, Br, I)	Ag: MVol.B7-65/6
$AgC_{14}H_{11}N_2O_2$	$Ag(C_6H_5CONNHCOC_6H_5)$	Ag: MVol.B6-331
$AgC_{14}H_{11}N_2O_3$	$Ag(C_6H_5NHCONHC_6H_4COO)$	Ag: MVol.B6-335
$AgC_{14}H_{11}N_4OS$	$[Ag(CH_3OC_7H_3NSN_3C_6H_5)]$	Ag: MVol.B6-303
$AgC_{14}H_{11}N_4O_2$	$Ag(C_6H_5C_3HN_3O_2) \cdot C_5H_5N$	Ag: MVol.B6-308/12
$AgC_{14}H_{11}N_4S$	$[Ag(CH_3C_7H_3NSN_3C_6H_5)]$	Ag: MVol.B6-302, 303
$AgC_{14}H_{11}N_6$	$[Ag(C_6H_5N_3CCN_3HC_6H_5)]$	Ag: MVol.B6-297
$AgC_{14}H_{11}N_8S_2$	$Ag(C_6H_5CN_4S) \cdot C_6H_5CHN_4S \cdot 2\ H_2O$	Ag: MVol.B7-57
$AgC_{14}H_{11}O_2S$	$[Ag(C_6H_5CH_2SC_6H_4COO)]$	Ag: MVol.B7-37
$AgC_{14}H_{12}^+$	$[AgC_2H_2(C_6H_5)_2]^+$	Ag: MVol.B5-102
–	$[AgC_{14}H_{12}]^+$	Ag: MVol.B5-109
$AgC_{14}H_{12}IN_2$	$Ag((CH_3)_2C_{12}H_6N_2)I$	Ag: MVol.B6-127

$AgC_{14}H_{12}IN_2OS$ ..... $Ag(C_6H_5C_8H_6N_2OS) \cdot HI$ .... Ag: MVol.B7-65/6
$AgC_{14}H_{12}NO_3$ ...... $AgNO_3 \cdot C_6H_5C_8H_7$ .... Ag: MVol.B5-106
$AgC_{14}H_{12}NS$ .... $[Ag(CH_3C_7H_4NSC_6H_5)]$ .... Ag: MVol.B7-78
$AgC_{14}H_{12}N_2^+$ ...... $[Ag((CH_3)_2C_{12}H_6N_2)]^+$ .... Ag: MVol.B6-126, 128
$AgC_{14}H_{12}N_3O_3$ ...... $[AgNO_3((CH_3)_2C_{12}H_6N_2)]$ .... Ag: MVol.B6-126
– .... $AgNO_3 \cdot C_2(C_6H_4NH_2)_2$ .... Ag: MVol.B6-65
$AgC_{14}H_{12}N_5O_3$ ...... $[Ag(CHNCHNC_5H_4)_2]NO_3$ .... Ag: MVol.B6-141
– .... $Ag(C_6H_3(NH_2)_2NNC_6H_4NHCOCOO) \cdot 3\ H_2O$ Ag: MVol.B6-287
– .... $[Ag(NHC_6H_4CHN)_2]NO_3$ .... Ag: MVol.B6-132
– .... $[Ag(NHC_6H_4NCH)_2]NO_3$ .... Ag: MVol.B6-144
$AgC_{14}H_{12}N_5O_3S$ .... $AgNO_3 \cdot NH_2C_2N_2SN(C_6H_5)_2$ .... Ag: MVol.B7-80
$AgC_{14}H_{12}N_5O_4$ ...... $Ag(ONC_4H_2N_2O_3) \cdot 2\ C_5H_5N$ .... Ag: MVol.B6-315
$AgC_{14}H_{13}NO_3P$ ..... $[Ag(CH_2CHP(C_6H_5)_2)]NO_3$ .... Ag: MVol.B7-222/3
$AgC_{14}H_{13}N_2O_3S_2$ .... $[Ag(SCH_2CONHC_6H_4SO_2C_6H_4NH_2)]$ .... Ag: MVol.B7-23/4
$AgC_{14}H_{13}N_4O_3$ ...... $AgNO_3 \cdot C_{10}H_7C_2N_3(CH_3)_2$ .... Ag: MVol.B6-183
$AgC_{14}H_{13}N_4O_3S$ .... $Ag(CH_3C_6H_4N_2C_6H_3(CH_3)N_2SO_3)$ .... Ag: MVol.B6-294
$AgC_{14}H_{14}^+$ .... $[AgC_2H_4(C_6H_5)_2]^+$ .... Ag: MVol.B5-102
$AgC_{14}H_{14}NO_3S$ ..... $AgNO_3 \cdot S(CH_2C_6H_5)_2$ .... Ag: MVol.B7-14
$AgC_{14}H_{14}NO_3S_2$ .... $AgNO_3 \cdot C_6H_5CH_2SSCH_2C_6H_5$ .... Ag: MVol.B7-15
$AgC_{14}H_{14}N_2O_4^+$ .... $[Ag(NC_5H_3(CH_3)COOH)_2]^+$ .... Ag: MVol.B6-102
– .... $[Ag(NC_5H_4COOCH_3)_2]^+$ .... Ag: MVol.B6-103
$AgC_{14}H_{14}N_3$ .... $[Ag(CH_3C_6H_4N_3C_6H_4CH_3)]$ .... Ag: MVol.B6-296
$AgC_{14}H_{14}N_3O$ ...... $[Ag(CH_3OC_6H_4N_3C_6H_4CH_3)]$ .... Ag: MVol.B6-296
$AgC_{14}H_{14}N_3O_2$ ...... $[Ag(CH_3OC_6H_4N_3C_6H_4OCH_3)]$ .... Ag: MVol.B6-296
$AgC_{14}H_{14}N_3O_3$ ...... $AgNO_3 \cdot C_2H_2(C_6H_4NH_2)_2$ .... Ag: MVol.B6-65
– .... $[AgNO_3(C_2H_3C_5H_4N)_2]$ .... Ag: MVol.B6-90
$AgC_{14}H_{14}N_3O_3S$ .... $Ag(CH_3C_6H_4N(CH_3)NNC_6H_4SO_3)$ .... Ag: MVol.B6-303
– .... $Ag((CH_3)_2NC_6H_4NNC_6H_4SO_3) \cdot 2\ H_2O$ ..... Ag: MVol.B6-290
$AgC_{14}H_{14}N_4^+$ ...... $[Ag(C_2H_4(NCHC_5H_4N)_2)]^+$ .... Ag: MVol.B6-280
$AgC_{14}H_{14}N_4O_2^+$ .... $[Ag(C_{10}H_2N_4O_2(CH_3)_4)]^+$ .... Ag: MVol.B6-173
$AgC_{14}H_{14}N_5O_3$ ...... $[AgNO_3 \cdot N_2(CCH_3C_5H_4N)_2]_2$ .... Ag: MVol.B6-295
$AgC_{14}H_{14}O_8S_2^-$ ..... $[Ag(CH_3OC_6H_4SO_3)_2]^-$ .... Ag: MVol.B6-218
$AgC_{14}H_{15}N^+$ .... $[Ag(NH(CH_2C_6H_5)_2)]^+$ .... Ag: MVol.B6-54
$AgC_{14}H_{15}N_2OS_2$ .... $[Ag((C_2H_5)_2NC_6H_4CHC_3NOS_2)]$ .... Ag: MVol.B7-75
$AgC_{14}H_{15}N_2O_3$ ...... $Ag(C_6H_3(CH_3)_2(OH)_2NNC_6H_4O) \cdot 2\ C_5H_5N$ .. Ag: MVol.B6-285
$AgC_{14}H_{16}^+$ .... $[Ag(C_6H_5CH_3)_2]^+$ .... Ag: MVol.B5-97
$AgC_{14}H_{16}^{2+}$ .... $[Ag(C_7H_8)_2]^{2+}$ .... Ag: MVol.B5-61
$AgC_{14}H_{16}IN_2$ .... $Ag((CH_3)_4C_{10}H_4N_2)I$ .... Ag: MVol.B6-121
$AgC_{14}H_{16}NO_3$ ...... $AgNO_3 \cdot C_{14}H_{16}$ .... Ag: MVol.B5-86
$AgC_{14}H_{16}NO_7$ ...... $AgNO_3 \cdot C_{10}H_{10}(COOCH_3)_2$ .... Ag: MVol.B5-90
$AgC_{14}H_{16}N_2^+$ ...... $[Ag(C_4H_8(C_5H_4N)_2)]^+$ .... Ag: MVol.B6-123
$AgC_{14}H_{16}N_3$ .... $[Ag(C_{10}H_7N_3C_4H_9)]$ .... Ag: MVol.B6-297
$AgC_{14}H_{16}N_3O_3$ ...... $Ag((CH_3)_4C_{10}H_4N_2)NO_3$ .... Ag: MVol.B6-121
– .... $AgNO_3 \cdot C_2H_4(C_6H_4NH_2)_2$ .... Ag: MVol.B6-64
$AgC_{14}H_{16}N_3O_5$ ...... $[AgC_{14}H_{16}N_3O_5]$ .... Ag: MVol.B6-269
$AgC_{14}H_{16}N_4O_3S^+$ ... $[Ag(HOOC(CH_2)_2CONHC_6H_4(CH)_3NNHCSNH_2)]^+$
Ag: MVol.B7-171
$AgC_{14}H_{16}N_5O_3S_2$ .... $[Ag(C_6H_5NHCSNH_2)_2]NO_3$ .... Ag: MVol.B7-158
$AgC_{14}H_{16}N_5O_7$ ...... $AgNO_3 \cdot 2\ (CH_3)_2(NO_2)C_5H_2N$ .... Ag: MVol.B6-101
$AgC_{14}H_{17}N_2OS$ ..... $[Ag(C_4HN_2S(CH_3)_3C_6H_4OCH_3)]$ .... Ag: MVol.B7-64

Formula	Compound	Reference
$AgC_{14}H_{17}N_2S$	$[Ag(C_4HN_2S(CH_3)_3CH_2C_6H_5)]$	Ag: MVol.B7-64
–	$[Ag(C_4HN_2S(CH_3)_3C_6H_4CH_3)]$	Ag: MVol.B7-64
$AgC_{14}H_{17}N_3O_3PS$	$[Ag((CH_3)_2NNHP(S)(C_6H_5)_2)]NO_3$	Ag: MVol.B7-254
$AgC_{14}H_{18}N_2^+$	$[Ag(CH_3C_6H_4NH_2)_2]^+$	Ag: MVol.B6-52
–	$[Ag((CH_3)_2C_5H_3N)_2]^+$	Ag: MVol.B6-87/8
–	$[Ag(C_2H_5C_5H_4N)_2]^+$	Ag: MVol.B6-86
–	$[Ag(C_6H_5CH_2NH_2)_2]^+$	Ag: MVol.B6-53/4
–	$[Ag(C_6H_5NHCH_3)_2]^+$	Ag: MVol.B6-53
$AgC_{14}H_{18}N_2O_2^+$	$[Ag((CH_3)_2C_5H_3NO)_2]^+$	Ag: MVol.B6-105
–	$[Ag(C_2H_5C_5H_4NO)_2]^+$	Ag: MVol.B6-105
$AgC_{14}H_{18}N_2O_8^{3-}$	$[Ag(C_6H_{10}(N(CH_2COO)_2)_2)]^{3-}$	Ag: MVol.B6-278
$AgC_{14}H_{18}N_3O_3$	$[AgNO_3((CH_3)_2C_5H_3N)_2]$	Ag: MVol.B6-88
$AgC_{14}H_{18}N_3O_{10}^{4-}$	$[Ag(OCOCH_2N(C_2H_4N(CH_2COO)_2)_2)]^{4-}$	Ag: MVol.B6-278
$AgC_{14}H_{19}N_2O$	$[Ag((CH_3)_2C_5H_3N)_2]OH$	Ag: MVol.B6-87/8
–	$[Ag(C_2H_5C_5H_4N)_2]OH$	Ag: MVol.B6-86
$AgC_{14}H_{19}N_4O_2$	$[Ag(NH_2(CH_3)C_5H_3N)_2CH_3COO]$	Ag: MVol.B6-96
$AgC_{14}H_{20}NO_8$	$AgNO_3 \cdot C_{14}H_{20}O_5$	Ag: MVol.B6-219
$AgC_{14}H_{20}N_2O_{10}^{3-}$	$[Ag(C_2H_4(OC_2H_4N(CH_2COO)_2)_2)]^{3-}$	Ag: MVol.B6-278
$AgC_{14}H_{20}N_4^+$	$[Ag(CH_3(NH_2CH_2)C_5H_3N)_2]^+$	Ag: MVol.B6-96/7
–	$[Ag(C_2H_5(CH_2CH)C_3H_2N_2)_2]^+$	Ag: MVol.B6-138/9
$AgC_{14}H_{20}N_4O_2^+$	$[Ag(HOC_2H_4(CH_2CH)C_3H_2N_2)_2]^+$	Ag: MVol.B6-138/9
$AgC_{14}H_{20}N_5O_3$	$[Ag(C_2H_5(CH_2CH)C_3H_2N_2)_2]NO_3$	Ag: MVol.B6-140
$AgC_{14}H_{20}O_8S_4^{3-}$	$[Ag(C_3H_6(SCH_2COO)_2)_2]^{3-}$	Ag: MVol.B7-39/40
$AgC_{14}H_{20}P$	$AgCCC_6H_5 \cdot P(C_2H_5)_3$	Ag: MVol.B5-18
$AgC_{14}H_{21}N_2O_{10}^{2-}$	$[Ag(C_2H_4(OC_2H_4N(CH_2COO)_2)_2H)]^{2-}$	Ag: MVol.B6-278
$AgC_{14}H_{22}^+$	$[AgC_6H_2(C_2H_5)_4]^+$	Ag: MVol.B5-98
–	$[AgC_6H_4(C_4H_9)_2]^+$	Ag: MVol.B5-98
$AgC_{14}H_{22}N_4^{3+}$	$[Ag(C_5H_5NC_2H_4NH_2)_2]^{3+}$	Ag: MVol.B6-97/8
$AgC_{14}H_{22}O_4S_2^-$	$[Ag(CH_2CH(CH_2)_3SCH_2COO)_2]^-$	Ag: MVol.B5-50/1
$AgC_{14}H_{22}O_4Se_2^-$	$[Ag(CH_2CH(CH_2)_3SeCH_2COO)_2]^-$	Ag: MVol.B5-50/1
		Ag: MVol.B7-191/3, 196
$AgC_{14}H_{23}N_6O_2S$	$Ag(NHC_6H_4SO_2NHC_4H_3N_2) \cdot 2\ C_2H_5NH_2$	Ag: MVol.B6-264
$AgC_{14}H_{23}O_2$	$AgCCCH(C_4H_9)(CH_2)_2OC_5H_9O$	Ag: MVol.B5-16
$AgC_{14}H_{23}O_8S_4$	$[Ag(C_7H_{11}O_4S_2)(C_7H_{12}O_4S_2)]$	Ag: MVol.B7-41
$AgC_{14}H_{24}N_2O_4^-$	$[Ag(C_5H_9CH(NH_2)COO)_2]^-$	Ag: MVol.B6-250
$AgC_{14}H_{24}N_2S_4$	$[Ag(SCSN(CH_2)_6)_2]$	Ag: MVol.B7-317
$AgC_{14}H_{24}N_8^+$	$[Ag((CH_2)_6CN_4)_2]^+$	Ag: MVol.B6-194
$AgC_{14}H_{24}N_9O_5$	$[Ag(C_7H_{12}N_4O)_2]NO_3$	Ag: MVol.B6-140
$AgC_{14}H_{24}N_9O_7$	$AgNO_3 \cdot 2\ NH_2C_3N_3(OC_2H_5)_2$	Ag: MVol.B6-189
$AgC_{14}H_{24}O_4S_2^+$	$[Ag(CH_2CH(CH_2)_3SCH_2COOH)_2]^+$	Ag: MVol.B5-50/1
$AgC_{14}H_{24}O_4Se_2^+$	$[Ag(CH_2CH(CH_2)_3SeCH_2COOH)_2]^+$	Ag: MVol.B5-50/1
		Ag: MVol.B7-193
$AgC_{14}H_{24}O_8S_4^+$	$[Ag(C_3H_6(SCH_2COOH)_2)_2]^+$	Ag: MVol.B7-40
$AgC_{14}H_{25}N_4O_4S$	$Ag(NHC_6H_4SO_2NH_2) \cdot 2\ NH(CH_2CH_2)_2O$	Ag: MVol.B6-263
$AgC_{14}H_{25}N_8O_2S$	$Ag(NHC_6H_4SO_2NHC_4H_3N_2) \cdot 2\ C_2H_8N_2$	Ag: MVol.B6-264
$AgC_{14}H_{26}N_2^+$	$[Ag(C_7H_{13}N)_2]^+$	Ag: MVol.B6-73
$AgC_{14}H_{26}N_3O_3$	$AgNO_3 \cdot 2\ C_7H_{13}N$	Ag: MVol.B6-73
$AgC_{14}H_{26}O_4S_2^-$	$[Ag(C_5H_{11}SCH_2COO)_2]^-$	Ag: MVol.B7-25/6
$AgC_{14}H_{28}NO_3S_2$	$AgNO_3 \cdot 2\ C_2H_5C_5H_9S$	Ag: MVol.B7-84
$AgC_{14}H_{28}N_2S_4$	$[Ag(SCSN(C_3H_7)_2)_2]$	Ag: MVol.B7-317/8

Formula	Compound	Reference
$AgC_{14}H_{28}N_5O_5S_2$	$[Ag((CH_3)_3C_4H_5N_2OS)_2]NO_3$	Ag: MVol.B7-64/5
$AgC_{14}H_{30}NO_3S$	$AgNO_3 \cdot S(C_7H_{15})_2$	Ag: MVol.B7-12
$AgC_{14}H_{30}N_2^+$	$[Ag(C_2H_5N(CH_2)_5)_2]^+$	Ag: MVol.B6-73
–	$[Ag(NHC_5H_8(CH_3)_2)_2]^+$	Ag: MVol.B6-73
–	$[Ag(NHC_5H_9C_2H_5)_2]^+$	Ag: MVol.B6-73
$AgC_{14}H_{32}NO_3S_2$	$AgNO_3 \cdot 2\ C_3H_7SC_4H_9$	Ag: MVol.B7-11
$AgC_{14}H_{32}N_2O_4^+$	$[Ag(C_2H_4(N(CH_2CHOHCH_3)_2)_2)]^+$	Ag: MVol.B6-230
$AgC_{15}ClCoH_{21}O_{10}$	$[Co(CH(COCH_3)_2)_3] \cdot AgClO_4$	Ag: MVol.B6-213
$AgC_{15}ClCrH_{21}O_{10}$	$[Cr(CH(COCH_3)_2)_3] \cdot AgClO_4$	Ag: MVol.B6-213
$AgC_{15}ClFeH_{21}O_{10}$	$[Fe(CH(COCH_3)_2)_3] \cdot AgClO_4$	Ag: MVol.B6-213
–	$[Fe(CH(COCH_3)_2)_3] \cdot AgClO_4 \cdot H_2O$	Ag: MVol.B6-213
$AgC_{15}ClH_{10}O_4$	$AgClO_4 \cdot C_{15}H_{10}$	Ag: MVol.B5-115
$AgC_{15}ClH_{11}N_3O_4$	$Ag(NC_5H_3(C_5H_4N)_2)ClO_4$	Ag: MVol.B6-122
–	$Ag(NC_5H_4CHNC_9H_6N)ClO_4$	Ag: MVol.B6-281
$AgC_{15}ClH_{12}N_4O_2$	$Ag(CH_3C_3N_3O_2C_6H_4Cl) \cdot C_5H_5N$	Ag: MVol.B6-309/12
$AgC_{15}ClH_{15}N_3O_7$	$Ag(C_5H_5NO)_2ClO_4 \cdot C_5H_5NO$	Ag: MVol.B6-104
$AgC_{15}ClH_{36}N_6O_4S_3$	$[Ag(C_2H_5NHCSNHC_2H_5)_3]ClO_4$	Ag: MVol.B7-155
$AgC_{15}Cl_2H_9N_4S$	$[Ag(NHC_3NS(C_6H_5)NNC_6H_3Cl_2)]$	Ag: MVol.B6-289
$AgC_{15}Cl_2H_{11}N_3O_6$	$Ag(NC_5H_3(C_5H_4N)_2)(ClO_3)_2$	Ag: MVol.B7-296
$AgC_{15}Cl_2H_{11}N_3O_8$	$Ag(NC_5H_3(C_5H_4N)_2)(ClO_4)_2$	Ag: MVol.B7-296
$AgC_{15}Cl_2H_{13}N_3O_9$	$[Ag(H_2O)(NC_5H_3(C_5H_4N)_2)](ClO_4)_2$	Ag: MVol.B7-296
$AgC_{15}F_6H_{11}N_3P$	$Ag(NC_5H_3(C_5H_4N)_2)PF_6$	Ag: MVol.B6-122
–	$Ag(NC_5H_4CHNC_9H_6N)PF_6$	Ag: MVol.B6-281
$AgC_{15}H_9N_2O_2$	$Ag(C_6H_5NNC_9H_4O_2) \cdot C_5H_5N$	Ag: MVol.B6-216
$AgC_{15}H_9N_2O_3$	$[Ag(C_9H_4O_3NNC_6H_5)]$	Ag: MVol.B6-288
–	$Ag(HOC_6H_4NNC_9H_4O_2) \cdot 2\ C_5H_5N$	Ag: MVol.B6-216
$AgC_{15}H_9N_4O_4$	$[Ag(C_6H_5C_3N_3O_2C_6H_4NO_2)]$	Ag: MVol.B6-308/12
$AgC_{15}H_{10}N_3O$	$[Ag(ONC_3N_2(C_6H_5)_2)]$	Ag: MVol.B6-317
$AgC_{15}H_{10}N_3O_2$	$[Ag(C_3N_3O_2(C_6H_5)_2)]$	Ag: MVol.B6-309/12
$AgC_{15}H_{10}N_3O_2S$	$Ag(C_6H_5CONC_2N_2OSC_6H_5)$	Ag: MVol.B7-82
$AgC_{15}H_{10}N_3O_3$	$[Ag((C_6H_5)_2C_3N_3O_3)]$	Ag: MVol.B6-188
$AgC_{15}H_{10}N_5O_2$	$Ag(C_6H_5N_2C_2N_3(C_6H_5)COO)$	Ag: MVol.B6-287
$AgC_{15}H_{11}N_2$	$Ag((C_6H_5)_2C_3HN_2)$	Ag: MVol.B6-142
$AgC_{15}H_{11}N_2O_2$	$Ag(C_7H_4N_2CH_2COOC_6H_5)$	Ag: MVol.B6-146
$AgC_{15}H_{11}N_3^+$	$[Ag(NC_5H_3(C_5H_4N)_2)]^+$	Ag: MVol.B6-122
–	$[Ag(NC_5H_4CHNC_9H_6N)]^+$	Ag: MVol.B6-281
$AgC_{15}H_{11}N_3O_6S_2$	$Ag(NC_5H_3(C_5H_4N)_2)S_2O_6$	Ag: MVol.B7-296
$AgC_{15}H_{11}N_3O_8S_2$	$Ag(NC_5H_3(C_5H_4N)_2)S_2O_8$	Ag: MVol.B7-296
–	$Ag(NC_5H_3(C_5H_4N)_2)S_2O_8 \cdot H_2O$	Ag: MVol.B7-296/7
$AgC_{15}H_{11}N_4O_3$	$Ag(NC_5H_3(C_5H_4N)_2)NO_3$	Ag: MVol.B6-122
–	$Ag((C_6H_5NN)_2CCOCOOH)$	Ag: MVol.B6-304
$AgC_{15}H_{11}N_4S$	$[Ag(NHC_3NS(C_6H_5)NNC_6H_5)]$	Ag: MVol.B6-289
$AgC_{15}H_{11}N_5O_6$	$Ag(NC_5H_3(C_5H_4N)_2)(NO_3)_2$	Ag: MVol.B7-296
$AgC_{15}H_{11}OS$	$[Ag(C_6H_5COCHCSC_6H_5)]$	Ag: MVol.B7-19/20
$AgC_{15}H_{11}O_2$	$[Ag(C_6H_5COCHCOC_6H_5)]$	Ag: MVol.B6-212
$AgC_{15}H_{12}NO_4$	$AgNO_3 \cdot C_8H_7COC_6H_5$	Ag: MVol.B5-118
		Ag: MVol.B6-210
$AgC_{15}H_{13}I_2N_4S$	$[Ag(CS(NNC_6H_3ICH_3)_2H)]$	Ag: MVol.B7-185
$AgC_{15}H_{13}NO_7^-$	$[Ag(HOC_{10}H_5O_2(CH_3)N(CH_2COO)_2)]^-$	Ag: MVol.B6-275
$AgC_{15}H_{13}N_2O_2$	$Ag(C_{10}H_7(CH_3)_2C_3N_2O_2)$	Ag: MVol.B6-148

Formula	Compound	Reference
$AgC_{15}H_{13}N_3O^{2+}$	$[Ag(H_2O)(NC_5H_3(C_5H_4N)_2)]^{2+}$	Ag: MVol.B7-296
$AgC_{15}H_{13}N_4$	$Ag(NC)_2CC_6H_4C(CN)NC_4H_9$	Ag: MVol.B6-356
$AgC_{15}H_{13}N_4O$	$Ag(CH_3COC(NNC_6H_5)_2)$	Ag: MVol.B6-304
$AgC_{15}H_{13}N_4S$	$[Ag(C_6H_5CH_2N_3C_7H_3NSCH_3)]$	Ag: MVol.B6-302
$AgC_{15}H_{14}NO_2$	$[Ag((C_6H_5)_2NCH_2CH_2COO)]$	Ag: MVol.B6-244
$AgC_{15}H_{14}NS_2$	$[Ag(N(CH_2C_6H_5)_2CSS)]$	Ag: MVol.B7-117
$AgC_{15}H_{14}N_5$	$[Ag(N(CH_2C_6H_5)_2CN_4)]$	Ag: MVol.B6-192/3
$AgC_{15}H_{14}O_2P$	$[Ag((C_6H_5)_2PCH_2CH_2COO)]$	Ag: MVol.B7-222
$AgC_{15}H_{15}N_2O_2$	$[Ag(CH(NC_6H_4OCH_3)_2)]$	Ag: MVol.B6-341
$AgC_{15}H_{15}N_4O_3$	$AgNO_3 \cdot 3\ C_5H_5N$	Ag: MVol.B6-76/7, 78
$AgC_{15}H_{15}N_4S$	$[Ag(CH_3C_6H_4NNCSNNHC_6H_4CH_3)]$	Ag: MVol.B7-184/5
$AgC_{15}H_{15}O_2P^+$	$[Ag((C_6H_5)_2PCH_2CH_2COOH)]^+$	Ag: MVol.B7-222
$AgC_{15}H_{16}N_3O$	$[Ag(C_6H_5N_3CH_2C(CH_3)(C_6H_5)OH)]$	Ag: MVol.B6-301
$AgC_{15}H_{17}N_2O_4$	$AgNO_3 \cdot (CH_3)_3C_{12}H_8NO$	Ag: MVol.B6-116
–	$Ag(C_6H_5C_3H_2N_2(COOC_2H_5)_2)$	Ag: MVol.B6-135
$AgC_{15}H_{18}N_2^+$	$[Ag(C_5H_{10}(C_5H_4N)_2)]^+$	Ag: MVol.B6-123
$AgC_{15}H_{18}N_3$	$[Ag(C_{10}H_7N_3C_5H_{11})]$	Ag: MVol.B6-297
$AgC_{15}H_{18}N_3O_5$	$[AgC_{15}H_{18}N_3O_5]$	Ag: MVol.B6-269
$AgC_{15}H_{18}N_6^+$	$[Ag(NH_2C_5H_4N)_3]^+$	Ag: MVol.B6-93
$AgC_{15}H_{18}N_7O_3$	$AgNO_3 \cdot 3\ NH_2C_5H_4N$	Ag: MVol.B6-93
$AgC_{15}H_{19}N_2^{2+}$	$[AgH(C_5H_{10}(C_5H_4N)_2)]^{2+}$	Ag: MVol.B6-123
$AgC_{15}H_{19}N_2O_5S$	$[AgC_{15}H_{19}N_2O_5S]$	Ag: MVol.B6-267/8
$AgC_{15}H_{20}NO_6S$	$[AgC_{15}H_{20}NSO_6]$	Ag: MVol.B6-263
$AgC_{15}H_{22}NO_2S_3$	$[Ag(C_6H_5SO_2NCSSC_8H_{17})]$	Ag: MVol.B7-104
$AgC_{15}H_{22}N_3O_3S$	$[Ag((C_2H_5)_2NC_2H_4OC_7H_5N_2S)]CH_3COO$	Ag: MVol.B7-52
$AgC_{15}H_{23}N_2$	$Ag((C_6H_{11})_2C_3HN_2)$	Ag: MVol.B6-142
$AgC_{15}H_{24}^+$	$[AgC_6H_3(C_3H_7)_3]^+$	Ag: MVol.B5-98
$AgC_{15}H_{24}MgN_{11}$	$MgAg(CN)_3 \cdot 2\ (CH_2)_6N_4 \cdot 9\ H_2O$	Ag: MVol.B6-200
$AgC_{15}H_{24}NO_3$	$AgNO_3 \cdot (CH_3)_2C_{10}H_{12}C(CH_3)_2$	Ag: MVol.B5-70/1
–	$AgNO_3 \cdot C_{15}H_{24}$	Ag: MVol.B5-80/1
$AgC_{15}H_{25}N_3S_4$	$Ag(SCSN(C_2H_5)_2)_2 \cdot C_5H_5N$	Ag: MVol.B7-318
$AgC_{15}H_{33}NPS_2$	$[Ag(S_2CN(C_4H_9)_2)P(C_2H_5)_3]$	Ag: MVol.B7-202/3
$AgC_{15}H_{33}N_2S$	$[Ag(SC(CH_3)_2CH_2NHC_3H_6N(C_4H_9)_2)]$	Ag: MVol.B7-17
$AgC_{15}H_{35}OP_2S_2$	$[(C_2H_5)_3P]_2Ag[S_2COC_2H_5]$	C: MVol.D4-257
$AgC_{15}H_{36}N_6S_3^+$	$[Ag(C_2H_5NHCSNHC_2H_5)_3]^+$	Ag: MVol.B7-155
$AgC_{15}H_{36}N_7O_3S_3$	$Ag((C_2H_5)_2NCSNH_2)_3NO_3$	Ag: MVol.B7-155
$AgC_{16}ClH_{10}O_4$	$AgClO_4 \cdot C_{16}H_{10}$	Ag: MVol.B5-116
$AgC_{16}ClH_{14}N_2O_4S_2$	$AgClO_4 \cdot 2\ CH_3C_7H_4NS$	Ag: MVol.B7-77
$AgC_{16}ClH_{14}O_4$	$AgClO_4 \cdot C_6H_5(CH)_4C_6H_5$	Ag: MVol.B5-104
$AgC_{16}ClH_{20}N_4O_4S_2$	$[Ag(CH_3C_6H_4NHCSNH_2)_2]ClO_4$	Ag: MVol.B7-160
$AgC_{16}ClH_{20}O_4$	$AgClO_4 \cdot 2\ C_6H_4(CH_3)_2$	Ag: MVol.B5-98/9
$AgC_{16}ClH_{24}N_4O_8$	$[Ag(C_4H_9OCOC_3H_3N_2)_2]ClO_4$	Ag: MVol.B6-140
$AgC_{16}ClH_{28}O_4$	$AgClO_4 \cdot 2\ C_8H_{14}$	Ag: MVol.B5-65
$AgC_{16}ClH_{36}N_4O_4$	$[Ag((CH_3)_6C_{10}H_{18}N_4)]ClO_4$	Ag: MVol.B6-201/2
$AgC_{16}ClH_{48}N_8O_{10}P_4$	$[Ag(O_3P_2(N(CH_3)_2)_4)_2]ClO_4$	Ag: MVol.B7-252
$AgC_{16}Cl_2H_{12}O_4S_2^-$	$[Ag(ClC_6H_4SCH_2COO)_2]^-$	Ag: MVol.B7-27/8
$AgC_{16}Cl_2H_{12}O_4Se_2^-$	$[Ag(ClC_6H_4SeCH_2COO)_2]^-$	Ag: MVol.B7-191/2, 194
$AgC_{16}Cl_2H_{16}IrN_{10}O_6$	$[Ir(C_4H_4N_2)_3(C_4H_4N_2Ag)Cl_2](NO_3)_2$	Ir: SVol.2-77
$AgC_{16}Cl_2H_{32}N_4O_8$	$Ag((CH_3)_6C_{10}H_{14}N_4)(ClO_4)_2$	Ag: MVol.B7-300/1
$AgC_{16}Cl_2H_{36}N_4^+$	$[AgCl_2((CH_3)_6C_{10}H_{18}N_4)]^+$	Ag: MVol.B7-326

$AgC_{16}Cl_2H_{36}N_4O_8$ ... $Ag((CH_3)_6C_{10}H_{18}N_4)(ClO_4)_2$ ... Ag: MVol.B7-300/1
$AgC_{16}Cl_3H_{36}N_4O_{12}$ .. $[Ag((CH_3)_6C_{10}H_{18}N_4)](ClO_4)_3$ ... Ag: MVol.B7-325/6
$AgC_{16}CrH_{20}N_8S_4$ .... $[Ag(NH_3)_2][Cr(NCS)_4(C_6H_5NH_2)_2]$ ... Ag: MVol.B6-31
$AgC_{16}CuH_8Mn_2O_6$ ... $CuAg[C_5H_4Mn(CO)_3]_2 \cdot C_6H_6$ ... Ag: MVol.B5-23
$AgC_{16}F_6H_{20}P$ ... $AgPF_6 \cdot 2\ C_6H_4(CH_3)_2$ ... Ag: MVol.B5-99
$AgC_{16}F_6H_{20}Sb$ ... $AgSbF_6 \cdot 2\ C_6H_4(CH_3)_2$ ... Ag: MVol.B5-99
$AgC_{16}H_8N_2O_{12}$ ... $Ag(NC_5H_2(COOH)_2COO)_2$ ... Ag: MVol.B7-285
– ... $Ag(NC_5H_2(COOH)_2COO)_2 \cdot H_2O$ ... Ag: MVol.B7-285
$AgC_{16}H_{10}K$ ... $K[Ag(CCC_6H_5)_2]$ ... Ag: MVol.B5-13, 17
$AgC_{16}H_{10}N_3O_3S$ ... $[Ag(ONC_4N_2O_2S(C_6H_5)_2)]$ ... Ag: MVol.B6-315
$AgC_{16}H_{10}N_3O_4$ ... $[Ag(ONC_4N_2O_3(C_6H_5)_2)]$ ... Ag: MVol.B6-315
$AgC_{16}H_{11}N_2OS$ ... $[Ag(C_6H_5CHC_3NOSNC_6H_5)]$ ... Ag: MVol.B7-71
$AgC_{16}H_{11}N_2O_4S$ ... $Ag(HOC_6H_4NNC_{10}H_6SO_3)$ ... Ag: MVol.B6-289
$AgC_{16}H_{11}N_4OS$ ... $Ag(C_6H_5CHNNC_3NS(C_6H_5)NO)$ ... Ag: MVol.B6-285
$AgC_{16}H_{11}N_4O_3$ ... $Ag(C_6H_5C_3N_2O(NNHC_6H_5)COO)$ ... Ag: MVol.B6-135
$AgC_{16}H_{11}N_4O_4$ ... $[AgH(ONC_8H_5NO)_2]$ ... Ag: MVol.B6-316
$AgC_{16}H_{12}NO_3$ ... $AgNO_3 \cdot C_{16}H_{12}$ ... Ag: MVol.B5-113
$AgC_{16}H_{12}N_2O_8S_2^-$ ... $[Ag(NO_2C_6H_4SCH_2COO)_2]^-$ ... Ag: MVol.B7-27/8
$AgC_{16}H_{12}N_2O_8Se_2^-$ .. $[Ag(NO_2C_6H_4SeCH_2COO)_2]^-$ ... Ag: MVol.B7-191/4
$AgC_{16}H_{12}N_3$ ... $[Ag(C_{10}H_7N_3C_6H_5)]$ ... Ag: MVol.B6-297
$AgC_{16}H_{12}N_3OS$ ... $[Ag(ONC_3NS(C_6H_5)NC_6H_4CH_3)]$ ... Ag: MVol.B7-71
$AgC_{16}H_{12}N_4O_4^+$ ... $[Ag(NO_2C_6H_4CH_2CN)_2]^+$ ... Ag: MVol.B6-350
$AgC_{16}H_{13}$ ... $AgCCCH(C_6H_5)CH_2C_6H_5$ ... Ag: MVol.B5-14
$AgC_{16}H_{13}N_2$ ... $Ag(CH_3(C_6H_5)_2C_3N_2)$ ... Ag: MVol.B6-142
$AgC_{16}H_{13}N_2O_2S$ ... $[Ag(CH_3C_6H_4SO_2NC_9H_6N)]$ ... Ag: MVol.B6-115
$AgC_{16}H_{13}N_2O_3$ ... $Ag(C_6H_5NHCONHC_6H_4C_2H_2COO)$ ... Ag: MVol.B6-335
$AgC_{16}H_{13}N_2O_4$ ... $Ag(N_2H(COC_6H_5)_2CH_2COO)$ ... Ag: MVol.B6-331
$AgC_{16}H_{13}N_2O_7$ ... $[Ag(N_2O(C_6H_4OCH_2COO)_2H)]$ ... Ag: MVol.B6-291/2
$AgC_{16}H_{13}N_4OS$ ... $[Ag(NHC_3NS(C_6H_5)NNC_6H_4OCH_3)]$ ... Ag: MVol.B6-289
$AgC_{16}H_{13}N_4O_3$ ... $AgNO_3 \cdot C_6H_5NNC_{10}H_6NH_2$ ... Ag: MVol.B6-286
– ... $Ag((C_6H_5NN)_2CCOCOOCH_3)$ ... Ag: MVol.B6-304
$AgC_{16}H_{13}OS$ ... $[Ag(C_6H_5COC(CH_3)CSC_6H_5)]$ ... Ag: MVol.B7-20
$AgC_{16}H_{14}NO_3$ ... $[Ag(C_6H_5C_2H_3(NHCOC_6H_5)COO)]$ ... Ag: MVol.B6-245
$AgC_{16}H_{14}N_2^+$ ... $[Ag(C_6H_5CH_2CN)_2]^+$ ... Ag: MVol.B6-350
$AgC_{16}H_{14}N_3$ ... $[Ag(C_2N_3(CH_2C_6H_5)_2)]$ ... Ag: MVol.B6-186
– ... $[Ag(C_2N_3(C_6H_4CH_3)_2)]$ ... Ag: MVol.B6-186
$AgC_{16}H_{14}N_3O_3$ ... $[Ag(CH_3C_6H_4NC)_2]NO_3$ ... Ag: MVol.B5-25
$AgC_{16}H_{14}N_3O_3S_2$ ... $AgNO_3 \cdot 2\ CH_3C_7H_4NS$ ... Ag: MVol.B7-77
$AgC_{16}H_{14}N_3O_4$ ... $AgNO_3 \cdot (CH_3C_6H_4)_2C_2N_2O$ ... Ag: MVol.B6-203
$AgC_{16}H_{14}N_3O_5S_2$ ... $AgNO_3 \cdot 2\ HOCH_2C_7H_4NS$ ... Ag: MVol.B7-77
$AgC_{16}H_{14}N_3O_7$ ... $[Ag(C_7H_4NOCH_2OH)_2]NO_3$ ... Ag: MVol.B6-202
$AgC_{16}H_{14}N_7O_3$ ... $AgNO_3 \cdot 2\ NCNC(C_6H_5)NH_2$ ... Ag: MVol.B6-340, 341
$AgC_{16}H_{14}N_7O_7$ ... $AgNO_3 \cdot 2\ HOC_6H_4CON_2H_2CN$ ... Ag: MVol.B6-332
$AgC_{16}H_{14}O_4S_2^-$ ... $[Ag(C_6H_5SCH_2COO)_2]^-$ ... Ag: MVol.B7-27/8
$AgC_{16}H_{14}O_4Se_2^-$ ... $[Ag(C_6H_5SeCH_2COO)_2]^-$ ... Ag: MVol.B7-191/3
$AgC_{16}H_{15}N_2OS$ ... $Ag(C_8H_5N_2S(C_6H_5)OC_2H_5) \cdot HNO_3$ ... Ag: MVol.B7-65/6
$AgC_{16}H_{15}N_2O_4S_3$ ... $[AgH(SO_2(C_6H_4NHCOCH_2S)_2)]$ ... Ag: MVol.B7-23/4
$AgC_{16}H_{15}N_2O_6S$ ... $Ag(C_{16}H_{15}N_2O_6S)$ ... Ag: MVol.B6-319
$AgC_{16}H_{15}N_8O_2S_2$ ... $Ag(CH_3OC_6H_4CN_4S) \cdot CH_3OC_6H_4CHN_4S \cdot 2\ H_2O$
Ag: MVol.B7-57/8

Formula	Compound	Reference
$AgC_{16}H_{15}N_8S_2$	$Ag(C_7H_7CN_4S) \cdot C_7H_7CHN_4S \cdot 2\ H_2O$	Ag: MVol.B7-57/8
$AgC_{16}H_{16}NO_3$	$AgNO_3 \cdot 2\ C_8H_8$	Ag: MVol.B5-63
–	$AgNO_3 \cdot C_{16}H_{16}$	Ag: MVol.B5-86/7
$AgC_{16}H_{16}N_2^+$	$[Ag(C_2H_4(NCHC_6H_5)_2)]^+$	Ag: MVol.B6-279
$AgC_{16}H_{16}N_2O_4S_2^-$	$[Ag(NH_2C_6H_4SCH_2COO)_2]^-$	Ag: MVol.B7-27/8
$AgC_{16}H_{16}N_3O_4S$	$Ag(C_8H_5N_2S(C_6H_5)OC_2H_5) \cdot HNO_3$	Ag: MVol.B7-65/6
$AgC_{16}H_{16}N_5O_3$	$[Ag(CHNC(CH_3)NC_5H_4)_2]NO_3$	Ag: MVol.B6-141
–	$[Ag(NHC_6H_4NC(CH_3))_2]NO_3$	Ag: MVol.B6-145
$AgC_{16}H_{16}N_5O_3S$	$AgNO_3 \cdot (CH_3C_6H_4)_2C_2N_2S(NH)_2$	Ag: MVol.B7-80
–	$AgNO_3 \cdot (CH_3C_6H_4)_2C_2N_2S(NH)_2 \cdot 0.5\ H_2O$	Ag: MVol.B7-80
$AgC_{16}H_{16}N_5O_5$	$[Ag(HOCH_2C_7H_5N_2)_2]NO_3$	Ag: MVol.B6-145
$AgC_{16}H_{16}N_5O_7$	$Ag(OC_6H_2(NO_2)_3) \cdot C_{10}H_{14}N_2$	Ag: MVol.B6-107
$AgC_{16}H_{16}O_2^+$	$[Ag(CH_3COC_6H_5)_2]^+$	Ag: MVol.B6-210
$AgC_{16}H_{18}N_3O_7$	$Ag(NC_5H_4COOC_2H_5)_2NO_3$	Ag: MVol.B6-103
$AgC_{16}H_{18}N_7O_3S_2$	$AgNO_3 \cdot 2\ C_6H_5CHNNHCSNH_2$	Ag: MVol.B6-283/4
$AgC_{16}H_{18}O_6S_4^-$	$[Ag(C_2H_5SC_6H_4SO_3)_2]^-$	Ag: MVol.B7-44
$AgC_{16}H_{19}N_4O_8S_2$	$Ag(C_8H_9N_2O_4S) \cdot C_8H_{10}N_2O_4S \cdot 2\ H_2O$	Ag: MVol.B6-263
$AgC_{16}H_{20}^+$	$[Ag(C_6H_4(CH_3)_2)_2]^+$	Ag: MVol.B5-100
–	$[Ag(C_6H_5C_2H_5)_2]^+$	Ag: MVol.B5-100
$AgC_{16}H_{20}N_2^+$	$[Ag(C_6H_{12}(C_5H_4N)_2)]^+$	Ag: MVol.B6-123
$AgC_{16}H_{20}N_2O_3P$	$AgNO_3 \cdot (CH_3)_3CNHP(C_6H_5)_2$	Ag: MVol.B7-241
$AgC_{16}H_{20}N_2O_6S_2^-$	$[Ag((CH_3)_2NC_6H_4SO_3)_2]^-$	Ag: MVol.B6-262
$AgC_{16}H_{20}N_4O_2^+$	$[Ag(HOCH_2C(NH)NHC_6H_5)_2]^+$	Ag: MVol.B6-339/40
$AgC_{16}H_{20}N_5O_3S_2$	$[Ag(CH_3C_6H_4NHCSNH_2)_2]NO_3$	Ag: MVol.B7-160
$AgC_{16}H_{20}N_5O_7$	$AgNO_3 \cdot 2\ (CH_3)_3(NO_2)C_5HN$	Ag: MVol.B6-102
$AgC_{16}H_{21}N_2^{2+}$	$[AgH(C_6H_{12}(C_5H_4N)_2)]^{2+}$	Ag: MVol.B6-123
$AgC_{16}H_{21}N_4O_3$	$[Ag(HOCH_2C(NH)NHC_6H_5)_2]OH$	Ag: MVol.B6-339/40
$AgC_{16}H_{22}IP_2$	$[AgI((CH_3)_2PC_6H_5)_2]$	Ag: MVol.B7-221
$AgC_{16}H_{22}N_2^+$	$[Ag(CH_3(C_2H_5)C_5H_3N)_2]^+$	Ag: MVol.B6-89
–	$[Ag((CH_3)_2C_6H_3NH_2)_2]^+$	Ag: MVol.B6-52/3
–	$[Ag((CH_3)_3C_5H_2N)_2]^+$	Ag: MVol.B6-89
–	$[Ag(C_3H_7C_5H_4N)_2]^+$	Ag: MVol.B6-87
–	$[Ag(C_6H_5NHC_2H_5)_2]^+$	Ag: MVol.B6-53
$AgC_{16}H_{22}N_2O_2^+$	$[Ag(C_3H_7C_5H_4NO)_2]^+$	Ag: MVol.B6-105
$AgC_{16}H_{22}N_4NaO_6$	$Na[Ag((C_2H_5)_2C_4HN_2O_3)_2]$	Ag: MVol.B6-154/5
$AgC_{16}H_{22}N_4O_6^-$	$[Ag((C_2H_5)_2C_4HN_2O_3)_2]^-$	Ag: MVol.B6-154
$AgC_{16}H_{22}N_7O_9S_2$	$AgNO_3 \cdot 2\ CH_3CONHC_6H_4SO_2NHNH_2$	Ag: MVol.B6-265
$AgC_{16}H_{23}N_2O$	$[Ag((CH_3)_3C_5H_2N)_2]OH$	Ag: MVol.B6-89
–	$[Ag(C_3H_7C_5H_4N)_2]OH$	Ag: MVol.B6-87
$AgC_{16}H_{24}NO_3$	$AgNO_3 \cdot 2\ C_8H_{12}$	Ag: MVol.B5-62/3
–	$AgNO_3 \cdot C_{16}H_{24}$	Ag: MVol.B5-74
$AgC_{16}H_{24}N_5O_3$	$[Ag(CH_3C_7H_9N_2)_2]NO_3$	Ag: MVol.B6-132
$AgC_{16}H_{24}O_8S_2^{3-}$	$[Ag(S(C_3H_6COO)_2)_2]^{3-}$	Ag: MVol.B7-39
$AgC_{16}H_{24}O_8S_4^{3-}$	$[Ag(C_2H_4(SC_2H_4COO)_2)_2]^{3-}$	Ag: MVol.B7-41
–	$[Ag(C_4H_8(SCH_2COO)_2)_2]^{3-}$	Ag: MVol.B7-39/40
$AgC_{16}H_{25}IN_2P$	$[AgI((C_5H_{10}N)_2PC_6H_5)]_4$	Ag: MVol.B7-242
$AgC_{16}H_{26}N_4^{3+}$	$[Ag(C_5H_5NC_3H_6NH_2)_2]^{3+}$	Ag: MVol.B6-97/8
$AgC_{16}H_{27}N_3S_4$	$Ag(SCSN(C_2H_5)_2)_2 \cdot CH_3C_5H_4N$	Ag: MVol.B7-318
$AgC_{16}H_{27}O_8S_4$	$H_3Ag(C_2H_4(SC_2H_4COO)_2)_2$	Ag: MVol.B7-41
$AgC_{16}H_{28}NO_3$	$AgNO_3 \cdot 2\ C_8H_{14}$	Ag: MVol.B5-62

Formula	Compound	Reference
$AgC_{16}H_{28}NO_3$	$AgNO_3 \cdot C_{16}H_{28}$	Ag: MVol.B5-74
$AgC_{16}H_{28}N_8^+$	$[Ag((CH_2)_7CN_4)_2]^+$	Ag: MVol.B6-194
–	$[Ag(C_6H_{11}CN_4CH_3)_2]^+$	Ag: MVol.B6-193
$AgC_{16}H_{28}O_8S_4^+$	$[Ag(C_2H_4(SC_2H_4COOH)_2)_2]^+$	Ag: MVol.B7-41
$AgC_{16}H_{30}O_4S_2^-$	$[Ag(C_6H_{13}SCH_2COO)_2]^-$	Ag: MVol.B7-25/6
$AgC_{16}H_{30}O_6^+$	$[Ag(C_{16}H_{30}O_6)]^+$	Ag: MVol.B6-219
$AgC_{16}H_{31}IN_2P$	$[AgI((C_5H_{10}N)_2PC_6H_{11})]_4$	Ag: MVol.B7-241
$AgC_{16}H_{31}N_4O_4S_4$	$[Ag(C_8H_{15}N_2O_2S_2)(C_8H_{16}N_2O_2S_2)]$	Ag: MVol.B7-189
$AgC_{16}H_{33}N_2S_2$	$[Ag((C_2H_5)_2N(CH_2)_{10}N(CH_3)CSS)]$	Ag: MVol.B7-115
$AgC_{16}H_{34}NO_3S$	$AgNO_3 \cdot S(C_8H_{17})_2$	Ag: MVol.B7-12/3
$AgC_{16}H_{34}N_2^+$	$[Ag(NHC_5H_9C_3H_7)_2]^+$	Ag: MVol.B6-73
$AgC_{16}H_{35}N_3S_4$	$Ag(SCSN(C_2H_5)_2)_2 \cdot (C_2H_5)_3N$	Ag: MVol.B7-318
$AgC_{16}H_{36}NO_3S_2$	$AgNO_3 \cdot 2\ CH_3SC_7H_{15}$	Ag: MVol.B7-11
–	$AgNO_3 \cdot 2\ C_2H_5SC_6H_{13}$	Ag: MVol.B7-11
–	$AgNO_3 \cdot 2\ C_3H_7SC_5H_{11}$	Ag: MVol.B7-11
–	$AgNO_3 \cdot 2\ S(C_4H_9)_2$	Ag: MVol.B7-12
$AgC_{16}H_{36}N_4^+$	$[Ag((CH_3)_6C_{10}H_{18}N_4)]^+$	Ag: MVol.B6-202
$AgC_{16}H_{36}N_4^{2+}$	$[Ag((CH_3)_6C_{10}H_{18}N_4)]^{2+}$	Ag: MVol.B6-202
$AgC_{16}H_{36}N_6O_6$	$Ag((CH_3)_6C_{10}H_{18}N_4)(NO_3)_2$	Ag: MVol.B7-300
$AgC_{16}H_{37}OP_2S_2$	$[(C_2H_5)_3P]_2Ag[S_2COC_3H_7]$	C: MVol.D4-257
$AgC_{16}H_{38}N_2^+$	$[Ag(C_5H_{11}C(CH_3)_2NH_2)_2]^+$	Ag: MVol.B6-46
–	$[Ag(NH(C_4H_9)_2)_2]^+$	Ag: MVol.B6-48
$AgC_{16}H_{38}N_2O_2^+$	$[Ag((CH(CH_3)_2)_2NC_2H_4OH)_2]^+$	Ag: MVol.B6-227
$AgC_{16}H_{38}N_2O_4^+$	$[Ag(NH(C_4H_8OH)_2)_2]^+$	Ag: MVol.B6-229
$AgC_{16}H_{40}N_4^+$	$[Ag(NH_2(CH_2)_8NH_2)_2]^+$	Ag: MVol.B6-63
$AgC_{17}ClH_{13}N_3O_4$	$Ag(C_{12}H_8N_2)(C_5H_5N)ClO_4$	Ag: MVol.B6-125
$AgC_{17}ClH_{20}N_4O_{10}$	$AgClO_4 \cdot (CH_3)_2C_{10}H_3N_4O_2C_5H_{11}O_4$	Ag: MVol.B6-177
$AgC_{17}H_7N_2O_5$	$Ag(C_{17}H_7N_2O_5)$	Ag: MVol.B6-328
$AgC_{17}H_{11}N_3O_4$	$Ag(NC_5H_3(COO)_2)(C_{10}H_8N_2) \cdot 3\ H_2O$	Ag: MVol.B7-299/300
$AgC_{17}H_{11}N_4S$	$[Ag(C_{11}H_6NSN_3C_6H_5)]$	Ag: MVol.B6-303
$AgC_{17}H_{11}OS_3$	$[Ag(C_5HOS_3(C_6H_5)_2)]$	Ag: MVol.B7-83/4
$AgC_{17}H_{13}N_6O_4$	$[Ag(NO_2C_6H_4N_3C_6H_4NO_2)(C_5H_5N)]$	Ag: MVol.B6-298, 299
$AgC_{17}H_{14}N_3O_2$	$[Ag(C_6H_5C_3N_3O_2C_2H_4C_6H_5)]$	Ag: MVol.B6-309/12
$AgC_{17}H_{15}N_2$	$Ag((CH_3C_6H_4)_2C_3HN_2)$	Ag: MVol.B6-142
–	$Ag(C_2H_5(C_6H_5)_2C_3N_2)$	Ag: MVol.B6-142
$AgC_{17}H_{15}N_2O_2$	$[Ag(CH(NC_6H_4COCH_3)_2)]$	Ag: MVol.B6-341
–	$Ag((CH_3OC_6H_4)_2C_3HN_2)$	Ag: MVol.B6-142
$AgC_{17}H_{15}N_2O_3$	$HOC_6H_4COOAg \cdot 2\ C_5H_5N$	Ag: MVol.B6-81
$AgC_{17}H_{16}NO_3$	$[Ag(C_6H_5C_2H_3(COO)NHCOCH_2C_6H_5)]$	Ag: MVol.B6-245
$AgC_{17}H_{17}N_2OS$	$[Ag(C_8H_5N_2S(C_6H_4CH_3)OC_2H_5)]$	Ag: MVol.B7-65/6
$AgC_{17}H_{18}N_3O_2$	$Ag(C_6H_5NNC_6H_3(N(C_2H_5)_2)COO)$	Ag: MVol.B6-287
$AgC_{17}H_{19}N_2O_2$	$[Ag(CH(NC_6H_4OC_2H_5)_2)]$	Ag: MVol.B6-341
$AgC_{17}H_{19}N_4O_6$	$[Ag((CH_3)_2C_{10}H_2N_4O_2C_5H_{11}O_4)]$	Ag: MVol.B6-173, 177
$AgC_{17}H_{20}N_4O_6^+$	$[Ag((CH_3)_2C_{10}H_3N_4O_2C_5H_{11}O_4)]^+$	Ag: MVol.B6-173, 177
$AgC_{17}H_{20}N_5O_7$	$[Ag(C_6H_5CO(NHCH_2CO)_4NHCH_2COO)]$	Ag: MVol.B6-270
$AgC_{17}H_{22}N_3O_5$	$[AgC_{17}H_{22}N_3O_5]$	Ag: MVol.B6-269
$AgC_{17}H_{26}N_5O_4S_2$	$Ag(NHC_6H_4SO_2NHC_3H_2NS) \cdot 2\ NH(CH_2CH_2)_2O$	Ag: MVol.B6-265
$AgC_{17}H_{34}NS_2$	$[Ag((C_8H_{17})_2NCSS)]$	Ag: MVol.B7-113
$AgC_{17}H_{39}OP_2S_2$	$[(C_2H_5)_3P]_2Ag[S_2COC_4H_9]$	C: MVol.D4-257

$AgC_{17.45}ClH_{20}N_4O_{10.45}$ ... $AgClO_4 \cdot C_{17}H_{20}N_4O_6(CO)_{0.45} \cdot 0.5\ H_2O$ ... Ag: MVol.B6-177
$AgC_{17.45}ClH_{20.9}N_4O_{10.9}$
$[AgC_{17}H_{20}N_4O_6]ClO_4 \cdot 0.45\ HCOOH$ ... Ag: MVol.B6-177
$AgC_{18}ClH_{12}N_6O_4$ ... $Ag(NCC_5H_4N)_3ClO_4$ ... Ag: MVol.B6-100
$AgC_{18}ClH_{14}N_2O_4$ ... $[Ag(C_9H_7N)_2]ClO_4$ ... Ag: MVol.B6-108,116
$AgC_{18}ClH_{14}N_2O_6$ ... $Ag(C_9H_7NO)_2ClO_4$ ... Ag: MVol.B6-112
$AgC_{18}ClH_{14}N_2O_8$ ... $Ag(C_6H_5C_3H_2NO_2)_2ClO_4$ ... Ag: MVol.B6-203
$AgC_{18}ClH_{14}O_4$ ... $AgClO_4 \cdot C_5H_4C(C_6H_5)_2$ ... Ag: MVol.B5-106
$AgC_{18}ClH_{15}NO_4$ ... $[Ag(C_6H_5)_3N]ClO_4$ ... Ag: MVol.B6-53
$AgC_{18}ClH_{15}N_3O_{10}$ ... $AgClO_4 \cdot 3\ C_6H_5NO_2$ ... Ag: MVol.B5-118
$AgC_{18}ClH_{15}P$ ... $[AgCl(P(C_6H_5)_3)]_4$ ... Ag: MVol.B7-208
$AgC_{18}ClH_{18}N_6O_4$ ... $[Ag(NC_5H_3(CH_3)C_3H_3N_2)_2]ClO_4 \cdot 0.5\ H_2O$ ... Ag: MVol.B6-141
$AgC_{18}ClH_{18}O_4$ ... $AgClO_4 \cdot 3\ C_6H_6$ ... Ag: MVol.B5-93
$AgC_{18}ClH_{21}N_3O_4$ ... $AgClO_4 \cdot 3\ C_6H_5NH_2$ ... Ag: MVol.B6-51
$AgC_{18}ClH_{21}N_3O_7$ ... $[Ag(CH_3C_5H_4NO)_2]ClO_4 \cdot CH_3C_5H_4NO$ ... Ag: MVol.B6-105
$AgC_{18}ClH_{38}N_4O_3S_2$ ... $[Ag((CH_3)_3C_4H_5N_2OS)_2]Cl \cdot (C_2H_5)_2O$ ... Ag: MVol.B7-65
$AgC_{18}ClH_{45}O_9P_3$ ... $[AgCl(P(OC_2H_5)_3)_3]$ ... Ag: MVol.B7-247
$AgC_{18}ClH_{45}O_{13}P_3$ ... $[Ag(P(OC_2H_5)_3)_3]ClO_4$ ... Ag: MVol.B7-246
$AgC_{18}Cl_4H_{10}N_3O_3$ ... $Ag(Cl_2C_9H_5N)_2NO_3$ ... Ag: MVol.B6-115
$AgC_{18}Cl_4H_{30}IrN_2S_2$ ... $[Ag(C_5H_5N)_2][Ir(S(C_2H_5)_2)_2Cl_4]$ ... Ir: SVol.2-166
$AgC_{18}CrH_{24}N_8S_4$ ... $[Ag(CH_3NH_2)_2][Cr(NCS)_4(C_6H_5NH_2)_2]$ ... Ag: MVol.B6-42
$AgC_{18}FH_{14}N_2O_2$ ... $Ag(C_9H_7NO)_2F$ ... Ag: MVol.B6-112
$AgC_{18}F_6H_{12}O_4S_2^-$ ... $[Ag(CF_3C_6H_4SCH_2COO)_2]^-$ ... Ag: MVol.B7-27/8
$AgC_{18}GeH_{25}IP$ ... $[AgI((C_2H_5)_3GeP(C_6H_5)_2)]_4$ ... Ag: MVol.B7-270
$AgC_{18}H_{11}N_4O_6$ ... $[AgH(ONC_3NO_2C_6H_5)_2]$ ... Ag: MVol.B6-313
$AgC_{18}H_{12}^+$ ... $[AgC_{18}H_{12}]^+$ ... Ag: MVol.B5-115
$AgC_{18}H_{12}KN_2O_2$ ... $K[Ag(C_9H_6NO)_2] \cdot 4\ H_2O$ ... Ag: MVol.B6-113
$AgC_{18}H_{12}N_2^+$ ... $[Ag(C_6H_5C_{12}H_7N_2)]^+$ ... Ag: MVol.B6-128
$AgC_{18}H_{12}N_2NaO_2$ ... $Na[Ag(C_9H_6NO)_2] \cdot 3\ H_2O$ ... Ag: MVol.B6-113
$AgC_{18}H_{12}N_2O_2^-$ ... $[Ag(C_9H_6NO)_2]^-$ ... Ag: MVol.B6-110
$AgC_{18}H_{12}N_2O_4S_2^-$ ... $[Ag(NCC_6H_4SCH_2COO)_2]^-$ ... Ag: MVol.B7-27/8
$AgC_{18}H_{12}N_5O_7$ ... $Ag(NO_2C_9H_6N)_2NO_3 \cdot x\ H_2O$ ... Ag: MVol.B6-117
$AgC_{18}H_{12}O_8Se_2^{3-}$ ... $[Ag(OCOC_6H_4SeCH_2COO)_2]^{3-}$ ... Ag: MVol.B7-191/3
$AgC_{18}H_{13}N_2O_2$ ... $AgC_9H_6NO \cdot C_9H_7NO$ ... Ag: MVol.B6-110/1
$AgC_{18}H_{13}N_2O_3$ ... $AgOH(C_9H_6NO)_2 \cdot 0.5\ C_5H_5N \cdot H_2O$ ... Ag: MVol.B7-326
$AgC_{18}H_{13}N_2S_2$ ... $AgC_9H_6NS \cdot C_9H_7NS \cdot 3\ H_2O$ ... Ag: MVol.B7-60
$AgC_{18}H_{13}N_6O_2$ ... $[Ag(NO_2C_6H_4N_3C_6H_4N_2C_6H_5)]$ ... Ag: MVol.B6-298/9
$AgC_{18}H_{14}NO_3$ ... $AgNO_3 \cdot C_{18}H_{14}$ ... Ag: MVol.B5-113
$AgC_{18}H_{14}N_2^+$ ... $[Ag(C_9H_7N)_2]^+$ ... Ag: MVol.B6-107/8, 116
$AgC_{18}H_{14}N_2O_2^+$ ... $[Ag(HOC_9H_6N)_2]^+$ ... Ag: MVol.B6-110
$AgC_{18}H_{14}N_3O_3$ ... $[AgNO_3(C_9H_7N)_2]$ ... Ag: MVol.B6-108, 116
$AgC_{18}H_{14}N_3O_3S$ ... $[Ag(ONC_4N_2O_2S(C_6H_4CH_3)_2)]$ ... Ag: MVol.B6-315/6
$AgC_{18}H_{14}N_3O_5$ ... $Ag(C_9H_7NO)_2NO_3$ ... Ag: MVol.B6-111/2
$AgC_{18}H_{14}N_5$ ... $[Ag(C_6H_5N_3C_6H_4N_2C_6H_5)]$ ... Ag: MVol.B6-297
$AgC_{18}H_{14}N_7O_3$ ... $AgNO_3 \cdot 2\ C_6H_5C_3H_2N_3$ ... Ag: MVol.B6-187
$AgC_{18}H_{14}O_3PS$ ... $[Ag((C_6H_5)_2PC_6H_4SO_3)]$ ... Ag: MVol.B7-233
$AgC_{18}H_{14}O_8S_2^-$ ... $[Ag(C_6H_4(COOH)SCH_2COO)_2]^-$ ... Ag: MVol.B7-27/8
$AgC_{18}H_{15}IP$ ... $[AgI(P(C_6H_5)_3)]_4$ ... Ag: MVol.B7-208/9
$AgC_{18}H_{15}NO_4P$ ... $AgNO_3 \cdot (C_6H_5)_3PO$ ... Ag: MVol.B7-244
$AgC_{18}H_{15}N_2O$ ... $Ag(C_9H_7N)_2OH$ ... Ag: MVol.B6-108

Formula	Compound	Reference
$AgC_{18}H_{15}N_2O_4S$	$Ag(HOC_{10}H_6NNC_6H_2(CH_3)_2SO_3)$	Ag: MVol.B6-289
$AgC_{18}H_{15}N_4O_4$	$[AgH(ONC_8H_4NOCH_3)_2]$	Ag: MVol.B6-316
$AgC_{18}H_{16}N_3O_3$	$AgNO_3 \cdot (CH_3)_2C_4N_2(C_6H_5)_2$	Ag: MVol.B6-160
$AgC_{18}H_{16}N_5O_3$	$Ag(NH_2C_9H_6N)_2NO_3$	Ag: MVol.B6-113/4
$AgC_{18}H_{16}N_5O_7S_2$	$[Ag(C_6H_5SO_2C_3H_3N_2)_2]NO_3$	Ag: MVol.B6-141
$AgC_{18}H_{17}N_2$	$Ag(C_3H_7(C_6H_5)_2C_3N_2)$	Ag: MVol.B6-142
$AgC_{18}H_{17}N_2O_4$	$[AgC_{18}H_{17}N_2O_4]$	Ag: MVol.B6-267/8
$AgC_{18}H_{17}N_2O_5P_2$	$Ag((C_6H_5O)_2(O)PNP(O)(NH_2)OC_6H_5)$	Ag: MVol.B7-252
$AgC_{18}H_{18}N_3O_7$	$[Ag(NC_5H_4COCH_2COCH_3)_2]NO_3$	Ag: MVol.B6-214
$AgC_{18}H_{18}N_3O_9S_3{}^{2-}$	$[Ag(NH_2C_6H_4SO_3)_3]^{2-}$	Ag: MVol.B6-262
$AgC_{18}H_{18}N_{11}O_5$	$AgNO_3 \cdot 2\ (HN)_2C_3H_2N_3OC_6H_5$	Ag: MVol.B6-188
$AgC_{18}H_{18}N_{11}O_9$	$AgNO_3 \cdot 2\ NHC_3H_4N_3OC_6H_4NO_2$	Ag: MVol.B6-190
$AgC_{18}H_{18}O_4S_2{}^-$	$[Ag(CH_3C_6H_4SCH_2COO)_2]^-$	Ag: MVol.B7-27/8
–	$[Ag(C_6H_5CH_2SCH_2COO)_2]^-$	Ag: MVol.B7-29
$AgC_{18}H_{18}O_4S_2Se_2{}^-$	$[Ag(CH_3SC_6H_4SeCH_2COO)_2]^-$	Ag: MVol.B7-191/2, 195
$AgC_{18}H_{18}O_4S_4{}^-$	$[Ag(CH_3SC_6H_4SCH_2COO)_2]^-$	Ag: MVol.B7-27/8
$AgC_{18}H_{18}O_4Se_2{}^-$	$[Ag(CH_3C_6H_4SeCH_2COO)_2]^-$	Ag: MVol.B7-191/2, 194
$AgC_{18}H_{18}O_6S_2{}^-$	$[Ag(CH_3OC_6H_4SCH_2COO)_2]^-$	Ag: MVol.B7-27/8
$AgC_{18}H_{18}O_6Se_2{}^-$	$[Ag(CH_3OC_6H_4SeCH_2COO)_2]^-$	Ag: MVol.B7-191/2, 195
$AgC_{18}H_{20}N_2O_4{}^-$	$[Ag(C_6H_5CH_2CH(NH_2)COO)_2]^-$	Ag: MVol.B6-244
$AgC_{18}H_{20}N_5O_3$	$[Ag((CH_3)_2C_7H_4N_2)_2]NO_3$	Ag: MVol.B6-145
–	$[Ag(C_2H_5C_7H_5N_2)_2]NO_3$	Ag: MVol.B6-141
$AgC_{18}H_{20}N_5O_3S$	$AgNO_3 \cdot ((CH_3)_2C_6H_3)_2C_2N_2S(NH)_2$	Ag: MVol.B7-80
$AgC_{18}H_{20}N_5O_5$	$[Ag(CH_3(CH_2OH)C_7H_4N_2)_2]NO_3$	Ag: MVol.B6-145
$AgC_{18}H_{21}N_3{}^+$	$[Ag(C_6H_5NH_2)_3]^+$	Ag: MVol.B6-50/1
$AgC_{18}H_{21}N_4O_3$	$[AgNO_3(CH_3C_5H_4N)_3]$	Ag: MVol.B6-85/6
$AgC_{18}H_{24}{}^+$	$[Ag(C_6H_3(CH_3)_3)_2]^+$	Ag: MVol.B5-100
$AgC_{18}H_{24}LiN_2$	$AgLi(C_6H_4CH_2N(CH_3)_2)_2$	Ag: MVol.B5-12
$AgC_{18}H_{24}N_4O_2{}^+$	$[Ag(CH_3C(C_6H_5)(OH)C(NH)NH_2)_2]^+$	Ag: MVol.B6-339/40
$AgC_{18}H_{24}N_4O_{12}{}^{5-}$	$[Ag(C_6H_{12}N_4(CH_2COO)_6)]^{5-}$	Ag: MVol.B6-279
$AgC_{18}H_{24}N_6P_2{}^+$	$[Ag(P(C_2H_4CN)_3)_2]^+$	Ag: MVol.B7-203
$AgC_{18}H_{25}N_4O_3$	$[Ag(CH_3C(C_6H_5)(OH)C(NH)NH_2)_2]OH$	Ag: MVol.B6-339/40
$AgC_{18}H_{25}N_4O_{12}{}^{4-}$	$[Ag(C_6H_{12}N_4(CH_2COO)_6H)]^{4-}$	Ag: MVol.B6-279
$AgC_{18}H_{26}N_2{}^+$	$[Ag((CH_3)_3C_6H_2NH_2)_2]^+$	Ag: MVol.B6-53
$AgC_{18}H_{26}N_4O_{12}{}^{3-}$	$[Ag(C_6H_{12}N_4(CH_2COO)_6H_2)]^{3-}$	Ag: MVol.B6-279
$AgC_{18}H_{27}N_6O_4S$	$Ag(NHC_6H_4SO_2NHC_4H_3N_2) \cdot 2\ NH(CH_2CH_2)O$	Ag: MVol.B6-264
$AgC_{18}H_{28}O_8S_4{}^{3-}$	$[Ag(C_5H_{10}(SCH_2COO)_2)_2]^{3-}$	Ag: MVol.B7-39/40
$AgC_{18}H_{30}{}^+$	$[AgC_6(C_2H_5)_6]^+$	Ag: MVol.B5-98
–	$[AgC_6H_3(C_4H_9)_3]^+$	Ag: MVol.B5-98
$AgC_{18}H_{30}N_4{}^{3+}$	$[Ag(C_5H_5NC_4H_8NH_2)_2]^{3+}$	Ag: MVol.B6-97/8
$AgC_{18}H_{32}NO_3$	$AgNO_3 \cdot 2\ C_9H_{16}$	Ag: MVol.B5-66
$AgC_{18}H_{32}NO_3S_2$	$AgNO_3 \cdot 2\ C_9H_{16}S$	Ag: MVol.B7-85
$AgC_{18}H_{33}P^+$	$[Ag(P(C_6H_{11})_3)]^+$	Ag: MVol.B7-207
$AgC_{18}H_{34}N_2{}^+$	$[Ag(C_9H_{17}N)_2]^+$	Ag: MVol.B6-116
$AgC_{18}H_{36}N_2S_4$	$[Ag(SCSN(C_4H_9)_2)_2]$	Ag: MVol.B7-317/8
$AgC_{18}H_{36}N_2S_4{}^+$	$[Ag((C_4H_9)_2NCSS)_2]^+$	Ag: MVol.B7-326
$AgC_{18}H_{36}N_2Se_4$	$[Ag((C_4H_9)_2NCSeSe)_2]$	Ag: MVol.B7-319
$AgC_{18}H_{36}N_3O_9$	$AgNO_3 \cdot C_{18}H_{36}N_2O_6$	Ag: MVol.B6-205
$AgC_{18}H_{36}N_{12}{}^+$	$[Ag((CH_2)_6N_4)_3]^+$	Ag: MVol.B6-197/8

Formula	Compound	Reference
$AgC_{18}H_{36}N_{13}O_3$	$AgNO_3 \cdot 2\ (C_2H_5NH)_3C_3N_3$	Ag: MVol.B6-188
$AgC_{18}H_{40}NO_3S_2$	$AgNO_3 \cdot 2\ C_4H_9SC_5H_{11}$	Ag: MVol.B7-12
$AgC_{18}H_{41}OP_2S_2$	$[(C_2H_5)_3P]_2Ag[S_2COC_5H_{11}]$	C: MVol.D4-257
$AgC_{18}H_{42}N_2O_6{}^+$	$[Ag(N(CH_2CHOHCH_3)_3)_2]^+$	Ag: MVol.B6-230
$AgC_{18}H_{45}IO_9P_3$	$[AgI(P(OC_2H_5)_3)_3]$	Ag: MVol.B7-247
$AgC_{18}H_{45}NO_{12}P_3$	$[Ag(P(OC_2H_5)_3)_3]NO_3$	Ag: MVol.B7-246
$AgC_{18}H_{45}N_3O_9{}^+$	$[Ag(N(C_2H_4OH)_3)_3]^+$	Ag: MVol.B6-229
$AgC_{18}H_{45}O_9P_3{}^+$	$[Ag(P(OC_2H_5)_3)_3]^+$	Ag: MVol.B7-246
$AgC_{18}H_{54}N_9O_3P_3{}^+$	$[Ag(OP(N(CH_3)_2)_3)_3]^+$	Ag: MVol.B7-250/1
$AgC_{19}ClH_{17}N_3O_4$	$[Ag((CH_3)_2C_{12}H_6N_2)(C_5H_5N)]ClO_4$	Ag: MVol.B6-127
$AgC_{19}ClH_{21}N_3O_4Si$	$[AgClO_4(CH_3Si(CH_2C_5H_4N)_3)]$	Ag: MVol.B7-269
$AgC_{19}H_{11}N_3O_4$	$Ag(NC_5H_3(COO)_2)(C_{12}H_8N_2) \cdot 3.5\ H_2O$	Ag: MVol.B7-299/300
$AgC_{19}H_{14}{}^+$	$[Ag(C_{18}H_{11}CH_3)]^+$	Ag: MVol.B5-115
$AgC_{19}H_{14}NO_2S$	$Ag(C_{13}H_9NSO_2C_6H_5)$	Ag: MVol.B6-343
$AgC_{19}H_{15}$	$AgC(C_6H_5)_3$	Ag: MVol.B5-4
$AgC_{19}H_{15}HgN_4S$	$[C_6H_5Hg(AgC_6H_5NNCSNNC_6H_5)]$	Ag: MVol.B7-182
$AgC_{19}H_{15}N_2$	$[AgNC_8H_4(CH_3)CHC_8H_4(CH_3)N]$	Ag: MVol.B6-70
$AgC_{19}H_{15}N_2O$	$Ag((C_6H_5)_2NCONC_6H_5) \cdot NH_3$	Ag: MVol.B6-335
$AgC_{19}H_{15}N_4O_3$	$Ag(ONC_3NO_2C_6H_5) \cdot 2\ C_5H_5N$	Ag: MVol.B6-313
$AgC_{19}H_{15}N_6$	$Ag(C_6H_5NNC(NNC_6H_5)_2)$	Ag: MVol.B6-304
$AgC_{19}H_{16}N_3$	$[Ag((C_6H_5)_2NC(NH)NC_6H_5)]$	Ag: MVol.B6-337
$AgC_{19}H_{18}N_3O$	$Ag(C_6H_5)_2NCONC_6H_5 \cdot NH_3$	Ag: MVol.B6-335
$AgC_{19}H_{23}N_6O_8$	$[Ag(C_6H_5CO(NHCH_2CO)_5NHCH_2COO)]$	Ag: MVol.B6-270
$AgC_{19}H_{28}N_5O_4S$	$Ag(NHC_6H_4SO_2NHC_5H_4N) \cdot 2\ NH(CH_2CH_2)_2O$	Ag: MVol.B6-264
$AgC_{19}H_{45}NO_9P_3$	$[AgCN(P(OC_2H_5)_3)_3]$	Ag: MVol.B7-247
$AgC_{19}H_{45}NO_9P_3S$	$[AgSCN(P(OC_2H_5)_3)_3]$	Ag: MVol.B7-247
$AgC_{20}ClH_{16}N_4O_4$	$[Ag(C_5H_5N)(NC_5H_4CHNC_9H_6N)]ClO_4$	Ag: MVol.B6-281
–	$Ag(NC_5H_4C_5H_4N)_2ClO_4$	Ag: MVol.B6-120
$AgC_{20}ClH_{16}N_4O_8$	$Ag(ONC_5H_4C_5H_4NO)_2ClO_4 \cdot 3\ H_2O$	Ag: MVol.B6-122
$AgC_{20}ClH_{18}N_2O_4$	$[Ag(CH_3C_9H_6N)_2]ClO_4$	Ag: MVol.B6-109
$AgC_{20}ClH_{18}N_2O_4S_2$	$[Ag(CH_3SC_9H_6N)_2]ClO_4$	Ag: MVol.B7-61/2
$AgC_{20}ClH_{20}N_4O_4$	$[Ag((CH_3)_2C_8H_4N_2)_2]ClO_4$	Ag: MVol.B6-129
–	$AgClO_4 \cdot 4\ C_5H_5N$	Ag: MVol.B6-80
$AgC_{20}ClH_{32}O_4$	$AgClO_4 \cdot 2\ C_7H_7(CH_3)_3$	Ag: MVol.B5-78
–	$AgClO_4 \cdot 2\ C_7H_8(CH_3)_2CH_2$	Ag: MVol.B5-78
$AgC_{20}ClH_{36}O_{16}P_4$	$[Ag(P(OCH_2)_3CCH_3)_4]ClO_4$	Ag: MVol.B7-248/9
$AgC_{20}ClH_{40}N_4O_4S_4$	$[Ag((C_2H_5)_2NCSCSN(C_2H_5)_2)_2]ClO_4$	Ag: MVol.B7-190
$AgC_{20}Cl_2H_{16}N_4O_6$	$Ag(NC_5H_4C_5H_4N)_2(ClO_3)_2$	Ag: MVol.B7-291
$AgC_{20}Cl_2H_{16}N_4O_8$	$Ag(NC_5H_4C_5H_4N)_2(ClO_4)_2$	Ag: MVol.B7-292
$AgC_{20}Cl_2H_{20}IrN_6O_6$	$[Ir(C_5H_5N)_4Cl_2][Ag(NO_3)_2]$	Ir: SVol.2-68
$AgC_{20}Cl_3H_{14}N_2O_4$	$Ag(C_9H_7NO)_2(CCl_3COO)$	Ag: MVol.B6-112
$AgC_{20}Cl_4H_{20}IrN_4$	$[Ir(C_5H_5N)_4Cl_2][AgCl_2]$	Ir: SVol.2-68
$AgC_{20}CrH_{28}N_8S_4$	$[Ag(C_2H_5NH_2)_2][Cr(NCS)_4(C_6H_5NH_2)_2]$	Ag: MVol.B6-43
$AgC_{20}F_{10}H_{20}NS_2$	$[N(C_2H_5)_4][Ag(SC_6F_5)_2]$	Ag: MVol.B7-7/8
$AgC_{20}H_{12}{}^+$	$[AgC_{20}H_{12}]^+$	Ag: MVol.B5-115
$AgC_{20}H_{12}KN_2O_4$	$K[Ag(OC_{10}H_6NO)_2]$	Ag: MVol.B6-307
$AgC_{20}H_{12}N_2NaO_4$	$Na[Ag(OC_{10}H_6NO)_2]$	Ag: MVol.B6-307
$AgC_{20}H_{12}N_2O_2S_4{}^-$	$[Ag(C_6H_5CHC_3NOS_2)_2]^-$	Ag: MVol.B7-73/4
$AgC_{20}H_{12}N_2O_4$	$Ag(C_9H_6NCOO)_2$	Ag: MVol.B7-286, 287

$AgC_{20}H_{12}N_2O_4$ ...... $Ag(C_9H_6NCOO)_2 \cdot 2\ H_2O$ .... Ag: MVol.B7-286
$AgC_{20}H_{13}N_2O_2$ ...... $Ag(C_7H_5N_2C_6H_4C_6H_4COO)$ .... Ag: MVol.B6-146
$AgC_{20}H_{13}N_2O_4$ ...... $H[Ag(OC_{10}H_6NO)_2]$ .... Ag: MVol.B6-307
$AgC_{20}H_{14}N_3O_2$ ...... $Ag(C_6H_5NNC_9H_4O_2) \cdot C_5H_5N$ .... Ag: MVol.B6-216
$AgC_{20}H_{14}N_3O_7$ ...... $[Ag(HOC_{10}H_6NO)_2]NO_3$ .... Ag: MVol.B6-307
$AgC_{20}H_{14}N_5O_3$ ...... $Ag(NC_5H_4(NC_5H_3)_2C_5H_4N)NO_3$ .... Ag: MVol.B6-123
$AgC_{20}H_{14}N_5O_3S_2$ .... $[Ag(C_7H_4NSNNC(C_6H_5)NNHC_6H_4SO_3)]$ ..... Ag: MVol.B6-304
$AgC_{20}H_{14}N_6O_6$ ...... $Ag(NC_5H_4(NC_5H_3)_2C_5H_4N)(NO_3)_2$ .... Ag: MVol.B6-123
$AgC_{20}H_{15}N_6O_7$ ...... $AgNO_3 \cdot NH(COC_8H_4N_2OCH_3)_2$ .... Ag: MVol.B6-152
$AgC_{20}H_{15}N_8O_8$ ...... $[HAg(CH_3C_3N_3O_2C_6H_4NO_2)_2]$ .... Ag: MVol.B6-308/12
$AgC_{20}H_{16}^+$ .... $[Ag(C_{18}H_{10}(CH_3)_2)]^+$ .... Ag: MVol.B5-115
$AgC_{20}H_{16}N_4^+$ ...... $[Ag(NC_5H_4C_5H_4N)_2]^+$ .... Ag: MVol.B6-118/9
$AgC_{20}H_{16}N_4^{2+}$ ...... $[Ag(NC_5H_4C_5H_4N)_2]^{2+}$ .... Ag: MVol.B7-287/8
$AgC_{20}H_{16}N_4O_8S_2$ .... $Ag(NC_5H_4C_5H_4N)_2S_2O_8$ .... Ag: MVol.B7-292/3
– .... $Ag(NC_5H_4C_5H_4N)_2S_2O_8 \cdot H_2O$ .... Ag: MVol.B7-293
$AgC_{20}H_{16}N_5O_3$ ...... $Ag(NC_5H_4C_5H_4N)_2NO_3$ .... Ag: MVol.B6-119
– .... $Ag(NC_5H_4C_5H_4N)_2NO_3 \cdot 2\ H_2O$ .... Ag: MVol.B6-119/20
$AgC_{20}H_{16}N_6O_6$ ...... $[AgNO_3(NC_5H_4C_5H_4N)_2]NO_3$ .... Ag: MVol.B7-289/91
– .... $[AgNO_3(NC_5H_4C_5H_4N)_2]NO_3 \cdot H_2O$ .... Ag: MVol.B7-289/91
$AgC_{20}H_{18}N_2^+$ ...... $[Ag(CH_3C_9H_6N)_2]^+$ .... Ag: MVol.B6-109, 117
– .... $[Ag(C_{10}H_7NH_2)_2]^+$ .... Ag: MVol.B6-54
$AgC_{20}H_{18}N_3O$ ...... $[Ag(C_6H_5N_3CH_2C(C_6H_5)_2OH)]$ .... Ag: MVol.B6-301
– .... $[Ag(C_6H_5N_3C_6H_4C(CH_3)(C_6H_5)OH)]$ .... Ag: MVol.B6-301
$AgC_{20}H_{18}N_3O_3$ ...... $Ag(CH_3C_9H_6N)_2NO_3$ .... Ag: MVol.B6-109, 117
$AgC_{20}H_{18}N_3O_3S_2$ .... $[Ag(CH_3SC_9H_6N)_2]NO_3 \cdot CH_3OH$ .... Ag: MVol.B7-61/2
$AgC_{20}H_{18}N_4O_2$ ...... $[Ag(C_3HN_2O(C_6H_5)CH_3)_2]$ .... Ag: MVol.B7-299
$AgC_{20}H_{18}N_4O_8S_2$ .... $Ag(NC_5H_4C_5H_4N)_2(HSO_4)_2$ .... Ag: MVol.B7-292
$AgC_{20}H_{19}N_4O_2$ ...... $AgC_{10}H_9N_2O \cdot C_{10}H_{10}N_2O$ .... Ag: MVol.B6-134
$AgC_{20}H_{20}N_4O_2^{2+}$ .... $[Ag(H_2O)_2(NC_5H_4C_5H_4N)_2]^{2+}$ .... Ag: MVol.B7-288
$AgC_{20}H_{20}N_4O_8S_2$ .... $Ag(C_5H_5N)_4S_2O_8$ .... Ag: MVol.B7-273/4
– .... $Ag(C_5H_5N)_4S_2O_8$ solid solutions
$(Ag_xCd_{1-x})(C_5H_5N)_4S_2O_8$ .... Ag: MVol.B7-274
$(Ag_xCu_{1-x})(C_5H_5N)_4S_2O_8$ .... Ag: MVol.B7-274
$AgC_{20}H_{20}N_5O_3$ ...... $AgNO_3 \cdot 4\ C_5H_5N$ .... Ag: MVol.B6-78
$AgC_{20}H_{20}N_5O_7S_2$ .... $[Ag(CH_3C_6H_4SO_2C_3H_3N_2)_2]NO_3$ .... Ag: MVol.B6-141
$AgC_{20}H_{20}N_6O_6$ ...... $Ag(C_5H_5N)_4(NO_3)_2$ .... Ag: MVol.B7-273
$AgC_{20}H_{22}N_4NaO_6$ ... $Na[Ag((C_3H_5)_2C_4HN_2O_3)_2]$ .... Ag: MVol.B6-154/5
$AgC_{20}H_{22}O_4S_2^-$ ..... $[Ag(C_6H_5CH_2SC_2H_4COO)_2]^-$ .... Ag: MVol.B7-30
$AgC_{20}H_{22}O_6Se_2^-$ .... $[Ag(C_2H_5OC_6H_4SeCH_2COO)_2]^-$ .... Ag: MVol.B7-191/2, 195
$AgC_{20}H_{24}NO_3$ ...... $AgNO_3 \cdot C_{16}H_{12}(CH_3)_4$ .... Ag: MVol.B5-114
$AgC_{20}H_{24}N_5O_3$ ...... $[Ag(CH_3(C_2H_5)C_7H_4N_2)_2]NO_3$ .... Ag: MVol.B6-141
$AgC_{20}H_{28}N_4^+$ ...... $[Ag(C_6H_5C_4H_9N_2)_2]^+$ .... Ag: MVol.B6-163/4
$AgC_{20}H_{28}N_4O_2^+$ .... $[Ag(NC_5H_4CON(C_2H_5)_2)_2]^+$ .... Ag: MVol.B6-103
$AgC_{20}H_{28}N_5O_3$ ...... $AgNO_3 \cdot 2\ NC_5H_4C_4H_7NCH_3$ .... Ag: MVol.B6-107
$AgC_{20}H_{30}IP_2$ .... $[AgI((C_2H_5)_2PC_6H_5)_2]$ .... Ag: MVol.B7-221
$AgC_{20}H_{30}N_2O_2^+$ .... $[Ag(C_5H_{11}C_5H_4NO)_2]^+$ .... Ag: MVol.B6-105
$AgC_{20}H_{32}IN_2P_2$ ..... $[AgI((CH_3)_2PC_6H_4N(CH_3)_2)_2]$ .... Ag: MVol.B7-232
$AgC_{20}H_{32}O_8S_4^{3-}$ .... $[Ag(C_2H_4(SC_3H_6COO)_2)_2]^{3-}$ .... Ag: MVol.B7-42
– .... $[Ag(C_6H_{12}(SCH_2COO)_2)_2]^{3-}$ .... Ag: MVol.B7-39/40
$AgC_{20}H_{34}N_4^{3+}$ ...... $[Ag(C_5H_5NC_5H_{10}NH_2)_2]^{3+}$ .... Ag: MVol.B6-97/8

Formula	Compound	Reference
$AgC_{20}H_{36}NO_3$	$AgNO_3 \cdot 2\ C_{10}H_{18}$	Ag: MVol.B5-67/8
–	$AgNO_3 \cdot C_{20}H_{36}$	Ag: MVol.B5-74
$AgC_{20}H_{36}NO_{15}P_4$	$[Ag(P(OCH_2)_3CCH_3)_4]NO_3$	Ag: MVol.B7-248/9
$AgC_{20}H_{36}N_8^+$	$[Ag((C_3H_7)(CH_3)C_5H_8CN_4)_2]^+$	Ag: MVol.B6-194
–	$[Ag(C_4H_9CH(CH_2)_4CN_4)_2]^+$	Ag: MVol.B6-194
$AgC_{20}H_{36}O_6^+$	$[Ag(C_{20}H_{36}O_6)]^+$	Ag: MVol.B6-219
$AgC_{20}H_{36}O_8S_4^+$	$[Ag(C_2H_4(SC_3H_6COOH)_2)_2]^+$	Ag: MVol.B7-42
$AgC_{20}H_{36}O_{12}P_4^+$	$[Ag(P(OCH_2)_3CCH_3)_4]^+$	Ag: MVol.B7-248
$AgC_{20}H_{42}NO_3S$	$AgNO_3 \cdot S(C_{10}H_{21})_2$	Ag: MVol.B7-13
$AgC_{20}H_{42}NO_4$	$N(C_4H_9)_4Ag(CH_3COO)_2$	Ag: MVol.B5-147
$AgC_{20}H_{42}NS$	$[Ag(SC(CH_3)_2CH_2NHC_{16}H_{33})]$	Ag: MVol.B7-17
$AgC_{20}H_{42}S_2^-$	$[Ag(SC_{10}H_{21})_2]^-$	Ag: MVol.B7-6
$AgC_{20}H_{46}N_2O_4^+$	$[Ag(NH(C_5H_{10}OH)_2)_2]^+$	Ag: MVol.B6-229
$AgC_{20}H_{48}N_9O_3S_4$	$Ag((C_2H_5)_2NCSNH_2)_4NO_3$	Ag: MVol.B7-155
$AgC_{20.5}H_{15.5}N_{2.5}O_3$	$AgOH(C_9H_6NO)_2 \cdot 0.5\ C_5H_5N \cdot H_2O$	Ag: MVol.B7-326
$AgC_{21}ClH_{27}N_3O_{10}$	$Ag(C_2H_5OC_5H_4NO)_2ClO_4 \cdot C_2H_5OC_5H_4NO$	Ag: MVol.B6-105/6
$AgC_{21}ClH_{42}N_6O_3S_3$	$[Ag((CH_3)_3C_4H_5N_2OS)_3]Cl$	Ag: MVol.B7-64/5
$AgC_{21}F_3H_{16}N_4O_3S$	$Ag(NC_5H_4C_5H_4N)_2SO_3CF_3$	Ag: MVol.B6-120
$AgC_{21}H_{13}N_2O_4$	$[AgC_{18}H_8N_2O_2(COOC_2H_5)]$	Ag: MVol.B6-71
$AgC_{21}H_{15}N_2$	$Ag((C_6H_5)_3C_3N_2)$	Ag: MVol.B6-142
$AgC_{21}H_{15}N_4O_6S$	$Ag(NC_5H_4COO)(C_{15}H_{11}N_3)(S_2O_8)_{0.5} \cdot 2\ H_2O$	Ag: MVol.B7-299/300
$AgC_{21}H_{16}^+$	$[Ag(C_{20}H_{13}CH_3)]^+$	Ag: MVol.B5-115
$AgC_{21}H_{16}N_5S$	$[Ag(C_6H_5C_3HN_3S(NC_6H_5)_2)]$	Ag: MVol.B7-68
$AgC_{21}H_{18}N_6S_3^+$	$[Ag(C_7H_6N_2S)_3]^+$	Ag: MVol.B7-52
$AgC_{21}H_{19}IP$	$[AgI(CH_2CHCH_2C_6H_4P(C_6H_5)_2)]$	Ag: MVol.B7-231
$AgC_{21}H_{19}NO_3P$	$[AgNO_3(CH_2CHCH_2C_6H_4P(C_6H_5)_2)]$	Ag: MVol.B7-231
$AgC_{21}H_{21}IP$	$[AgI(P(C_6H_4CH_3)_3)]$	Ag: MVol.B7-227
$AgC_{21}H_{22}N_3O_4S_2$	$[Ag(CH_3SC_9H_6N)_2]NO_3 \cdot CH_3OH$	Ag: MVol.B7-61/2
$AgC_{21}H_{27}N_2O_2$	$[Ag(CH(NC_6H_4OC_4H_9)_2)]$	Ag: MVol.B6-341
$AgC_{21}H_{27}N_3^+$	$[Ag(CH_3C_6H_4NH_2)_3]^+$	Ag: MVol.B6-52
$AgC_{21}H_{29}O_5$	$[Ag(C_5H_2O_3(COC_4H_9)(COC_5H_9)C_5H_9)]$	Ag: MVol.B6-217
$AgC_{21}H_{30}NP_2S$	$[AgSCN((C_2H_5)_2PC_6H_5)_2]$	Ag: MVol.B7-221
$AgC_{21}H_{42}N_7O_6S_3$	$[Ag((CH_3)_3C_4H_5N_2OS)_3]NO_3$	Ag: MVol.B7-64/5
$AgC_{22}ClH_{15}N_4O_8$	$[Ag(C_7H_4NO_4)(C_{15}H_{11}N_3)]ClO_4 \cdot 2\ H_2O$	Ag: MVol.B7-299
$AgC_{22}ClH_{18}N_4O_4$	$[Ag(C_2H_4(NCHC_9H_6N)_2)]ClO_4$	Ag: MVol.B6-280
$AgC_{22}ClH_{22}N_2O_8$	$AgClO_4 \cdot 2\ C_3HNO_2(CH_3)CH_2C_6H_5$	Ag: MVol.B6-203
$AgC_{22}CoH_{18}N_{10}O_6$	$Co(NC_5H_4CHNNC_5H_4N)_2NO_3 \cdot AgNO_3$	Ag: MVol.B6-281/2
$AgC_{22}CuH_{14}Mn_2O_6$	$CuAg[C_5H_4Mn(CO)_3]_2 \cdot C_6H_6$	Ag: MVol.B5-23
$AgC_{22}F_6H_{10}N_4O_6S_2$	$Ag(NC_5H_4C_5H_4N)_2(CF_3SO_3)_2$	Ag: MVol.B7-293/4
$AgC_{22}F_6H_{28}IP_2$	$[AgI((C_2H_5)_2PC_6H_4CF_3)_2]$	Ag: MVol.B7-232/3
$AgC_{22}H_{14}^+$	$[AgC_{22}H_{14}]^+$	Ag: MVol.B5-115
$AgC_{22}H_{14}N_4O_4$	$Ag(C_5H_3N(COO)_2)(C_{15}H_{11}N_3) \cdot 4\ H_2O$	Ag: MVol.B7-299/300
$AgC_{22}H_{15}N_4O_3$	$AgNO_3 \cdot (C_{10}H_7)_2C_2HN_3$	Ag: MVol.B6-186
$AgC_{22}H_{15}N_4O_8S$	$Ag(C_7H_4NO_4)(C_{15}H_{11}N_3)(S_2O_8)_{0.5}$	Ag: MVol.B7-299/300
–	$Ag(C_7H_4NO_4)(C_{15}H_{11}N_3)(S_2O_8)_{0.5} \cdot H_2O$	Ag: MVol.B7-299/300
–	$Ag(C_7H_4NO_4)(C_{15}H_{11}N_3)(S_2O_8)_{0.5} \cdot 2\ H_2O$	Ag: MVol.B7-299/300
$AgC_{22}H_{16}N_5O_3S$	$AgNO_3 \cdot (C_{10}H_7)_2C_2N_2S(NH)_2$	Ag: MVol.B7-80
$AgC_{22}H_{16}N_9O_6$	$[Ag(CH_3C_{10}H_5N_4O_2)_2]NO_2 \cdot 4\ H_2O$	Ag: MVol.B6-174/6
$AgC_{22}H_{16}N_9O_{6.45}$	$[Ag(C_{11}H_8N_4O_2)_2](NO_2)_{0.55}(NO_3)_{0.45} \cdot 4\ H_2O$	Ag: MVol.B6-174/6
$AgC_{22}H_{20}NO_4$	$AgNO_3 \cdot C_8H_8CHC(C_6H_5)_2OH$	Ag: MVol.B5-119

Formula	Compound	Reference
$AgC_{22}H_{20}N_4^+$	$[Ag(CH_2(C_5H_4N)_2)_2]^+$	Ag: MVol.B6-123
$AgC_{22}H_{21}N_2O_2S_2$	$[Ag(CH_3SC_9H_6N)_2]CH_3COO$	Ag: MVol.B7-61/2
$AgC_{22}H_{21}N_4O_3$	$Ag(C_2H_5C_4HN_2O_3C_6H_5) \cdot 2\ C_5H_5N$	Ag: MVol.B6-154
$AgC_{22}H_{22}N_3O_5$	$Ag(CH_3O(CH_3)C_9H_5N)_2NO_3$	Ag: MVol.B6-110
$AgC_{22}H_{22}N_7O_3$	$AgNO_3 \cdot 2\ C_3N_3(CH_3)_2C_6H_5$	Ag: MVol.B6-187
$AgC_{22}H_{30}N_3O$	$[Ag(C_{10}H_{21}OC_6H_4N_3C_6H_5)]$	Ag: MVol.B6-296
$AgC_{22}H_{34}N_4NaO_6$	$Na[Ag(C_2H_5(C_5H_{11})C_4HN_2O_3)_2]$	Ag: MVol.B6-155
$AgC_{22}H_{40}P$	$AgCH_2C(CH_3)_2C_6H_5 \cdot P(C_4H_9)_3$	Ag: MVol.B5-4
–	$[C_6H_5C(CH_3)_2CH_2Ag(P(C_4H_9)_3)]_n$	Ag: MVol.B7-206
$AgC_{22}H_{44}N_2S_4$	$[Ag(SCSN(C_5H_{11})_2)_2]$	Ag: MVol.B7-317
$AgC_{22}H_{48}N_2S_2$	$Ag(SCH(CH_3)CH_2N(C_4H_9)_2)_2$	Ag: MVol.B7-18
$AgC_{23}F_3H_{19}O_2P$	$[Ag(CF_3COCHCOCH_3)(P(C_6H_5)_3)]$	Ag: MVol.B7-209
$AgC_{23}F_6H_{16}O_2P$	$[Ag(CH(COCF_3)_2)(P(C_6H_5)_3)]$	Ag: MVol.B7-209
$AgC_{23}H_{18}N_3O_2$	$AgC_9H_6NO \cdot C_9H_7NO \cdot C_5H_5N$	Ag: MVol.B6-111
$AgC_{23}H_{18}N_7O_4$	$HAg(C_6H_5C_3HN_3O_2)_2 \cdot C_5H_5N$	Ag: MVol.B6-308/12
$AgC_{23}H_{22}NO_2S$	$[Ag(SC_2H_3(NHC(C_6H_5)_3)COOCH_3)]$	Ag: MVol.B6-257
$AgC_{23}H_{22}O_2P$	$[Ag(CH(COCH_3)_2)(P(C_6H_5)_3)]$	Ag: MVol.B7-209
$AgC_{23}H_{25}NPS_2$	$[Ag(S_2CN(C_2H_5)_2)(P(C_6H_5)_3)]$	Ag: MVol.B7-209
$AgC_{23}H_{27}HgN_6O_3$	$[Ag(CH_3C_6H_4NH_2)_3][HgNO_3(CN)_2]$	Ag: MVol.B6-52
$AgC_{23}H_{28}N_3O_2$	$(C_{18}H_{23}N_2O_2)Ag(NC_5H_5)$	Ag: MVol.B6-228
$AgC_{24}ClH_{16}N_4O_4$	$[Ag(C_{12}H_8N_2)_2]ClO_4$	Ag: MVol.B6-125
$AgC_{24}ClH_{22}N_6O_4$	$[Ag(C_6H_5CHNNHC_5H_4N)_2]ClO_4 \cdot 0.5\ H_2O$	Ag: MVol.B6-282
$AgC_{24}ClH_{32}O_4$	$AgClO_4 \cdot 2\ C_6H_5C_6H_{11}$	Ag: MVol.B5-105
$AgC_{24}ClH_{33}N_3O_7$	$[Ag(C_3H_7C_5H_4NO)_2]ClO_4 \cdot C_3H_7C_5H_4NO$	Ag: MVol.B6-105
$AgC_{24}ClH_{36}O_4$	$AgClO_4 \cdot 2\ C_{12}H_{18}$	Ag: MVol.B5-74
$AgC_{24}ClH_{36}O_{16}P_4$	$[Ag(C_6H_9O_3P)_4]ClO_4$	Ag: MVol.B7-249
$AgC_{24}ClH_{54}N_2$	$AgCl \cdot 2\ C_{12}H_{25}NH_2$	Ag: MVol.B6-46
$AgC_{24}ClH_{54}O_6P_2$	$AgCl \cdot 2\ (C_4H_9O)_3P$	Ag: MVol.B7-248
$AgC_{24}ClH_{60}O_{12}P_4$	$[Ag(P(OC_2H_5)_3)_4]Cl$	Ag: MVol.B7-247
$AgC_{24}ClH_{60}O_{16}P_4$	$[Ag(P(OC_2H_5)_3)_4]ClO_4$	Ag: MVol.B7-247
$AgC_{24}Cl_2H_{14}N_4^+$	$[Ag(ClC_{12}H_7N_2)_2]^+$	Ag: MVol.B6-128
$AgC_{24}Cl_2H_{16}N_4O_6$	$Ag(C_{12}H_8N_2)_2(ClO_3)_2$	Ag: MVol.B7-294/5
$AgC_{24}Cl_2H_{16}N_4O_8$	$Ag(C_{12}H_8N_2)_2(ClO_4)_2$	Ag: MVol.B7-294/5
$AgC_{24}Cl_8H_8O_4S_2^{3-}$	$[Ag(S(C_6H_2(O)Cl_2)_2)_2]^{3-}$	Ag: MVol.B7-13
$AgC_{24}Cr_2H_{16}N_4O_7$	$Ag(C_{12}H_8N_2)_2Cr_2O_7$	Ag: MVol.B7-294/5
$AgC_{24}F_2H_{16}N_4$	$Ag(C_{12}H_8N_2)_2F_2$	Ag: MVol.B7-294/5
$AgC_{24}F_6H_{60}O_{12}P_5$	$[Ag(P(OC_2H_5)_3)_4]PF_6$	Ag: MVol.B7-247
$AgC_{24}H_{14}N_3O_6S_3$	$Ag(C_8H_4NO_2S) \cdot 2\ C_8H_5NO_2S$	Ag: MVol.B7-69
$AgC_{24}H_{14}N_7O_7$	$[Ag(NO_2C_{12}H_7N_2)_2]NO_3$	Ag: MVol.B6-128
$AgC_{24}H_{16}N_2^+$	$[Ag((C_6H_5)_2C_{12}H_6N_2)]^+$	Ag: MVol.B6-128
$AgC_{24}H_{16}N_4^+$	$[Ag(C_{12}H_8N_2)_2]^+$	Ag: MVol.B6-124
$AgC_{24}H_{16}N_4^{2+}$	$[Ag(C_{12}H_8N_2)_2]^{2+}$	Ag: MVol.B7-295
$AgC_{24}H_{16}N_4O_8S_2$	$Ag(C_{12}H_8N_2)_2S_2O_8$	Ag: MVol.B7-294/5
$AgC_{24}H_{16}N_5O_3$	$[Ag(C_{12}H_8N_2)_2]NO_3$	Ag: MVol.B6-124/5
–	$[Ag(C_{12}H_8N_2)_2]NO_3 \cdot H_2O$	Ag: MVol.B6-125
$AgC_{24}H_{16}N_6O_6$	$Ag(C_{12}H_8N_2)_2(NO_3)_2$	Ag: MVol.B7-294/5
$AgC_{24}H_{16}O_{12}S_6^{3-}$	$[Ag(S(C_6H_4SO_3)_2)_2]^{3-}$	Ag: MVol.B7-45
$AgC_{24}H_{18}N_4O_8S_2$	$Ag(C_{12}H_8N_2)_2(HSO_4)_2$	Ag: MVol.B7-294/5
$AgC_{24}H_{18}N_7O_3$	$[Ag(NC_5H_4C_7H_5N_2)_2]NO_3$	Ag: MVol.B6-145
$AgC_{24}H_{18}O_6S_2Se_2^-$	$[Ag(C_6H_5SeC_6H_4SO_3)_2]^-$	Ag: MVol.B7-191

Formula	Compound	Reference
$AgC_{24}H_{18}O_6S_4^-$	$[Ag(C_6H_5SC_6H_4SO_3)_2]^-$	Ag: MVol.B7-44/5
$AgC_{24}H_{20}NO_3S_2$	$AgNO_3 \cdot 2\ S(C_6H_5)_2$	Ag: MVol.B7-13
$AgC_{24}H_{20}NO_6P_2$	$AgN(P(O)(OC_6H_5)_2)_2$	Ag: MVol.B7-252
$AgC_{24}H_{20}N_5O_5$	$[Ag(NC_5H_4C(C_6H_5)NOH)_2]NO_3$	Ag: MVol.B6-306
–	$Ag(NC_5H_4CONHC_6H_5)_2NO_3$	Ag: MVol.B6-103
$AgC_{24}H_{22}N_4NaO_6$	$Na[Ag(C_2H_5(C_6H_5)C_4HN_2O_3)_2]$	Ag: MVol.B6-155
–	$Na[Ag(C_2H_5(C_6H_5)C_4HN_2O_3)_2] \cdot H_2O$	Ag: MVol.B6-154/5
$AgC_{24}H_{22}N_4O_2S_4^-$	$[Ag((CH_3)_2NC_6H_4CHC_3NOS_2)_2]^-$	Ag: MVol.B7-74
$AgC_{24}H_{22}N_4O_6^-$	$[Ag(C_2H_5(C_6H_5)C_4HN_2O_3)_2]^-$	Ag: MVol.B6-155
$AgC_{24}H_{23}N_6O$	$AgOH \cdot 2\ C_6H_5NNC_6H_4NH_2$	Ag: MVol.B6-286
$AgC_{24}H_{24}K_2N_3O_6S_9$	$K_2[Ag(C_6H_5SO_2NCSSCH_3)_3]$	Ag: MVol.B7-104
$AgC_{24}H_{24}N_4^+$	$[Ag(C_2H_4(C_5H_4N)_2)_2]^+$	Ag: MVol.B6-123
$AgC_{24}H_{24}N_4O_4^+$	$[Ag(C_2H_4(NCHC_4H_3O)_2)_2]^+$	Ag: MVol.B6-280
$AgC_{24}H_{24}N_4O_{12}S_4^{3-}$	$[Ag(NH_2C_6H_4SO_3)_4]^{3-}$	Ag: MVol.B6-262
$AgC_{24}H_{24}N_4S_4^+$	$[Ag(C_2H_4(NCHC_4H_3S)_2)_2]^+$	Ag: MVol.B6-280
$AgC_{24}H_{24}N_5O_3$	$AgNO_3 \cdot 2\ NH_2C_6H_4C_6H_4NH_2$	Ag: MVol.B6-64
$AgC_{24}H_{24}N_5O_3S_4$	$[Ag(C_2H_4(NCHC_4H_3S)_2)_2]NO_3$	Ag: MVol.B6-280
$AgC_{24}H_{24}N_5O_7$	$[Ag(C_2H_4(NCHC_4H_3O)_2)_2]NO_3$	Ag: MVol.B6-280
$AgC_{24}H_{25}N_4O_3$	$Ag((CH_3)_2C_6H_3(OH)_2NNC_6H_4O) \cdot 2\ C_5H_5N$	Ag: MVol.B6-285
$AgC_{24}H_{26}N_4O_4$	$[Ag(C_5H_5N)_4](CH_3COO)_2$	Ag: MVol.B5-148
$AgC_{24}H_{26}N_6^+$	$[Ag(NH(CH_2C_5H_4N)_2)_2]^+$	Ag: MVol.B6-98
$AgC_{24}H_{27}N_6O_2S$	$Ag(NHC_6H_4SO_2NHC_4H_3N_2) \cdot 2\ C_6H_5CH_2NH_2$	Ag: MVol.B6-264
$AgC_{24}H_{27}O_9S_6^{2-}$	$[Ag(C_2H_5SC_6H_4SO_3)_3]^{2-}$	Ag: MVol.B7-44
$AgC_{24}H_{28}N_3O_6S$	$[AgC_{24}H_{28}N_3O_6S]$	Ag: MVol.B6-269
$AgC_{24}H_{28}N_4O_8S_2$	$Ag(CH_3C_5H_4N)_4S_2O_8$	Ag: MVol.B7-274/5
$AgC_{24}H_{30}N_3O_9S_3^{2-}$	$[Ag((CH_3)_2NC_6H_4SO_3)_3]^{2-}$	Ag: MVol.B6-262
$AgC_{24}H_{30}N_4NaO_6$	$Na[Ag(C_6H_9C_4N_2O_3(CH_3)_2)_2]$	Ag: MVol.B6-155/6
$AgC_{24}H_{33}N_3^+$	$[Ag((CH_3)_2C_6H_3NH_2)_3]^+$	Ag: MVol.B6-52/3
$AgC_{24}H_{36}NO_3$	$AgNO_3 \cdot 2\ C_{12}H_{18}$	Ag: MVol.B5-74
$AgC_{24}H_{36}NO_{15}P_4$	$[Ag(C_6H_9O_3P)_4]NO_3$	Ag: MVol.B7-249
$AgC_{24}H_{36}N_4^+$	$[Ag(CH_3(CH_3C_6H_4)C_4H_8N_2)_2]^+$	Ag: MVol.B6-163/4
$AgC_{24}H_{36}N_4O_2^+$	$[Ag(CH_3(CH_3OC_6H_4)C_4H_8N_2)_2]^+$	Ag: MVol.B6-163/4
$AgC_{24}H_{40}IN_2P_2$	$[AgI((C_2H_5)_2PC_6H_4N(CH_3)_2)_2]$	Ag: MVol.B7-232
$AgC_{24}H_{40}NO_3S_6$	$AgNO_3 \cdot 2\ (CH_3)_3CC_3S_3CH_2C(CH_3)_3$	Ag: MVol.B7-82
$AgC_{24}H_{40}O_8S_4^{3-}$	$[Ag(C_2H_4(SCH_2C(CH_3)_2COO)_2)_2]^{3-}$	Ag: MVol.B7-42
$AgC_{24}H_{44}NO_3$	$AgNO_3 \cdot 2\ C_{12}H_{22}$	Ag: MVol.B5-73
$AgC_{24}H_{44}NO_{15}P_4$	$[Ag(P(OCH_2)_3CC_2H_5)_4]NO_3$	Ag: MVol.B7-249
$AgC_{24}H_{44}O_8S_4^+$	$[Ag(C_2H_4(SCH_2C(CH_3)_2COOH)_2)_2]^+$	Ag: MVol.B7-42
$AgC_{24}H_{48}N_4O_6^+$	$[Ag(O(C_2H_4N(C_2H_4OC_2H_4)_2NC_2H_4)_2O)]^+$	Ag: MVol.B6-205
$AgC_{24}H_{48}N_5O_9$	$AgNO_3 \cdot C_{24}H_{48}N_4O_6$	Ag: MVol.B6-205
$AgC_{24}H_{52}NO_3S_2$	$AgNO_3 \cdot 2\ S(C_6H_{13})_2$	Ag: MVol.B7-12
$AgC_{24}H_{54}N_2^+$	$[Ag(N(C_4H_9)_3)_2]^+$	Ag: MVol.B6-49
$AgC_{24}H_{54}N_3O_3$	$[AgNO_3(C_{12}H_{25}NH_2)_2]$	Ag: MVol.B6-46
$AgC_{24}H_{54}P_2^+$	$[Ag(P(C_4H_9)_3)_2]^+$	Ag: MVol.B7-205
$AgC_{24}H_{60}IO_{12}P_4$	$[Ag(P(OC_2H_5)_3)_4]I$	Ag: MVol.B7-247
$AgC_{24}H_{60}NO_{15}P_4$	$[Ag(P(OC_2H_5)_3)_4]NO_3$	Ag: MVol.B7-247
$AgC_{24}H_{60}O_{12}P_4^+$	$[Ag(P(OC_2H_5)_3)_4]^+$	Ag: MVol.B7-247
$AgC_{24}H_{64}N_8^+$	$[Ag(C_2H_4(N(CH_3)_2)_2)_4]^+$	Ag: MVol.B6-60
$AgC_{24}H_{72}N_{12}O_4P_4^+$	$[Ag(OP(N(CH_3)_2)_3)_4]^+$	Ag: MVol.B7-250/1
$AgC_{25}ClH_{52}N_6O_4S_3$	$[Ag((CH_3)_3C_4H_5N_2OS)_3]Cl \cdot (C_2H_5)_2O$	Ag: MVol.B7-65

$AgC_{25}H_{17}N_2O_2$ ...... $Ag(C_6H_5C_{19}H_9NO_2) \cdot NH_3$ ...... Ag: MVol.B6-106/7
$AgC_{25}H_{19}N_4O_3$ ...... $Ag(HOC_6H_4NNC_9H_4O_2) \cdot 2\ C_5H_5N$ ...... Ag: MVol.B6-216
$AgC_{25}H_{19}N_4S$ ...... $[Ag(CS(NNC_6H_4C_6H_5)_2H)]$ ...... Ag: MVol.B7-185
$AgC_{25}H_{26}N_3O_2$ ...... $AgC_6H_5COO \cdot 3\ C_6H_5NH_2$ ...... Ag: MVol.B6-51
$AgC_{25}H_{26}N_3O_3$ ...... $AgC_6H_4(OH)COO \cdot 3\ C_6H_5NH_2$ ...... Ag: MVol.B6-51
$AgC_{25}H_{27}N_4O_2$ ...... $AgC_6H_4(NH_2)COO \cdot 3\ C_6H_5NH_2$ ...... Ag: MVol.B6-51
$AgC_{25}H_{27}N_4O_{10}$ ...... $[Ag((CH_3)_2C_{10}H_2N_4O_2C_{13}H_{19}O_8)]$ ...... Ag: MVol.B6-173
$AgC_{25}H_{32}N_3O_2$ ...... $(C_{20}H_{27}N_2O_2)Ag(NC_5H_5)$ ...... Ag: MVol.B6-228
$AgC_{25}H_{60}NO_{12}P_4$ ...... $[Ag(P(OC_2H_5)_3)_4]CN$ ...... Ag: MVol.B7-247
$AgC_{25}H_{60}NO_{12}P_4S$ ...... $[Ag(P(OC_2H_5)_3)_4]SCN$ ...... Ag: MVol.B7-247
$AgC_{26}ClH_{20}N_4O_4$ ...... $[Ag(CH_3C_{12}H_7N_2)_2]ClO_4$ ...... Ag: MVol.B6-126
$AgC_{26}ClH_{20}P_2$ ...... $[AgCl((C_6H_5)_2PCCP(C_6H_5)_2)]$ ...... Ag: MVol.B7-235/6
$AgC_{26}ClH_{26}P_2$ ...... $[AgCl(CH_3P(C_6H_5)_2)_2]$ ...... Ag: MVol.B7-220
$AgC_{26}H_{18}N_2^+$ ...... $[Ag(C_{13}H_9N)_2]^+$ ...... Ag: MVol.B6-116
$AgC_{26}H_{19}N_2O_4S_2$ ...... $Ag(C_6H_5SO_2NC_{14}H_8NHSO_2C_6H_5)$ ...... Ag: MVol.B6-343
$AgC_{26}H_{20}IP_2$ ...... $[AgI((C_6H_5)_2PCCP(C_6H_5)_2)]$ ...... Ag: MVol.B7-235/6
$AgC_{26}H_{20}NO_3P_2$ ...... $[AgNO_3((C_6H_5)_2PCCP(C_6H_5)_2)]$ ...... Ag: MVol.B7-235/6
$AgC_{26}H_{20}N_4^+$ ...... $[Ag(CH_3C_{12}H_7N_2)_2]^+$ ...... Ag: MVol.B6-126
$AgC_{26}H_{20}N_5O_3$ ...... $[Ag(CH_3C_{12}H_7N_2)_2]NO_3$ ...... Ag: MVol.B6-126
$AgC_{26}H_{21}N_2O_5$ ...... $Ag(C_9H_7NO)_2(C_6H_5CHOHCOO)$ ...... Ag: MVol.B6-112
$AgC_{26}H_{24}IP_2$ ...... $[AgI((C_6H_5)_2PC_2H_4P(C_6H_5)_2)]_n$ ...... Ag: MVol.B7-235
$AgC_{26}H_{24}NO_3P_2$ ...... $[AgNO_3((C_6H_5)_2PC_2H_4P(C_6H_5)_2)] \cdot 2\ H_2O$ ...... Ag: MVol.B7-234/5
$AgC_{26}H_{24}N_4S_2^+$ ...... $[Ag(C_6H_5NHCSNHC_6H_5)_2]^+$ ...... Ag: MVol.B7-158
$AgC_{26}H_{26}N_3O_2P_2$ ...... $[AgN_3((C_6H_5)_2POCH_3)_2]$ ...... Ag: MVol.B7-240
$AgC_{26}H_{26}N_3P_2$ ...... $[AgN_3(CH_3P(C_6H_5)_2)_2]$ ...... Ag: MVol.B7-220
$AgC_{26}H_{26}N_4NaO_6$ ...... $Na[Ag(C_2H_5(C_6H_5)C_4N_2O_3CH_3)_2]$ ...... Ag: MVol.B6-155
$AgC_{26}H_{26}N_6^+$ ...... $[Ag(C_6H_5NHC(NH)NHC_6H_5)_2]^+$ ...... Ag: MVol.B6-337
$AgC_{26}H_{26}N_7O_3$ ...... $[Ag(NHC(NHC_6H_5)_2)_2]NO_3$ ...... Ag: MVol.B6-337
$AgC_{26}H_{28}N_4^+$ ...... $[Ag(C_3H_6(C_5H_4N)_2)_2]^+$ ...... Ag: MVol.B6-123
$AgC_{27}F_6H_{36}P$ ...... $AgPF_6 \cdot 3\ C_6H_3(CH_3)_3$ ...... Ag: MVol.B5-99
$AgC_{27}H_{19}N_2$ ...... $Ag((C_{12}H_9)_2C_3HN_2)$ ...... Ag: MVol.B6-142
$AgC_{27}H_{20}NP_2S$ ...... $[AgSCN((C_6H_5)_2PCCP(C_6H_5)_2)]$ ...... Ag: MVol.B7-235/6
$AgC_{27}H_{20}N_3O_3P_2$ ...... $AgOCN \cdot 2\ (C_6H_5)_2PNCO$ ...... Ag: MVol.B7-241
$AgC_{27}H_{20}N_9O_6$ ...... $[H_2Ag(C_6H_5C_3HN_3O_2)_3]$ ...... Ag: MVol.B6-308/12
$AgC_{27}H_{26}NP_2S$ ...... $[AgSCN(CH_3P(C_6H_5)_2)_2]$ ...... Ag: MVol.B7-220
$AgC_{27}H_{54}N_3S_6$ ...... $[Ag((C_4H_9)_2NCSS)_3]$ ...... Ag: MVol.B7-326
$AgC_{28}ClH_{20}O_4$ ...... $AgClO_4 \cdot 2\ C_{14}H_{10}$ ...... Ag: MVol.B5-111
$AgC_{28}ClH_{24}N_4O_4$ ...... $[Ag((CH_3)_2C_{12}H_6N_2)_2]ClO_4$ ...... Ag: MVol.B6-127
$AgC_{28}ClH_{26}P_2$ ...... $[AgCl(C_6H_4(CH_2)_2PC_6H_5)_2]$ ...... Ag: MVol.B7-222
$AgC_{28}ClH_{28}P_2S$ ...... $[AgCl((C_6H_5)_2PCH_2CH_2)_2S]_2$ ...... Ag: MVol.B7-237/9
$AgC_{28}ClH_{32}N_4O_4$ ...... $Ag((CH_3)_4C_{10}H_4N_2)_2ClO_4$ ...... Ag: MVol.B6-121/2
$AgC_{28}H_{19}N_4O_4$ ...... $[AgH(ONC_8H_4NOC_6H_5)_2]$ ...... Ag: MVol.B6-316
$AgC_{28}H_{21}N_6$ ...... $Ag(C_6H_5)_2C_2N_3 \cdot (C_6H_5)_2C_2HN_3$ ...... Ag: MVol.B6-185/6
$AgC_{28}H_{24}N_4^+$ ...... $[Ag((CH_3)_2C_{12}H_6N_2)_2]^+$ ...... Ag: MVol.B6-126, 128
$AgC_{28}H_{24}N_5O_3$ ...... $[Ag((CH_3)_2C_{12}H_6N_2)_2]NO_3$ ...... Ag: MVol.B6-127
– ...... $[Ag(C_6H_5CH_2C_7H_5N_2)_2]NO_3$ ...... Ag: MVol.B6-145
$AgC_{28}H_{24}O_2P$ ...... $[Ag(C_6H_5COCHCOCH_3)(P(C_6H_5)_3)]$ ...... Ag: MVol.B7-208
$AgC_{28}H_{26}IP_2$ ...... $[AgI(C_6H_4(CH_2)_2PC_6H_5)_2]$ ...... Ag: MVol.B7-222
$AgC_{28}H_{26}NO_3P_2$ ...... $[Ag(CH_2CHP(C_6H_5)_2)_2]NO_3$ ...... Ag: MVol.B7-222/3
– ...... $[AgNO_3(C_6H_4(CH_2)_2PC_6H_5)_2]$ ...... Ag: MVol.B7-222

$AgC_{28}H_{28}IP_2$ .... $[AgI((C_6H_5)_2P(CH_2)_4P(C_6H_5)_2)]_n$ .... Ag: MVol.B7-237
$AgC_{28}H_{28}IP_2S$ ...... $[AgI((C_6H_5)_2PCH_2CH_2)_2S]_2$ .... Ag: MVol.B7-237/9
$AgC_{28}H_{28}NO_3S_2$ .... $AgNO_3 \cdot 2\ S(CH_2C_6H_5)_2$ .... Ag: MVol.B7-14
$AgC_{28}H_{28}N_8^+$ ...... $[Ag(C_2H_4(NCHC_5H_4N)_2)_2]^+$ .... Ag: MVol.B6-280
$AgC_{28}H_{29}N_2O_6S$ .... $Ag(C_9H_7NO)_2C_{10}H_{15}O_4S$ .... Ag: MVol.B6-112
$AgC_{28}H_{30}N_2^+$ ...... $[Ag(NH(CH_2C_6H_5)_2)_2]^+$ .... Ag: MVol.B6-54
$AgC_{28}H_{32}N_4^+$ ...... $[Ag(C_4H_8(C_5H_4N)_2)_2]^+$ .... Ag: MVol.B6-123
$AgC_{28}H_{32}N_5O_3$ ...... $Ag((CH_3)_4C_{10}H_4N_2)_2NO_3$ .... Ag: MVol.B6-121
$AgC_{28}H_{35}N_6O_6S$ .... $Ag(NHC_6H_4SO_2NHC_4H_3N_2)$
$\cdot\ 2\ HOC_6H_4CHOHCH_2NHCH_3$ .... Ag: MVol.B6-264
$AgC_{28}H_{44}NO_3S_2$ .... $AgNO_3 \cdot 2\ C_6H_5SC_8H_{17}$ .... Ag: MVol.B7-14
$AgC_{28}H_{52}NO_3$ ...... $AgNO_3 \cdot 2\ C_{10}H_{14}(CH_3)_4$ .... Ag: MVol.B5-68
$AgC_{29}H_{15}N_2O_4$ ...... $[Ag(CH(NC_6H_3(CO)_2C_6H_4)_2)]$ .... Ag: MVol.B6-341
$AgC_{29}H_{28}NP_2S_2$ ..... $[AgSCN((C_6H_5)_2PCH_2CH_2)_2S]_2$ .... Ag: MVol.B7-237/9
$AgC_{29}H_{30}NO_2P_2S$ ... $[AgSCN((C_6H_5)_2POC_2H_5)_2]$ .... Ag: MVol.B7-241
$AgC_{30}ClH_{34}N_2O_4S_2$ .. $[Ag(C_6H_{11}SC_9H_6N)_2]ClO_4$ .... Ag: MVol.B7-62
$AgC_{30}ClH_{34}P_2S^+$ .... $[AgCl((C_6H_5)_2(CH_3)PCH_2CH_2)_2S]^+$ .... Ag: MVol.B7-238/9
$AgC_{30}Cl_2H_{24}N_6O_6$ ... $Ag(ClO_3)_2 \cdot 3\ NC_5H_4C_5H_4N$ .... Ag: MVol.B7-287, 291/2
$AgC_{30}Cl_2H_{24}N_6O_8$ ... $Ag(ClO_4)_2 \cdot 3\ NC_5H_4C_5H_4N$ .... Ag: MVol.B7-287, 292
$AgC_{30}H_{21}N_4O_6S_3$ .... $[Ag(SO_2(C_6H_4SO_2NC_9H_6N)_2H)]$ .... Ag: MVol.B6-115
$AgC_{30}H_{22}N_6O_8S_2$ .... $Ag(NC_5H_3(C_5H_4N)_2)_2S_2O_8 \cdot 3\ H_2O$ .... Ag: MVol.B7-297
$AgC_{30}H_{23}N_4O_4$ ...... $[AgH(ONC_8H_4NOCH_2C_6H_5)_2]$ .... Ag: MVol.B6-316
$AgC_{30}H_{24}N_8O_6$ ...... $Ag(NO_3)_2 \cdot 3\ NC_5H_4C_5H_4N$ .... Ag: MVol.B7-287, 291
$AgC_{30}H_{28}N_2S_4$ ...... $[Ag(SCSN(CH_2C_6H_5)_2)_2]$ .... Ag: MVol.B7-317
$AgC_{30}H_{28}N_6O_2$ ...... $Ag(C_3HN_2O(C_6H_5)CH_3)_2 \cdot 2\ C_5H_5N$ .... Ag: MVol.B7-299
$AgC_{30}H_{28}O_4P_2^-$ ..... $[Ag((C_6H_5)_2PCH_2CH_2COO)_2]^-$ .... Ag: MVol.B7-222
$AgC_{30}H_{30}^+$ .... $[Ag(C_{10}H_{10})_3]^+$ .... Ag: MVol.B5-84
$AgC_{30}H_{30}N_7O_3$ ...... $AgNO_3 \cdot 6\ C_5H_5N$ .... Ag: MVol.B6-76/7, 78
$AgC_{30}H_{32}N_5O_3$ ...... $AgNO_3 \cdot 2\ CH_3C_6H_4NCHNHC_6H_4CH_3$ ..... Ag: MVol.B6-340, 342
$AgC_{30}H_{33}N_2O_4S_3$ .... $[Ag(CH_3SC_9H_6N)_2]C_{10}H_{15}O_4S \cdot 2.5\ H_2O$ ... Ag: MVol.B7-61/2
$AgC_{30}H_{33}OP_2S_2$ ..... $[Ag(SC_2H_4OH)((C_6H_5)_2PC_2H_4)_2S]_2$ .... Ag: MVol.B7-237/9
$AgC_{30}H_{34}IO_2P_2$ ..... $[AgI((C_2H_5)(C_6H_5)PC_6H_4OCH_3)_2]$ .... Ag: MVol.B7-232
$AgC_{30}H_{34}IP_2S^+$ ..... $[AgI((C_6H_5)_2(CH_3)PCH_2CH_2)_2S]^+$ .... Ag: MVol.B7-238/9
$AgC_{30}H_{36}K_2N_3O_6S_9$ .. $K_2[Ag(C_6H_5SO_2NCSSC_3H_7)_3]$ .... Ag: MVol.B7-104
$AgC_{30}H_{52}IP_4$ .... $[Ag(CH_3C_6H_3(P(C_2H_5)_2)_2)_2]I$ .... Ag: MVol.B7-239
$AgC_{32}ClCu_2H_{28}N_4O_8$. $Ag(CuC_2H_4(NCHC_6H_4O)_2)_2ClO_4 \cdot 2\ H_2O$ ... Ag: MVol.B6-279/80
$AgC_{32}ClH_{26}N_2O_4S_2$ .. $[Ag(C_6H_5CH_2SC_9H_6N)_2]ClO_4$ .... Ag: MVol.B7-62
$AgC_{32}ClH_{28}N_4O_4$ .... $[Ag(CH_3C_6H_4NC)_4]ClO_4$ .... Ag: MVol.B5-25
$AgC_{32}H_{16}N_8$ .... $Ag(C_{32}H_{16}N_8)$ .... Ag: MVol.B7-315/6
$AgC_{32}H_{17}N_8$ .... $Ag(C_{32}H_{17}N_8)$ .... Ag: MVol.B7-315/6
$AgC_{32}H_{24}N_5O_7$ ...... $AgNO_3 \cdot 2\ C_6H_5COC_3H_2N_2OC_6H_5$ .... Ag: MVol.B6-147
$AgC_{32}H_{26}N_3O_3S_2$ .... $[Ag(C_6H_5CH_2SC_9H_6N)_2]NO_3$ .... Ag: MVol.B7-62
$AgC_{32}H_{28}N_5O_3$ ...... $[Ag(CH_3C_6H_4NC)_4]NO_3 \cdot H_2O$ .... Ag: MVol.B5-25
$AgC_{32}H_{32}N_4^+$ ...... $[Ag(C_2H_4(NCHC_6H_5)_2)_2]^+$ .... Ag: MVol.B6-279
$AgC_{32}H_{32}N_4O_4$ ...... $[Ag(C_{20}H_6N_4(CH_3)_4(C_2H_4COOCH_3)_2)]$ ...... Ag: MVol.B7-304, 310
$AgC_{32}H_{32}N_5O_3$ ...... $[Ag(C_2H_4(NCHC_6H_5)_2)_2]NO_3$ .... Ag: MVol.B6-279
$AgC_{32}H_{36}N_4$ .... $[Ag(C_{20}H_4N_4(CH_3)_4(C_2H_5)_4)]$ .... Ag: MVol.B7-304, 309
$AgC_{32}H_{36}O_{12}S_8^{3-}$ .... $[Ag(C_2H_5SC_6H_4SO_3)_4]^{3-}$ .... Ag: MVol.B7-44
$AgC_{32}H_{40}N_4O_{12}S_4^{3-}$ .. $[Ag((CH_3)_2NC_6H_4SO_3)_4]^{3-}$ .... Ag: MVol.B6-262
$AgC_{32}H_{45}N_2O_5S$ .... $Ag(C_{17}H_{35}C_3H_2N_2OC_6H_3(SO_3)OC_6H_5)$ ...... Ag: MVol.B6-134

$AgC_{32}H_{58}IP_2$	$[AgI((C_6H_{11})_2PCH_2CH_2CHCH_2)_2]$	Ag: MVol.B7-207/8
$AgC_{32}H_{68}NO_3S_2$	$AgNO_3 \cdot 2\ S(C_8H_{17})_2$	Ag: MVol.B7-12/3
$AgC_{33}ClH_{26}N_3O_4P$	$Ag(NC_5H_3(C_5H_4N)_2)(P(C_6H_5)_3)ClO_4$	Ag: MVol.B6-122/3
–	$[AgP(C_6H_5)_3(NC_5H_4CHNC_9H_6N)]ClO_4$	Ag: MVol.B6-281
$AgC_{33}ClH_{36}N_6O_7$	$[Ag(C_6H_5C_3HN_2O(CH_3)_2)_3]ClO_4$	Ag: MVol.B6-134
$AgC_{33}H_{34}N_4O_4$	$[Ag(C_2H_5C_{20}H_4N_4(CH_3)_5(C_2H_4COOH)_2)]$	Ag: MVol.B7-305, 311
$AgC_{34}CrH_{28}N_8S_4$	$[Ag(C_9H_7N)_2][Cr(NCS)_4(C_6H_5NH_2)_2]$	Ag: MVol.B6-108/9, 116/7
$AgC_{34}H_{28}N_3O_3P_2$	$[AgNO_3(NC_5H_4P(C_6H_5)_2)_2]$	Ag: MVol.B7-242
$AgC_{34}H_{30}N_4O_5$	$[Ag(C_{34}H_{30}N_4O_5)]$	Ag: MVol.B7-306, 312
$AgC_{34}H_{31}N_4O_4S_2$	$AgC_{17}H_{15}N_2O_2S \cdot C_{17}H_{16}N_2O_2S$	Ag: MVol.B6-115
$AgC_{34}H_{34}N_4O_6$	$[Ag(C_{34}H_{34}N_4O_6)]$	Ag: MVol.B7-306, 312
$AgC_{34}H_{36}N_4O_6$	$[Ag(C_{34}H_{36}N_4O_6)]$	Ag: MVol.B7-304, 310
$AgC_{34}H_{38}N_8NaO_{12}$	$Na[Ag((CH_3)_2C_{10}H_2N_4O_2C_5H_{11}O_4)_2]$	Ag: MVol.B6-178
$AgC_{34}H_{38}N_8O_{12}^-$	$[Ag((CH_3)_2C_{10}H_2N_4O_2C_5H_{11}O_4)_2]^-$	Ag: MVol.B6-173, 177
$AgC_{35}H_{25}$	$AgC_5(C_6H_5)_5$	Ag: MVol.B5-54
$AgC_{35}H_{35}OP_2S_2$	$[Ag(SC_6H_4OCH_3)((C_6H_5)_2PC_2H_4)_2S]_2$	Ag: MVol.B7-237/9
$AgC_{35}H_{38}N_4O_4$	$[Ag(C_2H_5C_{20}H_4N_4(CH_3)_5(C_2H_4COOCH_3)_2)]$	Ag: MVol.B7-305, 311/2
$AgC_{36}ClH_{28}O_4$	$AgClO_4 \cdot 2\ C_5H_4C(C_6H_5)_2$	Ag: MVol.B5-106
$AgC_{36}ClH_{30}N_6O_{16}$	$AgClO_4 \cdot 6\ C_6H_5NO_2$	Ag: MVol.B5-118
$AgC_{36}ClH_{30}O_4P_2Se_2$	$[Ag((C_6H_5)_3PSe)_2]ClO_4$	Ag: MVol.B7-254/5
$AgC_{36}ClH_{42}N_6O_4$	$AgClO_4 \cdot 6\ C_6H_5NH_2$	Ag: MVol.B6-51
$AgC_{36}ClH_{44}N_4O_4$	$[Ag(C_{20}H_4N_4(C_2H_5)_8)]ClO_4$	Ag: MVol.B7-304, 309
$AgC_{36}Cl_2H_{82}O_{12}P_3$	$HAgCl_2 \cdot 3\ (C_4H_9O)_3PO \cdot x\ H_2O$	Ag: MVol.B7-250
$AgC_{36}CrH_{32}N_8S_4$	$[Ag(C_9H_7N)_2][Cr(NCS)_4(CH_3C_6H_4NH_2)_2]$	Ag: MVol.B6-108/9, 116/7
$AgC_{36}GeH_{30}P$	$[AgGe(C_6H_5)_3(P(C_6H_5)_3)]$	Ag: MVol.B7-219
$AgC_{36}H_{24}N_4^+$	$[Ag(C_6H_5C_{12}H_7N_2)_2]^+$	Ag: MVol.B6-128
–	$[Ag(NC_9H_6C_9H_6N)_2]^+$	Ag: MVol.B6-115
$AgC_{36}H_{24}N_5O_3$	$Ag(NC_9H_6C_9H_6N)_2NO_3 \cdot x\ H_2O$	Ag: MVol.B6-115
$AgC_{36}H_{24}O_{18}S_9^{5-}$	$[Ag(S(C_6H_4SO_3)_2)_3]^{5-}$	Ag: MVol.B7-45
$AgC_{36}H_{27}O_9S_3Se_3^{2-}$	$[Ag(C_6H_5SeC_6H_4SO_3)_3]^{2-}$	Ag: MVol.B7-191
$AgC_{36}H_{27}O_9S_6^{2-}$	$[Ag(C_6H_5SC_6H_4SO_3)_3]^{2-}$	Ag: MVol.B7-44/5
$AgC_{36}H_{28}N_5O_3$	$Ag(C_9H_7N)_4NO_3$	Ag: MVol.B6-108
$AgC_{36}H_{28}O_6P_2S_2^-$	$[Ag((C_6H_5)_2PC_6H_4SO_3)_2]^-$	Ag: MVol.B7-233
$AgC_{36}H_{30}NO_8P_2S_2$	$[Ag((C_6H_5O)_3PS)_2]NO_2$	Ag: MVol.B7-253
$AgC_{36}H_{30}NO_9P_2S_2$	$[Ag((C_6H_5O)_3PS)_2]NO_3$	Ag: MVol.B7-253
$AgC_{36}H_{30}N_3P_2$	$[AgN_3(P(C_6H_5)_3)_2]$	Ag: MVol.B7-209/10
$AgC_{36}H_{30}P_2^+$	$[Ag(P(C_6H_5)_3)_2]^+$	Ag: MVol.B7-213
$AgC_{36}H_{34}N_5O_3P_2$	$AgNO_3 \cdot 2\ NC_5H_3(CH_3)NHP(C_6H_5)_2$	Ag: MVol.B7-242
$AgC_{36}H_{36}N_4O_4$	$[Ag(C_{36}H_{36}N_4O_4)]$	Ag: MVol.B7-305, 311
$AgC_{36}H_{36}N_4O_6$	$[Ag(C_{36}H_{36}N_4O_6)]$	Ag: MVol.B7-304, 310
$AgC_{36}H_{36}N_4O_8$	$[Ag(C_{20}H_4N_4(CH_3)_4(C_2H_4COOH)_4)]$	Ag: MVol.B7-305, 311
$AgC_{36}H_{40}N_4O_4$	$[Ag(C_{20}H_2N_4(CH_3)_6(C_2H_5)_2(C_2H_4COOH)_2)]$	Ag: MVol.B7-305, 312
–	$[Ag(C_{20}H_4N_4(CH_3)_4(C_2H_5)_2(C_2H_4COOCH_3)_2)]$	Ag: MVol.B7-304, 310
$AgC_{36}H_{40}N_4O_6$	$[Ag(C_{36}H_{40}N_4O_6)]$	Ag: MVol.B7-305, 310
$AgC_{36}H_{44}N_4$	$[Ag(C_{20}H_4N_4(CH_3)_4(C_3H_7)_4)]$	Ag: MVol.B7-304, 310
–	$[Ag(C_{20}H_4N_4(C_2H_5)_8)]$	Ag: MVol.B7-304, 309
$AgC_{36}H_{54}N_3O_7P_4$	$[Ag(CH_3C_6H_3(P(C_2H_5)_2)_2)_2]OC_6H_2(NO_2)_3$	Ag: MVol.B7-239
$AgC_{36}H_{66}P_2^+$	$[Ag(P(C_6H_{11})_3)_2]^+$	Ag: MVol.B7-207
$AgC_{36}H_{81}P_3^+$	$[Ag(P(C_4H_9)_3)_3]^+$	Ag: MVol.B7-205
$AgC_{37}H_{30}NOP_2$	$[AgCNO(P(C_6H_5)_3)_2]$	Ag: MVol.B7-211

Formula	Compound	Reference
$AgC_{37}H_{30}NP_2S$	$[AgSCN(P(C_6H_5)_3)_2]$	Ag: MVol.B7-211/2
$AgC_{37}H_{30}NSSb_2$	$[AgSCN((C_6H_5)_3Sb)_2]$	Ag: MVol.B7-269
$AgC_{37}H_{30}N_3P_2S_2$	$[Ag(SCSN_3)(P(C_6H_5)_3)_2]$	Ag: MVol.B7-213
$AgC_{37}H_{40}N_4O_6$	$[Ag(C_{37}H_{40}N_4O_6)]$	Ag: MVol.B7-306, 312
$AgC_{37}H_{66}NOP_2$	$[AgCNO(P(C_6H_{11})_3)_2]$	Ag: MVol.B7-207
$AgC_{37}H_{66}NP_2S$	$[AgSCN(P(C_6H_{11})_3)_2]$	Ag: MVol.B7-207
$AgC_{38}ClH_{34}P_2$	$[AgCl((C_6H_5)_2PC_6H_4CH_3)_2]$	Ag: MVol.B7-230
$AgC_{38}F_3H_{30}N_4P_2$	$[Ag(CF_3CN_4)(P(C_6H_5)_3)_2]_2$	Ag: MVol.B7-213
$AgC_{38}H_{34}N_3P_2$	$[AgN_3((C_6H_5)_2PC_6H_4CH_3)_2]$	Ag: MVol.B7-230
$AgC_{38}H_{36}N_{13}O_{21}$	$AgC_6H_2N_3O_7 \cdot 2\ (C_6H_3N_3O_7 \cdot C_{10}H_{14}N_2)$	Ag: MVol.B6-107
$AgC_{38}H_{44}N_4O_6$	$[Ag(C_{38}H_{44}N_4O_6)]$	Ag: MVol.B7-305, 310
$AgC_{39}H_{34}NP_2S$	$[AgSCN((C_6H_5)_2PC_6H_4CH_3)_2]$	Ag: MVol.B7-230/1
$AgC_{39}H_{34}N_3P_2S_2$	$[AgSCSN_3((C_6H_5)_2PC_6H_4CH_3)_2]$	Ag: MVol.B7-230/1
$AgC_{39}H_{36}NP_2S_2$	$[Ag(S_2CN(CH_3)_2)(P(C_6H_5)_3)_2]$	Ag: MVol.B7-212
$AgC_{39}H_{54}K_2N_3O_6S_9$	$K_2[Ag(C_6H_5SO_2NCSSC_6H_{13})_3]$	Ag: MVol.B7-104
$AgC_{40}H_{42}N_6O_{12}$	$[Ag(C_{40}H_{42}N_6O_{12})]$	Ag: MVol.B7-306, 312
$AgC_{40}H_{44}N_4O_8$	$[Ag(C_{20}H_4N_4(CH_3)_4(C_2H_4COOCH_3)_4)]$	Ag: MVol.B7-305, 311
$AgC_{40}H_{67}O_{10}$	$[Ag(C_{40}H_{67}O_{10})]$	Ag: MVol.B6-223
$AgC_{40}H_{67}O_{11}$	$[Ag(C_{40}H_{67}O_{11})]$	Ag: MVol.B6-223/4
$AgC_{41}ClH_{35}NP_2$	$[AgCl(P(C_6H_5)_3)_2(C_5H_5N)]$	Ag: MVol.B7-215
$AgC_{41}F_6H_{31}O_2P_2$	$[Ag(CH(COCF_3)_2)(P(C_6H_5)_3)_2]$	Ag: MVol.B7-211
$AgC_{41}H_{40}NP_2S_2$	$[Ag(S_2CN(C_2H_5)_2)(P(C_6H_5)_3)_2]$	Ag: MVol.B7-212
$AgC_{42}ClH_{38}P_2$	$[AgCl(CH_2CHCH_2C_6H_4P(C_6H_5)_2)_2]$	Ag: MVol.B7-231
$AgC_{42}ClH_{39}O_4P_3$	$[Ag(C_6H_4(CH_2)_2PC_6H_5)_3]ClO_4$	Ag: MVol.B7-222
$AgC_{42}ClH_{42}O_4P_2$	$[Ag(P(C_6H_4CH_3)_3)_2]ClO_4$	Ag: MVol.B7-223/4
$AgC_{42}ClH_{42}P_2$	$[AgCl(P(C_6H_4CH_3)_3)_2]$	Ag: MVol.B7-227/8
$AgC_{42}FH_{42}P_2$	$[AgF(P(C_6H_4CH_3)_3)_2]$	Ag: MVol.B7-227/8
$AgC_{42}F_2H_{42}P_3S_2$	$[AgS_2PF_2(P(C_6H_4CH_3)_3)_2]$	Ag: MVol.B7-227, 228/9
$AgC_{42}F_5H_{30}P_2S$	$[AgSC_6F_5(P(C_6H_5)_3)_2]$	Ag: MVol.B7-212
$AgC_{42}F_6H_{42}P_3$	$[Ag(P(C_6H_4CH_3)_3)_2]PF_6$	Ag: MVol.B7-224
$AgC_{42}H_{35}O_2P_2S$	$[Ag(O_2SC_6H_5)(P(C_6H_5)_3)_2]$	Ag: MVol.B7-212
$AgC_{42}H_{38}IP_2$	$[AgI(CH_2CHCH_2C_6H_4P(C_6H_5)_2)_2]$	Ag: MVol.B7-231
$AgC_{42}H_{39}P_3{}^+$	$[Ag(C_6H_4(CH_2)_2PC_6H_5)_3]^+$	Ag: MVol.B7-222
$AgC_{42}H_{42}NO_3P_2$	$[AgNO_3(P(C_6H_4CH_3)_3)_2]$	Ag: MVol.B7-227
$AgC_{42}H_{42}P_2{}^+$	$[Ag(P(C_6H_4CH_3)_3)_2]^+$	Ag: MVol.B7-223/4
$AgC_{42}H_{60}N_{12}O_2$	$[Ag(C_{42}H_{60}N_{12}O_2)]$	Ag: MVol.B7-305, 311
$AgC_{43}H_{42}NOP_2$	$[AgNCO(P(C_6H_4CH_3)_3)_2]$	Ag: MVol.B7-227
$AgC_{43}H_{42}NP_2$	$[AgCN(P(C_6H_4CH_3)_3)_2]$	Ag: MVol.B7-227/8
$AgC_{43}H_{42}NP_2S$	$[AgSCN(P(C_6H_4CH_3)_3)_2]$	Ag: MVol.B7-227
$AgC_{43}H_{44}NP_2S_2$	$[Ag(S_2CN(C_3H_7)_2)(P(C_6H_5)_3)_2]$	Ag: MVol.B7-212
$AgC_{43}H_{45}NP_3S$	$[AgSCN(C_2H_5P(C_6H_5)_2)_3]$	Ag: MVol.B7-221
$AgC_{44}F_3H_{42}O_2P_2$	$[AgOOCCF_3(P(C_6H_4CH_3)_3)_2]$	Ag: MVol.B7-227/8
$AgC_{44}H_{28}N_4$	$[Ag(C_{20}H_8N_4(C_6H_5)_4)]$	Ag: MVol.B7-304, 308/9
$AgC_{44}H_{28}N_4O_{12}S_4$	$[Ag(C_{20}H_8N_4(C_6H_4SO_3H)_4)]$	Ag: MVol.B7-306, 312
$AgC_{44}H_{29}N_4$	$[Ag(C_{20}H_9N_4(C_6H_5)_4)]$	Ag: MVol.B7-304, 308
$AgC_{44}H_{30}N_4$	$[Ag(C_{20}H_{10}N_4(C_6H_5)_4)]$	Ag: MVol.B7-306, 312
$AgC_{45}H_{48}NP_2S_2$	$[Ag(S_2CN(C_4H_9)_2)(P(C_6H_5)_3)_2]$	Ag: MVol.B7-212
$AgC_{45}H_{66}K_2N_3O_6S_9$	$K_2[Ag(C_6H_5SO_2NCSSC_8H_{17})_3]$	Ag: MVol.B7-104
$AgC_{46}H_{39}O_2P_2$	$[Ag(C_6H_5COCHCOCH_3)(P(C_6H_5)_3)_2]$	Ag: MVol.B7-211
$AgC_{48}H_{28}N_4O_8$	$Ag(C_{20}H_8N_4(C_7H_5O_2)_4) \cdot C_5H_5N$	Ag: MVol.B7-306, 312

Formula	Compound	Reference
$AgC_{48}H_{32}N_4^+$	$[Ag((C_6H_5)_2C_{12}H_6N_2)_2]^+$	Ag: MVol.B6-128
$AgC_{48}H_{36}O_{12}S_4Se_4^{3-}$	$[Ag(C_6H_5SeC_6H_4SO_3)_4]^{3-}$	Ag: MVol.B7-191
$AgC_{48}H_{36}O_{12}S_8^{3-}$	$[Ag(C_6H_5SC_6H_4SO_3)_4]^{3-}$	Ag: MVol.B7-44/5
$AgC_{48}H_{51}N_5O_{18}$	$[Ag(C_{48}H_{51}N_5O_{18})]$	Ag: MVol.B7-306, 313
$AgC_{48}H_{52}N_4O_{16}$	$[Ag(C_{20}H_4N_4(CH_3)_4(C_2H_3(COOCH_3)_2)_4)]$	Ag: MVol.B7-304, 308
$AgC_{48}H_{53}N_4O_{16}$	$[Ag(C_{20}H_5N_4(CH_3)_4(C_2H_3(COOCH_3)_2)_4)]$	Ag: MVol.B7-304, 307
$AgC_{48}H_{60}N_7O_3$	$[Ag((CH_3)_2C_2H_2(C_6H_4NH_2)_2)_3]NO_3$	Ag: MVol.B6-64
$AgC_{48}H_{104}O_{12}P_3$	$AgO_2P(OC_8H_{17})_2 \cdot 2\ HO_2P(OC_8H_{17})_2$	Ag: MVol.B7-250
$AgC_{50}H_{54}N_8O_{20}^-$	$[Ag((CH_3)_2C_{10}H_2N_4O_2C_{13}H_{19}O_8)_2]^-$	Ag: MVol.B6-173
$AgC_{52}F_6H_{40}P_5$	$[Ag((C_6H_5)_2PCCP(C_6H_5)_2)_2]_n(PF_6)_n$	Ag: MVol.B7-235
$AgC_{53}H_{33}N_5O_8$	$Ag(C_{20}H_8N_4(C_7H_5O_2)_4) \cdot C_5H_5N$	Ag: MVol.B7-306, 312
$AgC_{54}ClH_{45}P_3$	$[AgCl(P(C_6H_5)_3)_3]$	Ag: MVol.B7-214/5
$AgC_{54}Cl_3H_{45}P_3Sn$	$[AgSnCl_3(P(C_6H_5)_3)_3]$	Ag: MVol.B7-220
$AgC_{54}H_{42}O_9P_3S_3^{2-}$	$[Ag((C_6H_5)_2PC_6H_4SO_3)_3]^{2-}$	Ag: MVol.B7-233
$AgC_{54}H_{45}IP_3$	$[AgI(P(C_6H_5)_3)_3]$	Ag: MVol.B7-215
$AgC_{55}ClH_{72}MgN_4O_9$	$[Mg(C_{55}H_{72}N_4O_5)] \cdot AgClO_4$	Ag: MVol.B7-314
$AgC_{55}H_{45}NOP_3$	$[AgNCO(P(C_6H_5)_3)_3]$	Ag: MVol.B7-216
$AgC_{55}H_{45}NP_3S$	$[AgSCN(P(C_6H_5)_3)_3]$	Ag: MVol.B7-216
$AgC_{55}H_{72}N_4O_5$	$[Ag(C_{55}H_{72}N_4O_5)]$	Ag: MVol.B7-314/5
$AgC_{63}ClH_{63}O_4P_3$	$[Ag(P(C_6H_4CH_3)_3)_3]ClO_4$	Ag: MVol.B7-224
$AgC_{63}ClH_{63}P_3$	$[AgCl(P(C_6H_4CH_3)_3)_3]$	Ag: MVol.B7-229
$AgC_{63}FH_{63}P_3$	$[AgF(P(C_6H_4CH_3)_3)_3]$	Ag: MVol.B7-229
$AgC_{63}F_2H_{63}P_4S_2$	$[AgS_2PF_2(P(C_6H_4CH_3)_3)_3]$	Ag: MVol.B7-229/30
$AgC_{63}F_6H_{63}P_4$	$[Ag(P(C_6H_4CH_3)_3)_3]PF_6$	Ag: MVol.B7-224/5
$AgC_{63}H_{63}IP_3$	$[AgI(P(C_6H_4CH_3)_3)_3]$	Ag: MVol.B7-229
$AgC_{63}H_{63}NO_3P_3$	$[Ag(P(C_6H_4CH_3)_3)_3]NO_3$	Ag: MVol.B7-224
$AgC_{63}H_{63}P_3^+$	$[Ag(P(C_6H_4CH_3)_3)_3]^+$	Ag: MVol.B7-224
$AgC_{64}H_{63}NOP_3$	$[AgNCO(P(C_6H_4CH_3)_3)_3]$	Ag: MVol.B7-229/30
$AgC_{64}H_{63}NP_3$	$[AgCN(P(C_6H_4CH_3)_3)_3]$	Ag: MVol.B7-229
$AgC_{64}H_{63}NP_3S$	$[AgSCN(P(C_6H_4CH_3)_3)_3]$	Ag: MVol.B7-229
$AgC_{65}F_3H_{63}O_2P_3$	$[AgOOCCF_3(P(C_6H_4CH_3)_3)_3]$	Ag: MVol.B7-229/30
$AgC_{72}ClH_{60}O_4P_4$	$[Ag(P(C_6H_5)_3)_4]ClO_4$	Ag: MVol.B7-216/7
$AgC_{72}ClH_{60}O_4Sb_4$	$[Ag((C_6H_5)_3Sb)_4]ClO_4$	Ag: MVol.B7-268/9
$AgC_{72}F_6H_{60}O_4P_4Sb$	$AgSbF_6 \cdot 4\ (C_6H_5)_3PO$	Ag: MVol.B7-244
$AgC_{72}GeH_{60}P_3$	$[AgGe(C_6H_5)_3(P(C_6H_5)_3)_3]$	Ag: MVol.B7-219
$AgC_{72}H_{60}NO_3P_4$	$[Ag(P(C_6H_5)_3)_4]NO_3$	Ag: MVol.B7-216/7
$AgC_{74}CuH_{60}O_2P_4S_2$	$[(P(C_6H_5)_3)_2AgC_2O_2S_2Cu(P(C_6H_5)_3)_2]$	Ag: MVol.B7-217
$AgC_{80}GeH_{80}O_4P_3$	$[AgGe(C_6H_5)_3(P(C_6H_5)_3)_3] \cdot 2\ C_2H_4(OCH_3)_2$	Ag: MVol.B7-219
$AgC_{84}ClH_{84}O_4P_4$	$[Ag(P(C_6H_4CH_3)_3)_4]ClO_4$	Ag: MVol.B7-225/6
$AgC_{84}ClH_{84}P_4$	$[Ag(P(C_6H_4CH_3)_3)_4]Cl$	Ag: MVol.B7-225/6
$AgC_{84}FH_{84}P_4$	$[Ag(P(C_6H_4CH_3)_3)_4]F$	Ag: MVol.B7-225/6
$AgC_{84}F_2H_{84}P_5S_2$	$[Ag(P(C_6H_4CH_3)_3)_4]S_2PF_2$	Ag: MVol.B7-225, 227
$AgC_{84}F_6H_{84}P_5$	$[Ag(P(C_6H_4CH_3)_3)_4]PF_6$	Ag: MVol.B7-225, 227
$AgC_{84}H_{84}IP_4$	$[Ag(P(C_6H_4CH_3)_3)_4]I$	Ag: MVol.B7-225/6
$AgC_{84}H_{84}NO_3P_4$	$[Ag(P(C_6H_4CH_3)_3)_4]NO_3$	Ag: MVol.B7-225
$AgC_{84}H_{84}P_4^+$	$[Ag(P(C_6H_4CH_3)_3)_4]^+$	Ag: MVol.B7-224, 225
$AgC_{85}H_{84}NP_4$	$[Ag(P(C_6H_4CH_3)_3)_4]CN$	Ag: MVol.B7-225/6
$AgC_{86}F_3H_{84}O_2P_4$	$[Ag(P(C_6H_4CH_3)_3)_4]OOCCF_3$	Ag: MVol.B7-225, 227
$AgC_{90}ClH_{75}O_9P_5$	$AgClO_4 \cdot 5\ (C_6H_5)_3PO$	Ag: MVol.B7-244
$AgC_{90}F_6H_{75}O_5P_6$	$AgPF_6 \cdot 5\ (C_6H_5)_3PO$	Ag: MVol.B7-244

Formula	Compound	Reference
$AgC_{96}H_{135}K_2N_6O_{15}S_3$	$K_2[Ag(C_{32}H_{45}N_2SO_5)_3]$	Ag: MVol.B6-134
$AgC_{96}H_{135}N_6O_{15}S_3^{2-}$	$[Ag(C_{17}H_{35}C_3H_2N_2OC_6H_3(SO_3)OC_6H_5)_3]^{2-}$	Ag: MVol.B6-134
AgCl	AgCl systems	
	$AgCl-C_5H_5N$	Ag: MVol.B6-78/9
	$AgCl-NH_3$	Ag: MVol.B6-19
	$AgCl-NH_3-H_2O$	Ag: MVol.B6-19
	$AgCl-UCl_4$	U: SVol.C9-24
$AgClH_3N$	$[AgCl(NH_3)]$	Ag: MVol.B6-8, 10
–	$AgCl \cdot NH_3$	Ag: MVol.B6-19/20
$AgClH_3NO_2$	$AgClO_2 \cdot NH_3$	Ag: MVol.B6-21
$AgClH_6N_2$	$[AgCl(NH_3)_2]$	Ag: MVol.B6-8, 10
–	$[Ag(NH_3)_2]Cl$	Ag: MVol.B6-20
$AgClH_6N_2O_2$	$AgClO_2 \cdot 2\ NH_3$	Ag: MVol.B6-21
$AgClH_6N_2O_3$	$AgClO_3 \cdot 2\ NH_3$	Ag: MVol.B6-21
$AgClH_6N_2O_4$	$[Ag(NH_3)_2]ClO_4$	Ag: MVol.B6-13, 21
$AgClH_7N_3O_3$	$[Ag(NH_3)_2]HNClO_3$	Ag: MVol.B6-22
$AgClH_9N_3$	$AgCl \cdot 3\ NH_3$	Ag: MVol.B6-19, 20/1
$AgClH_9N_3O_2$	$AgClO_2 \cdot 3\ NH_3$	Ag: MVol.B6-21
$AgClH_9N_3O_3$	$AgClO_3 \cdot 3\ NH_3$	Ag: MVol.B6-21
$AgClH_9N_3O_4$	$AgClO_4 \cdot 3\ NH_3$	Ag: MVol.B6-21/2
$AgClH_{10}N_2O_2$	$[Ag(NH_3)_2(H_2O)_2]Cl$	Ag: MVol.B6-20
$AgClO_3$	$AgClO_3$ systems	
	$AgClO_3-(CH_2)_6N_4-CH_3OH$	Ag: MVol.B6-198
	$AgClO_3-(CH_2)_6N_4-H_2O$	Ag: MVol.B6-198
$AgClO_4$	$AgClO_4$ systems	
	$AgClO_4-CH_3CCC_2H_5$	Ag: MVol.B5-39
$AgCl_2H_3N^-$	$[AgCl_2(NH_3)]^-$	Ag: MVol.B6-8, 10
$AgCl_2H_6N_2^-$	$[AgCl_2(NH_3)_2]^-$	Ag: MVol.B6-10
$AgCl_6H_6N_2Sb$	$[Ag(NH_3)_2]SbCl_6$	Ag: MVol.B6-30
$AgCl_6U$	$AgUCl_6$	U: SVol.C9-60
$AgCl_9H_6N_2W_2$	$[Ag(NH_3)_2]W_2Cl_9$	Ag: MVol.B6-31
$AgCrH_6N_2O_4^-$	$[AgCrO_4(NH_3)_2]^-$	Ag: MVol.B6-8
$AgCr_2H_6N_2O_8^{3-}$	$[Ag(CrO_4)_2(NH_3)_2]^{3-}$	Ag: MVol.B6-8
$AgFH_3N$	$AgF \cdot NH_3$	Ag: MVol.B6-19
$AgFH_6N_2$	$AgF \cdot 2\ NH_3 \cdot 2\ H_2O$	Ag: MVol.B6-19
AgF_2NOS	$Ag[NS(O)F_2]$	S: S-N-Verb. 1-61
$AgF_2NO_4S_2$	$AgN(SO_2F)_2$	S: S-N-Verb. 1-50
$AgF_2N_3O_4S_3$	$Ag[SO_2(NS(O)F)_2N]$	S: S-N-Verb. 1-11
AgF_6Ir	$Ag[IrF_6]$	Ir: SVol.2-109
AgF_6Pa	$AgPaF_6$	Pa: SVol.2-44/5
$AgHKNO_3S$	$KAg[NHSO_3] \cdot H_2O$	S: S-N-Verb. 1-96
$AgH_{1.5}IN_{0.5}$	$AgI \cdot 0.5\ NH_3$	Ag: MVol.B6-23
AgH_3IN	$AgI \cdot NH_3$	Ag: MVol.B6-23
AgH_3IP	$AgI \cdot PH_3$	Ag: MVol.B7-199
AgH_3N^+	$[AgNH_3]^+$	Ag: MVol.B6-7/10
AgH_3N^{2+}	$[Ag(NH_3)]^{2+}$	Ag: MVol.B7-272
AgH_3NO^+	$[Ag(NH_2OH)]^+$	Ag: MVol.B6-36
$AgH_3NO_3S^-$	$[AgSO_3(NH_3)]^-$	Ag: MVol.B6-10
$AgH_3NO_4P^{2-}$	$[AgPO_4(NH_3)]^{2-}$	Ag: MVol.B6-8

$AgH_3N_2O_2$	$AgNO_2 \cdot NH_3 = AgNO_2 \cdot Ag(NH_3)_2NO_2$	Ag: MVol.B6-16
$AgH_3N_2O_3$	$AgNO_3 \cdot NH_3$	Ag: MVol.B6-17
$AgH_3N_3O_2^-$	$[AgN_2O_2(NH_3)]^-$	Ag: MVol.B6-8, 10
AgH_4N	$[(NH_3)AgH]$	Ag: MVol.B6-14
$AgH_4N_3O_3$	$AgNO_3 \cdot N_2H_4$	Ag: MVol.B6-35
$AgH_4N_3O_4S_2$	$Ag[N(SO_2NH_2)_2] \cdot 2\ H_2O$	S: S-N-Verb. 1-160
$AgH_4O_2^+$	$[Ag(H_2O)_2]^+$	Ag: MVol.B6-6
$AgH_{4.5}IN_{1.5}$	$AgI \cdot 1.5\ NH_3$	Ag: MVol.B6-23
AgH_6IN_2	$AgI \cdot 2\ NH_3$	Ag: MVol.B6-23/4
–	$AgI \cdot 2\ NH_3 \cdot H_2O$	Ag: MVol.B6-24
$AgH_6IN_2O_3$	$AgIO_3 \cdot 2\ NH_3$	Ag: MVol.B6-24
$AgH_6I_2N_2O_6^-$	$[Ag(IO_3)_2(NH_3)_2]^-$	Ag: MVol.B6-10
$AgH_6MnN_2O_4$	$AgMnO_4 \cdot 2\ NH_3 = [Ag(NH_3)_2]MnO_4$	Ag: MVol.B6-31
–	$AgMnO_4 \cdot 2\ NH_3 \cdot H_2O$	Ag: MVol.B6-31
$AgH_6N_2^+$	$[Ag(NH_3)_2]^+$	Ag: MVol.B6-7/16
$AgH_6N_2O_2^+$	$[Ag(NH_2OH)_2]^+$	Ag: MVol.B6-36
$AgH_6N_2O_3S^-$	$[AgSO_3(NH_3)_2]^-$	Ag: MVol.B6-10
$AgH_6N_2O_7P_2^{3-}$	$[AgP_2O_7(NH_3)_2]^{3-}$	Ag: MVol.B6-8, 10, 14
$AgH_6N_3O_2$	$Ag(NH_3)_2NO_2$	Ag: MVol.B6-16
$AgH_6N_3O_3$	$[Ag(NH_3)_2]NO_3$	Ag: MVol.B6-12/3, 17/8
–	$[Ag(ND_3)_2]NO_3$	Ag: MVol.B6-17
$AgH_6N_4O_4P$	$[Ag(PO(NH_2)_3)]NO_3$	Ag: MVol.B7-250
AgH_7N_2O	$[Ag(NH_3)_2]OH$	Ag: MVol.B6-8, 13/5, 18
$AgH_8N_3O_3S$	$[Ag(NH_3)_2]SO_3NH_2$	Ag: MVol.B6-26
$AgH_8N_3O_3Se$	$[Ag(NH_3)_2]SeO_3NH_2$	Ag: MVol.B6-27
$AgH_8N_5O_3$	$[Ag(N_2H_4)_2]NO_3$	Ag: MVol.B6-35
$AgH_8N_7O_2P_2S$	$Ag[NSO_2][NP(NH_2)_2]_2$	S: S-N-Verb.1-37
AgH_9IN_3	$AgI \cdot 3\ NH_3$	Ag: MVol.B6-23/4
$AgH_9MnN_3O_4$	$AgMnO_4 \cdot 3\ NH_3$	Ag: MVol.B6-31
$AgH_9N_3^+$	$[Ag(NH_3)_3]^+$	Ag: MVol.B6-10
$AgH_9N_4O_2$	$AgNO_2 \cdot 3\ NH_3$	Ag: MVol.B6-16
$AgH_9N_4O_3$	$AgNO_3 \cdot 3\ NH_3$	Ag: MVol.B6-18/9
$AgH_{28}N_8O_{12}S_3$	$Ag(NH_3)_4SO_4 \cdot 2\ (NH_4)_2SO_4 \cdot 2\ H_2O$	Ag: MVol.B7-272
$AgH_{2x}O_x^+$	$[Ag(H_2O)_x]^+$	Ag: MVol.B6-6/7, 15
$AgH_{3x}N_x^+$	$[Ag(NH_3)_x]^+$	Ag: MVol.B6-15
$AgH_{3x}N_x^{2+}$	$[Ag(NH_3)_x]^{2+}$	Ag: MVol.B6-14
AgI	AgI systems	
	AgI-NH_3	Ag: MVol.B6-23
	AgI-TeI_4	Te: SVol.B3-62
$AgMn_3N$	Mn_3AgN	Mn: MVol.C3-253/4
$AgNO_3$	$AgNO_3$ systems	
	$AgNO_3$-$CO(NH_2)_2$-H_2O	Ag: MVol.B6-333/4
	$AgNO_3$-CH_3CONH_2	Ag: MVol.B6-323
	$AgNO_3$-CH_2CHCH_3	Ag: MVol.B5-31
	$AgNO_3$-CH_2CHCH_3-C_3H_8	Ag: MVol.B5-31
	$AgNO_3$-CH_2CHCH_3-H_2O	Ag: MVol.B5-31
	$AgNO_3$-$(CH_2)_2(CN)_2$-H_2O	Ag: MVol.B6-350
	$AgNO_3$-$CH_2CHC_2H_5$	Ag: MVol.B5-33
	$AgNO_3$-C_5H_5N	Ag: MVol.B6-76
	$AgNO_3$-$CH_3CCC_2H_5$	Ag: MVol.B5-38

Formula	Compound	Reference
$AgNO_3$	$AgNO_3$ systems	
	$AgNO_3$-C_6H_5OH	Ag: MVol.B5-118
	$AgNO_3$-$(CH_2)_6N_4$-CH_3OH	Ag: MVol.B6-197
	$AgNO_3$-$(CH_2)_6N_4$-H_2O	Ag: MVol.B6-197
	$AgNO_3$-KBr-$(CH_2)_6N_4$-H_2O	Ag: MVol.B6-198
	$AgNO_3$-KI-$(CH_2)_6N_4$-H_2O	Ag: MVol.B6-199
	$AgNO_3$-K_2CrO_4-$(CH_2)_6N_4$-H_2O	Ag: MVol.B6-200
	$AgNO_3$-NaCl-$(CH_2)_6N_4$-H_2O	Ag: MVol.B6-198
$Ag_{1.55}C_8H_6N_{3.55}O_{1.65}S$	$0.55\ AgNO_3 \cdot Ag(C_6H_5C_2HN_3S)$	Ag: MVol.B7-56
$Ag_{1.61}C_8Cl_{0.61}H_6N_3S$	$0.61\ AgCl \cdot Ag(C_6H_5C_2HN_3S)$	Ag: MVol.B7-56
$Ag_{1.61}C_8H_6N_{3.61}O_{1.83}S$	$0.61\ AgNO_3 \cdot Ag(C_6H_5C_2HN_3S)$	Ag: MVol.B7-56
$Ag_{1.67}C_8Cl_{0.67}H_6N_3S$	$0.67\ AgCl \cdot Ag(C_6H_5C_2HN_3S)$	Ag: MVol.B7-56
$Ag_{1.75}C_4H_6N_{1.75}O_{5.25}$	$1.75\ AgNO_3 \cdot CH_3CCCH_3$	Ag: MVol.B5-36
$Ag_{1.8}C_3H_4N_{3.8}O_{2.4}S$	$0.8\ AgNO_3 \cdot Ag(CH_3C_2HN_3S)$	Ag: MVol.B7-56
$Ag_{1.84}C_3Cl_{0.84}H_4N_3S$	$0.84\ AgCl \cdot Ag(CH_3C_2HN_3S)$	Ag: MVol.B7-56
$Ag_{1.84}C_5Cl_{0.84}H_8N_3S$	$0.84\ AgCl \cdot Ag(CH_3C_2N_3SC_2H_5)$	Ag: MVol.B7-56
$Ag_{1.98}C_9H_8N_{3.98}O_{2.94}S$	$0.98\ AgNO_3 \cdot Ag(C_6H_5C_2N_3SCH_3)$	Ag: MVol.B7-56
$Ag_2AsCF_3O_3$	$CF_3As(O)(OAg)_2$	F: PerFHalOrg.3-197, 198
$Ag_2AsC_2F_5O_3$	$C_2F_5As(O)(OAg)_2$	F: PerFHalOrg.3-197
$Ag_2AsC_6H_{25}N_{12}O_4S_6$	$[Ag(NH_2CSNH_2)_3]_2HAsO_4$	Ag: MVol.B7-147
$Ag_2AsC_{11}H_{13}O_4S^{2+}$	$[Ag_2(CH_3SC_6H_4As(CH_2COOH)_2)]^{2+}$	Ag: MVol.B7-262
$Ag_2AsC_{11}H_{15}N_2O_6$	$2\ AgNO_3 \cdot C_3H_5C_6H_4As(CH_3)_2$	Ag: MVol.B5-119
$Ag_2As_2CF_{12}HN$	$2\ AgAsF_6 \cdot HCN$	Ag: MVol.B6-346
$Ag_2As_2C_{10}H_{16}N_2O_6$	$[Ag_2(NO_3)_2((CH_3)_2AsC_6H_4As(CH_3)_2)]$	Ag: MVol.B7-266
$Ag_2As_2C_{12}Cl_2H_{14}N_4O_8$	$2\ AgNO_3 \cdot As_2(C_6H_3(OH)NH_2)_2 \cdot 2\ HCl$	Ag: MVol.B7-263
$Ag_2As_2C_{16}Cl_2H_{24}N_2$	$[Ag(NC_5H_4CH_2As(CH_3)_2)_2][AgCl_2]$	Ag: MVol.B7-255
$Ag_2As_2C_{20}F_8H_{20}N_4O_2$	$[Ag(C_5H_5N)_2]_2As_2F_8O_2$	Ag: MVol.B6-82
$Ag_2As_2C_{28}Cl_2H_{48}P_2$	$[Ag((C_2H_5)_2PC_6H_4As(C_2H_5)_2)_2][AgCl_2]$	Ag: MVol.B7-268
$Ag_2As_2C_{28}H_{48}I_2P_2$	$[Ag((C_2H_5)_2PC_6H_4As(C_2H_5)_2)_2][AgI_2]$	Ag: MVol.B7-268
$Ag_2As_4Br_2C_{20}H_{32}$	$[Ag(C_6H_4(As(CH_3)_2)_2)_2][AgBr_2]$	Ag: MVol.B7-265/6
$Ag_2As_4C_{20}Cl_2H_{32}$	$[Ag(C_6H_4(As(CH_3)_2)_2)_2][AgCl_2]$	Ag: MVol.B7-265/6
$Ag_2As_4C_{20}H_{32}I_2$	$[Ag(C_6H_4(As(CH_3)_2)_2)_2][AgI_2]$	Ag: MVol.B7-265/6
$Ag_2As_4C_{20}H_{32}N_2O_4$	$[Ag(C_6H_4(As(CH_3)_2)_2)_2][Ag(NO_2)_2]$	Ag: MVol.B7-265/6
$Ag_2As_4C_{56}H_{64}N_2O_{10}$	$[Ag(C_4H_8(As(C_6H_5)_2)_2)_2(H_2O)_2]$ $[Ag(NO_3)_2(H_2O)_2]$	Ag: MVol.B7-265
$Ag_2As_6C_{38}FeH_{46}O_4$	$(CO)_4FeAg_2(CH_3As(C_6H_4As(CH_3)_2)_2)_2$	Ag: MVol.B7-267
		Fe: Org.Verb.B2-167
$Ag_2As_6C_{78}H_{72}I_2$	$[Ag_2I_2((C_6H_5)_2AsC_2H_4As(C_6H_5)_2)_3]$	Ag: MVol.B7-264/5
$Ag_2B_2C_3F_8H_{12}N_6S_3$	$[Ag_2(NH_2CSNH_2)_3](BF_4)_2$	Ag: MVol.B7-138
$Ag_2B_2C_7H_{22}I_3N_3$	$[(CH_3)_3NBH_2]_2CNAg_2I_3$	B: B-Verb.15-17
$Ag_2B_2C_8F_8H_8$	$2\ AgBF_4 \cdot C_8H_8$	Ag: MVol.B5-83
$Ag_2B_2C_8H_{14}N_8O_6$	$2\ AgNO_3 \cdot B_2(CN)_4 \cdot 2\ NH(CH_3)_2$	Ag: MVol.B7-269
$Ag_2B_2C_{12}F_8H_{18}$	$2\ AgBF_4 \cdot 3\ C_4H_6$	Ag: MVol.B5-35/6
$Ag_2B_2C_{15}F_8H_{18}$	$2\ AgBF_4 \cdot 3\ C_5H_6$	Ag: MVol.B5-54
$Ag_2B_2C_{16}F_8H_{10}$	$2\ AgBF_4 \cdot C_{16}H_{10}$	Ag: MVol.B5-116
$Ag_2B_2C_{18}F_8H_{14}$	$2\ AgBF_4 \cdot C_5H_4C(C_6H_5)_2$	Ag: MVol.B5-106
$Ag_2B_2C_{18}F_8H_{18}$	$2\ AgBF_4 \cdot 3\ C_6H_6$	Ag: MVol.B5-93
$Ag_2B_2C_{18}F_8H_{24}$	$2\ AgBF_4 \cdot 3\ C_6H_8$	Ag: MVol.B5-60
$Ag_2B_2C_{20}F_7H_{21}N_4O$	$[Ag(C_5H_5N)_2]BF_3OH \cdot [Ag(C_5H_5N)_2]BF_4$	Ag: MVol.B6-81
		B: B-Verb.8-74

Formula	Compound	Reference
$Ag_2B_2C_{21}F_8H_{24}$	$2\ AgBF_4 \cdot 3\ C_6H_5CH_3$	Ag: MVol.B5-97
–	$2\ AgBF_4 \cdot 3\ C_7H_8$	Ag: MVol.B5-78/9
–	$2\ AgBF_4 \cdot 3\ C_7H_8 \cdot y\ H_2O$	Ag: MVol.B5-78/9
$Ag_2B_2C_{28}F_8H_{30}$	$2\ AgBF_4 \cdot C_{12}H_{10} \cdot C_6H_6 \cdot C_6H_5CH_2CH(CH_3)_2$	Ag: MVol.B5-112
$Ag_2B_2C_{30}F_8H_{24}$	$2\ AgBF_4 \cdot 3\ C_{10}H_8$	Ag: MVol.B5-109
$Ag_2B_2C_{36}F_8H_{24}$	$2\ AgBF_4 \cdot 3\ C_{12}H_8$	Ag: MVol.B5-113
$Ag_2B_2C_{36}F_8H_{30}$	$2\ AgBF_4 \cdot 3\ (C_6H_5)_2$	Ag: MVol.B5-103
$Ag_2B_2C_{36}F_8H_{30}N_6$	$2\ AgBF_4 \cdot 3\ C_6H_5NNC_6H_5$	Ag: MVol.B6-285
$Ag_2B_2C_{39}F_8H_{30}$	$2\ AgBF_4 \cdot 3\ C_{13}H_{10}$	Ag: MVol.B5-111
$Ag_2B_2C_{42}F_8H_{30}$	$2\ AgBF_4 \cdot 3\ C_{14}H_{10}$	Ag: MVol.B5-111
$Ag_2B_2C_{42}F_8H_{36}$	$2\ AgBF_4 \cdot 3\ (C_6H_5)_2CCH_2$	Ag: MVol.B5-103
$Ag_2B_2C_{45}F_8H_{30}$	$2\ AgBF_4 \cdot 3\ C_{15}H_{10}$	Ag: MVol.B5-115
$Ag_2B_2O_4$	$B_2O_3 \cdot Ag_2O = AgBO_2$	B: B-Verb.7-74
–	$B_2O_3 \cdot Ag_2O \cdot 4\ H_2O = Ag[B(OH)_4]$	B: B-Verb.7-149
$Ag_2B_4H_4O_9$	$Ag_2[B_4O_5(OH)_4] = 2\ B_2O_3 \cdot Ag_2O \cdot 2\ H_2O$	B: B-Verb.7-148/9
$Ag_2B_4O_7$	$2\ B_2O_3 \cdot Ag_2O = Ag_2B_4O_7$	B: B-Verb.7-73
–	$2\ B_2O_3 \cdot Ag_2O \cdot 2\ H_2O = Ag_2[B_4O_5(OH)_4]$	B: B-Verb.7-148/9
$Ag_2B_6H_6$	$Ag_2[B_6H_6]$	B: B-Verb.20-78
$Ag_2B_8O_{13}$	$4\ B_2O_3 \cdot Ag_2O = Ag_2B_8O_{13}$	B: B-Verb.7-73/4
$Ag_2B_{10}Br_{10}$	$Ag_2[B_{10}Br_{10}]$	B: B-Verb.19-331
$Ag_2B_{10}Cl_{10}$	$Ag_2[B_{10}Cl_{10}]$	B: B-Verb.19-329
$Ag_2B_{10}H_{10}$	$Ag_2[B_{10}H_{10}]$	B: B-Verb.20-182, 188
$Ag_2B_{12}Br_{10}H_2$	$Ag_2[Br_{10}B_{12}H_2]$	B: B-Verb.14-304
$Ag_2B_{12}Br_{12}$	$Ag_2[B_{12}Br_{12}]$	B: B-Verb.19-334
$Ag_2B_{12}CH_{12}O_2$	$Ag_2[(HOCO)B_{12}H_{11}]$	B: B-Verb.20-238
$Ag_2B_{12}C_2H_{12}O_4$	$Ag_2[(HOCO)_2B_{12}H_{10}]$	B: B-Verb.20-238
$Ag_2B_{12}Cl_{12}$	$Ag_2[B_{12}Cl_{12}]$	B: B-Verb.19-333
$Ag_2B_{12}H_{12}$	$Ag_2[B_{12}H_{12}]$	B: B-Verb.20-220, 222/3
$Ag_2B_{12}I_{12}$	$Ag_2[B_{12}I_{12}]$	B: B-Verb.19-336
$Ag_2B_{18}O_{28}$	$9\ B_2O_3 \cdot Ag_2O = Ag_2B_{18}O_{28}$	B: B-Verb.7-73/4
$Ag_2B_{20}H_{18}$	$Ag_2[B_{20}H_{18}]$	B: B-Verb.20-277
–	$Ag_2[B_{20}H_{18}] \cdot 5\ H_2O$	B: B-Verb.20-277
$Ag_2BaC_{32}H_{20}$	$Ba[Ag(CCC_6H_5)_2]_2$	Ag: MVol.B5-17
$Ag_2BaC_{32}H_{24.5}N_{1.5}$	$Ba[Ag(CCC_6H_5)_2]_2 \cdot 1.5\ NH_3$	Ag: MVol.B5-17
$Ag_2BrC_6H_5^{2+}$	$[Ag_2C_6H_5Br]^{2+}$	Ag: MVol.B5-117
$Ag_2BrC_8H_7^{2+}$	$[Ag_2(C_6H_5CHCHBr)]^{2+}$	Ag: MVol.B5-117
$Ag_2BrC_{16}CuH_{20}N_2$	$2\ AgC_6H_4N(CH_3)_2 \cdot CuBr$	Ag: MVol.B5-11
$Ag_2Br_2CH_5N_3S$	$[Ag_2(NH_2CSNHNH_2)]Br_2$	Ag: MVol.B7-165
$Ag_2Br_2C_2H_2^{2+}$	$[Ag_2CHBrCHBr]^{2+}$	Ag: MVol.B5-46
$Ag_2Br_2C_2H_8N_2$	$(AgBr)_2C_2H_8N_2$	Ag: MVol.B6-58
$Ag_2Br_2C_3H_{15}N_9S_3$	$Ag_2(NH_2CSNHNH_2)_3Br_2$	Ag: MVol.B7-168/9
$Ag_2Br_2C_4N_2O_2$	$Ag_2C_4Br_2N_2O_2$	Ag: MVol.B6-148
$Ag_2Br_2C_6H_4^{2+}$	$[Ag_2C_6H_4Br_2]^{2+}$	Ag: MVol.B5-117
$Ag_2Br_2C_{19}H_6O_8S^{2-}$	$[Ag_2(C_{19}H_6Br_2O_8S)]^{2-}$	Ag: MVol.B6-207/8
$Ag_2Br_2C_{56}H_{56}P_4S_2$	$[AgBr((C_6H_5)_2PC_2H_4)_2S]_2$	Ag: MVol.B7-237/9
$Ag_2Br_2H_9N_3$	$2\ AgBr \cdot 3\ NH_3 = AgBr \cdot 1.5\ NH_3$	Ag: MVol.B6-22
$Ag_2Br_4C_{68}H_{38}N_8O_5$	$[Ag(C_{12}H_8N_2)_2]_2C_{20}H_6Br_4O_5$	Ag: MVol.B6-124

Formula	Compound	Reference
$Ag_2CF_3H_3N_2O_2S$	$CF_3SO_2NAg_2 \cdot NH_3$	Ag: MVol.B6-26
$Ag_2CHKO_6S_3$	$KAg[Ag(SCH(SO_3)_2)] \cdot H_2O$	Ag: MVol.B7-47
Ag_2CHNO_2	Ag_2NHCOO	Ag: MVol.B6-29
$Ag_2CH_2N_2O$	$Ag_2(CO(NH)_2)$	Ag: MVol.B6-334
$Ag_2CH_2N_4O_2$	$[Ag_2(NO_2NC(NH)NH)]$	Ag: MVol.B6-337
$Ag_2CH_3N_3$	$[Ag_2(NHC(NH)NH)]$	Ag: MVol.B6-337
–	$Ag_2(NHC(NH)NH) \cdot H_2O$	Ag: MVol.B6-337
$Ag_2CH_3N_3O_3S$	$Ag_2(NHCONHSO_2NH)$	Ag: MVol.B6-336 S: S-N-Verb.1-167
$Ag_2CH_4I_2N_2S$	$Ag_2(NH_2CSNH_2)I_2$	Ag: MVol.B7-136
$Ag_2CH_4N_4O_6S$	$Ag_2(NH_2CSNH_2)(NO_3)_2$	Ag: MVol.B7-136
$Ag_2CH_4N_4O_7$	$2\ AgNO_3 \cdot CO(NH_2)_2$	Ag: MVol.B6-333/4
$Ag_2CH_5I_2N$	$2\ AgI \cdot CH_3NH_2 = AgI \cdot 0.5\ CH_3NH_2$	Ag: MVol.B6-42
$Ag_2CH_5NO_6S_3$	$NH_4Ag[Ag(SCH(SO_3)_2)] \cdot H_2O$	Ag: MVol.B7-47
$Ag_2CH_5N_3O$	$Ag_2O \cdot NH_2C(NH)NH_2$	Ag: MVol.B6-337
$Ag_2CH_{12}IrN_7O_9S_2$	$Ir(NCS)(SO_3)(NH_3)_4 \cdot 2\ AgNO_3$	Ir: SVol.2-163
$Ag_2CH_{12}N_4O_3$	$[Ag(NH_3)_2]_2CO_3$	Ag: MVol.B6-28
–	$[Ag(NH_3)_2]_2CO_3 \cdot H_2O$	Ag: MVol.B6-28
$Ag_2CH_{12}N_4O_4$	$AgO \cdot Ag(NH_3)_4CO_3 \cdot 2\ H_2O$	Ag: MVol.B7-272
Ag_2CN_2	Ag_2CN_2	Ag: MVol.B5-22/3
Ag_2CN_4O	$[Ag_2(OCN_4)]$	Ag: MVol.B6-193
Ag_2CO_5S	$Ag_2SO_4 \cdot CO$	Ag: MVol.B5-25
$Ag_2C_2Cl_2H_2^{2+}$	$[Ag_2CHClCHCl]^{2+}$	Ag: MVol.B5-46
$Ag_2C_2Cl_2H_8N_2$	$(AgCl)_2C_2H_8N_2$	Ag: MVol.B6-58
$Ag_2C_2F_2H_2^{2+}$	$[Ag_2CHFCHF]^{2+}$	Ag: MVol.B5-46
$Ag_2C_2HN_3O_4S$	$Ag_2[HNSO_2][NCO]_2 \cdot 3\ H_2O$	Ag: MVol.B7-86 S: S-N-Verb.1-30
$Ag_2C_2HN_3S_2$	$[Ag_2(C_2HN_3S_2)]$	Ag: MVol.B7-56
$Ag_2C_2HN_5O_6$	$Ag_2(NH(CONNO_2)_2)$	Ag: MVol.B6-336
$Ag_2C_2H_2I_2^{2+}$	$[Ag_2CHICHI]^{2+}$	Ag: MVol.B5-46
$Ag_2C_2H_2N_2S_2$	$[Ag_2(NHCSCSNH)]$	Ag: MVol.B7-186
$Ag_2C_2H_2N_4$	$[Ag_2(NHC(NH)NCN)]$	Ag: MVol.B6-338
$Ag_2C_2H_2N_4S$	$Ag_2(NH_2C_2N_3S)$	Ag: MVol.B7-55
–	$[Ag_2(NH_2C_2N_3S)]_x$	Ag: MVol.B7-55
$Ag_2C_2H_2N_4S_2$	$[Ag_2(C_2H_2N_4S_2)]$	Ag: MVol.B7-68
–	$[Ag_2(NH_2C_2N_3S_2)]$	Ag: MVol.B7-56
$Ag_2C_2H_2O_2S$	$[Ag_2(SCH_2COO)]$	Ag: MVol.B7-21
$Ag_2C_2H_3N_3O_2$	$Ag_2(NHCONHCONH)$	Ag: MVol.B6-336
$Ag_2C_2H_3O_2^{+}$	$[Ag_2CH_3COO]^{+}$	Ag: MVol.B5-139, 146/7
$Ag_2C_2H_4^{2+}$	$[Ag_2C_2H_4]^{2+}$	Ag: MVol.B5-28
$Ag_2C_2H_4N_2O_6$	$2\ AgNO_3 \cdot C_2H_4$	Ag: MVol.B5-28/9
$Ag_2C_2H_4N_6O_3$	$[Ag_2(ONC(NH_2)N(O)NC(NH_2)NO)]$	Ag: MVol.B6-318
$Ag_2C_2H_5OS^{+}$	$[Ag_2(SCH_2CH_2OH)]^{+}$	Ag: MVol.B7-6
$Ag_2C_2H_6I_2Se$	$2\ AgI \cdot (CH_3)_2Se$	Ag: MVol.B7-191
$Ag_2C_2H_6I_2Te$	$2\ AgI \cdot (CH_3)_2Te$	Ag: MVol.B7-198
$Ag_2C_2H_6N_4$	$[Ag(NH_3)_2][Ag(CN)_2]$	Ag: MVol.B6-32
$Ag_2C_2H_6N_4O_6S$	$2\ AgNO_3 \cdot CH_3NHCSNH_2$	Ag: MVol.B7-150
$Ag_2C_2H_8I_2N_2$	$(AgI)_2C_2H_8N_2$	Ag: MVol.B6-58
$Ag_2C_2H_8N_2^{2+}$	$[Ag_2(C_2H_8N_2)]^{2+}$	Ag: MVol.B6-55/7
$Ag_2C_2H_8N_4O_4S_2W$	$Ag_2(NH_2CSNH_2)_2WO_4 \cdot H_2O$	Ag: MVol.B7-137

$Ag_2C_2H_8N_4O_4S_3$ .... $Ag_2(NH_2CSNH_2)_2SO_4$ .... Ag: MVol.B7-137
$Ag_2C_2H_8N_8O_6$ ...... 2 $AgNO_3$ · $(NH_2NH)_2C_2(NH)_2$ .... Ag: MVol.B6-332
$Ag_2C_2H_{10}N_6O_4S_3$ .... $(AgNH_2CSNHNH_2)_2SO_4$ .... Ag: MVol.B7-167
$Ag_2C_2H_{12}N_4O_4$ ...... $[Ag(NH_3)_2]_2C_2O_4$ .... Ag: MVol.B6-29
$Ag_2C_2H_{4x}N_{2x}O_4$ ..... $Ag_2C_2O_4$ · x N_2H_4 .... Ag: MVol.B6-36
$Ag_2C_2N_2S_2$ .... $[Ag_2(NCNCSS)]$ .... Ag: MVol.B7-104
$Ag_2C_2N_2S_3$ .... $[Ag_2(C_2N_2S_3)]$ .... Ag: MVol.B7-80
Ag_2C_2O .... Ag_2C_2O .... Ag: MVol.B5-21
$Ag_2C_2O_4$ .... $Ag_2C_2O_4$
Chemical reactions .... Ag: MVol.B5-160/84
Electrochemical behavior .... Ag: MVol.B5-160
Formation and preparation .... Ag: MVol.B5-148/50
Physical properties .... Ag: MVol.B5-150/60
Solubility .... Ag: MVol.B5-184/6
– .... $Ag_2C_2O_4$ solutions
$Ag_2C_2O_4$-H_2O .... Ag: MVol.B5-186
$Ag_2C_3Cl_2H_{12}N_6O_8S_3$. $[Ag_2(NH_2CSNH_2)_3](ClO_4)_2$ .... Ag: MVol.B7-138
$Ag_2C_3CrH_{12}N_6O_4S_3$ .. $[Ag_2(NH_2CSNH_2)_3]CrO_4$ .... Ag: MVol.B7-138
– .... $[Ag_2(NH_2CSNH_2)_3]CrO_4$ · H_2O .... Ag: MVol.B7-138
$Ag_2C_3HN_3O_2$ .... $[Ag_2(C_3HN_3O_2)]$ .... Ag: MVol.B6-308/12
$Ag_2C_3HN_3O_3$ .... $[Ag_2(O_2C_3N_3OH)]$ .... Ag: MVol.B6-189/90
– .... $Ag_2(O_2C_3N_3OH)$ · H_2O .... Ag: MVol.B6-189/90
$Ag_2C_3H_2N_2OS$ ...... $Ag_2C_3H_2N_2OS$ · 0.5 H_2O .... Ag: MVol.B7-71
$Ag_2C_3H_2N_4O_5S$ ..... $AgC_3H_2N_3O_2S$ · $AgNO_3$ · H_2O .... Ag: MVol.B7-72
$Ag_2C_3H_2S_6$ .... $[Ag_2(CH_2(SCSS)_2)]$ .... Ag: MVol.B7-95/6
$Ag_2C_3H_3N_3S_2$ ...... $[Ag_2(C_3H_3N_3S_2)]$ .... Ag: MVol.B7-68
$Ag_2C_3H_3N_5O_5$ ...... $AgNO_3$ · $AgNHC_3N_3(OH)_2$ .... Ag: MVol.B6-189
$Ag_2C_3H_3N_5S$ .... $[Ag_2(C_3H_3N_5S)]$ .... Ag: MVol.B7-68
C: MVol.D6-44
$Ag_2C_3H_4N_2O_3S_2$ .... $AgNO_3$ · $AgC_3H_4NS_2$ .... Ag: MVol.B7-73
$Ag_2C_3H_4N_6$ .... $Ag_2(C_3N_3(NH)_2NH_2)$ .... Ag: MVol.B6-188
$Ag_2C_3H_4O_2S$ .... $[Ag_2(SCH_2CH_2COO)]$ .... Ag: MVol.B7-30
$Ag_2C_3H_5NO_2$ .... $Ag_2(NCOOC_2H_5)$ .... Ag: MVol.B6-330
$Ag_2C_3H_5N_3O_2$ ...... $[Ag_2(NHC(NH)NHCH_2COO)]$ .... Ag: MVol.B6-339
$Ag_2C_3H_5NaO_3S_3$ .... $Na[Ag_2(SCH_2CH(S)CH_2SO_3)]$ .... Ag: MVol.B7-46/7
$Ag_2C_3H_5O_3S_3^-$ ...... $[Ag_2(SCH_2CH(S)CH_2SO_3)]^-$ .... Ag: MVol.B7-46
$Ag_2C_3H_6^{2+}$ .... $[Ag_2(CH_2CHCH_3)]^{2+}$ .... Ag: MVol.B5-30
$Ag_2C_3H_6N_4O_2$ ...... $[Ag_2(NH_2C(NO)CH_2C(NO)NH_2)]$ .... Ag: MVol.B6-318
$Ag_2C_3H_6N_8O_6$ ...... 2 $AgNO_3$ · $C_3N_3(NH_2)_3$ .... Ag: MVol.B6-188
$Ag_2C_3H_6OS_2$ .... $[Ag_2(SCH(CH_2OH)CH_2S)]$ .... Ag: MVol.B7-8/9
$Ag_2C_3H_6S_3^{2+}$ .... $[Ag_2(S(CH_2S)_2CH_2)]_n^{2n+}$ .... Ag: MVol.B7-88
$Ag_2C_3H_7N_5O_3$ ...... $Ag_2(O_2C_3N_3OH)$ · 2 NH_3 .... Ag: MVol.B6-189/90
$Ag_2C_3H_8N_4O_6S$ ..... 2 $AgNO_3$ · $CH_3NHCSNHCH_3$ .... Ag: MVol.B7-152
– .... 2 $AgNO_3$ · $(CH_3)_2NCSNH_2$ .... Ag: MVol.B7-152
$Ag_2C_3H_{10}N_2^{2+}$ ...... $[Ag_2(NH_2C_3H_6NH_2)]^{2+}$ .... Ag: MVol.B6-62
$Ag_2C_3H_{11}N_3^{2+}$ ...... $[Ag_2(NH_2CH(CH_2NH_2)_2)]^{2+}$ .... Ag: MVol.B6-65
$Ag_2C_3H_{12}I_2N_6O_6S_3$ ... $[Ag_2(NH_2CSNH_2)_3](IO_3)_2$ .... Ag: MVol.B7-138
$Ag_2C_3H_{12}N_6O_3S_5$ .... $[Ag_2(NH_2CSNH_2)_3]S_2O_3$ .... Ag: MVol.B7-138
$Ag_2C_3H_{12}N_6O_4S_4$ .... $[Ag_2(NH_2CSNH_2)_3]SO_4$ · H_2O .... Ag: MVol.B7-138

Formula	Compound	Reference
$Ag_2C_3H_{12}N_6S_3^{2+}$	$[Ag_2(NH_2CSNH_2)_3]^{2+}$	Ag: MVol.B7-133/4
$Ag_2C_3H_{12}N_8O_6S_3$	$[Ag_2(NH_2CSNH_2)_3](NO_3)_2$	Ag: MVol.B7-137/8
$Ag_2C_3H_{15}N_9S_3^{2+}$	$[Ag_2(NH_2CSNHNH_2)_3]^{2+}$	Ag: MVol.B7-165
$Ag_2C_3H_{15}N_{11}O_6S_3$	$Ag_2(NH_2CSNHNH_2)_3(NO_3)_2$	Ag: MVol.B7-167/8
Ag_2C_4	$Ag[CC]_2Ag$	Ag: MVol.B5-13
$Ag_2C_4Cl_2H_{16}N_4$	$[Ag_2(C_2H_8N_2)_2]Cl_2$	Ag: MVol.B6-58
$Ag_2C_4Cl_4H_{14}N_2Pt$	$[Ag(C_2H_5NH_2)]_2[PtCl_4]$	Ag: MVol.B6-43
$Ag_2C_4Cl_4H_{16}N_4Pt$	$[Ag(C_2H_8N_2)]_2PtCl_4$	Ag: MVol.B6-59
$Ag_2C_4Cl_5H_4IrN_2$	$Ag_2[Ir(NCHCHCHCHN)Cl_5]$	Ir: SVol.2-79
–	$Ag_2[Ir(NCHCHNCHCH)Cl_5]$	Ir: SVol.2-78
$Ag_2C_4CuH_{18}N_{10}O_4$	$[Ag(NH_3)_2]_2[Cu(NH(CONH)_2)_2]$	Ag: MVol.B6-32
$Ag_2C_4FHN_2O_2$	$Ag_2(C_4HFN_2O_2)$	Ag: MVol.B6-152
$Ag_2C_4F_6H_7NS_2$	$2\ AgSCF_3 \cdot NH(CH_3)_2$	Ag: MVol.B7-4
$Ag_2C_4HN_3O_4$	$[Ag_2(C_2HN_3(COO)_2)]$	Ag: MVol.B6-181
–	$[Ag_2(OCONC_3HN_2O_2)]$	Ag: MVol.B6-148
–	$[Ag_2(ONC_4HN_2O_3)]$	Ag: MVol.B6-314
$Ag_2C_4H_2N_2OS$	$[Ag_2(C_4H_2N_2OS)]$	Ag: MVol.B7-63
$Ag_2C_4H_2N_2S_2$	$[Ag_2(C_4H_2N_2S_2)]$	Ag: MVol.B7-63
$Ag_2C_4H_2N_4O_4$	$[Ag_2(C_2H_2N_4(COO)_2)]$	Ag: MVol.B6-196
$Ag_2C_4H_2N_4O_7$	$Ag(ONC_4H_2N_2O_3) \cdot AgNO_3$	Ag: MVol.B6-314
$Ag_2C_4H_2N_6O$	$[Ag_2(C_4H_2N_6O)]$	Ag: MVol.B6-183
$Ag_2C_4H_3N_3O_2$	$[Ag_2(CH_3C_3N_3O_2)]$	Ag: MVol.B6-308/12
–	$Ag_2(CH_3C_3N_3O_2) \cdot 2.5\ H_2O$	Ag: MVol.B6-189
$Ag_2C_4H_3N_5Pt$	$Ag_2Pt(CN)_4 \cdot NH_3$	Ag: MVol.B6-33
$Ag_2C_4H_4N_2^{2+}$	$[Ag(NC(CH_2)_2CN)Ag]^{2+}$	Ag: MVol.B6-351
$Ag_2C_4H_4N_2O_2$	$[Ag_2C_4H_4N_2O_2]$	Ag: MVol.B6-267
$Ag_2C_4H_4N_4O_6$	$[Ag_2(NCC_2H_4CN)](NO_3)_2$ $= 2\ AgNO_3 \cdot NCC_2H_4CN$	Ag: MVol.B6-350, 351
$Ag_2C_4H_4O_4S$	$Ag_2(S(CH_2COO)_2)$	Ag: MVol.B7-38
–	$Ag[Ag(HSC_2H_3(COO)_2)] \cdot H_2O$	Ag: MVol.B7-32
$Ag_2C_4H_5NO_4$	$Ag[Ag(NH(CH_2COO)_2)]$	Ag: MVol.B6-275
–	$[Ag_2(NH_2C_2H_3(COO)_2)]$	Ag: MVol.B6-252
$Ag_2C_4H_5NO_4S_2$	$Ag_2(C_3H_5CSNOSO_3)$	Ag: MVol.B7-126
$Ag_2C_4H_6N_2O_5S$	$AgNO_3 \cdot Ag(HOCH_2C_3H_3NOS)$	Ag: MVol.B7-53
$Ag_2C_4H_6N_2O_6$	$2\ AgNO_3 \cdot C_4H_6$	Ag: MVol.B5-35
$Ag_2C_4H_6N_6Pt$	$Ag_2Pt(CN)_4 \cdot 2\ NH_3$	Ag: MVol.B6-33
$Ag_2C_4H_6N_6S_2$	$[Ag_2(C_2H_2(NNCSNH_2)_2)]$	Ag: MVol.B6-282
$Ag_2C_4H_6O_2^{2+}$	$[Ag_2CH_2CHCOOCH_3]^{2+}$	Ag: MVol.B5-49
–	$[Ag_2CH_2CHOOCCH_3]^{2+}$	Ag: MVol.B5-48
$Ag_2C_4H_7N_3O_2$	$[Ag_2(NHC(NH)N(CH_3)CH_2COO)]$	Ag: MVol.B6-339
$Ag_2C_4H_8^{2+}$	$[Ag_2C_4H_8]^{2+}$	Ag: MVol.B5-32/3
$Ag_2C_4H_8N_2O_6S_2$	$2\ AgNO_3 \cdot C_4H_8S_2$	Ag: MVol.B7-86, 87
$Ag_2C_4H_8N_4$	$(AgCN)_2C_2H_8N_2$	Ag: MVol.B6-58
$Ag_2C_4H_8N_4O_6S$	$2\ AgNO_3 \cdot CH_2CHCH_2NHCSNH_2$	Ag: MVol.B7-156
$Ag_2C_4H_8N_4S_2$	$(AgSCN)_2C_2H_8N_2$	Ag: MVol.B6-58/9
$Ag_2C_4H_8N_4Se_2$	$(AgSeCN)_2C_2H_8N_2$	Ag: MVol.B6-59
$Ag_2C_4H_8O^{2+}$	$[Ag_2CH_2CHOC_2H_5]^{2+}$	Ag: MVol.B5-48
$Ag_2C_4H_{10}I_2N_2$	$Ag_2(C_4H_{10}N_2)I_2 = Ag(C_4H_{10}N_2)_{0.5}I$	Ag: MVol.B6-164/5
$Ag_2C_4H_{10}N_2^{2+}$	$[Ag_2(NH(CH_2CH_2)_2NH)]^{2+}$	Ag: MVol.B6-162/3
$Ag_2C_4H_{10}N_4O_{10}$	$[Ag(NO_3)(NH_2CH_2COOH)]_2$	Ag: MVol.B6-236/8

Formula	Compound	Reference
$Ag_2C_4H_{12}N_8Pd$	$[Ag(NH_3)_2]_2[Pd(CN)_4]$	Ag: MVol.B6-32
$Ag_2C_4H_{12}N_8Pt$	$Ag_2Pt(CN)_4 \cdot 4\ NH_3$	Ag: MVol.B6-33
$Ag_2C_4H_{12}N_8PtS_4$	$Ag_2Pt(SCN)_4 \cdot 4\ NH_3$	Ag: MVol.B6-33
$Ag_2C_4H_{13}N_3^{2+}$	$[Ag_2(NH(C_2H_4NH_2)_2)]^{2+}$	Ag: MVol.B6-65
$Ag_2C_4H_{14}N_4O_4$	$Ag_2(NH_2C_2H_3(COO)_2) \cdot 3\ NH_3$	Ag: MVol.B6-252
–	$[Ag(NH_3)_2]_2C_2H_2(COO)_2$	Ag: MVol.B6-29
$Ag_2C_4H_{16}N_4^{2+}$	$[Ag_2(C_2H_8N_2)_2]^{2+}$	Ag: MVol.B6-55/6
$Ag_2C_4H_{16}N_4O_4S$	$[Ag_2(C_2H_8N_2)_2]SO_4$	Ag: MVol.B6-58
$Ag_2C_4H_{17}N_5O_4$	$Ag_2(NH_2C_2H_3(COO)_2) \cdot 4\ NH_3$	Ag: MVol.B6-252
$Ag_2C_4H_{18}N_{10}NiO_4$	$[Ag(NH_3)_2]_2[Ni(NH(CONH)_2)_2] \cdot 6\ H_2O$	Ag: MVol.B6-32
$Ag_2C_4H_{18}N_{10}PtS_4$	$Ag_2Pt(SCN)_4 \cdot 6\ NH_3$	Ag: MVol.B6-33
$Ag_2C_4H_{20}N_{12}O_4S_5$	$[Ag(NH_2CSNHNH_2)_2]_2SO_4$	Ag: MVol.B7-169
$Ag_2C_4H_{22}N_4O_{42}SiW_{12}$	$[Ag_2(C_2H_8N_2)_2][Si(W_2O_7)_6H_6]$	Ag: MVol.B6-59
$Ag_2C_4N_4O_2$	$[Ag_2(NCC(NO)C(NO)CN)]$	Ag: MVol.B6-318
$Ag_2C_4N_6S_4$	$[Ag_2(C_2N_2S_2NNC_2N_2S_2)]$	Ag: MVol.B6-289
$Ag_2C_5H_2N_4S$	$[Ag_2(C_5H_2N_4S)]$	Ag: MVol.B7-66/7
$Ag_2C_5H_3NO_3$	$AgCCCCCH_3 \cdot AgNO_3$	Ag: MVol.B5-14
$Ag_2C_5H_3NO_6$	$Ag_2(OOCCOCONHCH_2COO)$	Ag: MVol.B6-327
$Ag_2C_5H_4N_2OS$	$[Ag_2(CH_3C_4HN_2OS)]$	Ag: MVol.B7-63
$Ag_2C_5H_5N_3S_2$	$[Ag_2(C_3H_5NC_2N_2S_2)]$	Ag: MVol.B7-81
$Ag_2C_5H_7NO_4$	$[Ag_2(NH_2C_3H_5(COO)_2)]$	Ag: MVol.B6-253
$Ag_2C_5H_7NO_5$	$[Ag_2(NH_2C_3H_4(OH)(COO)_2)]$	Ag: MVol.B6-254
$Ag_2C_5H_7N_3O$	$[Ag_2(C_3H_7C_2N_3O)]$	Ag: MVol.B6-187
$Ag_2C_5H_9O_3S_4^-$	$[Ag_2(SCH_2CH(S)CH_2SC_2H_4SO_3)]^-$	Ag: MVol.B7-46
$Ag_2C_5H_9O_4S_3^-$	$[Ag_2(SCH_2CH(S)CH_2OC_2H_4SO_3)]^-$	Ag: MVol.B7-45
$Ag_2C_5H_9O_5S_4^-$	$[Ag_2(SCH_2CH(S)CH_2SO_2C_2H_4SO_3)]^-$	Ag: MVol.B7-46
$Ag_2C_5H_{11}O_3S^+$	$[Ag_2(SCH_2C(CH_2OH)_3)]^+$	Ag: MVol.B7-7
$Ag_2C_5H_{12}N_4O_6S$	$2\ AgNO_3 \cdot (C_2H_5)_2NCSNH_2$	Ag: MVol.B7-155
$Ag_2C_5H_{19}N_5O_4$	$Ag_2(NH_2C_3H_5(COO)_2) \cdot 4\ NH_3$	Ag: MVol.B6-253
$Ag_2C_5H_{20}N_{10}O_4S_6$	$Ag_2SO_4 \cdot 5\ (NH_2)_2CS \cdot 2\ H_2O$	Ag: MVol.B7-144
Ag_2C_6	$Ag[CC]_3Ag$	Ag: MVol.B5-13
$Ag_2C_6ClH_5^{2+}$	$[Ag_2C_6H_5Cl]^{2+}$	Ag: MVol.B5-117
$Ag_2C_6Cl_2H_8O_8$	$2\ AgClO_4 \cdot C_6H_8$	Ag: MVol.B5-60
$Ag_2C_6Cl_2H_{12}N_4O_6$	$2\ AgClO_3 \cdot (CH_2)_6N_4$	Ag: MVol.B6-198
$Ag_2C_6CrH_{12}N_4O_4$	$Ag_2CrO_4 \cdot (CH_2)_6N_4$	Ag: MVol.B6-200
$Ag_2C_6CrH_{24}N_{12}O_4S_6$	$[Ag(NH_2CSNH_2)_3]_2CrO_4 \cdot 2\ H_2O$	Ag: MVol.B7-147/8
$Ag_2C_6FH_5^{2+}$	$[Ag_2C_6H_5F]^{2+}$	Ag: MVol.B5-117
$Ag_2C_6H_2N_2O_4$	$[Ag_2(C_4H_2N_2(COO)_2)]$	Ag: MVol.B6-160/1
$Ag_2C_6H_2O_3S_3$	$[Ag_2(C_5H_2OS_3COO)]$	Ag: MVol.B7-83/4
$Ag_2C_6H_2O_4$	$[Ag_2(C_6H_2O_4)]$	Ag: MVol.B6-217/8
$Ag_2C_6H_4I_2^{2+}$	$[Ag_2C_6H_4I_2]^{2+}$	Ag: MVol.B5-117
$Ag_2C_6H_4N_2O_4S$	$[Ag_2(SO_3C_6H_4NNO)]$	Ag: MVol.B6-293
$Ag_2C_6H_4N_4O_4$	$[Ag_2(C_6H_4(N(NO)O)_2)]$	Ag: MVol.B6-321
$Ag_2C_6H_4OS_3$	$[Ag_2(CH_3C_5HOS_3)]$	Ag: MVol.B7-83/4
$Ag_2C_6H_5I^{2+}$	$[Ag_2C_6H_5I]^{2+}$	Ag: MVol.B5-117
$Ag_2C_6H_5NO_3S$	$[Ag_2(C_6H_5SO_2NO)]$	Ag: MVol.B6-307
$Ag_2C_6H_5N_3$	$Ag_2CN_2 \cdot C_5H_5N$	Ag: MVol.B5-22/3
$Ag_2C_6H_6^{2+}$	$[Ag_2C_6H_6]^{2+}$	Ag: MVol.B5-91
$Ag_2C_6H_6N_2O_6$	$Ag_2(COCO(NHCH_2COO)_2)$	Ag: MVol.B6-327
$Ag_2C_6H_6N_6O_6S_2$	$2\ AgNO_3 \cdot CH_3C_2N_2SC_2N_2SCH_3$	Ag: MVol.B7-81

Formula	Compound	Reference
$Ag_2C_6H_6N_{12}O_2$	$[Ag_2(C_2H_5OCOCH(NNCN_4)_2)]$	Ag: MVol.B6-288
$Ag_2C_6H_6Na_4O_6S_3$	$2\ Na[Ag(SCH_2COO)] \cdot Na_2(SCH_2COO)$	Ag: MVol.B7-22
$Ag_2C_6H_6O^{2+}$	$[Ag_2(C_6H_5OH)]^{2+}$	Ag: MVol.B5-118
$Ag_2C_6H_7N^{2+}$	$[Ag_2(C_6H_5NH_2)]^{2+}$	Ag: MVol.B6-51
$Ag_2C_6H_7NO_6$	$[Ag_2(N(CH_2COO)_2CH_2COOH)]$	Ag: MVol.B6-276
$Ag_2C_6H_7N_3O_2$	$Ag_2(C_3H_3N_2C_2H_3(NH)COO) \cdot H_2O$	Ag: MVol.B6-246/7
$Ag_2C_6H_7N_3O_3$	$[Ag_2C_6H_7N_3O_3]$	Ag: MVol.B6-269
$Ag_2C_6H_7N_3O_3S$	$AgNO_3 \cdot Ag((CH_3)_2C_4HN_2S)$	Ag: MVol.B7-63
$Ag_2C_6H_8{}^{2+}$	$[Ag_2(C_6H_8)]^{2+}$	Ag: MVol.B5-40/1
$Ag_2C_6H_8N_2{}^{2+}$	$[Ag_2(NH_2CH_2C_5H_4N)]^{2+}$	Ag: MVol.B6-96
$Ag_2C_6H_8N_2O_6$	$2\ AgNO_3 \cdot C_6H_8$	Ag: MVol.B5-58
$Ag_2C_6H_8N_2S_4$	$[Ag_2(SCSN(CH_2CH_2)_2NCSS)]$	Ag: MVol.B7-118
$Ag_2C_6H_8N_3O_2{}^+$	$[Ag_2(C_3H_3N_2C_2H_3(NH_2)COO)]^+$	Ag: MVol.B6-246
$Ag_2C_6H_8N_8O_4$	$Ag_2(C_2(COO)_2(NNHC(NH)NH_2)_2) \cdot 2\ H_2O$	Ag: MVol.B6-345
$Ag_2C_6H_8N_8O_8$	$2\ AgNO_3 \cdot C_2H_4(CON_2H_2CN)_2$	Ag: MVol.B6-333
$Ag_2C_6H_8O_4S_2$	$[Ag_2(C_2H_4(SCH_2COO)_2)]$	Ag: MVol.B7-39/40
$Ag_2C_6H_9O_2S^+$	$[Ag_2CH_2CH(CH_2)_2SCH_2COO]^+$	Ag: MVol.B5-50/1
$Ag_2C_6H_9O_2Se^+$	$[Ag_2(CH_2CHC_2H_4SeCH_2COO)]^+$	Ag: MVol.B5-50/1
		Ag: MVol.B7-193
$Ag_2C_6H_{10}{}^{2+}$	$[Ag_2(C_6H_{10})]^{2+}$	Ag: MVol.B5-40/1
$Ag_2C_6H_{10}N_2O_2S_2$	$[Ag_2(HOC_2H_4NCSCSNC_2H_4OH)]$	Ag: MVol.B7-188
$Ag_2C_6H_{10}N_2O_4S_2$	$[Ag_2(S_2(CH_2CH(NH_2)COO)_2)]$	Ag: MVol.B6-258
$Ag_2C_6H_{10}O_2Se^{2+}$	$[Ag_2(CH_2CHC_2H_4SeCH_2COOH)]^{2+}$	Ag: MVol.B5-50/1
		Ag: MVol.B7-193
$Ag_2C_6H_{10}O_4S_2{}^{2+}$	$[Ag_2(C_2H_4(SCH_2COOH)_2)]^{2+}$	Ag: MVol.B7-40
$Ag_2C_6H_{10}O_6S_6{}^{4-}$	$[Ag_2(SCH_2CH(S)CH_2SO_3)_2]^{4-}$	Ag: MVol.B7-46
$Ag_2C_6H_{12}N_4OS_2$	$[Ag(C_3H_6N_2S)]_2O$	Ag: MVol.B7-49
$Ag_2C_6H_{12}N_4O_6S_4$	$[Ag_2(SS(CSN(CH_3)_2)_2)](NO_3)_2$	Ag: MVol.B7-120
$Ag_2C_6H_{12}N_6O_6$	$2\ AgNO_3 \cdot (CH_2)_6N_4$	Ag: MVol.B6-197
$Ag_2C_6H_{15}N_{11}O_6S_6$	$[Ag_2(NH_2CSNHCSNH_2)_3](NO_3)_2$	Ag: MVol.B7-164
$Ag_2C_6H_{18}N_4{}^{2+}$	$[Ag_2(C_2H_4(NHC_2H_4NH_2)_2)]^{2+}$	Ag: MVol.B6-66
–	$[Ag_2(N(C_2H_4NH_2)_3)]^{2+}$	Ag: MVol.B6-65
$Ag_2C_6H_{18}N_{12}S_3{}^{2+}$	$[Ag_2(NH_2C(NH)NHCSNH_2)_3]^{2+}$	Ag: MVol.B7-162
$Ag_2C_6H_{24}N_{12}O_4S_7$	$[Ag(NH_2CSNH_2)_3]_2SO_4$	Ag: MVol.B7-146
$Ag_2C_6H_{24}N_{12}S_6{}^{2+}$	$[Ag_2(NH_2CSNH_2)_6]^{2+}$	Ag: MVol.B7-145
$Ag_2C_6H_{25}N_{12}O_4PS_6$	$[Ag(NH_2CSNH_2)_3]_2HPO_4$	Ag: MVol.B7-147
$Ag_2C_6H_{42}N_{20}PtS_6$	$Ag_2Pt(SCN)_6 \cdot 14\ NH_3$	Ag: MVol.B6-33
$Ag_2C_7Cl_3H_5IrNO_4$	$Ag_2[Ir(C_2O_4)(C_5H_5N)Cl_3]$	Ir: SVol.2-36
$Ag_2C_7F_3NO_4$	$NC_5F_3(COOAg)_2$	F: PerFHalOrg.5-193, 207
$Ag_2C_7F_6H_5NS_2$	$2\ AgSCF_3 \cdot C_5H_5N$	Ag: MVol.B7-4
$Ag_2C_7F_6H_{12}N_6O_4S_3$	$[Ag_2(NH_2CSNH_2)_3](CF_3COO)_2$	Ag: MVol.B7-138
$Ag_2C_7FeH_3N_6$	$[CH_3NCFe(CN)_5]Ag_2$	Fe: Org.Verb.B4-10
$Ag_2C_7H_3NO_4$	$Ag_2(NC_5H_3(COO)_2)$	Ag: MVol.B6-104
–	$Ag_2(NC_5H_3(COO)_2) \cdot H_2O$	Ag: MVol.B6-104
$Ag_2C_7H_4N_2S$	$[Ag_2(C_7H_4N_2S)]$	Ag: MVol.B7-77
$Ag_2C_7H_4O_3S_3$	$[Ag_2(C_5H_2OS_3CH_2COO)]$	Ag: MVol.B7-83/4
$Ag_2C_7H_5NO$	$Ag_2C_2O \cdot C_5H_5N$	Ag: MVol.B5-21
		Ag: MVol.B6-81
$Ag_2C_7H_5NO_4S$	$Ag_2(NHSO_2C_6H_4COO)$	Ag: MVol.B6-344
$Ag_2C_7H_5N_3$	$[AgC_5H_5N][Ag(CN)_2]$	Ag: MVol.B6-81

Formula	Compound		Reference
$Ag_2C_7H_6N_2S_2$	$[Ag_2(C_6H_5NHNCSS)]$	Ag:	MVol.B7-119
$Ag_2C_7H_6OS_3$	$[Ag_2(C_2H_5C_5HOS_3)]$	Ag:	MVol.B7-83/4
–	$[Ag_2(C_5OS_3(CH_3)_2)]$	Ag:	MVol.B7-83/4
$Ag_2C_7H_7NO_5$	$Ag_2(OCOCHC(CH_3)CONHCH_2COO)$	Ag:	MVol.B6-241
$Ag_2C_7H_7N_3O_3S$	$[Ag_2(OCONHC_4HN_2OSC_2H_5)]$	Ag:	MVol.B7-63
$Ag_2C_7H_8{}^{2+}$	$[Ag_2C_6H_5CH_3]^{2+}$	Ag:	MVol.B5-96
$Ag_2C_7H_8N_2OS^{2+}$	$[Ag_2(HOC_6H_4NHCSNH_2)]^{2+}$	Ag:	MVol.B7-159
$Ag_2C_7H_8N_2O_6$	$2\ AgNO_3 \cdot C_7H_8$	Ag:	MVol.B5-77/8
$Ag_2C_7H_9N^{2+}$	$[Ag_2(CH_3C_6H_4NH_2)]^{2+}$	Ag:	MVol.B6-52
–	$[Ag_2(C_6H_5NHCH_3)]^{2+}$	Ag:	MVol.B6-53
$Ag_2C_7H_{10}N_2{}^{2+}$	$[Ag_2(CH_3(NH_2CH_2)C_5H_3N)]^{2+}$	Ag:	MVol.B6-96/7
$Ag_2C_7H_{10}N_2O_6S_3$	$2\ AgNO_3 \cdot (CH_2)_4C_3H_2S_3$	Ag:	MVol.B7-83
$Ag_2C_7H_{10}O_4S_2$	$[Ag_2(CH_3C_2H_3(SCH_2COO)_2)]$	Ag:	MVol.B7-41
$Ag_2C_7H_{11}NO_4$	$AgCCCH(C_4H_9)OH \cdot AgNO_3$	Ag:	MVol.B5-15
$Ag_2C_7H_{11}O_2S^+$	$[Ag_2CH_2CH(CH_2)_3SCH_2COO]^+$	Ag:	MVol.B5-50/1
$Ag_2C_7H_{11}O_2Se^+$	$[Ag_2(CH_2CH(CH_2)_3SeCH_2COO)]^+$	Ag:	MVol.B5-50/1
		Ag:	MVol.B7-193
$Ag_2C_7H_{12}N_2O_8S$	$AgNO_3 \cdot Ag(C_4H_5(OH)_4C_3H_3NOS)$	Ag:	MVol.B7-53
$Ag_2C_7H_{12}O_2Se^{2+}$	$[Ag_2(CH_2CH(CH_2)_3SeCH_2COOH)]^{2+}$	Ag:	MVol.B5-50/1
		Ag:	MVol.B7-193
$Ag_2C_8ClH_7{}^{2+}$	$[Ag_2(ClC_6H_4CHCH_2)]^{2+}$	Ag:	MVol.B5-117
$Ag_2C_8Cl_2H_8O_8$	$2\ AgClO_4 \cdot C_8H_8$	Ag:	MVol.B5-83
$Ag_2C_8Cl_4H_{28}N_4Pt$	$[Ag(C_2H_5NH_2)_2]_2[PtCl_4]$	Ag:	MVol.B6-44
$Ag_2C_8CuH_{22}N_{10}O_4$	$[Ag(C_2H_8N_2)]_2[Cu(NH(CONH)_2)_2] \cdot 2\ H_2O$	Ag:	MVol.B6-59
$Ag_2C_8F_6H_6O_6S_2$	$2\ AgCF_3SO_3 \cdot C_6H_6$	Ag:	MVol.B5-93
$Ag_2C_8FeH_5N_6$	$[C_2H_5NCFe(CN)_5]Ag_2$	Fe:	Org.Verb.B4-10
$Ag_2C_8FeH_{10}N_2O_6S_2$	$(CO)_2Fe(OOCCH(NH_2)CH_2SAg)_2 \cdot 2\ H_2O$	Fe:	Org.Verb.B1-116
$Ag_2C_8H_4N_2O_2$	$[Ag_2(C(O)NNC(O)C_6H_4)]$	Ag:	MVol.B6-179/80
–	$[Ag_2(NC(NO)C(O)C_6H_4)]$	Ag:	MVol.B6-316
$Ag_2C_8H_4N_6O_8$	$[Ag_2(ONC_4H_2N_2O_3)_2]$	Ag:	MVol.B6-314
$Ag_2C_8H_4N_8S_2$	$[Ag_2(SCN_4C_6H_4CN_4S)]$	Ag:	MVol.B7-58
$Ag_2C_8H_4O_2S_2$	$[Ag_2(C_6H_4(COS)_2)]$	Ag:	MVol.B7-95
$Ag_2C_8H_5NO_3$	$AgCCC_6H_5 \cdot AgNO_3$	Ag:	MVol.B5-17
$Ag_2C_8H_5NO_4$	$AgCCC_6H_4OH \cdot AgNO_3$	Ag:	MVol.B5-18
$Ag_2C_8H_5N_3O$	$[Ag_2(C_6H_5C_2N_3O)]$	Ag:	MVol.B6-187
$Ag_2C_8H_5N_3OS$	$[Ag_2(C_6H_5C_2N_3OS)]$	Ag:	MVol.B6-184
$Ag_2C_8H_5N_3O_2$	$[Ag_2(C_6H_5C_2N_3O_2)]$	Ag:	MVol.B6-184
$Ag_2C_8H_6N_4O_3S$	$AgNO_3 \cdot Ag(C_6H_5C_2HN_3S)$	Ag:	MVol.B7-56
$Ag_2C_8H_6N_4O_6$	$2\ AgNO_3 \cdot C_8H_6N_2$	Ag:	MVol.B6-129
$Ag_2C_8H_7NO_2{}^{2+}$	$[Ag_2(CH_2CHC_6H_4NO_2)]^{2+}$	Ag:	MVol.B5-118
$Ag_2C_8H_7NO_4$	$Ag_2(C_2H_5NC_4H_2(COO)_2)$	Ag:	MVol.B6-68
$Ag_2C_8H_7NO_5$	$AgNO_3 \cdot C_5H_7CCCOOAg$	Ag:	MVol.B5-89
$Ag_2C_8H_8{}^{2+}$	$[Ag_2C_6H_5CHCH_2]^{2+}$	Ag:	MVol.B5-101
$Ag_2C_8H_8N_2O_6$	$2\ AgNO_3 \cdot C_8H_8$	Ag:	MVol.B5-83
$Ag_2C_8H_8O^{2+}$	$[Ag_2(C_6H_5COCH_3)]^{2+}$	Ag:	MVol.B5-118
–	$[Ag_2(C_6H_5OCHCH_2)]^{2+}$	Ag:	MVol.B5-118
$Ag_2C_8H_8OS_3$	$[Ag_2(C_3H_7C_5HOS_3)]$	Ag:	MVol.B7-83/4
$Ag_2C_8H_{10}{}^{2+}$	$[Ag_2C_6H_4(CH_3)_2]^{2+}$	Ag:	MVol.B5-97
–	$[Ag_2C_6H_5C_2H_5]^{2+}$	Ag:	MVol.B5-97
$Ag_2C_8H_{10}N_2O_3$	$Ag_2((C_2H_5)_2C_4N_2O_3)$	Ag:	MVol.B6-154, 156

Formula	Compound	Reference
$Ag_2C_8H_{10}N_6O_2$	$[Ag_2(C_4H_8(CONNHCN)_2)]$	Ag: MVol.B6-333
$Ag_2C_8H_{11}N^{2+}$	$[Ag_2((CH_3)_2C_6H_3NH_2)]^{2+}$	Ag: MVol.B6-53
$Ag_2C_8H_{11}NO_6$	$[Ag_2(C_2H_5OCONHC_3H_5(COO)_2)]$	Ag: MVol.B6-253
$Ag_2C_8H_{12}{}^{2+}$	$[Ag_2C_8H_{12}]^{2+}$	Ag: MVol.B5-63
$Ag_2C_8H_{12}N_2O_6$	$2\ AgNO_3 \cdot C_8H_{12}$	Ag: MVol.B5-63
$Ag_2C_8H_{12}N_4O_4$	$Ag_2C_2O_4 \cdot (CH_2)_6N_4$	Ag: MVol.B6-199
–	$Ag_2C_2O_4 \cdot (CH_2)_6N_4 \cdot 2\ H_2O$	Ag: MVol.B6-199
$Ag_2C_8H_{12}N_6O_2$	$2\ AgNCO \cdot (CH_2)_6N_4$	Ag: MVol.B6-200
$Ag_2C_8H_{12}N_6S_2$	$2\ AgSCN \cdot (CH_2)_6N_4$	Ag: MVol.B6-200
$Ag_2C_8H_{12}N_8O_8$	$2\ AgNO_3 \cdot C_4H_8(CONHNHCN)_2$	Ag: MVol.B6-333
$Ag_2C_8H_{12}O_3$	$AgCCCH(CH_3)OH \cdot AgC_3H_7COO$	Ag: MVol.B5-15
$Ag_2C_8H_{12}O_4S_3$	$[Ag_2(S(C_2H_4SCH_2COO)_2)]$	Ag: MVol.B7-42/3
$Ag_2C_8H_{14}N_2O_2S_2$	$[Ag_2(CSCS(NCH_2CH(OH)CH_3)_2)]$	Ag: MVol.B7-189
$Ag_2C_8H_{14}N_6OS_2$	$Ag_2(C_2H_5OC_4H_5(NNCSNH_2)_2) \cdot HNO_3$	Ag: MVol.B6-283
$Ag_2C_8H_{14}O_4S_3{}^{2+}$	$[Ag_2(S(C_2H_4SCH_2COOH)_2)]^{2+}$	Ag: MVol.B7-43
$Ag_2C_8H_{15}N_7O_4S_2$	$Ag_2(C_2H_5OC_4H_5(NNCSNH_2)_2) \cdot HNO_3$	Ag: MVol.B6-283
$Ag_2C_8H_{16}N_4O_4$	$[Ag(NH_3)_2]_2C_6H_4(COO)_2$	Ag: MVol.B6-29
$Ag_2C_8H_{20}P_2$	$(CH_3)_2P(CH_2AgCH_2)_2P(CH_3)_2$	Ag: MVol.B5-24
		Ag: MVol.B7-201
$Ag_2C_8H_{24}N_{12}O_4S_6$	$[Ag(NH_2CSNH_2)_3]_2C_2O_4$	Ag: MVol.B7-147
–	$[Ag(NH_2CSNH_2)_3]_2C_2O_4 \cdot 2\ H_2O$	Ag: MVol.B7-147
$Ag_2C_8H_{28}N_{20}O_{12}S_3$	$[Ag(NH(C(NH_2)NH)_2)_2]_2(SO_4)_3 \cdot 9\ H_2O$	Ag: MVol.B7-320
$Ag_2C_8H_{32}N_8O_{39}SiW_{12}$	$Ag_2O \cdot 4\ C_2H_8N_2 \cdot SiO_2 \cdot 12\ WO_3$	Ag: MVol.B6-59
$Ag_2C_8N_4S_4{}^{2-}$	$[Ag_2(NCC(S)C(S)CN)_2]^{2-}$	Ag: MVol.B7-33
$Ag_2C_9H_4NO_3S_4{}^{-}$	$[Ag_2(HOC_6H_3(COO)N(CSS)_2)]^{-}$	Ag: MVol.B7-117
$Ag_2C_9H_4O_5S$	$[Ag_2(SO_3C_9H_4O_2)]$	Ag: MVol.B6-216
$Ag_2C_9H_5N_3O_2$	$[Ag_2(C_6H_5C_3N_3O_2)]$	Ag: MVol.B6-308/12
$Ag_2C_9H_5N_3O_6$	$Ag(ONC_3NO_2C_6H_5) \cdot AgNO_3$	Ag: MVol.B6-313
$Ag_2C_9H_6N_2O$	$Ag_2(CO(C_4H_3N)_2)$	Ag: MVol.B6-69
$Ag_2C_9H_6O_5S_3$	$[Ag_2(C_5OS_3(COOCH_3)_2)]$	Ag: MVol.B7-83/4
$Ag_2C_9H_7NO_4$	$AgCCC_6H_4OCH_3 \cdot AgNO_3$	Ag: MVol.B5-19
$Ag_2C_9H_7N_3O$	$[Ag_2(C_6H_5CH_2C_2N_3O)]$	Ag: MVol.B6-187
$Ag_2C_9H_8N_2O_4$	$[Ag_2(OC_6H_3(NO)C_2H_3(NH_2)COO)]$	Ag: MVol.B6-307
$Ag_2C_9H_9NO_4S$	$[Ag_2(C_6H_5SO_2NCH(CH_3)COO)]$	Ag: MVol.B6-242
–	$[Ag_2(C_6H_5SO_2NCH_2CH_2COO)]$	Ag: MVol.B6-244
$Ag_2C_9H_{10}{}^{2+}$	$[Ag_2CH_3C_6H_4CHCH_2]^{2+}$	Ag: MVol.B5-102
–	$[Ag_2C_6H_5CHCHCH_3]^{2+}$	Ag: MVol.B5-102
$Ag_2C_9H_{10}N_2OS$	$Ag_2(C_5H_{10}CC_3N_2OS)$	Ag: MVol.B7-52
$Ag_2C_9H_{10}N_2O_6$	$2\ AgNO_3 \cdot C_9H_{10}$	Ag: MVol.B5-67
$Ag_2C_9H_{10}O^{2+}$	$[Ag_2(CH_2CHC_6H_4OCH_3)]^{2+}$	Ag: MVol.B5-118
$Ag_2C_9H_{10}OS_3$	$[Ag_2(C_5HOS_3(C_4H_9))]$	Ag: MVol.B7-83/4
–	$[Ag_2(C_5OS_3(CH_3)C_3H_7)]$	Ag: MVol.B7-83/4
–	$[Ag_2(C_5OS_3(C_2H_5)_2)]$	Ag: MVol.B7-83/4
$Ag_2C_9H_{10}O_2{}^{2+}$	$[Ag_2(C_6H_5COOC_2H_5)]^{2+}$	Ag: MVol.B5-118
$Ag_2C_9H_{11}N_3O_4S$	$AgNO_3 \cdot Ag(C_5H_{10}CC_3HN_2OS)$	Ag: MVol.B7-52
$Ag_2C_9H_{12}{}^{2+}$	$[Ag_2C_6H_3(CH_3)_3]^{2+}$	Ag: MVol.B5-97
–	$[Ag_2C_6H_5C_3H_7]^{2+}$	Ag: MVol.B5-97
$Ag_2C_9H_{14}O_3$	$AgCCC(CH_3)_2OH \cdot AgC_3H_7COO$	Ag: MVol.B5-15
$Ag_2C_9H_{18}N_6O_8S_5$	$[Ag_2(C_3H_6N_2S)_3]S_2O_8$	Ag: MVol.B7-50
$Ag_2C_9H_{18}N_8O_6S_3$	$[Ag_2(C_3H_6N_2S)_3](NO_3)_2$	Ag: MVol.B7-50
$Ag_2C_{10}ClH_{16}N_3O_{10}S$	$[Ag_2C_{10}H_{16}N_3O_6S]ClO_4$	Ag: MVol.B6-260

Formula	Compound	Reference
$Ag_2C_{10}Cl_2CoH_{14}O_{12}$	$[Co(CH(COCH_3)_2)_2] \cdot 2\,AgClO_4$	Ag: MVol.B6-213
$Ag_2C_{10}Cl_2CuH_{14}O_{12}$	$[Cu(CH(COCH_3)_2)_2] \cdot 2\,AgClO_4$	Ag: MVol.B6-213
$Ag_2C_{10}Cl_2H_8O_8$	$2\,AgClO_4 \cdot C_{10}H_8$	Ag: MVol.B5-108
$Ag_2C_{10}Cl_2H_{14}NiO_{12}$	$[Ni(CH(COCH_3)_2)_2] \cdot 2\,AgClO_4$	Ag: MVol.B6-213
$Ag_2C_{10}Cl_2H_{14}O_{12}Pd$	$[Pd(CH(COCH_3)_2)_2] \cdot 2\,AgClO_4$	Ag: MVol.B6-213
$Ag_2C_{10}F_6H_6O_4$	$2\,AgCF_3COO \cdot C_6H_6$	Ag: MVol.B5-93
$Ag_2C_{10}F_6H_{10}O_6S_2$	$2\,AgCF_3SO_3 \cdot C_6H_4(CH_3)_2$	Ag: MVol.B5-100
$Ag_2C_{10}H_5NOS_3$	$[Ag_2(NC_5H_4C_5HOS_3)]$	Ag: MVol.B7-83/4
$Ag_2C_{10}H_5N_3O_3S$	$[Ag_2(NO_2C_6H_4CHC_3N_2OS)]$	Ag: MVol.B7-51
$Ag_2C_{10}H_8^{2+}$	$[Ag_2C_{10}H_8]^{2+}$	Ag: MVol.B5-107
$Ag_2C_{10}H_9N_3O_3S$	$AgNO_3 \cdot Ag(CH_3C_3HN_2SC_6H_5) \cdot H_2O$	Ag: MVol.B7-48
$Ag_2C_{10}H_{10}N_2O_3$	$Ag_2((C_3H_5)_2C_4N_2O_3)$	Ag: MVol.B6-154, 156
$Ag_2C_{10}H_{10}N_2O_6$	1) $2\,AgNO_3 \cdot C_{10}H_{10}$	Ag: MVol.B5-83
–	2) $2\,AgNO_3 \cdot C_{10}H_{10}$	Ag: MVol.B5-83
$Ag_2C_{10}H_{10}N_6O_{10}S$	$Ag(C_3H_5NHCSNH_2)OC_6H_2(NO_2)_3 \cdot AgNO_3$	Ag: MVol.B7-157
$Ag_2C_{10}H_{11}N_5O_4S$	$Ag(CH_3CONHC_6H_4CHNNCSNH_2) \cdot AgNO_3$	Ag: MVol.B6-283
$Ag_2C_{10}H_{12}N_2O_3$	$Ag_2(C_3H_7(C_3H_5)C_4N_2O_3)$	Ag: MVol.B6-156
$Ag_2C_{10}H_{12}N_2O_6$	$Ag_2(C_4H_4N_2O_2(C_2H_4COO)_2)$	Ag: MVol.B6-165/6
–	$2\,AgNO_3 \cdot C_8H_6(CH_3)_2$	Ag: MVol.B5-65
–	$2\,AgNO_3 \cdot C_8H_7C_2H_5$	Ag: MVol.B5-65
–	$2\,AgNO_3 \cdot C_{10}H_{12}$	Ag: MVol.B5-69/70
$Ag_2C_{10}H_{12}N_2O_8^{2-}$	$[Ag_2(C_2H_4(N(CH_2COO)_2)_2)]^{2-}$	Ag: MVol.B6-276
$Ag_2C_{10}H_{12}N_4O_4$	$[Ag_2((CH_3)_2CHC_6H_2(CH_3)(N(NO)O)_2)]$	Ag: MVol.B6-321
$Ag_2C_{10}H_{12}OS_3$	$[Ag_2(C_5H_{11}C_5HOS_3)]$	Ag: MVol.B7-83/4
$Ag_2C_{10}H_{14}^{2+}$	$[Ag_2C_6H_2(CH_3)_4]^{2+}$	Ag: MVol.B5-97/8
–	$[Ag_2C_6H_5C_4H_9]^{2+}$	Ag: MVol.B5-97
$Ag_2C_{10}H_{14}N_2O_3$	$Ag_2(C_2H_5(C_4H_9)C_4N_2O_3)$	Ag: MVol.B6-156
$Ag_2C_{10}H_{14}N_2O_8$	$[Ag_2(C_2H_4(N(CH_2COO)_2)_2H_2)]$	Ag: MVol.B6-276/7
$Ag_2C_{10}H_{14}N_3O_6S^-$	$[Ag_2C_{10}H_{14}N_3O_6S]^-$	Ag: MVol.B6-260
$Ag_2C_{10}H_{15}N_3O_6S$	$[Ag_2C_{10}H_{15}N_3O_6S]$	Ag: MVol.B6-260
–	$Ag_2(C_{10}H_{15}N_3O_6S) \cdot 2\,H_2O$	Ag: MVol.B6-260
$Ag_2C_{10}H_{20}N_2S_4^{2+}$	$[Ag_2(SS(CSN(C_2H_5)_2)_2)]^{2+}$	Ag: MVol.B7-120
$Ag_2C_{10}H_{27}N_7O_6S$	$Ag_2(C_{10}H_{15}N_3O_6S) \cdot 4\,NH_3$	Ag: MVol.B6-260
$Ag_2C_{10}H_{33}N_9O_6S$	$Ag_2(C_{10}H_{15}N_3O_6S) \cdot 6\,NH_3$	Ag: MVol.B6-260
$Ag_2C_{10}N_6$	$Ag_2C(C(CN)_2)_2C(CN)_2$	Ag: MVol.B6-355
$Ag_2C_{10}N_8$	$Ag_2(N_2(C(CN)C(CN)_2)_2)$	Ag: MVol.B6-356
$Ag_2C_{11}Cl_2H_{10}O_8$	$2\,AgClO_4 \cdot C_{10}H_7CH_3$	Ag: MVol.B5-109
$Ag_2C_{11}F_6H_8O_4$	$2\,AgCF_3COO \cdot C_6H_5CH_3$	Ag: MVol.B5-97
$Ag_2C_{11}F_{18}GeO_{10}$	$Ag_2[CF_3Ge(OCOCF_3)_5]$	F: PerFHalOrg.4-166
$Ag_2C_{11}H_6N_2O_4S$	$[Ag_2(OCOCOC_3HNOSNC_6H_5)]$	Ag: MVol.B7-72
$Ag_2C_{11}H_6OS_3$	$[Ag_2(C_6H_5C_5HOS_3)]$	Ag: MVol.B7-83/4
$Ag_2C_{11}H_{10}N_2^{2+}$	$[Ag_2(CH_2(C_5H_4N)_2)]^{2+}$	Ag: MVol.B6-123
$Ag_2C_{11}H_{10}N_4OS$	$[Ag_2((CH_3)_2NC_6H_4NC_3N_2OS)]$	Ag: MVol.B7-51
$Ag_2C_{11}H_{12}O_2^{2+}$	$[Ag_2(C_6H_5C_2H_2COOC_2H_5)]^{2+}$	Ag: MVol.B5-118
$Ag_2C_{11}H_{14}N_2O_3$	$Ag_2(C_2H_5(CCH_3CHC_2H_5)C_4N_2O_3)$	Ag: MVol.B6-156
$Ag_2C_{11}H_{14}N_2O_6$	$2\,AgNO_3 \cdot C_8H_7C_3H_7$	Ag: MVol.B5-65
$Ag_2C_{11}H_{14}OS_3$	$[Ag_2(C_5OS_3(C_3H_7)_2)]$	Ag: MVol.B7-83/4
–	$[Ag_2(C_6H_{13}C_5HOS_3)]$	Ag: MVol.B7-83/4
$Ag_2C_{11}H_{16}^{2+}$	$[Ag_2C_6H_5C_5H_{11}]^{2+}$	Ag: MVol.B5-97
$Ag_2C_{11}H_{16}N_2O_3$	$Ag_2(C_2H_5(C_5H_{11})C_4N_2O_3)$	Ag: MVol.B6-156

$Ag_2C_{11}H_{16}N_2O_8$ ..... $[Ag_2(CH_3C_2H_3(N(CH_2COO)_2)_2H_2)]$ .... Ag: MVol.B6-278
$Ag_2C_{11}H_{17}NO_3$ ...... $AgCC(CH_2)_4CHCHC_3H_7 \cdot AgNO_3$ .... Ag: MVol.B5-14
$Ag_2C_{12}Cl_2H_6N_2O_3S$ .. $[Ag_2(CH_3COOC_6H_2Cl_2CHC_3N_2OS)]$ .... Ag: MVol.B7-51
$Ag_2C_{12}Cl_2H_8O_8$ ..... $2\ AgClO_4 \cdot C_{12}H_8$ .... Ag: MVol.B5-113
$Ag_2C_{12}Cl_2H_9N_3O_3S$ .. $Ag_2(CH_3COOC_6H_2Cl_2CHC_3N_2OS) \cdot NH_3$ ... Ag: MVol.B7-51
$Ag_2C_{12}Cl_2H_{10}O_8$ .... $2\ AgClO_4 \cdot C_{12}H_{10}$ .... Ag: MVol.B5-112
$Ag_2C_{12}Cl_2H_{24}O_8S_6$ ... $2\ AgClO_4 \cdot 3\ S(CH_2CH_2)_2S$ .... Ag: MVol.B7-87
$Ag_2C_{12}Cl_2H_{27}N$ ..... $4\ AgCl \cdot 2\ C_{12}H_{25}NH_2$ .... Ag: MVol.B6-46
$Ag_2C_{12}Cl_2H_{30}O_8S_6$ ... $2\ AgClO_4 \cdot 3\ C_2H_4(SCH_3)_2$ .... Ag: MVol.B7-14
$Ag_2C_{12}Cl_2H_{36}P_4$ ..... $[AgCl(P(CH_3)_3)_2]_2$ .... Ag: MVol.B7-200/1
$Ag_2C_{12}CoH_4N_4O_8$ ... $[Ag_2Co(C_4H_2N_2(COO)_2)_2]$ .... Ag: MVol.B6-161
$Ag_2C_{12}CrH_{24}N_8O_4$ ... $Ag_2CrO_4 \cdot 2\ (CH_2)_6N_4$ .... Ag: MVol.B6-200
– .... $Ag_2CrO_4 \cdot 2\ (CH_2)_6N_4 \cdot 5\ H_2O$ .... Ag: MVol.B6-200
$Ag_2C_{12}CuH_4N_4O_8$ ... $[Ag_2Cu(C_4H_2N_2(COO)_2)_2]$ .... Ag: MVol.B6-160/1
$Ag_2C_{12}F_3H_5K_2N_6O_6$ .. $K_2[Ag_2(C_4H_2FN_2O_2)_2(C_4HFN_2O_2)]$ .... Ag: MVol.B6-152
$Ag_2C_{12}F_6H_{10}O_4$ ..... $2\ AgCF_3COO \cdot C_6H_4(CH_3)_2$ .... Ag: MVol.B5-100
$Ag_2C_{12}F_8S_2$ .... $[Ag_2(SC_6F_4C_6F_4S)]$ .... Ag: MVol.B7-9
$Ag_2C_{12}H_4N_4NiO_8$ .... $[Ag_2Ni(C_4H_2N_2(COO)_2)_2]$ .... Ag: MVol.B6-160/1
$Ag_2C_{12}H_5NO_5$ ...... $[Ag_2(C_{12}H_5NO_5)]$ .... Ag: MVol.B6-204
$Ag_2C_{12}H_6N_4O_{10}S_2$ ... $Ag_2(N_2(C_6H_3(NO_2)SO_3)_2) \cdot 2\ H_2O$ .... Ag: MVol.B6-290
$Ag_2C_{12}H_8N_2O_3S$ .... $[Ag_2(CH_3COOC_6H_4CHC_3N_2OS)]$ .... Ag: MVol.B7-51
$Ag_2C_{12}H_8N_2O_6S_2$ .... $Ag_2(SO_3C_6H_4NNC_6H_4SO_3)$ .... Ag: MVol.B6-289
– .... $Ag_2(SO_3C_6H_4NNC_6H_4SO_3) \cdot H_2O$ .... Ag: MVol.B6-289
$Ag_2C_{12}H_8OS_3$ ...... $[Ag_2(CH_3C_5OS_3C_6H_5)]$ .... Ag: MVol.B7-83/4
– .... $[Ag_2(C_6H_5CH_2C_5HOS_3)]$ .... Ag: MVol.B7-83/4
$Ag_2C_{12}H_{10}{}^{2+}$ .... $[Ag_2C_6H_5C_6H_5]^{2+}$ .... Ag: MVol.B5-102
$Ag_2C_{12}H_{10}N_2O_3$ ..... $Ag_2(C_2H_5(C_6H_5)C_4N_2O_3)$ .... Ag: MVol.B6-155/6
$Ag_2C_{12}H_{10}N_4$ .... $[Ag(C_5H_5N)_2][Ag(CN)_2]$ .... Ag: MVol.B6-81
$Ag_2C_{12}H_{10}N_4O_6S_2$ ... $Ag_2(N_2(C_6H_3(NH_2)SO_3)_2)$ .... Ag: MVol.B6-290
$Ag_2C_{12}H_{10}N_4S_2$ ..... $[Ag_2(C_{10}H_7NCSNHNCSNH_2)]$ .... Ag: MVol.B7-169/70
$Ag_2C_{12}H_{10}N_6O_6$ ..... $2\ AgNO_3 \cdot N_2(CHC_5H_4N)_2$ .... Ag: MVol.B6-295
$Ag_2C_{12}H_{12}N_2{}^{2+}$ ..... $[Ag_2(C_2H_4(C_5H_4N)_2)]^{2+}$ .... Ag: MVol.B6-123
$Ag_2C_{12}H_{12}N_2O_3$ ..... $Ag_2(C_3H_5(C_5H_7)C_4N_2O_3)$ .... Ag: MVol.B6-156
$Ag_2C_{12}H_{12}N_2O_4S$ .... $Ag_2SO_4 \cdot NH_2C_6H_4C_6H_4NH_2$ .... Ag: MVol.B6-64
$Ag_2C_{12}H_{14}N_2O_3$ ..... $Ag_2(C_2H_5(C_6H_9)C_4N_2O_3)$ .... Ag: MVol.B6-156
$Ag_2C_{12}H_{14}N_2O_7$ ..... $2\ AgNO_3 \cdot C_{12}H_{14}O$ .... Ag: MVol.B5-90
$Ag_2C_{12}H_{16}N_2O_3$ ..... $Ag_2(C_5H_{11}(C_3H_5)C_4N_2O_3)$ .... Ag: MVol.B6-156
$Ag_2C_{12}H_{16}N_4{}^{2+}$ ..... $[Ag_2(NH_2CH_2C_5H_4N)_2]^{2+}$ .... Ag: MVol.B6-96
$Ag_2C_{12}H_{16}N_6O_4$ ..... $[Ag(C_3H_3N_2C_2H_3(NH_2)COO)]_2$ .... Ag: MVol.B6-246
$Ag_2C_{12}H_{18}N_2O_6$ ..... $2\ AgNO_3 \cdot C_6H_7(CH_3)(CHCH_2)C(CH_2)CH_3$ .. Ag: MVol.B5-58/9
– .... $2\ AgNO_3 \cdot C_{12}H_{18}$ .... Ag: MVol.B5-73/4
$Ag_2C_{12}H_{18}N_6O_7S_2$ ... $Ag_2S_2O_7 \cdot 6\ CH_3CN$ .... Ag: MVol.B6-348
$Ag_2C_{12}H_{19}NO_5$ ...... $Ag_2(OCOC_4H_8CONHC_5H_{10}COO)$ .... Ag: MVol.B6-328
$Ag_2C_{12}H_{20}N_4O_8$ ..... $[Ag_2(C_2H_4(NCCH_3CH_2COCH_3)_2)](NO_3)_2$ .... Ag: MVol.B6-279
$Ag_2C_{12}H_{22}N_2O_6$ ..... $2\ AgNO_3 \cdot CH_2CH(CH_2)_8CHCH_2$ .... Ag: MVol.B5-45
$Ag_2C_{12}H_{23}N_4O_4S_4{}^{+}$ .. $[Ag_2(C_6H_{11}N_2O_2S_2)(C_6H_{12}N_2O_2S_2)]^{+}$ .... Ag: MVol.B7-187/8
$Ag_2C_{12}H_{24}MoN_8O_4$ .. $Ag_2MoO_4 \cdot 2\ (CH_2)_6N_4 \cdot 2\ H_2O$ .... Ag: MVol.B6-200/1
$Ag_2C_{12}H_{24}N_4O_4S_4{}^{2+}$. $[Ag_2(HOC_2H_4NHCSCSNHC_2H_4OH)_2]^{2+}$ ..... Ag: MVol.B7-187/8
$Ag_2C_{12}H_{24}N_6O_6$ ..... $2\ AgNO_3 \cdot (CH_3)_3C_9H_{15}N_4 \cdot 2\ H_2O$ .... Ag: MVol.B6-201
– .... $2\ AgNO_3 \cdot (CH_3)_3C_9H_{15}N_4 \cdot 3\ H_2O$ .... Ag: MVol.B6-201

$Ag_2C_{12}H_{24}N_8O_4Se$ ... $Ag_2SeO_4 \cdot 2\,(CH_2)_6N_4 \cdot 12\,H_2O$ .... Ag: MVol.B6-199
$Ag_2C_{12}H_{24}N_8O_4W$ ... $Ag_2WO_4 \cdot 2\,(CH_2)_6N_4 \cdot 2\,H_2O$ .... Ag: MVol.B6-201
$Ag_2C_{12}H_{24}N_8O_8S_2$ ... $Ag_2S_2O_8 \cdot 2\,(CH_2)_6N_4 \cdot 2\,H_2O$ .... Ag: MVol.B6-199
$Ag_2C_{12}H_{27}N_{11}O_6S_3$ .. $2\,AgNO_3 \cdot 3\,(CH_3)_2CNNHCSNH_2$ .... Ag: MVol.B6-283
$Ag_2C_{12}H_{30}S_6^{2+}$ ..... $[Ag_2(C_2H_4(SCH_3)_2)_3]^{2+}$ .... Ag: MVol.B7-14
$Ag_2C_{12}H_{32}N_{16}O_{16}S_3$ .. $[Ag(NH_2C(NH)NC(OCH_3)NH_2)_2]_2(SO_4)_3$ .... Ag: MVol.B7-325
$Ag_2C_{12}H_{32}N_{20}O_{12}S_3$ .. $[Ag(C_2H_4(C_2H_6N_5)_2)]_2(SO_4)_3 \cdot 7\,H_2O$ ...... Ag: MVol.B7-321/4
$Ag_2C_{12}H_{32}N_{20}O_{24}S_6$ .. $[Ag(C_2H_4(C_2H_6N_5)_2)]_2(S_2O_8)_3$ .... Ag: MVol.B7-321, 324
$Ag_2C_{12}H_{36}N_2O_{18}P_4$ .. $[AgNO_3(P(OCH_3)_3)_2]_2$ .... Ag: MVol.B7-244
$Ag_2C_{12}H_{36}N_{20}O_{12}S_3$ .. $[Ag(CH_3C_2H_6N_5)_2]_2(SO_4)_3 \cdot 12.5\,H_2O$ ..... Ag: MVol.B7-321
$Ag_2C_{13}H_7NO_7S_2$ .... $[Ag_2(C_6H_5C_7NO(SO_3H)_2)]$ .... Ag: MVol.B7-53
$Ag_2C_{13}H_9NO_4$ ...... $C_6H_4(COOAg)_2 \cdot C_5H_5N$ .... Ag: MVol.B6-81
$Ag_2C_{13}H_{10}N_4S$ ...... $[Ag_2(C_6H_5NNCSNNC_6H_5)]$ .... Ag: MVol.B7-173/5
$Ag_2C_{13}H_{11}N_3O_3S$ .... $[Ag_2(CH_3C_5H_3N_2OSNHC_6H_4COO)]$ .... Ag: MVol.B7-65
$Ag_2C_{13}H_{12}^{2+}$ .... $[Ag_2CH_2(C_6H_5)_2]^{2+}$ .... Ag: MVol.B5-102
$Ag_2C_{13}H_{12}N_2O_6$ ..... $Ag_2C_6H_5CONHCH_2CONHC_2H_3(COO)_2 \cdot 1.5\,H_2O$
Ag: MVol.B6-267/8
$Ag_2C_{13}H_{15}N_3O_4$ ..... $Ag_2(NHCONN(COC_6H_5)C_2H_4COOC_2H_5)$ .... Ag: MVol.B6-336
$Ag_2C_{14}Cl_2H_{10}O_8$ .... $2\,AgClO_4 \cdot C_{14}H_{10}$ .... Ag: MVol.B5-110
$Ag_2C_{14}Cl_2H_{12}O_8$ .... $2\,AgClO_4 \cdot C_6H_5CHCHC_6H_5$ .... Ag: MVol.B5-103
$Ag_2C_{14}F_6H_9NO_5$ ..... $[Ag_2(C_3H_3(CF_3)_2(NHCOC_6H_5)(COO)_2)]$ ..... Ag: MVol.B6-253/4
$Ag_2C_{14}F_{14}H_6O_4$ ..... $2\,AgC_3F_7COO \cdot C_6H_6$ .... Ag: MVol.B5-93
$Ag_2C_{14}H_6N_4O_8$ ...... $Ag_2(N_2(C_6H_3(NO_2)COO)_2)$ .... Ag: MVol.B6-287
$Ag_2C_{14}H_7N_3O_6$ ...... $Ag_2(NO_2C_6H_3(COO)NNC_6H_4COO)$ .... Ag: MVol.B6-287
$Ag_2C_{14}H_8N_2O_4$ ...... $Ag_2(C_6H_4(COO)NNC_6H_4COO)$ .... Ag: MVol.B6-287
$Ag_2C_{14}H_8N_2O_5$ ...... $Ag_2(C_6H_4(COO)NNC_6H_3(OH)COO)$ .... Ag: MVol.B6-287
$Ag_2C_{14}H_9NO_5$ ...... $Ag(C_{13}H_6N(OH)_2COO) \cdot AgOH$ .... Ag: MVol.B6-116
$Ag_2C_{14}H_9N_3O_4$ ...... $Ag_2(OCOC_6H_4N_3HC_6H_4COO)$ .... Ag: MVol.B6-302
$Ag_2C_{14}H_{10}^{2+}$ .... $[Ag_2C_{14}H_{10}]^{2+}$ .... Ag: MVol.B5-109
$Ag_2C_{14}H_{10}N_2O_3S_4$ ... $[Ag_2(C_2H_5C_4N_2O_3(CSS)_2C_6H_5)]$ .... Ag: MVol.B7-118
$Ag_2C_{14}H_{12}^{2+}$ .... $[Ag_2C_2H_2(C_6H_5)_2]^{2+}$ .... Ag: MVol.B5-102
– .... $[Ag_2C_{14}H_{12}]^{2+}$ .... Ag: MVol.B5-109
$Ag_2C_{14}H_{12}N_2O_6S_2$ ... $Ag_2(N_2(C_6H_4CH_2SO_3)_2) \cdot H_2O$ .... Ag: MVol.B6-289
$Ag_2C_{14}H_{12}N_4S_2$ ..... $[Ag_2(C_6H_5NCSNHNHCSNC_6H_5)]$ .... Ag: MVol.B7-170
$Ag_2C_{14}H_{14}^{2+}$ .... $[Ag_2C_2H_4(C_6H_5)_2]^{2+}$ .... Ag: MVol.B5-102
$Ag_2C_{14}H_{14}N_2O_6$ ..... $2\,AgNO_3 \cdot C_{14}H_{14}$ .... Ag: MVol.B5-81
$Ag_2C_{14}H_{14}N_2O_{10}$ .... $2\,AgNO_3 \cdot C_{10}H_8(COOCH_3)_2$ .... Ag: MVol.B5-90
$Ag_2C_{14}H_{14}N_6O_6$ ..... $2\,AgNO_3 \cdot N_2(CCH_3C_5H_4N)_2$ .... Ag: MVol.B6-295
$Ag_2C_{14}H_{16}N_6O_7$ ..... $2\,AgNO_3 \cdot C_2H_5OC_6H_4NNC_6H_3(NH_2)_2$ ..... Ag: MVol.B6-286
$Ag_2C_{14}H_{20}N_4^{2+}$ ..... $[Ag_2(CH_3(NH_2CH_2)C_5H_3N)_2]^{2+}$ .... Ag: MVol.B6-96/7
$Ag_2C_{15}F_{14}H_8O_4$ ..... $2\,AgC_3F_7COO \cdot C_6H_5CH_3$ .... Ag: MVol.B5-97
$Ag_2C_{15}H_8N_2Na_2O_5S_2$. $Na_2[Ag_2(OC(NHC_6H_3(S)COO)_2)]$ .... Ag: MVol.B7-35
$Ag_2C_{15}H_{10}N_4O_3$ ..... $Ag_2((C_6H_5NN)_2CCOCOO)$ .... Ag: MVol.B6-304
$Ag_2C_{15}H_{14}N_4S$ ...... $[Ag_2(CS(NNC_6H_4CH_3)_2)]$ .... Ag: MVol.B7-184
$Ag_2C_{15}H_{20}N_2O_8$ ..... $2\,AgNO_3 \cdot C_{15}H_{20}O_2$ .... Ag: MVol.B5-90
$Ag_2C_{15}H_{21}NNiO_9$ .... $Ag[Ni(CH(COCH_3)_2)_3] \cdot AgNO_3 \cdot H_2O$ .... Ag: MVol.B6-213/4
$Ag_2C_{15}H_{24}N_2O_6$ ..... $2\,AgNO_3 \cdot C_{11}H_{12}(CH_3)_4$ .... Ag: MVol.B5-72
$Ag_2C_{15}H_{30}N_{12}O_{11}S_5$ .. $[Ag_2(CH_3CONHCSNH_2)_5](NO_3)_2$ .... Ag: MVol.B7-161
$Ag_2C_{16}ClH_{18}N_5O_6S$ .. $2\,AgNO_3 \cdot C_{16}H_{18}ClN_3S$ .... Ag: MVol.B7-85
$Ag_2C_{16}Cl_2H_{10}O_8$ .... $2\,AgClO_4 \cdot C_{16}H_{10}$ .... Ag: MVol.B5-116

$Ag_2C_{16}Cl_2H_{14}N_6$	$[Ag(NC_5H_4C_3H_3N_2)_2][AgCl_2]$	Ag:	MVol.B6-141
$Ag_2C_{16}FeH_8N_2O_4$	$Ag_2(C_{12}H_8N_2)Fe(CO)_4$	Ag:	MVol.B6-125
		Fe:	Org.Verb.B2-173
$Ag_2C_{16}FeH_{16}N_8$	$Ag_2H_2[Fe(CN)_6] \cdot C_{10}H_{14}N_2 \cdot 2\ H_2O$	Ag:	MVol.B6-107
$Ag_2C_{16}H_{12}N_2O_4$	$Ag_2(N_2(C_6H_4CH_2COO)_2)$	Ag:	MVol.B6-287
$Ag_2C_{16}H_{12}N_2O_6$	$Ag_2(N_2(C_6H_4OCH_2COO)_2) \cdot 3\ H_2O$	Ag:	MVol.B6-287
$Ag_2C_{16}H_{12}N_2O_7$	$[Ag_2(N_2O(C_6H_4OCH_2COO)_2)]$	Ag:	MVol.B6-291/2
$Ag_2C_{16}H_{15}N_5O_6S$	$2\ AgNO_3 \cdot N(CC_6H_5N)_2SC_2H_5$	Ag:	MVol.B7-86
$Ag_2C_{16}H_{16}N_2O_6$	$2\ AgNO_3 \cdot C_{16}H_{16}$	Ag:	MVol.B5-87
$Ag_2C_{16}H_{16}N_2O_6S_2$	$Ag_2(N_2(C_6H_2(CH_3)_2SO_3)_2)$	Ag:	MVol.B6-289
$Ag_2C_{16}H_{28}N_2O_6$	$2\ AgNO_3 \cdot C_{16}H_{28}$	Ag:	MVol.B5-74
$Ag_2C_{16}H_{28}N_2O_8$	$2\ AgNO_3 \cdot C_{16}H_{28}O_2$	Ag:	MVol.B5-90
$Ag_2C_{16}H_{40}N_{24}O_{12}S_3$	$[Ag(C_4H_8N_2(C_2H_6N_5)_2)]_2(SO_4)_3$	Ag:	MVol.B7-325
$Ag_2C_{17}F_{14}H_{12}O_4$	$2\ AgC_3F_7COO \cdot C_6H_3(CH_3)_3$	Ag:	MVol.B5-100
$Ag_2C_{17}H_{10}OS_3$	$[Ag_2(C_5OS_3(C_6H_5)_2)]$	Ag:	MVol.B7-83/4
$Ag_2C_{17}H_{24}O_4$	$[Ag_2(C_{11}H_{23}C_6HO_4)]$	Ag:	MVol.B6-218
$Ag_2C_{18}Cl_2H_{12}O_8$	$2\ AgClO_4 \cdot C_{18}H_{12}$	Ag:	MVol.B5-116
$Ag_2C_{18}Cl_2H_{30}O_8$	$2\ AgClO_4 \cdot 3\ CH_2CH(CH_2)_2CHCH_2$	Ag:	MVol.B5-41
$Ag_2C_{18}Cl_2H_{36}N_6O_8S_6$	$[Ag_2((CH_3)_2NCSCSN(CH_3)_2)_3](ClO_4)_2$	Ag:	MVol.B7-190
$Ag_2C_{18}Cl_2H_{36}N_6O_{14}S_6$	$[Ag_2(CSCS(NHC_2H_4OH)_2)_3](ClO_4)_2$	Ag:	MVol.B7-188
$Ag_2C_{18}CrH_{27}N_3O_7$	$[Ag_2(H_2O)_3(C_6H_5NH_2)_3]CrO_4$	Ag:	MVol.B6-51
$Ag_2C_{18}H_{12}{}^{2+}$	$[Ag_2C_{18}H_{12}]^{2+}$	Ag:	MVol.B5-115
$Ag_2C_{18}H_{12}N_2O_2$	$[AgC_9H_6NO]_2$	Ag:	MVol.B6-109
$Ag_2C_{18}H_{12}O_6Sn$	$Ag_2[Sn(C_6H_4O_2)_3]$	Sn:	MVol.C5-77
$Ag_2C_{18}H_{16}N_2S_2$	$[Ag_2(C_{18}H_{16}N_2S_2)]$	Ag:	MVol.B7-78
$Ag_2C_{18}H_{24}N_4O_{12}{}^{4-}$	$[Ag_2(C_6H_{12}N_4(CH_2COO)_6)]^{4-}$	Ag:	MVol.B6-279
$Ag_2C_{18}H_{24}P_2$	$CH_3C_6H_5P(CH_2AgCH_2)_2PCH_3C_6H_5$	Ag:	MVol.B5-24
$Ag_2C_{18}H_{25}N_4O_{12}{}^{3-}$	$[Ag_2(C_6H_{12}N_4(CH_2COO)_6H)]^{3-}$	Ag:	MVol.B6-279
$Ag_2C_{18}H_{26}N_4O_{12}{}^{2-}$	$[Ag_2(C_6H_{12}N_4(CH_2COO)_6H_2)]^{2-}$	Ag:	MVol.B6-279
$Ag_2C_{18}H_{30}{}^{2+}$	$[Ag_2(C_6H_{10})_3]^{2+}$	Ag:	MVol.B5-41
$Ag_2C_{18}H_{35}N_6O_6S_6{}^{+}$	$[Ag_2(C_6H_{11}N_2O_2S_2)(C_6H_{12}N_2O_2S_2)_2]^{+}$	Ag:	MVol.B7-187/8
$Ag_2C_{18}H_{36}N_6O_6S_6{}^{2+}$	$[Ag_2(HOC_2H_4NHCSCSNHC_2H_4OH)_3]^{2+}$	Ag:	MVol.B7-187/8
$Ag_2C_{19}H_8O_8S^{2-}$	$[Ag_2(C_{19}H_8O_8S)]^{2-}$	Ag:	MVol.B6-207/8
$Ag_2C_{19}H_{14}{}^{2+}$	$[Ag_2(C_{18}H_{11}CH_3)]^{2+}$	Ag:	MVol.B5-115
$Ag_2C_{20}Cl_2H_{12}O_8$	$2\ AgClO_4 \cdot C_{20}H_{12}$	Ag:	MVol.B5-116
$Ag_2C_{20}H_6I_4O_5$	$Ag_2(OCOC_6I_4C_{13}H_6O_3)$	Ag:	MVol.B6-207/8
$Ag_2C_{20}H_{12}{}^{2+}$	$[Ag_2C_{20}H_{12}]^{2+}$	Ag:	MVol.B5-115
$Ag_2C_{20}H_{12}N_2O_4$	$Ag[Ag(OC_{10}H_6NO)_2]$	Ag:	MVol.B6-307
$Ag_2C_{20}H_{12}N_4O_4$	$Ag_2((C_6H_5NN)_2C_6H_2(COO)_2)$	Ag:	MVol.B6-287
$Ag_2C_{20}H_{16}{}^{2+}$	$[Ag_2(C_{18}H_{10}(CH_3)_2)]^{2+}$	Ag:	MVol.B5-115
$Ag_2C_{20}H_{16}N_2O_2$	$[Ag(CH_3C_9H_5NO)]_2$	Ag:	MVol.B6-109
$Ag_2C_{20}H_{17}N_8O_{12}$	$Ag(NO_3)_2 \cdot 2\ NC_5H_4C_5H_4N \cdot AgNO_3 \cdot HNO_3$	Ag:	MVol.B7-291
$Ag_2C_{20}H_{20}N_2O_4$	$Ag_2(N_2(C_6H_3(C_3H_7)COO)_2)$	Ag:	MVol.B6-287
$Ag_2C_{20}H_{20}N_4O_{12}S$	$Ag_2(C_4H_3OCHNOH)_4SO_4$	Ag:	MVol.B6-305
$Ag_2C_{20}H_{21}N_5O_2$	$2\ AgNCO \cdot 3\ CH_3C_5H_4N$	Ag:	MVol.B6-86
$Ag_2C_{20}H_{28}N_4O_4$	$[Ag_2(C_{20}H_{28}N_4O_4)]$	Ag:	MVol.B6-321
$Ag_2C_{20}H_{36}N_2O_6$	$2\ AgNO_3 \cdot C_{20}H_{36}$	Ag:	MVol.B5-74
$Ag_2C_{20}H_{48}I_2P_4$	$[Ag((C_2H_5)_2PC_2H_4P(C_2H_5)_2)_2][AgI_2]$	Ag:	MVol.B7-234
$Ag_2C_{20}H_{48}N_{16}O_{16}S_3$	$[Ag(NH_2C(NH)NC(OC_3H_7)NH_2)_2]_2(SO_4)_3$	Ag:	MVol.B7-325

$Ag_2C_{21}H_{16}^{2+}$	$[Ag_2(C_{20}H_{13}CH_3)]^{2+}$	Ag: MVol.B5-115
$Ag_2C_{21}H_{18}N_6S_3^{2+}$	$[Ag_2(C_7H_6N_2S)_3]^{2+}$	Ag: MVol.B7-52
$Ag_2C_{22}H_{14}^{2+}$	$[Ag_2C_{22}H_{14}]^{2+}$	Ag: MVol.B5-115
$Ag_2C_{22}H_{20}N_4O_3$	$Ag_2(C_2H_5C_4N_2O_3C_6H_5) \cdot 2\ C_5H_5N$	Ag: MVol.B6-154
$Ag_2C_{22}H_{30}I_2N_2P_2$	$[Ag_2I_2(PP((NC_5H_{10})C_6H_5)_2)]$	Ag: MVol.B7-242/3
$Ag_2C_{22}H_{42}I_2N_2P_2$	$[Ag_2I_2(PP((NC_5H_{10})C_6H_{11})_2)]$	Ag: MVol.B7-242
$Ag_2C_{23}H_{43}I_2NP_2$	$[Ag_2I_2(C_5H_{10}NP(C_6H_{11})P(C_6H_{11})_2)]$	Ag: MVol.B7-242
$Ag_2C_{24}Cl_2H_{20}N_8O_8$	$[AgClO_4 \cdot N_2(CHC_5H_4N)_2]_2$	Ag: MVol.B6-295
$Ag_2C_{24}H_{20}N_{10}O_6$	$[AgNO_3 \cdot N_2(CHC_5H_4N)_2]_2$	Ag: MVol.B6-295
$Ag_2C_{24}H_{27}N_{11}O_6S_3$	$2\ AgNO_3 \cdot 3\ C_6H_5CHNNHCSNH_2$	Ag: MVol.B6-283/4
$Ag_2C_{24}H_{40}N_{18}O_6$	$[AgNO_3((CH_2)_5CN_4)_2]_2$	Ag: MVol.B6-195
$Ag_2C_{24}H_{44}I_2P_2$	$2\ AgI \cdot (C_6H_{11})_2PP(C_6H_{11})_2$	Ag: MVol.B7-233
$Ag_2C_{24}H_{48}N_6O_{12}$	$2\ AgNO_3 \cdot C_{24}H_{48}N_4O_6$	Ag: MVol.B6-205
$Ag_2C_{24}H_{56}N_{16}O_{16}S_3$	$[Ag(NH_2C(NH)NC(OC_4H_9)NH_2)_2]_2(SO_4)_3$	Ag: MVol.B7-325
$Ag_2C_{25}Cl_2H_{20}N_2P_2$	$2\ AgCl \cdot C(NP(C_6H_5)_2)_2 \cdot 0.5\ C_6H_6$	Ag: MVol.B7-243
$Ag_2C_{25}H_{22}N_2O_6P_2$	$[Ag_2(NO_3)_2((C_6H_5)_2PCH_2P(C_6H_5)_2)]$	Ag: MVol.B7-233/4
$Ag_2C_{25}H_{25}Mn_2N_5O_8$	$2\ AgMnO_4 \cdot 5\ C_5H_5N$	Ag: MVol.B6-83
$Ag_2C_{25}H_{30}N_{10}O_4S$	$Ag_2SO_4 \cdot 5\ NH_2C_5H_4N$	Ag: MVol.B6-93/4
$Ag_2C_{25}H_{45}I_2N_2P_2$	$[Ag_2I_2(PP((NC_5H_{10})C_6H_{11})_2)] \cdot 0.5\ C_6H_6$	Ag: MVol.B7-242
$Ag_2C_{26}Cl_2H_{20}O_8$	$2\ AgClO_4 \cdot (C_6H_5)_2CC(C_6H_5)_2$	Ag: MVol.B5-104
$Ag_2C_{26}H_{20}HgN_8S_2$	$[Hg(AgC_6H_5NNCSNNC_6H_5)_2]$	Ag: MVol.B7-182
$Ag_2C_{26}H_{22}N_2O_6P_2$	$[Ag_2(NO_3)_2((C_6H_5)_2PCHCHP(C_6H_5)_2)]$	Ag: MVol.B7-235
$Ag_2C_{26}H_{24}N_2O_6P_2$	$[Ag_2(NO_3)_2((C_6H_5)_2PC_2H_4P(C_6H_5)_2)]$	Ag: MVol.B7-234
$Ag_2C_{26}H_{25}N_3O_4$	$Ag_2C_6H_4(COO)_2 \cdot 3\ C_6H_5NH_2$	Ag: MVol.B6-51
$Ag_2C_{26}H_{26}N_4O_6$	$Ag_2(C_2H_5(C_6H_5)C_4N_2O_3CH_3)_2$	Ag: MVol.B6-155
$Ag_2C_{27}H_{26}N_2O_6P_2$	$[Ag_2(NO_3)_2((C_6H_5)_2P(CH_2)_3P(C_6H_5)_2)]$	Ag: MVol.B7-236/7
$Ag_2C_{28}Cl_2H_{23}N_2P_2$	$2\ AgCl \cdot C(NP(C_6H_5)_2)_2 \cdot 0.5\ C_6H_6$	Ag: MVol.B7-243
$Ag_2C_{28}Cl_2H_{28}N_8O_8$	$[AgClO_4 \cdot N_2(C(CH_3)C_5H_4N)_2]_2$	Ag: MVol.B6-295
$Ag_2C_{28}F_2H_{24}N_4$	$[Ag((CH_3)_2C_{12}H_6N_2)_2][AgF_2]$ $= Ag((CH_3)_2C_{12}H_6N_2)F$	Ag: MVol.B6-127
$Ag_2C_{28}H_{24}N_4O_4S$	$(Ag(CH_3)_2C_{12}H_6N_2)_2SO_4 \cdot 0.5\ H_2O$	Ag: MVol.B6-128
–	$(Ag(CH_3)_2C_{12}H_6N_2)_2SO_4 \cdot 2\ H_2O$	Ag: MVol.B6-127/8
$Ag_2C_{28}H_{28}N_{10}O_6$	$(AgNO_3 \cdot N_2(C(CH_3)C_5H_4N)_2)_2$	Ag: MVol.B6-295
$Ag_2C_{28}H_{28}P_2$	$(C_6H_5)_2P(CH_2AgCH_2)_2P(C_6H_5)_2$	Ag: MVol.B5-24
$Ag_2C_{30}CdH_{30}I_4N_6$	$2\ AgI \cdot CdI_2 \cdot 6\ C_5H_5N$	Ag: MVol.B6-80
$Ag_2C_{30}Cl_2H_{26}O_8$	$2\ AgClO_4 \cdot C_5H_4C(C_6H_5)_2 \cdot 2\ C_6H_6$	Ag: MVol.B5-106
$Ag_2C_{30}H_{28}N_6S_3$	$2\ AgC_{10}H_9N_2S \cdot C_{10}H_{10}N_2S$	Ag: MVol.B7-48
$Ag_2C_{30}H_{52}I_2P_4$	$[Ag(CH_3C_6H_3(P(C_2H_5)_2)_2)_2][AgI_2]$	Ag: MVol.B7-239
$Ag_2C_{32}Cl_2H_{30}O_8$	$2\ AgClO_4 \cdot C_5H_4C(C_6H_5)_2 \cdot 2\ C_6H_5CH_3$	Ag: MVol.B5-106
$Ag_2C_{32}H_{22}N_6O_6S_2$	$[Ag_2(C_{12}H_8(NNC_{10}H_5(NH_2)SO_3)_2)]$	Ag: MVol.B6-290
$Ag_2C_{32}H_{32}N_8O_8S$	$[Ag_2(HOCH_2C_7H_5N_2)_4]SO_4$	Ag: MVol.B6-145
$Ag_2C_{34}H_{32}N_4O_2$	$Ag_2((C_6H_5)_2C_2(NNCOC_6H_2(CH_3)_3)_2)$	Ag: MVol.B6-282
$Ag_2C_{36}Cl_2H_{30}N_6O_8$	$2\ AgClO_4 \cdot 3\ C_6H_5NNC_6H_5$	Ag: MVol.B6-285
$Ag_2C_{36}F_6H_{36}N_6Si$	$Ag_2SiF_6 \cdot 3\ NH_2C_6H_4C_6H_4NH_2 \cdot 8\ H_2O$	Ag: MVol.B6-64
$Ag_2C_{36}H_{30}Pb_2^{2+}$	$[Ag_2((C_6H_5)_3PbPb(C_6H_5)_3)]^{2+}$	Ag: MVol.B5-119
$Ag_2C_{36}H_{36}N_4O_8$	$[Ag_2(C_{20}H_4N_4(CH_3)_4(COOC_2H_5)_4)]$	Ag: MVol.B7-304, 307
$Ag_2C_{38}H_{32}N_2O_6$	$2\ Ag(C_6H_5COC_6H_4COO) \cdot C_{10}H_{14}N_2$	Ag: MVol.B6-107
$Ag_2C_{40}H_{36}N_4O_4S_5$	$[Ag(CH_3SC_9H_6N)_2]_2SO_4 \cdot 10\ H_2O$	Ag: MVol.B7-61/2
$Ag_2C_{42}Cl_2H_{36}O_8$	$2\ AgClO_4 \cdot 3\ (C_6H_5)_2CCH_2$	Ag: MVol.B5-103
$Ag_2C_{42}Cl_2H_{42}N_{12}O_8$	$2\ AgClO_4 \cdot 3\ N_2(CCH_3C_5H_4N)_2$	Ag: MVol.B6-295

$Ag_2C_{42}H_{30}N_8O_{12}S_2$ .. $Ag_2(NC_5H_4COO)_2(C_{15}H_{11}N_3)_2(S_2O_8)\cdot 4\,H_2O$ Ag: MVol.B7-299/300
$Ag_2C_{44}H_{28}N_4$ .... $[Ag_2(C_{20}H_8N_4(C_6H_5)_4)]$ .... Ag: MVol.B7-304, 308
$Ag_2C_{44}H_{30}N_8O_{16}S_2$ .. $Ag_2(C_7H_4NO_4)_2(C_{15}H_{11}N_3)_2(S_2O_8)\cdot n\,H_2O$ .. Ag: MVol.B7-299/300
$Ag_2C_{48}H_{36}N_{12}O_8S_2$ .. $[Ag(NC_5H_4C_7H_5N_2)_2]_2S_2O_8$ .... Ag: MVol.B6-145
$Ag_2C_{50}H_{40}N_{10}O_{16}S_4$ .. $Ag_2(S_2O_8)_2\cdot 5\,NC_5H_4C_5H_4N$ .... Ag: MVol.B7-287, 293
$Ag_2C_{52}H_{56}N_2O_{10}P_4$ .. $[Ag(H_2O)_2((C_6H_5)_2PC_2H_4P(C_6H_5)_2)_2]$
$[Ag(NO_3)_2(H_2O)_2]$ .... Ag: MVol.B7-235
$Ag_2C_{53}H_{48}N_2O_6P_4$ ... $[Ag_2(NO_3)_2C(CH_2P(C_6H_5)_2)_4]$ .... Ag: MVol.B7-239/40
$Ag_2C_{53}H_{48}P_4^{2+}$ ..... $[Ag_2C(CH_2P(C_6H_5)_2)_4]^{2+}$ .... Ag: MVol.B7-239/40
$Ag_2C_{54}H_{99}I_2P_3$ ...... $[AgI(P(C_6H_{11})_3)_2][AgI(P(C_6H_{11})_3)]$ .... Ag: MVol.B7-207
$Ag_2C_{56}Cl_2H_{56}P_4S_2$ ... $[AgCl((C_6H_5)_2PC_2H_4)_2S]_2$ .... Ag: MVol.B7-237/9
$Ag_2C_{56}H_{56}I_2P_4S_2$ .... $[AgI((C_6H_5)_2PC_2H_4)_2S]_2$ .... Ag: MVol.B7-237/9
$Ag_2C_{58}H_{56}N_2P_4S_4$ ... $[AgSCN((C_6H_5)_2PC_2H_4)_2S]_2$ .... Ag: MVol.B7-237/9
$Ag_2C_{60}H_{66}O_2P_4S_4$ ... $[Ag(SC_2H_4OH)((C_6H_5)_2PC_2H_4)_2S]_2$ .... Ag: MVol.B7-237/9
$Ag_2C_{68}H_{106}O_{16}$ ..... $[Ag_2(C_{34}H_{53}O_8)_2]\cdot 2\,(CH_3)_2CO$ .... Ag: MVol.B6-224/5
$Ag_2C_{70}H_{70}O_2P_4S_4$ ... $[Ag(SC_6H_4OCH_3)((C_6H_5)_2PC_2H_4)_2S]_2$ ...... Ag: MVol.B7-237/9
$Ag_2C_{74}H_{60}O_2P_4S_2$ ... $[(P(C_6H_5)_3)_2AgC_2O_2S_2Ag(P(C_6H_5)_3)_2]$ ...... Ag: MVol.B7-213
$Ag_2C_{74}H_{118}O_{18}$ ..... $[Ag_2(C_{34}H_{53}O_8)_2]\cdot 2\,CH_3COCH_3$ .... Ag: MVol.B6-224/5
$Ag_2C_{76}F_6H_{60}N_8P_4$ ... $[Ag(CF_3CN_4)(P(C_6H_5)_3)_2]_2$ .... Ag: MVol.B7-213
$Ag_2C_{76}H_{60}NiO_4P_4S_4$ .. $[(Ag(P(C_6H_5)_3)_2)_2Ni(C_2O_2S_2)_2]$ .... Ag: MVol.B7-219
$Ag_2C_{78}Cl_2H_{60}P_6$ ..... $[Ag_2Cl_2((C_6H_5)_2PCCP(C_6H_5)_2)_3]$ .... Ag: MVol.B7-236
$Ag_2C_{78}H_{72}N_6P_6$ ..... $[Ag_2(N_3)_2((C_6H_5)_2PC_2H_4P(C_6H_5)_2)_3]$
$\cdot\,2\,HCON(CH_3)_2$ .... Ag: MVol.B7-234
$Ag_2C_{80}H_{60}N_4NiP_4S_4$ .. $[Ag(P(C_6H_5)_3)_2]_2[Ni(S_2CC(CN)_2)_2]$ .... Ag: MVol.B7-213/4
– .... $[Ag(P(C_6H_5)_3)_2]_2[Ni(S_2C_2(CN)_2)_2]$ .... Ag: MVol.B7-213/4
$Ag_2C_{84}H_{86}N_8O_2P_6$ ... $[Ag_2(N_3)_2((C_6H_5)_2PC_2H_4P(C_6H_5)_2)_3]$
$\cdot\,2\,HCON(CH_3)_2$ .... Ag: MVol.B7-234
$Ag_2C_{85}Cl_2H_{70}N_5P_5$ ... $[Ag_2Cl_2(NC_5H_4P(C_6H_5)_2)_5]$ .... Ag: MVol.B7-242
$Ag_2C_{148}H_{120}NiO_4P_8S_4$ $[(Ag(P(C_6H_5)_3)_4)_2Ni(C_2O_2S_2)_2]$ .... Ag: MVol.B7-219

$Ag_2ClH_{12}N_5O_3$ ...... $[Ag(NH_3)_2]_2NClO_3$ .... Ag: MVol.B6-22
$Ag_2Cl_2H_4N_2$ .... $2\,AgCl\cdot N_2H_4$ .... Ag: MVol.B6-36
$Ag_2Cl_2H_9N_3$ .... $2\,AgCl\cdot 3\,NH_3$ .... Ag: MVol.B6-19, 20
– .... $[Ag_2(NH_3)_3]Cl_2$ .... Ag: MVol.B6-20
$Ag_2Cl_4H_{4.5}N_{1.5}Pt$ .... $Ag_2PtCl_4\cdot 1.5\,NH_3$ .... Ag: MVol.B6-32
$Ag_2Cl_4H_{12}N_4Pt$ ..... $Ag_2PtCl_4\cdot 4\,NH_3 = [Ag(NH_3)_2]_2PtCl_4$ .... Ag: MVol.B6-32/3
$Ag_2Cl_4H_{18}N_6Pt$ ..... $Ag_2PtCl_4\cdot 6\,NH_3$ .... Ag: MVol.B6-32/3
$Ag_2Cl_6H_6N_2Pt$ ...... $Ag_2PtCl_6\cdot 2\,NH_3$ .... Ag: MVol.B6-33
$Ag_2Cl_6H_{24}N_8Pt$ ..... $Ag_2PtCl_6\cdot 8\,NH_3$ .... Ag: MVol.B6-33
Ag_2Cl_6Ir .... $Ag_2[IrCl_6]$ .... Ir: SVol.2-136
Ag_2Cl_6U .... Ag_2UCl_6 .... U: SVol.C9-54/5
$Ag_2CrH_6N_2O_4$ ..... $[AgNH_3]_2CrO_4$ .... Ag: MVol.B6-31
$Ag_2CrH_{12}N_4O_4$ ...... $[Ag(NH_3)_2]_2CrO_4$ .... Ag: MVol.B6-8, 31
$Ag_2H_2N_2O_2S$ .... $Ag_2[SO_2N_2H_2]$ .... S: S-N-Verb.1-166
$Ag_2H_3I_2N$ .... $2\,AgI\cdot NH_3 = AgI\cdot 0.5\,NH_3$ .... Ag: MVol.B6-23
$Ag_2H_3I_2P$ .... $2\,AgI\cdot PH_3$ .... Ag: MVol.B7-199
$Ag_2H_3NO_3Se$ .... $Ag_2SeO_3\cdot NH_3$ .... Ag: MVol.B6-27
$Ag_2H_3NO_4P^-$ .... $[Ag_2PO_4(NH_3)]^-$ .... Ag: MVol.B6-8
$Ag_2H_4I_2N_2$ .... $2\,AgI\cdot N_2H_4$ .... Ag: MVol.B6-36
$Ag_2H_4N_4O_6S_3$ ...... $SO_2(NAgSO_2NH_2)_2$ .... S: S-N-Verb.1-179
$Ag_2H_6N_2O$ .... $Ag_2O\cdot 2\,NH_3$ .... Ag: MVol.B6-16

Formula	Compound		Reference
$Ag_2H_6N_2O_3Si$	$Ag_2SiO_3 \cdot 2\ NH_3 \cdot 2\ H_2O$	Ag:	MVol.B6-30
$Ag_2H_6N_4O_4$	$AgNO_2 \cdot Ag(NH_3)_2NO_2 = AgNO_2 \cdot NH_3$	Ag:	MVol.B6-16
$Ag_2H_9I_2N_3$	$2\ AgI \cdot 3\ NH_3 = AgI \cdot 1.5\ NH_3$	Ag:	MVol.B6-23
$Ag_2H_{12}N_4O_3S$	$Ag_2SO_3 \cdot 4\ NH_3$	Ag:	MVol.B6-24
$Ag_2H_{12}N_4O_3Se$	$Ag_2SeO_3 \cdot 4\ NH_3$	Ag:	MVol.B6-27
$Ag_2H_{12}N_4O_4S$	$[Ag(NH_3)_2]_2SO_4$	Ag:	MVol.B6-12, 24/6
–	$[Ag(ND_3)_2]_2SO_4$	Ag:	MVol.B6-12/3, 24
$Ag_2H_{12}N_4O_4Se$	$[Ag(NH_3)_2]_2SeO_4$	Ag:	MVol.B6-27
$Ag_2H_{12}N_4O_6S_2$	$Ag_2S_2O_6 \cdot 4\ NH_3 \cdot 2\ H_2O$	Ag:	MVol.B6-26
$Ag_2H_{12}N_6O_4S$	$Ag_2SO_4 \cdot 3\ N_2H_4$	Ag:	MVol.B6-36
$Ag_2N_2O_3$	$[Ag_2(ONN(O)O)]$	Ag:	MVol.B6-321
Ag_2O	Ag_2O glasses		
	Ag_2O-B_2O_3	B:	B-Verb.7-74/5
–	Ag_2O systems		
	Ag_2O-B_2O_3	B:	B-Verb.7-73
	Ag_2O-B_2O_3-H_2O	B:	B-Verb.7-148
$Ag_2O_7W_2$	$Ag_2W_2O_7$	W:	SVol.B3-155
Ag_2Pu	$PuAg_2$	Np:	TrU.B2-40
		Np:	TrU.B3-99/104
$Ag_{2.19}C_5H_8N_{4.19}O_{3.57}S$	$1.19\ AgNO_3 \cdot Ag(CH_3C_2N_3SC_2H_5)$	Ag:	MVol.B7-56
$Ag_{2.29}C_9Cl_{1.29}H_8N_3S$	$1.29\ AgCl \cdot Ag(C_6H_5C_2N_3SCH_3)$	Ag:	MVol.B7-56
$Ag_{2.3}C_{25.5}Cl_{2.3}H_{20.75}N_{2.25}P_2$			
	$2.3\ AgCl \cdot C(NP(C_6H_5)_2)_2 \cdot 0.25\ CH_3CN$	Ag:	MVol.B7-243
$Ag_{2.5}Br_{2.5}C_{26}Cl_2H_{22}N_2P_2$			
	$2.5\ AgBr \cdot C(NP(C_6H_5)_2)_2 \cdot CH_2Cl_2$	Ag:	MVol.B7-243
$Ag_{2.5}C_{25.5}Cl_{3.5}H_{21}N_2P_2$	$2.5\ AgCl \cdot C(NP(C_6H_5)_2)_2 \cdot 0.5\ CH_2Cl_2$	Ag:	MVol.B7-243
$Ag_3AlC_{114}H_{90}O_6P_6S_6$	$[(Ag(P(C_6H_5)_3)_2)_3Al(C_2O_2S_2)_3]$	Ag:	MVol.B7-217/9
$Ag_3AlC_{116}Cl_6H_{92}O_6P_6S_6$			
	$[(Ag(P(C_6H_5)_3)_2)_3Al(C_2O_2S_2)_3] \cdot 2\ CHCl_3$	Ag:	MVol.B7-217
$Ag_3AsC_6H_{24}N_{12}O_4S_6$	$[Ag(NH_2CSNH_2)_2]_3AsO_4$	Ag:	MVol.B7-143
$Ag_3AsH_{12}N_4O_4$	$Ag_3AsO_4 \cdot 4\ NH_3$	Ag:	MVol.B6-30
$Ag_3BC_6H_6O_9$	$Ag_3[BO_2(C_6H_6O_7)] \cdot 2\ H_2O$	B:	B-Verb.8-135
$Ag_3B_3C_{32}F_{12}H_{20}$	$3\ AgBF_4 \cdot 2\ C_{16}H_{10}$	Ag:	MVol.B5-116
$Ag_3BrC_{16}H_{20}N_2$	$2\ AgC_6H_4N(CH_3)_2 \cdot AgBr$	Ag:	MVol.B5-11
$Ag_3BrC_{18}H_{22}N_2$	$2\ AgC_6H_4N(CH_3)_2 \cdot AgBr \cdot 0.33\ C_6H_6$	Ag:	MVol.B5-11
$Ag_3BrC_{18}H_{24}N_2$	$2\ AgC_6H_4CH_2N(CH_3)_2 \cdot AgBr$	Ag:	MVol.B5-12
$Ag_3Br_3C_6H_{12}N_4$	$3\ AgBr \cdot (CH_2)_6N_4$	Ag:	MVol.B6-199
$Ag_3Br_3C_{27}H_{63}O_9P_3$	$[AgBr((i\text{-}C_3H_7O)_3P)]_3$	Ag:	MVol.B7-248
$Ag_3Br_3H_4N_2$	$3\ AgBr \cdot N_2H_4$	Ag:	MVol.B6-36
Ag_3Br_6Ir	$Ag_3[IrBr_6]$	Ir:	SVol.2-150
$Ag_3CHO_6S_3$	$Ag_2[Ag(SCH(SO_3)_2)] \cdot 2\ H_2O$	Ag:	MVol.B7-47
$Ag_3C_2Cl_3H_8N_4O_{12}S_2$	$Ag_3(NH_2CSNH_2)_2(ClO_4)_3$	Ag:	MVol.B7-136
$Ag_3C_2H_6NO_3$	$2\ AgCH_3 \cdot AgNO_3$	Ag:	MVol.B5-4
–	$2\ AgCD_3 \cdot AgNO_3$	Ag:	MVol.B5-4
$Ag_3C_2N_3OS$	$[Ag_3(C_2N_3OS)]$	Ag:	MVol.B7-56
$Ag_3C_2N_3S_2$	$[Ag_3(C_2N_3S_2)]$	Ag:	MVol.B7-56
$Ag_3C_{2.5}H_3I_3KNO_{0.5}Se$	$KAg_3I_3SeCN \cdot 0.5\ (CH_3)_2CO$	Ag:	MVol.B6-209
$Ag_3C_3H_2N_3O_3S$	$AgC_3H_2N_3O_2S \cdot Ag_2O$	Ag:	MVol.B7-72
$Ag_3C_3H_2N_3O_8S$	$2\ AgNO_3 \cdot AgC_3H_2NO_2S$	Ag:	MVol.B7-70

Formula	Compound	Reference
$Ag_3C_3H_2N_4O_6S$	$AgC_3H_2N_3O_2S \cdot AgNO_3 \cdot AgO$	Ag: MVol.B7-72
$Ag_3C_3H_5O_3S_3$	$Ag[Ag_2(SCH_2CH(S)CH_2SO_3)]$	Ag: MVol.B7-46
$Ag_3C_3H_5S_3$	$[Ag_3(SCH(CH_2S)_2)]$	Ag: MVol.B7-9
$Ag_3C_3N_3O_3$	$[Ag_3(C_3N_3O_3)]$	Ag: MVol.B6-190
$Ag_3C_3O_2$	$Ag_3C_3O_2$	Ag: MVol.B5-21/2, 128
$Ag_3C_3O_3$	$Ag_3C_3O_3$	Ag: MVol.B5-22
$Ag_3C_4HN_6O$	$[Ag_3(C_4HN_6O)]$	Ag: MVol.B6-183
$Ag_3C_4H_4NaO_4S_2$	$Na[Ag_3(SCH_2COO)_2]$	Ag: MVol.B7-22
$Ag_3C_4H_{10}NO_3S_2$	$2\ AgSC_2H_5 \cdot AgNO_3$	Ag: MVol.B7-5
$Ag_3C_4IrN_2O_{12}$	$Ag_3[Ir(NO_2)_2(C_2O_4)_2]$	Ir: SVol.2-61
$Ag_3C_5Cl_3H_8O_{12}$	$3\ AgClO_4 \cdot CH_3CCC_2H_5$	Ag: MVol.B5-39
$Ag_3C_5H_2N_3O_4$	$Ag_3(C_3N_3(O)_2CH_2COO) \cdot H_2O$	Ag: MVol.B6-189
$Ag_3C_5H_2N_3O_5$	$Ag_3(C_3N_3O_3CH_2COO) \cdot H_2O$	Ag: MVol.B6-190
$Ag_3C_5H_4NO_4S_2$	$[Ag_3(N(CH_2COO)_2CSS)]$	Ag: MVol.B7-115
$Ag_3C_5H_9O_3S_4$	$Ag[Ag_2(SCH_2CH(S)CH_2SC_2H_4SO_3)]$	Ag: MVol.B7-46
$Ag_3C_5H_9O_4S_3$	$Ag[Ag_2(SCH_2CH(S)CH_2OC_2H_4SO_3)]$	Ag: MVol.B7-45
$Ag_3C_5H_9O_5S_4$	$Ag[Ag_2(SCH_2CH(S)CH_2SO_2C_2H_4SO_3)]$	Ag: MVol.B7-46
$Ag_3C_6Cl_3H_{10}O_{12}$	$3\ AgClO_4 \cdot C_2H_5CCC_2H_5$	Ag: MVol.B5-41
$Ag_3C_6CoH_{21}N_{15}O_3$	$Ag_3Co(N(CNHNH_2)_2)_3(OH)_3 \cdot x\ H_2O$	Ag: MVol.B6-339
$Ag_3C_6CrO_{12}$	$Ag_3[Cr(C_2O_4)_3] \cdot 6\ H_2O$	Ag: MVol.B5-149
$Ag_3C_6H_3N_6O_4S_2$	$Ag_3C_6H_3N_6O_4S_2 \cdot 3\ H_2O$	Ag: MVol.B7-72
$Ag_3C_6H_6NO_6$	$[Ag_3(N(CH_2COO)_3)]$	Ag: MVol.B6-276
$Ag_3C_6H_6N_{12}O_2^+$	$[Ag_3(C_2H_5OCOCH(NNCN_4)_2)]^+$	Ag: MVol.B6-288
$Ag_3C_6H_8NO_7$	$Ag_3C_6H_5O_7 \cdot NH_3 \cdot 1.5\ H_2O$	Ag: MVol.B6-29
$Ag_3C_6H_9N_{12}$	$Ag_3(C_3N_3(NCN)_3) \cdot 3\ NH_3$	Ag: MVol.B6-188
$Ag_3C_6H_{10}N_2O_4S_2^-$	$[Ag_3(SCH_2CH(NH_2)COO)_2]^-$	Ag: MVol.B6-256
$Ag_3C_6H_{12}I_3N_4$	$3\ AgI \cdot (CH_2)_6N_4$	Ag: MVol.B6-199
$Ag_3C_6H_{12}N_3O_9S_3$	$3\ AgNO_3 \cdot S(CHCH_3S)_2CHCH_3$	Ag: MVol.B7-92
$Ag_3C_6H_{21}N_6O_6$	$[Ag(NH_3)_2]_3C_6H_3O_6$	Ag: MVol.B6-29
$Ag_3C_6H_{23}N_6O_7$	$[Ag(NH_3)_2]_3C_6H_5O_7$	Ag: MVol.B6-29
$Ag_3C_6IrN_6$	$Ag_3[Ir(CN)_6]$	Ir: SVol.2-183
$Ag_3C_6N_3O_6$	$[Ag_3(C_3N_3(COO)_3)]$	Ag: MVol.B6-190
$Ag_3C_6N_9$	$[Ag_3(C_3N_3(NCN)_3)]$	Ag: MVol.B6-188
$Ag_3C_7HN_2O_6$	$Ag_3(C_4HN_2(COO)_3) \cdot H_2O$	Ag: MVol.B6-161
$Ag_3C_7H_4NO_4S$	$Ag_3(NCOC_6H_4SO_3)$	Ag: MVol.B6-344
$Ag_3C_7H_{28}N_{17}O_9S_7$	$3\ AgNO_3 \cdot 7\ (NH_2)_2CS$	Ag: MVol.B7-143
$Ag_3C_8Fe_2H_2NO_{11}$	$[HFe(CO)_4Ag]_2AgNO_3$	Fe: Org.Verb.B2-50
$Ag_3C_8H_6Na_3O_8S_2$	$Na_3[Ag_3(SC_2H_3(COO)_2)_2]$	Ag: MVol.B7-32
$Ag_3C_8H_{10}N_3O_9$	$3\ AgNO_3 \cdot C_8H_{10}$	Ag: MVol.B5-63
$Ag_3C_8H_{12}O_4S_3^+$	$[Ag_3(S(C_2H_4SCH_2COO)_2)]^+$	Ag: MVol.B7-42/3
$Ag_3C_9H_7NO_9S_2$	$AgO \cdot 2\ AgSO_4 \cdot C_9H_7N$	Ag: MVol.B7-285/6
$Ag_3C_9H_{12}N_3O_9$	$3\ AgNO_3 \cdot C_9H_{12}$	Ag: MVol.B5-66/7
$Ag_3C_9H_{12}N_7$	$3\ AgCN \cdot (CH_2)_6N_4$	Ag: MVol.B6-199
$Ag_3C_{10}ClH_{10}N_4O_4$	$[Ag(NHC_5H_4N)]_2 \cdot AgClO_4$	Ag: MVol.B6-95
$Ag_3C_{10}H_5N_8S_2$	$Ag_2(C_5H_2N_4S) \cdot Ag(C_5H_3N_4S)$	Ag: MVol.B7-66/7
$Ag_3C_{10}H_6N_5O_9$	$3\ AgNO_3 \cdot C_8H_6(CN)_2$	Ag: MVol.B5-90
$Ag_3C_{10}H_8^{3+}$	$[Ag_3C_{10}H_8]^{3+}$	Ag: MVol.B5-107
$Ag_3C_{10}H_9N_8O_8$	$Ag_2C_5H_4N_4O_4 \cdot AgC_5H_5N_4O_4 \cdot 1.5\ H_2O$	Ag: MVol.B6-189
$Ag_3C_{10}H_{10}N_3O_6$	$[Ag_3(C_3N_3(O)_2C(COOC_2H_5)_2)]$	Ag: MVol.B6-189
$Ag_3C_{10}H_{10}N_4^+$	$[Ag_3(NHC_5H_4N)_2]^+$	Ag: MVol.B6-95

Formula	Compound	Reference
$Ag_3C_{10}H_{10}N_4O_2S_{0.5}$	$[Ag(NHC_5H_4N)]_2 \cdot 0.5\ Ag_2SO_4$	Ag: MVol.B6-95
$Ag_3C_{10}H_{10}N_5O_3$	$[Ag(NHC_5H_4N)]_2 \cdot AgNO_3$	Ag: MVol.B6-95
$Ag_3C_{10}H_{14}N_3O_6S$	$[Ag_3C_{10}H_{14}N_3O_6S]$	Ag: MVol.B6-260
$Ag_3C_{12}ClH_{12}N_{10}O_4$	$Ag_3(CH_3(NH)C_5H_2N_4)_2ClO_4$	Ag: MVol.B6-166/7
$Ag_3C_{12}FeH_{12}O_{12}S_3$	$[Ag_3Fe(HSC_2H_3(COO)_2)_3] \cdot 3\ H_2O$	Ag: MVol.B7-32
$Ag_3C_{12}FeH_{24}N_{18}S_6$	$[Ag(NH_2CSNH_2)_2]_3[Fe(CN)_6] \cdot 2\ H_2O$	Ag: MVol.B7-143
$Ag_3C_{12}H_4N_4O_8$	$Ag_3(C_4H_2N_2(COO)_2)_2$	Ag: MVol.B7-298
$Ag_3C_{12}H_{10}NO_3$	$2\ AgC_6H_5 \cdot AgNO_3$	Ag: MVol.B5-9
$Ag_3C_{12}H_{12}N_6O_4S_2^+$	$[Ag_3(CH_3SC_4HN_2(NH_2)COO)_2]^+$	Ag: MVol.B7-37/8
$Ag_3C_{12}H_{14}N_6NaO_4$	$Na[Ag_3(C_3H_3N_2C_2H_3(NH)COO)_2] \cdot 2\ H_2O$	Ag: MVol.B6-247
$Ag_3C_{12}H_{18}N_3O_9$	$3\ AgNO_3 \cdot C_{12}H_{18}$	Ag: MVol.B5-73
$Ag_3C_{12}H_{20}N_3O_9S_2$	$3\ AgNO_3 \cdot 2\ S(CH_2CHCH_2)_2$	Ag: MVol.B7-13
$Ag_3C_{12}H_{24}N_{11}O_9$	$3\ AgNO_3 \cdot 2\ (CH_2)_6N_4$	Ag: MVol.B6-197
$Ag_3C_{12}H_{24}N_{12}O_9P_3$	$3\ AgNO_3 \cdot N_3P_3(N(CH_2)_2)_6$	Ag: MVol.B7-243
$Ag_3C_{13}Fe_2H_6N_{11}$	$[(CH_3NC)_2Fe_2(CN)_9]Ag_3$	Fe: Org.Verb.B4-10
$Ag_3C_{14}H_8NaO_4S_2$	$Na[Ag_3(SC_6H_4COO)_2]$	Ag: MVol.B7-34, 36
$Ag_3C_{14}H_{10}^{3+}$	$[Ag_3C_{14}H_{10}]^{3+}$	Ag: MVol.B5-109
$Ag_3C_{14}H_{11}N_8O_2$	$[Ag_3(C_6H_5N_3CONNHCON_3C_6H_5)]$	Ag: MVol.B6-301/2
$Ag_3C_{14}H_{15}N_6S_4$	$3\ AgSCN \cdot HSCN \cdot C_{10}H_{14}N_2$	Ag: MVol.B6-107
$Ag_3C_{14}H_{21}O_4$	$2\ AgCCC(C_2H_5)(CH_3)OH \cdot AgCH_3COO$	Ag: MVol.B5-15
–	$2\ AgCCCH(C_3H_7)OH \cdot AgCH_3COO$	Ag: MVol.B5-15
$Ag_3C_{15}Cl_3CoH_{21}O_{18}$	$[Co(CH(COCH_3)_2)_3] \cdot 3\ AgClO_4$	Ag: MVol.B6-213
$Ag_3C_{15}Cl_3CrH_{21}O_{18}$	$[Cr(CH(COCH_3)_2)_3] \cdot 3\ AgClO_4$	Ag: MVol.B6-213
$Ag_3C_{15}Cl_3FeH_{21}O_{18}$	$[Fe(CH(COCH_3)_2)_3] \cdot 3\ AgClO_4$	Ag: MVol.B6-213
$Ag_3C_{15}H_{21}N_2NiO_{12}$	$Ag[Ni(CH(COCH_3)_2)_3] \cdot 2\ AgNO_3 \cdot H_2O$	Ag: MVol.B6-213/4
$Ag_3C_{16}H_{10}NO_3$	$2\ AgCCC_6H_5 \cdot AgNO_3$	Ag: MVol.B5-16/7
$Ag_3C_{16}H_{16}N_3O_9$	$3\ AgNO_3 \cdot 2\ C_8H_8$	Ag: MVol.B5-63
$Ag_3C_{16}H_{17}N_4O_8S_2$	$Ag_3(C_8H_9N_2O_4S)C_8H_8N_2O_4S$	Ag: MVol.B6-262/3
$Ag_3C_{16}H_{25}O_4$	$2\ AgCCC(C_2H_5)_2OH \cdot AgCH_3COO$	Ag: MVol.B5-15
$Ag_3C_{16}H_{32}I_4LiO_4$	$LiAg_3I_4 \cdot 4\ C_4H_8O$	Ag: MVol.B6-220
$Ag_3C_{18}H_{13}N_{12}O_4$	$Ag_2C_9H_6N_6O_2 \cdot AgC_9H_7N_6O_2$	Ag: MVol.B6-302
$Ag_3C_{18}H_{18}N_9O_3S_2$	$AgNO_3 \cdot 2\ Ag(NH_2C_2N_3SC_6H_4CH_3)$	Ag: MVol.B7-55
$Ag_3C_{18}H_{20}N_3O_9$	$3\ AgNO_3 \cdot 2\ C_8H_7CH_3$	Ag: MVol.B5-64
$Ag_3C_{18}H_{24}N_4O_{12}^{3-}$	$[Ag_3(C_6H_{12}N_4(CH_2COO)_6)]^{3-}$	Ag: MVol.B6-279
$Ag_3C_{18}H_{35}N_6O_6S_6^{2+}$	$[Ag_3(C_6H_{11}N_2O_2S_2)(C_6H_{12}N_2O_2S_2)_2]^{2+}$	Ag: MVol.B7-187/8
$Ag_3C_{18}H_{36}N_6O_6S_6^{3+}$	$[Ag_3(HOC_2H_4NHCSCSNHC_2H_4OH)_3]^{3+}$	Ag: MVol.B7-187/8
$Ag_3C_{20}ClH_{24}N_{10}O_{10}$	$Ag_3(C_{10}H_{12}N_5O_3)_2ClO_4$	Ag: MVol.B6-169
$Ag_3C_{20}H_{24}N_3O_9$	$3\ AgNO_3 \cdot 2\ C_8H_6(CH_3)_2$	Ag: MVol.B5-64
–	$3\ AgNO_3 \cdot 2\ C_8H_7C_2H_5$	Ag: MVol.B5-64
$Ag_3C_{20}H_{24}N_8^{3+}$	$[Ag_3(NH_2C_5H_4N)_4]^{3+}$	Ag: MVol.B6-93
$Ag_3C_{20}H_{24}N_{11}O_9$	$3\ AgNO_3 \cdot 4\ NH_2C_5H_4N$	Ag: MVol.B6-92/3
$Ag_3C_{22}F_9H_{20}O_6$	$3\ AgCF_3COO \cdot 2\ C_6H_4(CH_3)_2$	Ag: MVol.B5-100
$Ag_3C_{22}H_{28}N_3O_9$	$3\ AgNO_3 \cdot 2\ C_8H_7C_3H_7$	Ag: MVol.B5-64
$Ag_3C_{23}Cl_3H_{22}O_{12}$	$3\ AgClO_4 \cdot C_6H_5(CH)_4C_6H_5 \cdot C_6H_5CH_3$	Ag: MVol.B5-104
$Ag_3C_{24}Cl_3H_{24}O_{12}$	$3\ AgClO_4 \cdot 4\ C_6H_6$	Ag: MVol.B5-93
$Ag_3C_{24}H_{28}N_3O_9$	$3\ AgNO_3 \cdot 2\ C_{12}H_{14}$	Ag: MVol.B5-86
$Ag_3C_{24}H_{36}N_3O_9$	$3\ AgNO_3 \cdot C_{24}H_{36}$	Ag: MVol.B5-74
$Ag_3C_{24}H_{45}N_8O_8S_8$	$[Ag_3(C_6H_{11}N_2O_2S_2)_3(C_6H_{12}N_2O_2S_2)]$	Ag: MVol.B7-188
$Ag_3C_{24}H_{47}N_8O_8S_8^{2+}$	$[Ag_3(C_6H_{11}N_2O_2S_2)(C_6H_{12}N_2O_2S_2)_3]^{2+}$	Ag: MVol.B7-187/8
$Ag_3C_{24}H_{48}N_7O_{15}$	$3\ AgNO_3 \cdot C_{24}H_{48}N_4O_6$	Ag: MVol.B6-205/6

Formula	Compound	Reference
$Ag_3C_{24}H_{48}N_8O_8S_8^{3+}$	$[Ag_3(HOC_2H_4NHCSCSNHC_2H_4OH)_4]^{3+}$	Ag: MVol.B7-187/8
$Ag_3C_{26}Cl_3H_{20}O_{12}$	$3\ AgClO_4 \cdot (C_6H_5)_2CC(C_6H_5)_2$	Ag: MVol.B5-104
$Ag_3C_{27}Cl_3H_{63}O_9P_3$	$[AgCl((i\text{-}C_3H_7O)_3P)]_3$	Ag: MVol.B7-248
$Ag_3C_{27}H_{30}N_3O_3$	$[AgC(OCH_3)NC_6H_4CH_3]_3$	Ag: MVol.B5-22
$Ag_3C_{27}H_{63}I_3O_9P_3$	$[AgI((i\text{-}C_3H_7O)_3P)]_3$	Ag: MVol.B7-248
$Ag_3C_{28}F_{21}H_{20}O_6$	$3\ AgC_3F_7COO \cdot 2\ C_6H_4(CH_3)_2$	Ag: MVol.B5-100
$Ag_3C_{30}ClCo_2H_{60}N_6O_{16}S_6$	$[Ag_3Co_2(CH_3SC_3H_5(NH_2)COO)_6]ClO_4$	Ag: MVol.B6-259
$Ag_3C_{30}ClH_{60}N_6Ni_2O_{16}S_6$	$[Ag_3Ni_2(CH_3SC_3H_5(NH_2)COO)_6]ClO_4$	Ag: MVol.B6-259
$Ag_3C_{30}Cl_3Cr_2H_{60}N_6O_{24}S_6$	$[Ag_3Cr_2(CH_3SC_3H_5(NH_2)COO)_6](ClO_4)_3$	Ag: MVol.B6-259
$Ag_3C_{30}Cl_3Fe_2H_{60}N_6O_{24}S_6$	$[Ag_3Fe_2(CH_3SC_3H_5(NH_2)COO)_6](ClO_4)_3$	Ag: MVol.B6-259
$Ag_3C_{30}Fe_2H_{60}N_9O_{21}S_6$	$[Ag_3Fe_2(CH_3SC_3H_5(NH_2)COO)_6](NO_3)_3$	Ag: MVol.B6-259
$Ag_3C_{30}H_{30}N_9O_6S_3$	$(AgNSO_2)_3 \cdot 6\ C_5H_5N$	Ag: MVol.B6-80
$Ag_3C_{30}H_{36}N_3O_3$	$[AgC(OC_2H_5)NC_6H_4CH_3]_3$	Ag: MVol.B5-22
$Ag_3C_{30}H_{60}N_7Ni_2O_{15}S_6$	$[Ag_3Ni_2(CH_3SC_3H_5(NH_2)COO)_6]NO_3$	Ag: MVol.B6-259
$Ag_3C_{30}H_{60}N_9O_{21}Rh_2S_6$	$[Ag_3Rh_2(CH_3SC_3H_5(NH_2)COO)_6](NO_3)_3$	Ag: MVol.B6-259
$Ag_3C_{32}Cl_3H_{20}O_{12}$	$3\ AgClO_4 \cdot 2\ C_{16}H_{10}$	Ag: MVol.B5-115
$Ag_3C_{33}H_{42}N_3O_3$	$[AgC(OC_3H_7)NC_6H_4CH_3]_3$	Ag: MVol.B5-22
$Ag_3C_{34}H_{30}N_4O_4$	$[Ag_3(C_{20}H_4N_4(CH_3)_4(C_2H_3)_2(C_2H_4COO)_2)]$	Ag: MVol.B7-305, 310
$Ag_3C_{36}H_{39}N_4O_6$	$Ag_3(C_{36}H_{39}N_4O_6)$	Ag: MVol.B7-304, 310
$Ag_3C_{44}Cl_3H_{32}N_{16}O_{20}$	$3\ AgClO_4 \cdot 4\ CH_3C_{10}H_5N_4O_2$	Ag: MVol.B6-176
$Ag_3C_{48}H_{40}N_{19}O_9$	$Ag_3((CH_3)_2C_{10}H_4N_4)_4(NO_3)_3$	Ag: MVol.B6-180
$Ag_3C_{60}CrH_{42}N_{12}S_6$	$[Ag(C_9H_7N)_2]_3[Cr(NCS)_6]$	Ag: MVol.B6-108/9, 116/7
$Ag_3C_{76}ClH_{56}N_{16}O_{10}S_2$	$(Ag(C_{10}H_8N_2)_2)_3C_{16}H_8ClN_4O_{10}S_2$	Ag: MVol.B6-290
$Ag_3C_{88}ClH_{56}N_{16}O_{10}S_2$	$(Ag(C_{12}H_8N_2)_2)_3C_{16}H_8ClN_4O_{10}S_2$	Ag: MVol.B6-290
$Ag_3C_{114}CoH_{90}O_6P_6S_6$	$[(Ag(P(C_6H_5)_3)_2)_3Co(C_2O_2S_2)_3]$	Ag: MVol.B7-217/9
$Ag_3C_{114}CrH_{90}O_6P_6S_6$	$[(Ag(P(C_6H_5)_3)_2)_3Cr(C_2O_2S_2)_3]$	Ag: MVol.B7-217/9
$Ag_3C_{114}FeH_{90}O_6P_6S_6$	$[(Ag(P(C_6H_5)_3)_2)_3Fe(C_2O_2S_2)_3]$	Ag: MVol.B7-217/9
$Ag_3C_{167}Cl_{18}H_{140}P_9$	$3\ [AgCl(P(C_6H_5)_3)_3] \cdot 5\ CHCl_3$	Ag: MVol.B7-214/5
$Ag_3C_{222}CoH_{180}O_6P_{12}S_6$	$[(Ag(P(C_6H_5)_3)_4)_3Co(C_2O_2S_2)_3]$	Ag: MVol.B7-219
$Ag_3Cl_3H_4N_2$	$3\ AgCl \cdot N_2H_4$	Ag: MVol.B6-35
Ag_3Cl_6Ir	$Ag_3[IrCl_6]$	Ir: SVol.2-136
$Ag_3H_4N_3O_2S$	$Ag_3NH_4[SO_2N_2] \cdot H_2O$	S: S-N-Verb.1-166
$Ag_3H_9N_4O_3S$	$Ag_3NSO_3 \cdot 3\ NH_3 \cdot 2\ H_2O$	Ag: MVol.B6-26
$Ag_3H_9N_4O_3Se$	$AgSeO_3NAg_2 \cdot 3\ NH_3 \cdot 2\ H_2O$	Ag: MVol.B6-27/8
$Ag_3H_{18}N_7O_6S_2$	$Ag_3(SO_3NSO_3) \cdot 6\ NH_3$	Ag: MVol.B6-26
$Ag_3H_{18}N_7O_6Se_2$	$[Ag(NH_3)_2]_3(O_3SeNSeO_3)$	Ag: MVol.B6-28
Ag_3NO_3S	$Ag_3[NSO_3]$	S: S-N-Verb.1-96
–	$Ag_3[NSO_3] \cdot H_2O$	S: S-N-Verb.1-96
$Ag_3N_2NaO_2S$	$Ag_3Na[SO_2N_2] \cdot H_2O$	S: S-N-Verb.1-166
$Ag_3N_3O_6S_3$	$Ag_3[NSO_2]_3$	S: S-N-Verb.1-9
–	$Ag_3[NSO_2]_3 \cdot 3\ H_2O$	S: S-N-Verb.1-9
Ag_3Pu	$PuAg_3$	Np: TrU.B2-40
		Np: TrU.B3-98/104
$Ag_{3.6}Pu$	$PuAg_{3.6}$	Np: TrU.B3-99/104
$Ag_{3.75}C_5H_6N_{5.75}O_{11.25}$	$3.75\ AgNO_3 \cdot C_3H_6(CN)_2$	Ag: MVol.B6-352

Formula	Compound	Reference
$Ag_4As_4C_{24}H_{60}I_4$	$[AgI(As(C_2H_5)_3)]_4$	Ag: MVol.B7-255/6
$Ag_4As_4C_{36}H_{84}I_4$	$[AgI(As(C_3H_7)_3)]_4$	Ag: MVol.B7-256
$Ag_4As_4C_{48}H_{108}I_4$	$[AgI(As(C_4H_9)_3)]_4$	Ag: MVol.B7-256
$Ag_4B_2C_{14}H_8O_5$	$[AgCOC_6H_4B(OAg)]_2O$	B: B-Verb.16-87
$Ag_4B_4C_6H_4O_7$	$Ag_2C_6H_4 \cdot Ag_2B_4O_7$	Ag: MVol.B5-9
$Ag_4B_4C_{12}H_{10}O_7$	$2\ AgC_6H_5 \cdot Ag_2B_4O_7$	Ag: MVol.B5-9
$Ag_4B_4C_{42}F_{16}H_{36}$	$4\ AgBF_4 \cdot 3\ C_6H_5CHCHC_6H_5$	Ag: MVol.B5-103
$Ag_4B_{10}O_{17}$	$5\ B_2O_3 \cdot 2\ Ag_2O \cdot 5\ H_2O$	B: B-Verb.7-148/9
$Ag_4Br_4C_6H_{12}N_4$	$4\ AgBr \cdot (CH_2)_6N_4$	Ag: MVol.B6-198/9
$Ag_4Br_4C_{56}H_{52}P_4$	$[AgBr(C_6H_4(CH_2)_2PC_6H_5)]_4$	Ag: MVol.B7-221
$Ag_4CH_4I_4N_2S$	$Ag_4(NH_2CSNH_2)I_4$	Ag: MVol.B7-136
$Ag_4C_4H_4N_6O_{12}$	$4\ AgNO_3 \cdot C_2H_4(CN)_2$	Ag: MVol.B6-350
$Ag_4C_5CoH_{13.5}N_{9.5}O_3S_2$	$Ag_4Co(CN)_5S_2O_3 \cdot 4.5\ NH_3 \cdot 6.5\ H_2O$	Ag: MVol.B6-32
$Ag_4C_6Cl_4H_{12}N_4$	$4\ AgCl \cdot (CH_2)_6N_4$	Ag: MVol.B6-198
$Ag_4C_6FeH_6N_8$	$Ag_4Fe(CN)_6 \cdot 2\ NH_3 \cdot 6\ H_2O$	Ag: MVol.B6-32
$Ag_4C_6H_7N_5O_2S_2$	$2\ Ag_2C_3H_2N_2OS \cdot NH_3 \cdot 2.5\ H_2O$	Ag: MVol.B7-71
$Ag_4C_6H_{12}I_4N_4$	$4\ AgI \cdot (CH_2)_6N_4$	Ag: MVol.B6-199
$Ag_4C_6H_{12}N_2O_8S_3$	$(AgSCH_2CH(NH_2)COOH)_2 \cdot Ag_2SO_4$	Ag: MVol.B6-256
$Ag_4C_6H_{12}N_{16}O_{21}$	$4\ AgNO_3 \cdot 3\ (C_2H_3N_3 \cdot HNO_3)$	Ag: MVol.B6-185
$Ag_4C_8H_9MoN_{11}$	$Ag_4(NH_3)_3Mo(CN)_8$	Ag: MVol.B6-31
$Ag_4C_8H_{32}N_8O_{40}SiW_{12}$	$2\ Ag_2O \cdot 4\ C_2H_8N_2 \cdot SiO_2 \cdot 12\ WO_3$	Ag: MVol.B6-59
$Ag_4C_8N_4S_4$	$Ag_2[Ag_2(NCC(S)C(S)CN)_2]$	Ag: MVol.B7-33
$Ag_4C_{10}Cl_4H_8O_{16}$	$4\ AgClO_4 \cdot C_{10}H_8 \cdot 4\ H_2O$	Ag: MVol.B5-108
$Ag_4C_{10}H_{12}N_2O_8$	$[Ag_4(C_2H_4(N(CH_2COO)_2)_2)]$	Ag: MVol.B6-276/7
$Ag_4C_{10}H_{12}N_4O_8$	$2\ Ag_2C_2O_4 \cdot (CH_2)_6N_4$	Ag: MVol.B6-199
$Ag_4C_{10}K_2N_6S_4$	$Ag_2[Ag_2(NCC(S)C(S)CN)_2] \cdot 2\ KCN$	Ag: MVol.B7-33
$Ag_4C_{12}Cl_4H_{36}O_{12}P_4$	$[AgCl(P(OCH_3)_3)]_4$	Ag: MVol.B7-244
$Ag_4C_{12}Cl_4H_{36}P_4$	$[AgCl(P(CH_3)_3)]_4$	Ag: MVol.B7-200/1
$Ag_4C_{12}F_{12}H_{24}I_4P_4$	$[AgI(CF_3P(CH_3)_2)]_4$	Ag: MVol.B7-202
$Ag_4C_{12}FeH_{18}I_6N_6$	$[(CH_3NC)_6Fe]I_2 \cdot 4\ AgI$	Fe: Org.Verb.B4-101
$Ag_4C_{12}H_{16}N_8S_4$	$Ag_2[Ag_2(NCC(S)C(S)CN)_2] \cdot 2\ C_2H_8N_2$	Ag: MVol.B7-33
$Ag_4C_{12}H_{24}N_4O_{12}S_6$	$4\ AgNO_3 \cdot 3\ S(CH_2CH_2)_2S$	Ag: MVol.B7-87
$Ag_4C_{12}H_{30}N_{10}O_{12}S_3$	$4\ AgNO_3 \cdot 3\ (CH_3)_2NCSNHCH_3$	Ag: MVol.B7-153
$Ag_4C_{12}H_{36}I_4P_4$	$[AgI(P(CH_3)_3)]_4$	Ag: MVol.B7-201
$Ag_4C_{14}Cl_4H_{10}O_{16}$	$4\ AgClO_4 \cdot C_{14}H_{10} \cdot 4\ H_2O$	Ag: MVol.B5-110
$Ag_4C_{14}FeH_{32}N_{22}S_8$	$[Ag(NH_2CSNH_2)_2]_4[Fe(CN)_6] \cdot 2\ H_2O$	Ag: MVol.B7-143
$Ag_4C_{14}H_{20}N_6O_4$	$[Ag_4(C_6H_{12}(OC_2H_4C_2N_3O)_2)]$	Ag: MVol.B6-187
$Ag_4C_{15}H_{10}N_4O_2S$	$Ag_2O \cdot Ag_2((C_6H_5)_2NNC_3N_2OS)$	Ag: MVol.B7-51
$Ag_4C_{16}H_6N_2O_8$	$Ag_4(C_6H_3(COO)_2NNC_6H_3(COO)_2)$	Ag: MVol.B6-287
$Ag_4C_{16}H_{10}O$	$2\ AgCCC_6H_5 \cdot Ag_2O$	Ag: MVol.B5-16
$Ag_4C_{18}FeH_{24}N_{14}$	$Ag_4Fe(CN)_6 \cdot 2\ (CH_2)_6N_4$	Ag: MVol.B6-200
$Ag_4C_{18}FeH_{30}I_6N_6$	$[(C_2H_5NC)_6Fe]I_2 \cdot 4\ AgI$	Fe: Org.Verb.B4-103
$Ag_4C_{18}H_8N_4O_{10}S_2$	$Ag[(AgOC_9H_4N(NO)SO_3)_2Ag]$	Ag: MVol.B6-113, 317
$Ag_4C_{18}H_{36}N_{12}O_8S_2$	$2\ Ag_2SO_4 \cdot 3\ (CH_2)_6N_4 \cdot 6\ H_2O$	Ag: MVol.B6-199
$Ag_4C_{21}H_{20}N_4O_{12}$	$[Ag_4C_{21}H_{20}N_4O_{12}]$	Ag: MVol.B6-270
$Ag_4C_{24}Cl_4H_{54}N_2$	$4\ AgCl \cdot 2\ C_{12}H_{25}NH_2$	Ag: MVol.B6-46
$Ag_4C_{24}H_{60}I_4P_4$	$[AgI(P(C_2H_5)_3)]_4$	Ag: MVol.B7-202
$Ag_4C_{26}H_{40}N_8S_4$	$Ag_2[Ag_2(NC(CS)_2CN)_2] \cdot 2\ [N(C_2H_5)_4]CN$	Ag: MVol.B7-33
$Ag_4C_{32}H_{68}I_4P_4$	$[AgI(C_4H_7P(C_2H_5)_2)]_4$	Ag: MVol.B7-206/7

Formula	Compound	Reference
$Ag_4C_{33}H_{35}N_3O_{18}S_3$	$Ag_2C_{11}H_{11}NO_6S \cdot 2\ AgC_{11}H_{12}NO_6S$	Ag: MVol.B6-254
$Ag_4C_{35}F_{12}H_{36}O_8$	$4\ AgCF_3COO \cdot 3\ C_6H_3(CH_3)_3$	Ag: MVol.B5-100
$Ag_4C_{36}H_{84}I_4P_4$	$[AgI(P(CH(CH_3)_2)_3)]_4$	Ag: MVol.B7-205
–	$[AgI(P(C_3H_7)_3)]_4$	Ag: MVol.B7-203/4
$Ag_4C_{40}F_{28}H_{30}O_8$	$4\ AgC_3F_7COO \cdot 3\ C_6H_4(CH_3)_2$	Ag: MVol.B5-100
$Ag_4C_{40}H_{40}N_8O_{40}SiW_{12}$	$2\ Ag_2O \cdot 8\ C_5H_5N \cdot SiO_2 \cdot 12\ WO_3$	Ag: MVol.B6-83
$Ag_4C_{40}H_{60}I_4P_4$	$[AgI(C_6H_5P(C_2H_5)_2)]_4$	Ag: MVol.B7-221
$Ag_4C_{45}Cl_4H_{45}N_9O_{16}$	$4\ AgClO_4 \cdot 9\ C_5H_5N$	Ag: MVol.B6-79/80
$Ag_4C_{48}H_{80}I_4N_4P_4$	$[AgI((C_2H_5)_2PC_6H_4N(CH_3)_2)]_4$	Ag: MVol.B7-232
$Ag_4C_{48}H_{108}I_4P_4$	$[AgI(P(C_4H_9)_3)]_4$	Ag: MVol.B7-205/6
$Ag_4C_{56}Cl_4H_{52}P_4$	$[AgCl(C_6H_4(CH_2)_2PC_6H_5)]_4$	Ag: MVol.B7-221
$Ag_4C_{64}H_{100}I_4N_8P_4$	$[AgI((C_5H_{10}N)_2PC_6H_5)]_4$	Ag: MVol.B7-242
$Ag_4C_{64}H_{124}I_4N_8P_4$	$[AgI((C_5H_{10}N)_2PC_6H_{11})]_4$	Ag: MVol.B7-241
$Ag_4C_{72}Cl_4H_{60}P_4$	$[AgCl(P(C_6H_5)_3)]_4$	Ag: MVol.B7-208
$Ag_4C_{72}Ge_4H_{100}I_4P_4$	$[AgI((C_2H_5)_3GeP(C_6H_5)_2)]_4$	Ag: MVol.B7-270
$Ag_4C_{72}H_{60}I_4P_4$	$[AgI(P(C_6H_5)_3)]_4$	Ag: MVol.B7-208/9
$Ag_4Cl_8H_9N_3Pt_2$	$2\ Ag_2PtCl_4 \cdot 3\ NH_3 = Ag_2PtCl_4 \cdot 1.5\ NH_3$	Ag: MVol.B6-32
$Ag_4N_2O_2S$	$Ag_4[SO_2N_2]$	S: S-N-Verb.1-167
$Ag_4N_2O_8S_3$	$SO_2(NAgSO_3Ag)_2$	S: S-N-Verb.1-178
$Ag_4N_4O_8S_4$	$Ag_4[NSO_2]_4$	S: S-N-Verb.1-4
$Ag_5B_5C_{60}F_{20}H_{40}$	$5\ AgBF_4 \cdot 4\ C_{15}H_{10}$	Ag: MVol.B5-115
$Ag_5B_5C_{64}F_{20}H_{40}$	$5\ AgBF_4 \cdot 4\ C_{16}H_{10}$	Ag: MVol.B5-116
$Ag_5C_4H_6N_{11}O_9S_2$	$3\ AgNO_3 \cdot 2\ Ag(NH_2C_2HN_3S)$	Ag: MVol.B7-55
$Ag_5C_4NO_5$	$2\ Ag_2C_2O \cdot AgNO_3$	Ag: MVol.B5-21
$Ag_5C_6CoKO_{12}$	$KAg_5[Co(C_2O_4)_3] \cdot x\ H_2O$	Ag: MVol.B5-149
$Ag_5C_8H_{20}N_9O_{15}S_2$	$5\ AgNO_3 \cdot 2\ (CH_3)_2NCSNHCH_3$	Ag: MVol.B7-153
$Ag_5C_{11}H_{44}N_{27}O_{15}S_{11}$	$5\ AgNO_3 \cdot 11\ (NH_2)_2CS \cdot 2\ H_2O$	Ag: MVol.B7-143
$Ag_5C_{12}H_{30}N_{11}O_{15}S_3$	$5\ AgNO_3 \cdot 3\ (CH_3)_2NCSNHCH_3$	Ag: MVol.B7-153
$Ag_5C_{15}H_{15}N_8O_{15}$	$5\ AgNO_3 \cdot 3\ C_5H_5N \cdot HNO_3$	Ag: MVol.B6-76/7
$Ag_5C_{15}H_{16}N_9O_{18}$	$5\ AgNO_3 \cdot 3\ C_5H_5N \cdot HNO_3$	Ag: MVol.B6-76/7
$Ag_5C_{24}H_{24}N_{12}O_{20}S_5$	$Ag_5(N(CHCH)_2N)_6(SO_4)_5$	Ag: MVol.B7-297
$Ag_5C_{27}H_{27}N_{14}O_6S_3$	$2\ AgNO_3 \cdot 3\ Ag(CH_3C_6H_3(NH_2)C_2HN_3S)$	Ag: MVol.B7-55
$Ag_5C_{32}H_{28}N_{17}O_3S_4$	$AgNO_3 \cdot 4\ Ag(NH_2C_2N_3SC_6H_5)$	Ag: MVol.B7-55
$Ag_5C_{36}H_{36}N_{17}O_3S_4$	$AgNO_3 \cdot 4\ Ag(CH_3C_6H_3(NH_2)C_2HN_3S)$	Ag: MVol.B7-55
$Ag_5C_{45}Cl_3H_{45}N_{17}O_{15}S_3$	$[Ag_5(NH_2CSNHC_{14}H_{12}ClN_2)_3](NO_3)_5$	Ag: MVol.B7-162
$Ag_5C_{45}Cl_5H_{30}O_{20}$	$5\ AgClO_4 \cdot 3\ C_{15}H_{10}$	Ag: MVol.B5-115
$Ag_6AsC_{15}H_{30}N_3S_6$	$Ag_3As \cdot 3\ Ag(SCSN(C_2H_5)_2)$	Ag: MVol.B7-109
$Ag_6B_{10}S_{18}$	$Ag_6B_{10}S_{18}$	B: B-Verb.3-5/6, 15/6
$Ag_6C_2H_8N_4O_5$	$3\ Ag_2O \cdot 2\ CO(NH_2)_2$	Ag: MVol.B6-334
$Ag_6C_8N_6O_6S_4$	$Ag_2[Ag_2(NCC(S)C(S)CN)_2] \cdot 2\ AgNO_3$	Ag: MVol.B7-33
$Ag_6C_{12}Fe_2H_9N_{15}$	$2\ Ag_3Fe(CN)_6 \cdot 3\ NH_3 \cdot x\ H_2O$	Ag: MVol.B6-32
$Ag_6C_{12}H_{36}N_{12}O_{12}$	$[Ag(NH_3)_2]_6C_6(COO)_6$	Ag: MVol.B6-29
$Ag_6C_{14}H_{56}N_{14}O_{71}P_2W_{21}$	$3\ Ag_2O \cdot P_2O_5 \cdot 21\ WO_3 \cdot 7\ C_2H_8N_2$	Ag: MVol.B6-59
$Ag_6C_{30}Fe_2H_{36}N_{24}$	$2\ Ag_3Fe(CN)_6 \cdot 3\ (CH_2)_6N_4$	Ag: MVol.B6-200
$Ag_6C_{33}H_{60}N_{20}O_9$	$3\ Ag_2CO_3 \cdot 5\ (CH_2)_6N_4 \cdot 15\ H_2O$	Ag: MVol.B6-199
$Ag_6C_{42}H_{84}N_6O_6S_6$	$[Ag((C_3H_7)_2NCOS)]_6$	Ag: MVol.B7-100/1
$Ag_6C_{44}Cl_6H_{86}N_6S_{12}$	$[Ag((C_3H_7)_2NCSS)]_6 \cdot 2\ CHCl_3$	Ag: MVol.B7-111
$Ag_6C_{62}Cl_6H_{52}O_{24}$	$6\ AgClO_4 \cdot 3\ C_5H_4C(C_6H_5)_2 \cdot C_6H_5C_2H_5$	Ag: MVol.B5-106

$Ag_6C_{70}Cl_6H_{60}O_{24}$ .... 6 $AgClO_4$ · 5 $C_6H_5CHCHC_6H_5$ .... Ag: MVol.B5-103
$Ag_7C_{24}H_{36}N_3O_9$ ..... 4 $AgCCC_4H_9$ · 3 $AgNO_3$ · C_2H_5OH .... Ag: MVol.B5-14
$Ag_7C_{26}H_{42}N_3O_{10}$ .... 4 $AgCCC_4H_9$ · 3 $AgNO_3$ · C_2H_5OH .... Ag: MVol.B5-14
$Ag_7C_{30}H_{25}N_2O_6$ ..... 5 AgC_6H_5 · 2 $AgNO_3$ .... Ag: MVol.B5-9
$Ag_7C_{35}H_{35}N_7O_{35.5}SiW_{10}$
3.5 Ag_2O · 7 C_5H_5N · SiO_2 · 10 WO_3 .... Ag: MVol.B6-83
$Ag_7C_{54}H_{54}N_{25}O_3S_6$ .. $AgNO_3$ · 6 $Ag(NH_2C_2N_3SC_6H_4CH_3)$ .... Ag: MVol.B7-55
$Ag_8C_4H_8N_8O_{26}$ ...... 8 $AgNO_3$ · $C_4H_8O_2$ .... Ag: MVol.B6-220
$Ag_8C_{30}Cr_8H_{60}N_{20}O_{28}$. 4 $Ag_2Cr_2O_7$ · 5 $(CH_2)_6N_4$ · 4 H_2O .... Ag: MVol.B6-200
$Ag_8C_{30}H_{60}N_{20}O_{16}S_4$ .. 4 Ag_2SO_4 · 5 $(CH_2)_6N_4$ · 13 H_2O .... Ag: MVol.B6-199
$Ag_8H_{12}Ir_2O_{29}P_6$ ..... $Ag_8H_2[Ir_2O(OH)_4(HPO_4)_6]$ · 3 H_2O .... Ir: SVol.2-38
$Ag_{10}B_2S_8$ .... $Ag_{10}B_2S_8$ .... B: B-Verb.3-15
$Ag_{10}C_4H_{12}I_{12}N_2$ ..... $Ag_{10}(C_4H_{12}N_2)I_{12}$ · 4 $HCON(CH_3)_2$ .... Ag: MVol.B6-165
$Ag_{10}C_{16}H_{40}I_{12}N_6O_4$ ... $Ag_{10}(C_4H_{12}N_2)I_{12}$ · 4 $HCON(CH_3)_2$ .... Ag: MVol.B6-165
$Ag_{10}C_{45}H_{99}O_{27}S_9^+$ ... $[Ag_{10}(SCH_2C(CH_2OH)_3)_9]^+$ .... Ag: MVol.B7-7
$Ag_{10}C_{65}H_{65}HgI_{12}N_{13}$ .. 10 AgI · HgI_2 · 13 C_5H_5N .... Ag: MVol.B6-80
$Ag_{26}I_{18}O_{16}W_4$ .... $Ag_{26}I_{18}W_4O_{16}$ .... W: SVol.B3-155

$Al_{0.1}O_3W$ .... $Al_{0.1}WO_3$ .... W: SVol.B2-190
$Al_{0.25}Li_{1.25}Mn_{1.5}O_4$ ... $Li_{1.25}Al_{0.25}Mn_{1.5}O_4$ .... Mn: MVol.C3-7
$Al_{0.5}As_2H_{0.5}O_{12}U_2$ ... $(HAl)_{0.5}(UO_2AsO_4)_2$ · 10 H_2O .... U: SVol.C3-177
– .... $(HAl)_{0.5}(UO_2AsO_4)_2$ · 16 H_2O .... U: SVol.C3-177
$Al_{0.5}Cr_{1.5}Ge_3Mn_3O_{12}$. $Mn_3Al_{0.5}Cr_{1.5}Ge_3O_{12}$ .... Mn: MVol.C3-203/4
Al .... Aluminum systems
Al-$AlCl_3$-Sn-$SnCl_2$ .... Sn: MVol.C4-101/2
Al-Am .... Np: TrU.B3-270
Al-As-Ga-Sn .... Sn: MVol.C4-112
Al-Ir-Nb .... Ir: SVol.1-84/6
Al-Ir-Ti .... Ir: SVol.1-65
Al-Ir-Zr .... Ir: SVol.1-69
Al-Np .... Np: TrU.B2-14
Al-O-U .... U: SVol.C3-144/5
Al-Pu .... Np: TrU.B2-19, 23
Np: TrU.B3-154/94
$AlBBr_3C_4H_{12}N$ ...... $(CH_3)_2BN(CH_3)_2$ · $AlBr_3$ .... B: B-Verb.5-216
$AlBC_3Cl_3H_9N_4$ ...... $CH_3B[N(CH_3)N]_2$ · $AlCl_3$ .... B: B-Verb.4-308
$AlBC_4Cl_2H_{14}O$ ...... $AlCl_2[BH_4]$ · $(C_2H_5)_2O$ .... B: B-Verb.19-155
$AlBC_6F_3H_{15}$ .... $[(C_2H_5)_2Al][F_3B(C_2H_5)]$ .... B: B-Verb.8-91, 94
$AlBC_6H_{24}N_2$ .... $[((CH_3)_3N)_2AlH_2][BH_4]$ .... B: B-Verb.14-93
$AlBC_8Cl_4H_{24}N_2$ ..... $[(C_4H_9NH_2)_2BH_2][AlCl_4]$ .... B: B-Verb.10-48, 52
$AlBC_{11}F_3H_{22}NiO_2$ ... C_2H_5Ni · $BF_2(CH(COCH_3)_2)$ · $FAl(C_2H_5)_2$ .. Ni: Org.Verb.1-108
$AlBC_{20}Cl_4H_{26}O_4$ .... $[(C_4H_8O_2)_2B(C_6H_5)_2][AlCl_4]$ .... B: B-Verb.10-193, 195
$AlBC_{20}Cl_4H_{30}O_2$ .... $[(C_2H_5OC_2H_5)_2B(C_6H_5)_2][AlCl_4]$ .... B: B-Verb.10-193, 195
$AlBC_{22}Cl_4H_{20}N_2$ .... $[(C_5H_5N)_2B(C_6H_5)_2][AlCl_4]$ .... B: B-Verb.10-193, 195
$AlBC_{30}Cl_4H_{22}O_4$ .... $[(CH(COC_6H_5)_2)_2B][AlCl_4]$ .... B: B-Verb.10-146
$AlBCaO_4$ .... $CaAl(OBO_3)$ = B_2O_3 · Al_2O_3 · 2 CaO ... B: B-Verb.7-110
$AlBMgO_4$ .... $AlMg(OBO_3)$ = B_2O_3 · Al_2O_3 · 2 MgO ... B: B-Verb.7-110
$AlBO_3$ .... $AlBO_3$
see $Al_2B_2O_6$ ... B_2O_3 · Al_2O_3

Formula	Compound		Reference
AlB_2	AlB_2	B:	B-Verb.1-74
$AlB_2Br_2C_{16}H_{44}N$	$[(C_4H_9)_4N][AlBr_2][BH_4]_2$	B:	B-Verb.8-31
$AlB_2C_2H_{14}N$	$(CH_3)_2NAl[BH_4]_2$	B:	B-Verb.10-10, 29/30
$AlB_2C_4H_{17}O$	$C_4H_9OAl[BH_4]_2$	B:	B-Verb.10-10
$AlB_2C_8H_{29}N_4$	$[H_2B(N(CH_3)_2)_2]_2AlH$	B:	B-Verb.14-93
$AlB_2C_{16}Cl_2H_{44}N$	$[(C_4H_9)_4N][AlCl_2][BH_4]_2$	B:	B-Verb.8-31
$AlB_3Br_3C_6H_{18}N_3S_3$	$(BBrS)_3 \cdot Al(N(CH_3)_2)_3$	B:	B-Verb.3-70
$AlB_3C_2H_{14}$	$(CH_3)_2AlB_3H_8$	B:	B-Verb.3-185
$AlB_3C_2H_{18}O$	$Al[BH_4]_3 \cdot (CH_3)_2O$	B:	B-Verb.8-37
$AlB_3C_3H_{21}N$	$Al[BH_4]_3 \cdot N(CH_3)_3$	B:	B-Verb.8-37
$AlB_3C_4H_{20}N_2$	$[H_2BN(CH_3)_2]_2Al[BH_4]$	B:	B-Verb.10-17
$AlB_3C_4H_{20}O$	$Al[BH_4]_3 \cdot C_4H_8O$	B:	B-Verb.8-37
$AlB_3C_4H_{22}N_2$	$H_2B[N(CH_3)_2]_2Al[BH_4]_2$	B:	B-Verb.5-74
		B:	B-Verb.10-29/31
$AlB_3C_4H_{22}O$	$[(C_2H_5)_2O \cdot Al(BH_4)_3]$	B:	B-Verb.10-10
$AlB_3C_6H_{21}N_3S_6$	$(BSHS)_3 \cdot Al(N(CH_3)_2)_3$	B:	B-Verb.3-70
$AlB_3C_8H_{28}O_2$	$Al[BH_4]_3 \cdot 2\ C_4H_8O$	B:	B-Verb.8-37
$AlB_3C_9H_{39}N_3$	$Al[BH_4]_3 \cdot 3\ N(CH_3)_3$	B:	B-Verb.8-37
$AlB_3C_{96}H_{60}$	$Al[B(CCC_6H_5)_4]_3$	B:	B-Verb.8-166
AlB_3CaO_7	$CaAl(B_3O_7) = 3\ B_2O_3 \cdot Al_2O_3 \cdot 2\ CaO$	B:	B-Verb.7-110
AlB_3H_{12}	$Al[BH_4]_3$	B:	B-Verb.8-1, 35, 37/9
$AlB_3H_{30}N_6$	$[H_2Al(NH_3)_4][H_2B(NH_3)_2][BH_4]_2$	B:	B-Verb.10-39
AlB_3O_6	$Al(BO_2)_3$ *see* $Al_2B_6O_{12}$ ... $3\ B_2O_3 \cdot Al_2O_3$		
$AlB_4C_8H_{36}N$	$[(C_2H_5)_4N][Al(BH_4)_4]$	B:	B-Verb.8-30
$AlB_9C_3H_{14}$	$CH_3C_2AlB_9H_{11}$	B:	B-Verb.6-27/9, 130
$AlB_9C_4H_{16}$	$C_2H_5C_2AlB_9H_{11}$	B:	B-Verb.6-27/9, 130
$AlB_9C_4H_{18}$	$[Al(CH_3)_2]C_2B_9H_{12}$	B:	B-Verb.6-34/6, 139
$AlB_9C_6H_{22}$	$[Al(C_2H_5)_2]C_2B_9H_{12}$	B:	B-Verb.6-34/6, 139
$AlB_9C_{12}H_{32}O_2$	$C_2H_5AlC_2B_9H_{11} \cdot 2\ C_4H_8O$	B:	B-Verb.6-27/9
$AlB_{10}C_3H_{24}N$	$[(CH_3)_3NH][B_{10}H_{12}AlH_2] \cdot n\ (C_2H_5)_2O$	B:	B-Verb.3-196/7
$AlB_{10}H_{14}^-$	$[B_{10}H_{12}AlH_2]^-$	B:	B-Verb.3-196/7
AlB_{12}	AlB_{12}	B:	B-Verb.1-74
$AlBiC_6Cl_6H_6$	$BiCl_2[AlCl_4] \cdot C_6H_6$	Bi:	Org.Verb.-142
$AlBiC_{12}H_{30}$	$(C_2H_5)_3Bi \cdot Al(C_2H_5)_3$	Bi:	Org.Verb.-146
$AlBiC_{13}ClH_{33}$	$[(C_2H_5)_3BiCH_3][AlCl(C_2H_5)_3]$	Bi:	Org.Verb.-14
$AlBrCCl_3N$	$BrCN \cdot AlCl_3$	C:	MVol.D3-237
$AlBrC_5Cl_2H_{10}Ni$	$C_3H_5NiAl(C_2H_5)BrCl_2$	Ni:	Org.Verb.2-278
$AlBrC_5H_{10}I_2Ni$	$C_3H_5NiAl(C_2H_5)BrI_2$	Ni:	Org.Verb.2-278
$AlBrC_6Cl_3H_{15}Ti$	$C_2H_5TiCl_3 \cdot (C_2H_5)_2AlBr$	Ti:	Org.Verb.1-63
$AlBrC_{12}F_{10}$	$(C_6F_5)_2AlBr$	F:	PerFHalOrg.4-126/7
$AlBrC_{18}F_{15}Li$	$Li[Al(C_6F_5)_3Br]$	F:	PerFHalOrg.4-126/7
$AlBr_2C_5ClH_{10}Ni$	$C_3H_5NiAl(C_2H_5)Br_2Cl$	Ni:	Org.Verb.2-275
$AlBr_2C_5H_{10}INi$	$C_3H_5NiAl(C_2H_5)Br_2I$	Ni:	Org.Verb.2-281
$AlBr_2C_6F_5$	$C_6F_5AlBr_2$	F:	PerFHalOrg.4-126/7
$AlBr_2C_9F_5H_9N$	$C_6F_5AlBr_2 \cdot N(CH_3)_3$	F:	PerFHalOrg.4-127
$AlBr_2C_{10}F_5H_{10}O$	$C_6F_5AlBr_2 \cdot (C_2H_5)_2O$	F:	PerFHalOrg.4-127
$AlBr_2C_{12}F_{10}Li$	$LiAl(C_6F_5)_2Br_2$	F:	PerFHalOrg.4-126
$AlBr_2C_{23}Cl_3FeH_{30}O_3P_2$	$[(CO)_3Fe(P(C_2H_5)_2C_6H_5)_2Br]AlCl_3Br$	Fe:	Org.Verb.B1-184/5

Formula	Compound	Reference
$AlBr_2C_{39}Cl_3FeH_{30}O_3P_2$	$[(CO)_3Fe(P(C_6H_5)_3)_2Br]AlCl_3Br$	Fe: Org.Verb.B1-184/5
$AlBr_3$	$AlBr_3$ systems	
	$AlBr_3$-$MnBr_2$	Mn: MVol.C5-306
	$AlBr_3$-$SnBr_2$	Sn: MVol.C4-103
	$AlBr_3$-$SnBr_4$	Sn: MVol.C4-103/4
	$AlBr_3$-SnI_4	Sn: MVol.C4-105
$AlBr_3CClN$	$AlBr_3 \cdot ClCN$	C: MVol.D3-215
$AlBr_3C_5H_{10}Ni$	$C_3H_5NiAl(C_2H_5)Br_3$	Ni: Org.Verb.2-278
$AlBr_3C_9Fe_2O_9$	$Fe_2(CO)_9 \cdot AlBr_3$	Fe: Org.Verb.C2-66
$AlBr_3C_{12}FeH_{12}O$	$C_5H_5FeC_5H_4COCH_3 \cdot AlBr_3$	Fe: Org.Verb.A2-214
$AlBr_4CN$	$BrCN \cdot AlBr_3$	C: MVol.D3-237
$AlBr_4C_3H_5Ni$	$C_3H_5NiAlBr_4$	Ni: Org.Verb.2-4
$AlBr_4C_5H_5NiO_2$	$C_3H_5Ni(CO)_2Br \cdot AlBr_3$	Ni: Org.Verb.2-43
$AlBr_4C_9H_{11}Ni$	$[C_6H_6Ni(C_3H_5)]AlBr_4$	Ni: Org.Verb.2-221, 224/5
$AlBr_4C_{11}H_{13}Ni$	$[C_3H_5NiC_8H_8]AlBr_4$	Ni: Org.Verb.2-278
$AlBr_4C_{11}H_{17}Ni$	$[C_3H_5NiC_8H_{12}]AlBr_4$	Ni: Org.Verb.2-86
$AlBr_4C_{20}H_{20}NiP$	$[C_2H_5NiP(C_6H_5)_3][AlBr_4]$	Ni: Org.Verb.2-278
$AlBr_4C_{21}H_{50}NiP_3$	$[C_3H_5Ni(P(C_2H_5)_3)_3][AlBr_4]$	Ni: Org.Verb.1-108
$AlBr_4C_{22}H_{38}NiOP$	$C_3H_5Ni(CO)(P(C_6H_{11})_3)AlBr_4$	Ni: Org.Verb.2-44/5
$AlBr_4C_{23}H_{20}NiO_2P$	$[C_3H_5Ni(CO)_2P(C_6H_5)_3][AlBr_4]$	Ni: Org.Verb.1-108
$AlBr_4C_{24}H_{36}Ni$	$[((CH_3)_6C_6)_2Ni][AlBr_4] \cdot x\ AlBr_3$	Ni: Org.Verb.2-248
$AlBr_5C_8FeH_{18}O_2P_2$	$(CO)_2Fe(P(CH_3)_3)_2Br(BrAlBr_3)$	Fe: Org.Verb.B1-109
$AlBr_5C_{23}FeH_{30}O_3P_2$	$[(CO)_3Fe(P(C_2H_5)_2C_6H_5)_2Br]AlBr_4$	Fe: Org.Verb.B1-184/5
$AlBr_5C_{39}FeH_{30}O_3P_2$	$[(CO)_3Fe(P(C_6H_5)_3)_2Br]AlBr_4$	Fe: Org.Verb.B1-184/5
$AlBr_5Sn$	$SnBr_2 \cdot AlBr_3$	Sn: MVol.C4-103
$AlBr_6Sb$	$AlBr_3 \cdot SbBr_3$ melts	
	$AlBr_3 \cdot SbBr_3$-$SnBr_4$	Sn: MVol.C4-104/5
$AlBr_7Te$	$TeBr_4 \cdot AlBr_3 = TeBr_3[AlBr_4]$	Te: SVol.B3-12/3
$AlCCl_4N$	$AlCl_3 \cdot ClCN$	C: MVol.D3-214
$AlCCl_5O$	$AlCl_3 \cdot COCl_2$	C: MVol.D3-99
$AlCF_3H_2ILi$	$Li[Al(CF_3)H_2I]$	F: PerFHalOrg.4-125
$AlC_3Cl_4H_5Ni$	$C_3H_5NiAlCl_4$	Ni: Org.Verb.2-4
$AlC_3Cl_4H_9Sn$	$(CH_3)_3Sn[AlCl_4]$	Sn: Org.Verb.5-85
$AlC_3F_7HI_2Li$	$Li[Al(C_3F_7)HI_2]$	F: PerFHalOrg.4-125
$AlC_3F_7H_2ILi$	$Li[Al(CF(CF_3)_2)H_2I]$	F: PerFHalOrg.4-125/7
–	$Li[Al(C_3F_7)H_2I]$	F: PerFHalOrg.4-125/7
$AlC_4Cl_2H_{12}O_2Ti$	$(C_2H_5)_2TiCl_2Al(OH)_2$	Ti: Org.Verb.1-79
$AlC_4Cl_5H_{10}Sn$	$(C_2H_5)_2SnCl_2 \cdot AlCl_3$	Sn: Org.Verb.6-61
$AlC_4F_4HI_2Li$	$Li[C(CF_2)_2CAlHI_2]$	F: PerFHalOrg.4-125
$AlC_4F_4H_2ILi$	$Li[C(CF_2)_2CAlH_2I]$	F: PerFHalOrg.4-125
$AlC_5ClH_{10}I_2Ni$	$C_3H_5NiAl(C_2H_5)I_2Cl$	Ni: Org.Verb.2-275
$AlC_5Cl_2H_{10}INi$	$C_3H_5NiAl(C_2H_5)Cl_2I$	Ni: Org.Verb.2-281
$AlC_5Cl_3H_{10}Ni$	$C_3H_5NiAlCl_3(C_2H_5)$	Ni: Org.Verb.2-3
$AlC_5Cl_6H_5Ti$	$C_5H_5TiCl_3 \cdot AlCl_3$	Ti: Org.Verb.1-201
$AlC_5F_6HI_2Li$	$Li[C(CF_2)_3CAlHI_2]$	F: PerFHalOrg.4-125
$AlC_5F_6H_2ILi$	$Li[C(CF_2)_3CAlH_2I]$	F: PerFHalOrg.4-125
$AlC_5H_{10}I_3Ni$	$C_3H_5NiAl(C_2H_5)I_3$	Ni: Org.Verb.2-281
$AlC_6ClF_5H_2Li$	$Li[Al(C_6F_4Cl)H_2F]$	F: PerFHalOrg.4-126

Formula	Compound	Reference
$AlC_6ClF_{14}ILi$	$LiAl(C_3F_7)_2ClI$	F: PerFHalOrg.4-127
$AlC_6Cl_4H_{12}NO_3Sn$	$SnCl_4 \cdot Al(OC_2H_4)_3N$	Sn: MVol.C6-182
$AlC_6Cl_4H_{15}Sn$	$(C_2H_5)_3SnCl \cdot AlCl_3$	Sn: Org.Verb.5-106
$AlC_6F_6H_2Li$	$Li[Al(C_6F_5)H_2F]$	F: PerFHalOrg.4-126/7
$AlC_6F_{14}I_2Li$	$Li[Al(C_3F_7)_2I_2]$	F: PerFHalOrg.4-125, 127
$AlC_6F_{18}N_3O_3$	$[(CF_3)_2NO]_3Al$	F: PerFHalOrg.7-102, 119
$AlC_6FeH_9I_5O_3P$	$(CO)_3Fe(P(CH_3)_3)(IAlI_3)I$	Fe: Org.Verb.B1-182
$AlC_7Cl_5H_{10}Ti$	$C_5H_5TiCl_3 \cdot C_2H_5AlCl_2$	Ti: Org.Verb.1-201
$AlC_7F_6H_{10}N$	$(CF_2CF)_2AlH \cdot N(CH_3)_3$	F: PerFHalOrg.4-125/6
$AlC_7H_{17}Ni$	$C_3H_5NiCH_3 \cdot Al(CH_3)_3$	Ni: Org.Verb.2-40
$AlC_7H_{21}Ti$	$[(CH_3)_3Ti][Al(CH_3)_4]$	Ti: Org.Verb.1-86, 97
$AlC_8Cl_5FeH_{18}O_2P_2$	$(CO)_2Fe(P(CH_3)_3)_2Cl(ClAlCl_3)$	Fe: Org.Verb.B1-109
$AlC_8Cl_5H_{18}Sn$	$(C_4H_9)_2SnCl_2 \cdot AlCl_3$	Sn: Org.Verb.6-99
$AlC_8FeH_{10}O_8^-$	$[(CH_3O)_3AlOCHFe(CO)_4]^-$	Fe: Org.Verb.B4-167
$AlC_8FeH_{18}I_5O_2P_2$	$(CO)_2Fe(P(CH_3)_3)_2I(IAlI_3)$	Fe: Org.Verb.B1-109
$AlC_8H_{19}Ni$	$CH_3C_3H_4NiCH_3 \cdot Al(CH_3)_3$	Ni: Org.Verb.2-40
$AlC_8H_{20}Ni$	$Ni(C_2H_5)_2Al(C_2H_5)_2$	Ni: Org.Verb.1-108
$AlC_9Cl_4FeH_9O_4$	$[CH_3COC_4H_6Fe(CO)_3]AlCl_4$	Fe: Org.Verb.B5-94, 96
$AlC_9Cl_5FeH_{18}O_3P_2$	$[(CO)_3Fe(P(CH_3)_3)_2Cl]AlCl_4$	Fe: Org.Verb.B1-184/5
$AlC_9F_9H_9N$	$(CF_2CF)_3Al \cdot N(CH_3)_3$	F: PerFHalOrg.4-125/7
$AlC_9FeH_{12}NO_5$	$((CH_3)_2N)((CH_3)_2AlO)CFe(CO)_4$	Fe: Org.Verb.B4-122/3
$AlC_9H_{21}Ni$	$(CH_3)_2C_3H_3NiCH_3 \cdot Al(CH_3)_3$	Ni: Org.Verb.2-40
$AlC_{10}Cl_4FeH_{10}$	$[Fe(C_5H_5)_2]AlCl_4$	Fe: Org.Verb.A1-225
$AlC_{10}Cl_4FeH_{11}$	$[HFe(C_5H_5)_2]AlCl_4$	Fe: Org.Verb.A1-232
–	$[DFe(C_5D_5)_2]AlCl_4$	Fe: Org.Verb.A1-232
$AlC_{10}Cl_4H_{17}Ni$	$(CH_3)_4C_4NiCl_2AlCl_2C_2H_5$	Ni: Org.Verb.2-326/7
$AlC_{10}FeH_{15}NO_5^-$	$[(CH_3)_3AlO((CH_3)_2N)CFe(CO)_4]^-$	Fe: Org.Verb.B4-164, 171
$AlC_{11}Cl_2H_{20}O_3Ti$	$C_5H_5(C_2H_5O)Ti(OC_2H_5)_2AlCl_2$	Ti: Org.Verb.1-201/2
$AlC_{11}Cl_4FeH_{13}O_4$	$[(CH_3CO)(CH_3)_2C_4H_4Fe(CO)_3]AlCl_4$	Fe: Org.Verb.B5-94/7
$AlC_{11}FeH_{18}N_3O_5$	$[(CH_3)_2N][((CH_3)_2N)_2AlO]CFe(CO)_4$	Fe: Org.Verb.B4-122/3, 126
$AlC_{12}Cl_3FeH_{12}O$	$C_5H_5FeC_5H_4COCH_3 \cdot AlCl_3$	Fe: Org.Verb.A2-213
$AlC_{12}Cl_3H_{27}Ti$	$(i-C_4H_9)(Cl)TiCl_2Al(C_4H_9-i)_2$	Ti: Org.Verb.1-61, 80
$AlC_{12}Cl_4FeH_{15}O_4$	$[(CH_3)_2CC(CH_3)C(CH_3)_2Fe(CO)_4]AlCl_4$	Fe: Org.Verb.B4-325
$AlC_{12}Cl_4H_{27}Sn$	$(n-C_4H_9)_3Sn[AlCl_4]$	Sn: Org.Verb.5-133
–	$(i-C_4H_9)_3Sn[AlCl_4]$	Sn: Org.Verb.5-138
$AlC_{13}Cl_3FeH_{17}NO$	$[C_5H_5FeC_5H_4CHN(CH_3)_2][HOAlCl_3]$	Fe: Org.Verb.A2-189
$AlC_{13}Cl_4FeH_{15}O_5$	$[(CH_3)_4(CH_3CO)C_3Fe(CO)_4]AlCl_4$	Fe: Org.Verb.B5-102, 106
$AlC_{14}Cl_4FeH_{11}O_4$	$[C_6H_5C(O)C_4H_6Fe(CO)_3]AlCl_4$	Fe: Org.Verb.B5-94/7
$AlC_{14}H_{29}O_3Ti$	$C_5H_5Ti(CH_3)(OC_2H_5)_2 \cdot (CH_3)_2AlOC_2H_5$	Ti: Org.Verb.1-207
$AlC_{14}H_{33}O_2Ti^+$	$[C_2H_5Ti(C_4H_9O)_2Al(C_2H_5)_2]^+$	Ti: Org.Verb.1-60/1
$AlC_{14}H_{35}O_4Ti$	$C_2H_5Ti(OC_2H_5)_3 \cdot (C_2H_5)_2AlOC_2H_5$	Ti: Org.Verb.1-63
$AlC_{15}H_{23}N_2Ni$	$(CH_3)_2Ni(NC_5H_4C_5H_4N)Al(CH_3)_3$	Ni: Org.Verb.1-73
$AlC_{15}H_{30}O_3Ti$	$C_5H_5(C_2H_5O)Ti(OC_2H_5)_2Al(C_2H_5)_2$	Ti: Org.Verb.1-201/2
$AlC_{16}Cl_4H_{36}Ti^-$	$[(i-C_4H_9)_2Ti(Cl_2Al(C_4H_9-i)_2)Cl_2]^-$	Ti: Org.Verb.1-80
$AlC_{16}H_{29}Ni$	$C_{12}H_{18}NiHAl(C_2H_5)_2$	Ni: Org.Verb.2-231, 236
$AlC_{16}H_{38}O_2Ti$	$(C_2H_5)_2Al(C_2H_5)_2Ti(OC_4H_9-s)_2$	Ti: Org.Verb.1-79
$AlC_{17}Cl_3FeH_{19}O$	$[C_5H_5FeC_5H_4C_7H_{10}][OAlCl_3]$	Fe: Org.Verb.A2-69
$AlC_{17}Cl_4FeH_{17}O_4$	$[(CH_3)_2CC(C_6H_5)C(CH_3)_2Fe(CO)_4]AlCl_4$	Fe: Org.Verb.B4-325
$AlC_{17}FeH_{33}N_4O_5$	$[C(N(CH_3)_2)_3][(CH_3)_3AlO((CH_3)_2N)CFe(CO)_4]$	Fe: Org.Verb.B4-171
$AlC_{17}H_{34}Ti$	$C_5H_5(i-C_4H_9)TiH_2Al(C_4H_9-i)_2$	Ti: Org.Verb.1-207

$AlC_{17}H_{35}O_3Ti$ ...... $C_5H_5Ti(CH_3)(OC_3H_7)_2 \cdot (CH_3)_2AlOC_3H_7$ .... Ti: Org.Verb.1-207
$AlC_{18}Cl_4FeH_{17}O_5$ .... $[(CH_3)_4(C_6H_5CO)C_3Fe(CO)_4]AlCl_4$ .... Fe: Org.Verb.B5-102, 106
$AlC_{18}F_{15}$ .... $(C_6F_5)_3Al \cdot (C_2H_5)_2O$ .... F: PerFHalOrg.4-126/7
$AlC_{18}H_{25}Ni_2O_2$ ...... $(C_5H_5NiCO)_2 \cdot Al(C_2H_5)_3$ .... Ni: Org.Verb.2-350
$AlC_{18}H_{34}O_4Ti$ ...... $(C_2H_5)_2Al(C_2H_5)_2Ti(OCCH_3CHCCH_3O)_2$ .... Ti: Org.Verb.1-79
$AlC_{18}H_{42}O_3Ti$ ...... $(C_2H_5)(C_4H_9O)Ti(C_4H_9O)_2Al(C_2H_5)_2$ .... Ti: Org.Verb.1-62
– .... $(C_2H_5)_2Ti(C_4H_9O)_2Al(OC_4H_9)C_2H_5$ .... Ti: Org.Verb.1-79
$AlC_{19}Cl_3FeH_{21}NO$ ... $[C_5H_5FeC_5H_4C(C_6H_5)N(CH_3)_2][HOAlCl_3]$ ... Fe: Org.Verb.A2-189
$AlC_{20}ClH_{46}O_4Ti$ ..... $(C_2H_5)(C_4H_9O)_2Ti(C_4H_9O)ClAl(C_2H_5)OC_4H_9$ Ti: Org.Verb.1-64
– .... $(C_2H_5)(C_4H_9O)_2Ti(C_4H_9O)_2Al(C_2H_5)Cl$ ...... Ti: Org.Verb.1-64
$AlC_{20}H_{46}O_4Ti$ ...... $C_2H_5(C_4H_9O)Ti(C_4H_9O)_2Al(OC_4H_9)C_2H_5$ .... Ti: Org.Verb.1-61
$AlC_{21}Cl_4H_{21}Ti$ ...... $(C_6H_5CH_2)_3TiCl_2AlCl_2$ .... Ti: Org.Verb.1-86/7
$AlC_{22}Cl_5H_{21}N_4Ti$ .... $C_2H_5(Cl)Ti(NC_5H_4C_5H_4N)_2AlCl_4$ .... Ti: Org.Verb.1-61
$AlC_{22}F_{15}H_{10}O$ ...... $(C_6F_5)_3Al \cdot (C_2H_5)_2O$ .... F: PerFHalOrg.4-126/7
$AlC_{22}H_{51}O_4Ti$ ...... $C_2H_5Ti(OC_4H_9)_3 \cdot (C_2H_5)_2AlOC_4H_9$ .... Ti: Org.Verb.1-63
$AlC_{23}FeH_{34}N_3O_5$ .... $[(CH_3)_2N][((C_4H_7)_2N)_2AlO]CFe(CO)_4$ .... Fe: Org.Verb.B4-122, 124
$AlC_{23}H_{30}Ni_3O_2$ ...... $(C_5H_5)_3Ni_3(CO)_2 \cdot Al(C_2H_5)_3$ .... Ni: Org.Verb.2-382/3
$AlC_{24}ClH_{54}O_3Ti$ ..... $(i\text{-}C_4H_9)(C_4H_9O)_2Ti(C_4H_9O)(Cl)Al(C_4H_9\text{-}i)_2$ .. Ti: Org.Verb.1-64
$AlC_{24}H_{31}O_3Sn$ ...... $(C_6H_5)_3SnHAl(OC_2H_5)_3$ .... Sn: Org.Verb.5-171
$AlC_{24}H_{37}Ni_2O_2$ ...... $(C_5H_5NiCO)_2 \cdot Al(C_4H_9\text{-}i)_3$ .... Ni: Org.Verb.2-350
$AlC_{24}H_{49}O_3Sn$ ...... $(C_6H_{11})_3SnH \cdot Al(OC_2H_5)_3$ .... Sn: Org.Verb. 5-148
$AlC_{28}ClH_{62}O_5Ti$ ..... $(i\text{-}C_4H_9)(OC_4H_9)_2(CH_3COOC_2H_5)Ti$
$(OC_4H_9)Al(C_4H_9\text{-}i)_2Cl$ .... Ti: Org.Verb.1-65
$AlC_{28}Cl_3H_{28}Ti$ ...... $(C_6H_5CH_2)_3TiCl_2Al(CH_2C_6H_5)Cl$ .... Ti: Org.Verb.1-86/7
$AlC_{29}H_{42}Ni_3O_2$ ...... $(C_5H_5)_3Ni_3(CO)_2 \cdot Al(C_4H_9\text{-}i)_3$ .... Ni: Org.Verb.2-382/3
$AlC_{32}H_{68}I_6Mn_2O_8$ .... $Mn_2AlH_4I_6 \cdot 8C_4H_8O$ .... Mn: MVol.D1-144
$AlC_{34}H_{43}O_4Ti$ ...... $C_2H_5Ti(OCH_2C_6H_5)_3 \cdot (C_2H_5)_2AlOCH_2C_6H_5$. Ti: Org.Verb.1-63
$AlC_{42}Fe_3H_{39}O_6$ ..... $Al(C_5H_5FeC_5H_4COCHCOCH_3)_3$ .... Fe: Org.Verb.A3-37
$AlC_{42}H_{41}Ti$ .... $(C_6H_5CH_2)_3TiCH(C_6H_5)Al(CH_2C_6H_5)_2$ ...... Ti: Org.Verb.1-95, 102/3
$AlC_{47}H_{52}NNiP_2$ ..... $(i\text{-}C_4H_9)_2AlNC_3H_4Ni(P(C_6H_5)_3)_2$ .... Ni: Org.Verb.2-73
$AlC_{48}H_{54}NNiP_2$ ..... $(i\text{-}C_4H_9)_2AlNC_4H_6Ni(P(C_6H_5)_3)_2$ .... Ni: Org.Verb.2-73
$AlC_{60}Fe_3H_{75}O_3Si_6$ ... $(C_6H_5(CH_3)_2SiC_5H_4FeC_5H_4Si(CH_3)_2O)_3Al$ ... Fe: Org.Verb.A6-279
$AlCe_{0.13}HTh_{0.88}$ ..... $(Th_{0.875}Ce_{0.125})AlH$ .... Th: SVol.C1-33/4
$AlCe_{0.25}HTh_{0.75}$ ..... $(Th_{0.75}Ce_{0.25})AlH$ .... Th: SVol.C1-33/4
$AlCe_2$ .... Ce_2Al solid solutions
$(Th_xCe_{1-x})_2Al$ .... Th: SVol.C1-33/4
$AlCe_3MnS_7$ .... $MnCe_3AlS_7$ .... Mn: MVol.C6-62/3
$AlCl_3$ .... $AlCl_3$ melts
$AlCl_3$-NaCl-$SnCl_2$ .... Sn: MVol.C4-102
– .... $AlCl_3$ solutions
$AlCl_3$-$COCl_2$ .... C: MVol.D3-98/9
– .... $AlCl_3$ systems
$AlCl_3$-Al-Sn-$SnCl_2$ .... Sn: MVol.C4-101/2
$AlCl_3$-KCl-$MnCl_2$ .... Mn: MVol.C5-225
$AlCl_3$-$MnCl_2$ .... Mn: MVol.C5-223/4
$AlCl_3$-$MnCl_2$-NaCl .... Mn: MVol.C5-225
$AlCl_3$-NaCl-$TeCl_4$ .... Te: SVol.B2-96
$AlCl_3$-$SnCl_2$ .... Sn: MVol.C4-102
$AlCl_3$-$SnCl_4$ .... Sn: MVol.C4-102/3

Formula	Compound	Reference
$AlCl_3$	$AlCl_3$ systems	
	$AlCl_3$-$TeCl_4$	Te: SVol.B2-96
$AlCl_{3.5}Te$	$TeAlCl_{3.5}$	Te: SVol.B2-96
$AlCl_4K$	$KAlCl_4$ systems	
	$KAlCl_4$-LiCl-$SnCl_2$	Sn: MVol.C4-102
$AlCl_4Te_2$	Te_2AlCl_4	Te: SVol.B2-96
$AlCl_4Te_3$	Te_3AlCl_4	Te: SVol.B2-96
$AlCl_5Mn$	$MnAlCl_5 = MnCl_2 \cdot AlCl_3$	Mn: MVol.C5-223/4
$AlCl_5OS$	$SOCl_2 \cdot AlCl_3$	S: SVol.1-46, 52
$AlCl_5S$	$[SCl][AlCl_4]$	S: SVol.2-236
$AlCl_5S_2$	$AlCl_3 \cdot S_2Cl_2$	S: SVol.2-259
$AlCl_5Sn$	$SnCl_2 \cdot AlCl_3$	Sn: MVol.C4-102
$AlCl_7O_2S_2$	$2\ SOCl_2 \cdot AlCl_3$	S: SVol.1-46
$AlCl_7S$	$[SCl_3][AlCl_4]$	S: SVol.2-273, 290/1
$AlCl_7S_4$	$AlCl_3 \cdot 2\ S_2Cl_2$	S: SVol.2-259/60
$AlCl_7Te$	$TeCl_4 \cdot AlCl_3$	Te: SVol.B2-92, 96, 103, 112
$AlCl_7U$	$UCl_4 \cdot AlCl_3$	U: SVol.C9-53
$AlCl_8U$	$UCl_5 \cdot AlCl_3$	U: SVol.C9-60/3
$AlCl_{13}U_2$	$2\ UCl_5 \cdot AlCl_3$	U: SVol.C9-61
$AlCrGe_3Mn_3O_{12}$	$Mn_3AlCrGe_3O_{12}$	Mn: MVol.C3-203/4
$AlCrMgO_4$	$MgAlCrO_4$ solid solutions	
	$MgAlCrO_4$-Mn_3O_4	Mn: MVol.C3-201
$AlCsMo_2O_8$	$CsAl(MoO_4)_2$	Mo: SVol.B2-84
$AlDy_3MnS_7$	$MnDy_3AlS_7$	Mn: MVol.C6-62/3
AlF_3	AlF_3 systems	
	AlF_3-MnF_2	Mn: MVol.C4-242
AlF_5Mn	$MnAlF_5 = MnF_2 \cdot AlF_3$	Mn: MVol.C4-242/4
–	$MnAlF_5$ solid solutions	
	$MnAl_{1-x}Fe_xF_5$	Mn: MVol.C4-244
AlF_6LiMn	$LiMnAlF_6 = LiF \cdot MnF_2 \cdot AlF_3$	Mn: MVol.C4-244
AlF_6Na_3	$Na_3[AlF_6]$ systems	
	$Na_3[AlF_6]$-SnO_2	Sn: MVol.C4-101
$AlGd_3MnS_7$	$MnGd_3AlS_7$	Mn: MVol.C6-62/3
$AlHO_{13}U_2V_2$	$Al(UO_2)_2(VO_4)_2(OH) \cdot 8\ H_2O$	U: SVol.C3-145, 305/7
–	$Al(UO_2)_2(VO_4)_2(OH) \cdot 11\ H_2O$	U: SVol.C3-145, 305/7
$AlHTh_2$	$Th_2AlH = Th_8Al_4H_4$	Th: SVol.C1-26/33
AlH_2LiMnO_4	$(AlLi)(OH)_2MnO_2$	Mn: MVol.C3-9
AlH_2Th_2	$Th_2AlH_2 = Th_8Al_4H_8$	Th: SVol.C1-26/33
AlH_3Th_2	$Th_2AlH_3 = Th_8Al_4H_{12}$	Th: SVol.C1-28/33
$AlH_4NO_8S_2$	$NH_4Al(SO_4)_2$ solid solutions	
	$NH_4(Al,Mn)(SO_4)_2 \cdot 12\ H_2O$	Mn: MVol.C6-248
AlH_4Th_2	$Th_2AlH_4 = Th_8Al_4H_{16}$	Th: SVol.C1-26/33
AlI_3	AlI_3 systems	
	AlI_3-SnI_4	Sn: MVol.C4-105
$AlInMnO_4$	$MnAlInO_4$	Mn: MVol.C3-14
AlIr	IrAl	Ir: SVol.1-52
AlIrMn	IrMnAl	Ir: SVol.1-112

Formula	Compound	Reference
AlIrU	IrUAl	Ir: SVol.1-105
$AlIr_2Mn$	Ir_2MnAl	Ir: SVol.1-112
$AlKMo_2O_8$	$KAl(MoO_4)_2$	Mo: SVol.B2-81/3
$AlK_3Mo_6O_{21}$	$K_{3n}[(AlMo_6O_{21})_n]$	Mo: SVol.B2-83
$AlK_5Mo_4O_{16}$	$K_5Al(MoO_4)_4$	Mo: SVol.B2-81, 83
$AlLa_3MnS_7$	$MnLa_3AlS_7$	Mn: MVol.C6-62/3
$AlLiMnO_4$	$LiAlMnO_4$	Mn: MVol.C3-7
$AlLiMo_2O_8$	$LiAl(MoO_4)_2$	Mo: SVol.B2-79/81
$AlLi_3Mo_3O_{12}$	$Li_3Al(MoO_4)_3$	Mo: SVol.B2-79/81
$AlMnNaO_{12}S_3$	$NaAlMn(SO_4)_3$	
	$= Na_2SO_4 \cdot Al_2(SO_4)_3 \cdot 2MnSO_4$	Mn: MVol.C6-247/8
$AlMnNd_3S_7$	$MnNd_3AlS_7$	Mn: MVol.C6-62/3
$AlMnPr_3S_7$	$MnPr_3AlS_7$	Mn: MVol.C6-62/3
$AlMnS_7Sm_3$	$MnSm_3AlS_7$	Mn: MVol.C6-62/3
$AlMnS_7Y_3$	MnY_3AlS_7	Mn: MVol.C6-62/3
$AlMo_2NaO_8$	$NaAl(MoO_4)_2$	Mo: SVol.B2-81
$AlMo_2O_8Rb$	$RbAl(MoO_4)_2$	Mo: SVol.B2-83/4
$AlMo_4O_{16}Rb_5$	$Rb_5Al(MoO_4)_4$	Mo: SVol.B2-83/4
$AlMo_6O_{21}^{3-}$	$[(AlO_6Mo_6O_{15})_n]^{3n-}$	Mo: SVol.B2-83
AlN_3O_9	$Al(NO_3)_3$ systems	
	$Al(NO_3)_3-Ba(NO_3)_2-Mn(NO_3)_2-H_2O$	Mn: MVol.C3-301
	$Al(NO_3)_3-HNO_3-Mn(NO_3)_2-H_2O$	Mn: MVol.C3-301
	$Al(NO_3)_3-Mn(NO_3)_2-H_2O$	Mn: MVol.C3-301
AlO_4W	$AlWO_4$	W: SVol.B2-42
AlPu	PuAl	Np: TrU.B2-25
		Np: TrU.B3-155/61
$AlPu_3$	Pu_3Al	Np: TrU.B2-41, 51
		Np: TrU.B3-155/61, 169/70
AlTh	ThAl	Th: SVol.C1-33
$AlTh_2$	Th_2Al	Th: SVol.C1-26
–	Th_2Al solid solutions	
	$(Th_xCe_{1-x})_2Al$	Th: SVol.C1-33/4
–	Th_2Al systems	
	Th_2Al-H	Th: SVol.C1-26/7
$Al_{1.5}Cr_{0.5}Ge_3Mn_3O_{12}$	$Mn_3Al_{1.5}Cr_{0.5}Ge_3O_{12}$	Mn: MVol.C3-203/4
$Al_{1.71}F_{0.29}MgMn_{0.29}O_{3.71}$		
	$MgMn_{0.29}Al_{1.71}O_{3.71}F_{0.29}$	Mn: MVol.C4-269
$Al_2BC_6Cl_9H_7N$	$[(CH_3C_5H_4N)BCl_2][Al_2Cl_7]$	B: B-Verb.10-193, 196
$Al_2BC_{10}H_{32}N_3$	$Al_2B[N(CH_3)_2]_3(CH_3)_4H_2$	B: B-Verb.4-50, 100
$Al_2BCa_3HO_{16}Si_4$	$HCa_3Al_2B(SiO_4)_4$ solid solutions	
	$H(Ca,Mn,Fe)_3Al_2B(SiO_4)_4$	B: B-Verb.7-215
$Al_2BFe_3HO_{16}Si_4$	$HFe_3Al_2B(SiO_4)_4$ solid solutions	
	$H(Fe,Ca,Mn)_3Al_2B(SiO_4)_4$	B: B-Verb.7-215
$Al_2BHMn_3O_{16}Si_4$	$HMn_3Al_2B(SiO_4)_4$ solid solutions	
	$H(Mn,Ca,Fe)_3Al_2B(SiO_4)_4$	B: B-Verb.7-215
$Al_2B_2C_4H_{18}$	$[(CH_3)_2Al]_2(BH_3)_2$	B: B-Verb.3-184
$Al_2B_2C_8Cl_6H_{24}N_4$	$B_2[N(CH_3)_2]_4 \cdot 2\ AlCl_3$	B: B-Verb.4-272
$Al_2B_2CaO_7$	$B_2O_3 \cdot Al_2O_3 \cdot CaO$	B: B-Verb.7-110

Formula	Compound		Reference
$Al_2B_2Ca_2O_8$	$B_2O_3 \cdot Al_2O_3 \cdot 2\ CaO = CaAl(OBO_3)$	B:	B-Verb.7-110
$Al_2B_2F_2O_5$	$Al_2B_2O_5F_2$	B:	B-Verb.7-28
$Al_2B_2Li_4O_8$	$B_2O_3 \cdot Al_2O_3 \cdot 2\ Li_2O$	B:	B-Verb.7-109
$Al_2B_2Mg_2O_8$	$B_2O_3 \cdot Al_2O_3 \cdot 2\ MgO = AlMg(OBO_3)$	B:	B-Verb.7-110
$Al_2B_2O_6$	$B_2O_3 \cdot Al_2O_3 = AlBO_3$	B:	B-Verb.7-97
–	$B_2O_3 \cdot Al_2O_3 \cdot 3\ H_2O$	B:	B-Verb.7-195/6
$Al_2B_4C_4H_{28}N_2$	$[(CH_3)_2NAl(BH_4)_2]_2$	B:	B-Verb.4-100
		B:	B-Verb.10-10
$Al_2B_4Li_6O_{12}$	$2\ B_2O_3 \cdot Al_2O_3 \cdot 3\ Li_2O$	B:	B-Verb.7-109
$Al_2B_4O_9$	$2\ B_2O_3 \cdot Al_2O_3 \cdot 2.7\ H_2O$	B:	B-Verb.7-196
$Al_2B_6Ca_2O_{14}$	$3\ B_2O_3 \cdot Al_2O_3 \cdot 2\ CaO = CaAl(B_3O_7)$	B:	B-Verb.7-110
$Al_2B_6O_{12}$	$3\ B_2O_3 \cdot Al_2O_3 = Al(BO_2)_3$	B:	B-Verb.7-97
–	$3\ B_2O_3 \cdot Al_2O_3 \cdot 7\ H_2O$	B:	B-Verb.7-196
$Al_2B_7C_{16}H_{64}N$	$[(C_4H_9)_4N][BH_4] \cdot 2\ Al[BH_4]_3$	B:	B-Verb.8-31
$Al_2B_{10}C_6H_{34}N_2$	$[(CH_3)_3N \cdot AlH_2]_2[B_{10}H_{12}]$	B:	B-Verb.20-178
$Al_2B_{12}Na_4O_{23}$	$6\ B_2O_3 \cdot Al_2O_3 \cdot 2\ Na_2O \cdot 5\ H_2O$	B:	B-Verb.7-197
–	$6\ B_2O_3 \cdot Al_2O_3 \cdot 2\ Na_2O \cdot 13\ H_2O$	B:	B-Verb.7-197
$Al_2B_{36}H_{60}O_{12}$	$[Al(H_2O)_6]_2[B_{12}H_{12}]_3 \cdot 6\ H_2O$	B:	B-Verb.20-220, 223
$Al_2Bi_2O_6$	$Al_2O_3 \cdot Bi_2O_3$ solid solutions		
	$Al_2O_3 \cdot Bi_2O_3$-$BaSnO_3$	Sn:	MVol.C4-101
$Al_2Br_3C_{17}H_{31}Ni$	$[C_6H_6Ni(C_3H_5)]Al_2(C_2H_5)_4Br_3$	Ni:	Org.Verb.2-221, 225
$Al_2Br_4C_{10}Fe_2H_{12}N_2O_6$	$[(CO)_3Fe(AlN(CH_3)_2)Br_2]_2$	Fe:	Org.Verb.B2-65
		Fe:	Org.Verb.C1-31/2
$Al_2Br_4C_{11}Fe_2H_{12}N_2O_8$	$[Br_2(CO)_3Fe]_2[Al_2(N(CH_3)_2)_2CO_2]$	Fe:	Org.Verb.C1-31/2
$Al_2Br_4C_{34}Fe_2H_{66}N_2O_6P_2$			
	$[Br_2P(C_4H_9)_3(CO)_2Fe]_2[AlCON(CH_3)_2]_2$	Fe:	Org.Verb.C1-31/2
$Al_2Br_6C_{10}FeH_{10}$	$[Fe(C_5H_5)_2][AlBr_3]_2$	Fe:	Org.Verb.A1-225
$Al_2Br_6C_{10}H_{10}Ni$	$Ni(C_5H_5)_2 \cdot 2\ AlBr_3$	Ni:	Org.Verb.2-211
$Al_2Br_7C_9H_{11}Ni$	$[C_6H_6Ni(C_3H_5)]AlBr_4 \cdot AlBr_3$	Ni:	Org.Verb.2-221, 224/5
$Al_2Br_7C_{15}H_{23}Ni$	$[(CH_3)_6C_6Ni(C_3H_5)]AlBr_4 \cdot AlBr_3$	Ni:	Org.Verb.2-223, 225
$Al_2Br_8C_8FeH_{18}O_2P_2$	$(CO)_2Fe(P(CH_3)_3)_2(BrAlBr_3)_2$	Fe:	Org.Verb.B1-109
$Al_2Br_8C_{24}H_{36}Ni$	$[((CH_3)_6C_6)_2Ni][AlBr_4]_2$	Ni:	Org.Verb.2-248
–	$[((CH_3)_6C_6)_2Ni][AlBr_4]_2 \cdot x\ AlBr_3$	Ni:	Org.Verb.2-248
Al_2Br_8Mn	$2\ AlBr_3 \cdot MnBr_2$	Mn:	MVol.C5-306
Al_2Br_8Sn	$SnBr_2 \cdot 2\ AlBr_3$	Sn:	MVol.C4-103
$Al_2Br_{18}H_{24}O_{12}Sn_3$	$[Al(H_2O)_6]_2[SnBr_6]_3 \cdot 12\ H_2O$	Sn:	MVol.C4-104
Al_2CCl_8O	$CO(AlCl_4)_2$	C:	MVol.D3-93, 98/9
$Al_2C_5Cl_8H_5Ti$	$C_5H_5Ti(Cl_2AlCl_2)_2$	Ti:	Org.Verb.1-202
$Al_2C_6Cl_6H_{14}Ti^+$	$[(C_2H_4)Ti(Cl_2Al(C_2H_5)Cl)_2]^+$	Ti:	Org.Verb.1-124
$Al_2C_6FeH_9I_8O_3P$	$(CO)_3Fe(P(CH_3)_3)(IAlI_3)_2$	Fe:	Org.Verb.B1-182
$Al_2C_7H_{18}I_8NiOP_2$	$CONi(P(CH_3)_3)_2I_2 \cdot 2\ AlI_3$	Ni:	Org.Verb.1-121
$Al_2C_8Cl_8FeH_{18}O_2P_2$	$(CO)_2Fe(P(CH_3)_3)_2(ClAlCl_3)_2$	Fe:	Org.Verb.B1-109
$Al_2C_8FeH_{18}I_8O_2P_2$	$(CO)_2Fe(P(CH_3)_3)_2(IAlI_3)_2$	Fe:	Org.Verb.B1-109
$Al_2C_9Cl_6H_{15}Ti$	$C_5H_5Ti(Cl_2Al(C_2H_5)Cl)_2$	Ti:	Org.Verb.1-202/3
$Al_2C_{11}Cl_6FeH_{10}O_2$	$Fe(C_5H_5)_2 \cdot 2\ AlCl_3 \cdot CO_2$	Fe:	Org.Verb.A3-50
$Al_2C_{12}Cl_4H_{30}Ti$	$Cl_2Ti[(C_2H_5)(Cl)Al(C_2H_5)_2]_2$	Ti:	Org.Verb.1-80
$Al_2C_{12}Cl_7H_{19}Ni$	$(C_9H_{15})C_3H_4NiAlCl_4 \cdot AlCl_3$	Ni:	Org.Verb.2-4/5
$Al_2C_{14}H_{28}NiS_2$	$C_8H_{12}Ni(C_2H_4(SAl(CH_3)_2)_2)$	Ni:	Org.Verb.2-79
$Al_2C_{16}Cl_6FeH_{14}O_2$	$Fe(C_5H_5)_2 \cdot C_6H_4O_2 \cdot 2\ AlCl_3$	Fe:	Org.Verb.A1-237

Formula	Compound	Reference
$Al_2C_{17}Cl_3H_{31}Ni$	$[C_6H_6Ni(C_3H_5)]Al_2(C_2H_5)_4Cl_3$	Ni: Org.Verb.2-221, 225
$Al_2C_{20}H_{36}N_6Ni_2O_8$	$(CO)_3Ni(N(CH_3)_2COAl(N(CH_3)_2)_2)_2Ni(CO)_3$	Ni: Org.Verb.2-264/5
$Al_2C_{20}H_{44}I_4MnO_4$	$MnAl_2(CH_3)_4I_4 \cdot 4\ C_4H_8O$	Mn: MVol.D1-143/4
$Al_2C_{21}H_{47}O_2Ti$	$[(C_2H_5)_2Al(C_2H_5)_2]_2Ti(OCCH_3CHCCH_3O)$	Ti: Org.Verb.1-122
$Al_2C_{22}Fe_2H_{36}N_6O_{10}$	$[(CO)_4FeC(N(CH_3)_2)OAl(N(CH_3)_2)_2]_2$	Fe: Org.Verb.C2-71/2
$Al_2C_{22}H_{53}O_2Ti$	$C_2H_5Ti[(C_2H_5)(OC_4H_9\text{-}n)Al(C_2H_5)_2]_2$	Ti: Org.Verb.1-86
$Al_2C_{24}Cl_2H_{56}O_4Ti$	$C_2H_5(C_4H_9O)Ti[(C_4H_9O)(Cl)Al(C_2H_5)_2]$ $[(C_4H_9O)(Cl)Al(C_2H_5)OC_4H_9]$	Ti: Org.Verb.1-64
$Al_2C_{24}Cl_4H_{54}$	$(i\text{-}C_4H_9)_2Ti(Cl_2Al(C_4H_9\text{-}i)_2)_2$	Ti: Org.Verb.1-80
$Al_2C_{24}Cl_8H_{36}Ni$	$[((CH_3)_6C_6)_2Ni][AlCl_4]_2$	Ni: Org.Verb.2-248
$Al_2C_{24}H_{34}NiS_2$	$C_8H_{12}Ni(S(C_6H_5)Al(CH_3)_2)_2$	Ni: Org.Verb.2-79
$Al_2C_{24}H_{42}O_6Sn$	$(C_6H_5)_2Sn[HAl(OC_2H_5)_3]_2$	Sn: Org.Verb.6-160
$Al_2C_{24}H_{52}I_4MnO_4$	$MnAl_2(C_2H_5)_4I_4 \cdot 4\ C_4H_8O$	Mn: MVol.D1-143/4
$Al_2C_{24}H_{57}O_3Ti$	$C_4H_9OTi[(C_2H_5)(OC_4H_9)Al(C_2H_5)_2]_2$	Ti: Org.Verb.1-79/80
$Al_2C_{26}Cl_2H_{60}O_5Ti$	$C_2H_5(C_4H_9O)Ti[(C_4H_9O)(Cl)Al(C_2H_5)OC_4H_9]_2$	Ti: Org.Verb.1-64
$Al_2C_{26}H_{61}O_4Ti$	$C_2H_5Ti[(OC_4H_9\text{-}n)_2Al(C_2H_5)_2]_2$	Ti: Org.Verb.1-62/3
$Al_2C_{29}H_{45}Ni_3O_2$	$(C_5H_5)_3Ni_3(CO)_2 \cdot 2\ Al(C_2H_5)_3$	Ni: Org.Verb.2-382/3
$Al_2C_{32}Cl_2H_{72}O_3Ti$	$(i\text{-}C_4H_9)(C_4H_9O)Ti[(C_4H_9O)(Cl)Al(C_4H_9\text{-}i)_2]_2$	Ti: Org.Verb.1-64
$Al_2C_{40}Cl_2H_{88}O_7Ti$	$(i\text{-}C_4H_9)(OC_4H_9)(CH_3COOC_2H_5)_2Ti$ $[(OC_4H_9)Al(C_4H_9\text{-}i)_2Cl]_2$	Ti: Org.Verb.1-65
$Al_2C_{40}H_{52}NiP_2$	$C_8H_{12}Ni(P(C_6H_5)_2Al(C_2H_5)_2)_2$	Ni: Org.Verb.2-79
$Al_2C_{41}H_{69}Ni_3O_2$	$(C_5H_5)_3Ni_3(CO)_2 \cdot 2\ Al(C_4H_9\text{-}i)_3$	Ni: Org.Verb.2-382/3
$Al_2C_{54}H_{36}O_{18}Sn_3$	$Al_2[Sn(C_6H_4O_2)_3]_3$	Sn: MVol.C5-77
$Al_2C_{64}Cl_2H_{144}O_{11}Ti_3$	$(i\text{-}C_4H_9)(C_4H_9O)Ti[C_4H_9OTi(OC_4H_9)_3]_2$ $[C_4H_9OAl(C_4H_9\text{-}i)_2Cl]_2$	Ti: Org.Verb.1-64/5
$Al_2CdGe_3Mn_2O_{12}$	$CdMn_2Al_2Ge_3O_{12}$	Mn: MVol.C3-81/2
$Al_2Cd_2Ge_3MnO_{12}$	$Cd_2MnAl_2Ge_3O_{12}$	Mn: MVol.C3-81/2
$Al_2Cl_8H_{24}N_8O_2U$	$Al_2UO_2Cl_8 \cdot 8\ NH_3$	U: SVol.E1-12
Al_2Cl_8Mn	$Mn(AlCl_4)_2 = MnCl_2 \cdot 2\ AlCl_3$	Mn: MVol.C5-223/4
Al_2Cl_8OS	$2\ AlCl_3 \cdot SOCl_2$	S: SVol.1-40, 46
Al_2Cl_8Sn	$SnCl_2 \cdot 2\ AlCl_3$	Sn: MVol.C4-102
Al_2Cl_9U	$U(Al_2Cl_7)Cl_2$	U: SVol.C9-50
$Al_2Cl_{10}Te$	$TeCl_4 \cdot 2\ AlCl_3$	Te: SVol.B2-95
$Al_2Cl_{10}U$	$UCl_4 \cdot 2\ AlCl_3$	U: SVol.C9-53, 55, 58
$Al_2Cl_{12}O_6U_3$	$Al_2[UO_2Cl_4]_3 \cdot 2\ H_2O$	U: SVol.C9-93, 96
$Al_2Cl_{18}Sn_3$	$Al_2[SnCl_6]_3 \cdot 24\ H_2O$	Sn: MVol.C4-103
$Al_2F_{21}Pa_3$	$Al_2(PaF_7)_3$	Pa: SVol.2-45
Al_2FeO_4	$FeAl_2O_4$ solid solutions $FeAl_2O_4\text{-}MnV_2O_4$	Mn: MVol.C3-154
$Al_2Ge_3Mn_3O_{12}$	$Mn_3Al_2Ge_3O_{12}$	Mn: MVol.C3-80/1
–	$Mn_3Al_2Ge_3O_{12}$ solid solutions	
	$Mn_3Al_xCr_{2-x}Ge_3O_{12}$	Mn: MVol.C3-203/4
	$Mn_3Al_{2-x}Ga_xGe_3O_{12}$	Mn: MVol.C3-82/3
$Al_2H_6LiMn_3O_{12}$	$[Al_2Li(OH)_6][Mn_3O_6]$	Mn: MVol.C3-10
Al_2H_8Sn	$Sn[AlH_4]_2$	Sn: MVol.C4-100
$Al_2I_4Mn_4O_{25}$	$Al_2O_3 \cdot 4\ MnO_2 \cdot 2\ I_2O_7 \cdot 10\ H_2O$	Mn: MVol.C5-341
$Al_2K_6Mo_{12}O_{42}$	$K_6[(AlMo_6O_{21})_2]$	Mo: SVol.B2-83
$Al_2La_6S_{14}Sn$	$La_6Al_2SnS_{14}$	Sn: MVol.C4-155

$Al_2LiMn_3O_9$ . . . . $Al_2LiMn_3O_9 \cdot 3\ H_2O$ . . . . Mn: MVol.C3-9/10
Al_2MgO_4 . . . . $MgAl_2O_4$ solid solutions
$MgAl_{2-2x}Mn_{2x}O_4$ . . . . Mn: MVol.C3-7
$Mn_{1-x}Mg_xAl_2O_4$ . . . . Mn: MVol.C3-7
$MgAl_2O_4$-MnV_2O_4 . . . . Mn: MVol.C3-154
$(Mn_3O_4)_{1-x}(MgAl_2O_4)_x$ . . . . Mn: MVol.C3-7/9

– . . . . $MgAl_2O_4$ systems
$MgAl_2O_4$-UO_2 . . . . U: SVol.C3-145

$Al_2Mn_{0.5}O_4Zn_{0.5}$ . . . . $Mn_{0.5}Zn_{0.5}Al_2O_4$ . . . . Mn: MVol.C3-9
Al_2MnO_4 . . . . $MnAl_2O_4$ . . . . Mn: MVol.C3-1/6
– . . . . $MnAl_2O_4$ solid solutions
$Mn_{1-x}Mg_xAl_2O_4$ . . . . Mn: MVol.C3-7
$MnAl_2O_4$-$MnCr_2O_4$ . . . . Mn: MVol.C3-201
$(Mn,V)Al_2O_4$ . . . . Mn: MVol.C3-151/2

$Al_2MnO_{16}S_4$ . . . . $Al_2Mn(SO_4)_4 \cdot 22\ H_2O$
$= Al_2(SO_4)_3 \cdot MnSO_4 \cdot 22\ H_2O$ . . . . . . Mn: MVol.C6-244/6
Al_2MnS_4 . . . . $MnAl_2S_4$ . . . . Mn: MVol.C6-52/3
$Al_2Mn_2Na_2O_{24}S_6$ . . . . $Na_2SO_4 \cdot Al_2(SO_4)_3 \cdot 2\ MnSO_4$
$= NaAlMn(SO_4)_3$ . . . . Mn: MVol.C6-247/8
$Al_2Mn_5O_{32}S_8$ . . . . $Al_2Mn_5(SO_4)_8 = Al_2(SO_4)_3 \cdot 5\ MnSO_4$ . . . Mn: MVol.C6-244
– . . . . $Al_2Mn_5(SO_4)_8 \cdot 23\ H_2O$ . . . . Mn: MVol.C6-244/6
Al_2MoO_6 . . . . Al_2MoO_6 . . . . Mo: SVol.B2-76
$Al_2Mo_3O_{12}$ . . . . $Al_2(MoO_4)_3 = Al_2O_3 \cdot 3\ MoO_3$ . . . . Mo: SVol.B2-77/9
– . . . . $Al_2(MoO_4)_3$ systems
$Al_2(MoO_4)_3$-K_2MoO_4 . . . . Mo: SVol.B2-81
$Al_2(MoO_4)_3$-Li_2MoO_4 . . . . Mo: SVol.B2-79/80
$Al_2(MoO_4)_3$-Rb_2MoO_4 . . . . Mo: SVol.B2-83/4

$Al_2Mo_8O_{25}$ . . . . $Al_2Mo_8O_{25}$ . . . . Mo: SVol.B2-76
Al_2Np . . . . $NpAl_2$ . . . . Np: TrU.B2-13/4, 50
Al_2O_3 . . . . Al_2O_3 ceramics
Al_2O_3-BeO-MgO-ThO_2 . . . . Th: SVol.C2-5
Al_2O_3-BeO-UO_2 . . . . U: SVol.C3-145

– . . . . Al_2O_3 glasses
Al_2O_3-B_2O_3-BaO . . . . B: B-Verb.7-110
Al_2O_3-B_2O_3-CaO . . . . B: B-Verb.7-110
Al_2O_3-B_2O_3-CdO . . . . B: B-Verb.7-110
Al_2O_3-B_2O_3-Li_2O-SnO_2-ZnO . . . . Sn: MVol.C4-101
Al_2O_3-B_2O_3-MgO . . . . B: B-Verb.7-110
Al_2O_3-B_2O_3-PbO . . . . B: B-Verb.7-111
Al_2O_3-B_2O_3-SrO . . . . B: B-Verb.7-110
Al_2O_3-B_2O_3-ZnO . . . . B: B-Verb.7-110
Al_2O_3-BeO-ThO_2 . . . . Th: SVol.C2-5
Al_2O_3-Na_2O-SiO_2-SnO_2-ZrO_2 . . . . Sn: MVol.C4-220
Al_2O_3-SiO_2-SnO-ZnO . . . . Sn: MVol.C4-100
Al_2O_3-SiO_2-ThO_2 . . . . Th: SVol.C2-26/7
Al_2O_3-SiO_2-UO_2 . . . . U: SVol.C3-158/60
Al_2O_3-TeO_2 . . . . Te: SVol.B1-62

Al_2O_3 .... Al_2O_3 solid solutions
$(Mn_{1-x}Al_x)_2O_3$ .... Mn: MVol.C3-9
$(Al,Mo)_2O_3$ .... Mo: SVol.B2-76
Al_2O_3-SiO_2-SnO_2 .... Sn: MVol.C4-101
Al_2O_3-WO_3 .... W: SVol.B2-201
– .... Al_2O_3 systems
Al_2O_3-B_2O_3 .... B: B-Verb.7-96
Al_2O_3-B_2O_3-Li_2O .... B: B-Verb.7-109
Al_2O_3-B_2O_3-Li_2O-H_2O .... B: B-Verb.7-197
Al_2O_3-B_2O_3-Na_2O-SiO_2-H_2O .... B: B-Verb.7-212
Al_2O_3-B_2O_3-SiO_2 .... B: B-Verb.7-215
Al_2O_3-BeO-ThO_2 .... Th: SVol.C2-5
Al_2O_3-CaF_2-CaO .... Fe: SVol.GD5a-67
Fe: SVol.GD5b-44
Al_2O_3-CaO .... Fe: SVol.GD5a-67
Fe: SVol.GD5b-45
Al_2O_3-CaO-SiO_2 .... Fe: SVol.GD5a-67
Fe: SVol.GD5b-45
Al_2O_3-MnO .... Mn: MVol.C3-1/2
Al_2O_3-$MnTiO_3$ .... Mn: MVol.C3-122
Al_2O_3-Mn_2O_3 .... Mn: MVol.C3-1/2
Al_2O_3-Mo .... Mo: SVol.B2-76
Al_2O_3-MoO_3 .... Mo: SVol.B2-76
Al_2O_3-SnO_2 .... Sn: MVol.C4-100/1
Al_2O_3-SnO_2-TiO_2 .... Sn: MVol.C4-212
Al_2O_3-ThO_2 .... Th: SVol.C2-20
Al_2O_3-UO_2 .... U: SVol.C3-142/4
Al_2O_4Sn .... SnO · Al_2O_3 .... Sn: MVol.C4-100
Al_2O_4V .... VAl_2O_4 solid solutions
$(Mn,V)Al_2O_4$ .... Mn: MVol.C3-151/2
$Al_2O_9U_2$ .... Al_2O_3 · 2 UO_3 .... U: SVol.C3-144
$Al_2O_{12}S_3$ .... $Al_2(SO_4)_3$ systems
$Al_2(SO_4)_3$-Li_2SO_4-$MnSO_4$-H_2O .... Mn: MVol.C6-246/7
$Al_2(SO_4)_3$-$MnSO_4$ .... Mn: MVol.C6-244/5
$Al_2(SO_4)_3$-$MnSO_4$-Rb_2SO_4-H_2O .... Mn: MVol.C6-248/9
$Al_2(SO_4)_3$-Rb_2SO_4-H_2O .... Mn: MVol.C6-248
$Al_2O_{12}U_3$ .... Al_2O_3 · 3 UO_3 .... U: SVol.C3-142
$Al_2O_{16}Rb_2S_4$ .... Rb_2SO_4 · $Al_2(SO_4)_3$ · 24 H_2O .... Mn: MVol.C6-248/9
Al_2Pu .... $PuAl_2$ .... Np: TrU.B2-41, 51, 56, 60
Np: TrU.B3-155/61, 169/79
– .... $PuAl_2$ systems
$PuAl_2$-UAl_2 .... Np: TrU.B3-179
Al_2U .... UAl_2 .... Np: TrU.B2-51, 56, 60, 62
– .... UAl_2 systems
UAl_2-$PuAl_2$ .... Np: TrU.B3-179
$Al_{2.5}Li_{0.5}O_4$ .... $Li_{0.5}Al_{2.5}O_4$
see Al_5LiO_8 ... $LiAl_5O_8$

$Al_{2.7}Ir$ ..... $IrAl_{2.7}$ ..... Ir: SVol.1-52
Al_3BO_6 ..... $Al_3BO_6 = B_2O_3 \cdot 3\,Al_2O_3$ ..... B: B-Verb.7-96
$Al_3B_3C_{14}H_{47}N_7$ ..... $Al_3B_3[N(CH_3)_2]_7H_5$ ..... B: B-Verb.3-166/7
B: B-Verb.5-64, 74, 277
$Al_3B_4DyO_{12}$ ..... $DyAl_3(BO_3)_4 = 4\,B_2O_3 \cdot 3\,Al_2O_3 \cdot Dy_2O_3$ B: B-Verb.7-115/6
$Al_3B_4ErO_{12}$ ..... $ErAl_3(BO_3)_4 = 4\,B_2O_3 \cdot 3\,Al_2O_3 \cdot Er_2O_3$. B: B-Verb.7-115/6
$Al_3B_4EuO_{12}$ ..... $EuAl_3(BO_3)_4 = 4\,B_2O_3 \cdot 3\,Al_2O_3 \cdot Eu_2O_3$ B: B-Verb.7-115/6
$Al_3B_4GdO_{12}$ ..... $GdAl_3(BO_3)_4 = 4\,B_2O_3 \cdot 3\,Al_2O_3 \cdot Gd_2O_3$ B: B-Verb.7-115/6
$Al_3B_4HoO_{12}$ ..... $HoAl_3(BO_3)_4 = 4\,B_2O_3 \cdot 3\,Al_2O_3 \cdot Ho_2O_3$ B: B-Verb.7-115/6
$Al_3B_4NdO_{12}$ ..... $NdAl_3(BO_3)_4 = 4\,B_2O_3 \cdot 3\,Al_2O_3 \cdot Nd_2O_3$ B: B-Verb.7-115/6
$Al_3B_4O_{12}Sm$ ..... $SmAl_3(BO_3)_4 = 4\,B_2O_3 \cdot 3\,Al_2O_3 \cdot Sm_2O_3$
B: B-Verb.7-115/6
$Al_3B_4O_{12}Tb$ ..... $TbAl_3(BO_3)_4 = 4\,B_2O_3 \cdot 3\,Al_2O_3 \cdot Tb_2O_3$ B: B-Verb.7-115/6
$Al_3B_4O_{12}Tm$ ..... $TmAl_3(BO_3)_4 = 4\,B_2O_3 \cdot 3\,Al_2O_3 \cdot Tm_2O_3$ B: B-Verb.7-115
$Al_3B_4O_{12}Y$ ..... $YAl_3(BO_3)_4 = 4\,B_2O_3 \cdot 3\,Al_2O_3 \cdot Y_2O_3$ .. B: B-Verb.7-115/6
$Al_3B_4O_{12}Yb$ ..... $YbAl_3(BO_3)_4 = 4\,B_2O_3 \cdot 3\,Al_2O_3 \cdot Yb_2O_3$ B: B-Verb.7-115/6
$Al_3B_{11}C_{32}H_{116}N_2$ ..... $2\,[(C_4H_9)_4N][BH_4] \cdot 3\,Al[BH_4]_3$ ..... B: B-Verb.8-31
$Al_3C_{11}Mn$ ..... $MnAl_3C_{11} = MnCl_2 \cdot 3\,AlCl_3$ ..... Mn: MVol.C5-223/4
$Al_3C_{20}H_{50}Ni$ ..... $Ni(C_2H_5) \cdot 3\,Al(C_2H_5)_3$ ..... Ni: Org.Verb.1-108
$Al_3C_{24}H_{53}O_9Sn$ ..... $C_6H_5Sn[HAl(OC_2H_5)_3]_3$ ..... Sn: Org.Verb.6-275, 277
$Al_3H_9Mn_2O_{11}$ ..... $[Mn_2Al_2(OH)_6O_2]Al(OH)_3$ ..... Mn: MVol.C3-10
Al_3Ir ..... $IrAl_3$ ..... Ir: SVol.1-52
Al_3Np ..... $NpAl_3$ ..... Np: TrU.B2-13
$Al_3Pu_{0.95}$ ..... $Pu_{0.95}Al_3$ ..... Np: TrU.B3-155
Al_3Pu ..... $PuAl_3$ ..... Np: TrU.B2-41
Np: TrU.B3-154/61
Al_3U ..... UAl_3 ..... Np: TrU.B2-51, 56, 60, 62
$Al_4BC_{12}H_{36}N_3$ ..... $Al_4B[N(CH_3)_2]_3(CH_3)_6$ ..... B: B-Verb.4-272
$Al_4B_2Ca_4Fe_2O_{31}Si_8$ .. $B_2O_3 \cdot 8\,SiO_2 \cdot 2\,Al_2O_3 \cdot 2\,FeO \cdot 4\,CaO$
solid solutions
$B_2O_3 \cdot 8\,SiO_2 \cdot 2\,Al_2O_3 \cdot 2\,(Fe,Mn)O$
$\cdot\,4\,CaO \cdot H_2O$ ..... B: B-Verb.7-215
$Al_4B_2Ca_4Mn_2O_{31}Si_8$. $B_2O_3 \cdot 8\,SiO_2 \cdot 2\,Al_2O_3 \cdot 2\,MnO \cdot 4\,CaO$
solid solutions
$B_2O_3 \cdot 8\,SiO_2 \cdot 2\,Al_2O_3 \cdot 2\,(Fe,Mn)O$
$\cdot\,4\,CaO \cdot H_2O$ ..... B: B-Verb.7-215
$Al_4B_2O_9$ ..... $B_2O_3 \cdot 2\,Al_2O_3$ ..... B: B-Verb.7-96/7
$Al_4B_6Ca_3F_4H_6Na_2O_{20}$ $Na_2Ca_3Al_4B_6H_6O_{20}F_4$ ..... B: B-Verb.7-206
$Al_4B_6Li_4O_{17}$ ..... $3\,B_2O_3 \cdot 2\,Al_2O_3 \cdot 2\,Li_2O$ ..... B: B-Verb.7-109
$Al_4Br_2C_{24}Cl_{12}H_{36}Ni$ .. $[((CH_3)_6C_6)_2Ni][Al_4Cl_{12}Br_2]$ ..... Ni: Org.Verb.2-248
$Al_4C_{26}Cl_2Fe_2H_{34}$ ..... $[C_5H_5FeC_5H_3Al_2(CH_3)_3Cl]_2$ ..... Fe: Org.Verb.A6-217/8
$Al_4Cl_{15}U$ ..... $U(Al_2Cl_7)_2Cl$ ..... U: SVol.C9-50
$Al_4H_4Th_8$ ..... $Th_8Al_4H_4 = Th_2AlH$ ..... Th: SVol.C1-26/33
$Al_4H_6Mo_3O_{18}$ ..... $Al_2(MoO_4)_3 \cdot 2\,Al(OH)_3$ ..... Mo: SVol.B2-79
$Al_4H_8Th_8$ ..... $Th_8Al_4H_8 = Th_2AlH_2$ ..... Th: SVol.C1-26/33
$Al_4H_{12}Th_8$ ..... $Th_8Al_4H_{12} = Th_2AlH_3$ ..... Th: SVol.C1-28/33
$Al_4H_{16}Sn$ ..... $Sn[AlH_4]_4$ ..... Sn: MVol.C4-100
$Al_4H_{16}Th_8$ ..... $Th_8Al_4H_{16} = Th_2AlH_4$ ..... Th: SVol.C1-26/33
$Al_4Mn_2O_{36}S_9$ ..... $Al_4Mn_2(SO_4)_9 = 2\,Al_2(SO_4)_3 \cdot Mn_2(SO_4)_3$ Mn: MVol.C6-246

Al_4Np . . . . $NpAl_4$ . . . . Np: TrU.B2-13/4
$Al_4Pu_{0.9}$ . . . . $Pu_{0.9}Al_4$ . . . . Np: TrU.B3-155
Al_4Pu . . . . $PuAl_4$ . . . . Np: TrU.B2-41
Np: TrU.B3-154/61, 184
Al_4U . . . . UAl_4 . . . . Np: TrU.B2-56, 60
$Al_{4.3}C_{12}Cl_{12.9}H_{12}Ni$ . . $(C_6H_6)_2Ni(AlCl_3)_{4.3}$ . . . . Ni: Org.Verb.2-248
$Al_5C_{12}Cl_{15}H_{12}Ni$ . . . . $(C_6H_6)_2Ni(AlCl_3)_5$ . . . . Ni: Org.Verb.2-248, 249
$Al_5C_{14}Cl_{15}H_{16}Ni$ . . . . $(CH_3C_6H_5)_2Ni(AlCl_3)_5$ . . . . Ni: Org.Verb.2-248
$Al_5C_{18}Cl_{15}H_{24}Ni$ . . . . $((CH_3)_3C_6H_3)_2Ni(AlCl_3)_5$ . . . . Ni: Org.Verb.2-248
Al_5LiO_8 . . . . $LiAl_5O_8$ solid solutions
$Li_{0.5}Mn_xAl_{2.5-x}O_4$ . . . . Mn: MVol.C3-6/7
$Al_6B_2O_{12}$ . . . . $B_2O_3 \cdot 3\,Al_2O_3 = Al_3BO_6$ . . . . B: B-Verb.7-96
$Al_6B_4O_{15}$ . . . . $2\,B_2O_3 \cdot 3\,Al_2O_3$ . . . . B: B-Verb.7-97
$Al_6B_5F_3O_{15}$ . . . . $Al_6B_5O_{15}F_3$ . . . . B: B-Verb.7-28
$Al_6B_5H_3O_{18}$ . . . . $[Al_6(B_5O_{15}(OH)_3)]_n$ . . . . B: B-Verb.7-195
$Al_6B_8Dy_2O_{24}$ . . . . $4\,B_2O_3 \cdot 3\,Al_2O_3 \cdot Dy_2O_3 = DyAl_3(BO_3)_4$ B: B-Verb.7-115/6
$Al_6B_8Er_2O_{24}$ . . . . $4\,B_2O_3 \cdot 3\,Al_2O_3 \cdot Er_2O_3 = ErAl_3(BO_3)_4$ B: B-Verb.7-115/6
$Al_6B_8Eu_2O_{24}$ . . . . $4\,B_2O_3 \cdot 3\,Al_2O_3 \cdot Eu_2O_3 = EuAl_3(BO_3)_4$ B: B-Verb.7-115/6
$Al_6B_8Gd_2O_{24}$ . . . . $4\,B_2O_3 \cdot 3\,Al_2O_3 \cdot Gd_2O_3 = GdAl_3(BO_3)_4$ B: B-Verb.7-115/6
$Al_6B_8Ho_2O_{24}$ . . . . $4\,B_2O_3 \cdot 3\,Al_2O_3 \cdot Ho_2O_3 = HoAl_3(BO_3)_4$ B: B-Verb.7-115/6
$Al_6B_8Nd_2O_{24}$ . . . . $4\,B_2O_3 \cdot 3\,Al_2O_3 \cdot Nd_2O_3 = NdAl_3(BO_3)_4$ B: B-Verb.7-115/6
$Al_6B_8O_{24}Sm_2$ . . . . $4\,B_2O_3 \cdot 3\,Al_2O_3 \cdot Sm_2O_3 = SmAl_3(BO_3)_4$
B: B-Verb.7-115/6
$Al_6B_8O_{24}Tb_2$ . . . . $4\,B_2O_3 \cdot 3\,Al_2O_3 \cdot Tb_2O_3 = TbAl_3(BO_3)_4$ B: B-Verb.7-115/6
$Al_6B_8O_{24}Tm_2$ . . . . $4\,B_2O_3 \cdot 3\,Al_2O_3 \cdot Tm_2O_3 = TmAl_3(BO_3)_4$ B: B-Verb.7-115
$Al_6B_8O_{24}Y_2$ . . . . $4\,B_2O_3 \cdot 3\,Al_2O_3 \cdot Y_2O_3 = YAl_3(BO_3)_4$ . . B: B-Verb.7-115/6
$Al_6B_8O_{24}Yb_2$ . . . . $4\,B_2O_3 \cdot 3\,Al_2O_3 \cdot Yb_2O_3 = YbAl_3(BO_3)_4$ B: B-Verb.7-115/6
$Al_6C_{20}Cl_{16}Fe_2H_{16}$ . . . . $[C_5H_4FeC_5H_4Al]_2[AlCl_4]_4$ . . . . Fe: Org.Verb.A1-169
$Al_6Ca_4MoO_{16}$ . . . . $3\,CaO \cdot 3\,Al_2O_3 \cdot CaMoO_4$ . . . . Mo: SVol.B2-85
$Al_6Cd_2Mo_2Na_6O_{32}Si_6$. $Cd_2Na_6Al_6Si_6O_{24}(MoO_4)_2$ . . . . Mo: SVol.B2-85
$Al_6Cl_{21}U$ . . . . $U(Al_2Cl_7)_3$ . . . . U: SVol.C9-50
$Al_6MoO_{16}Sr_4$ . . . . $3\,SrO \cdot 3\,Al_2O_3 \cdot SrMoO_4$ . . . . Mo: SVol.B2-85
$Al_6O_{13}Sn_2$ . . . . $2\,SnO_2 \cdot 3\,Al_2O_3$ . . . . Sn: MVol.C4-100
Al_9Ir_2 . . . . Ir_2Al_9 . . . . Ir: SVol.1-52
$Al_{10}B_6O_{24}$ . . . . $3\,B_2O_3 \cdot 5\,Al_2O_3$ . . . . B: B-Verb.7-97
$Al_{12}B_{10}O_{33}$ . . . . $5\,B_2O_3 \cdot 6\,Al_2O_3 \cdot 3\,H_2O$ . . . . B: B-Verb.7-195
$Al_{12}Cd_8Mo_2O_{32}$ . . . . $Cd_8Al_{12}O_{24}(MoO_4)_2$ . . . . Mo: SVol.B2-85
$Al_{18}B_4O_{33}$ . . . . $2\,B_2O_3 \cdot 9\,Al_2O_3$ . . . . B: B-Verb.7-96

$Am_{0.5}BaO_3Pa_{0.5}$ . . . . $Ba(Am_{0.5}Pa_{0.5})O_3$ . . . . Pa: SVol.2-21
$Am_{0.5}O_2Pa_{0.5}$ . . . . $(^{241}Am_{0.5}Pa_{0.5})O_2$ . . . . Pa: SVol.2-31/2
Am . . . . Americium
Chemical reactions . . . . Np: TrU.B1-66
Chemistry in organic solvents
see Extraction
Complex chemistry
in aqueous solution . . . . Np: TrU.D1-129/76
Electrochemical behavior . . . . Np: TrU.D1-5/18, 21/6

Am .... Americium
Extraction
by combinations of
extracting agents .... Np: TrU.D2-360/72
by β-dicarbonyl
and related compounds .... Np: TrU.D2-236/48
by neutral organic
oxygen compounds .... Np: TrU.D2-232
by neutral organophosphorus
compounds .... Np: TrU.D2-177/220
by other neutral solvents .... Np: TrU.D2-235/6
by onium compounds .... Np: TrU.D2-325/44
by organophosphoric acids ..... Np: TrU.D2-249/311
by other acidic extractants
and solvents .... Np: TrU.D2-312/25

Ion exchange reactions .... Np: TrU.D2-373/437
Oxidation states .... Np: TrU.D1-5/7
Physical properties .... Np: TrU.B1-62/6
Precipitation from aqueous solutions Np: TrU.D1-113/6, 120/6
Preparation .... Np: TrU.B1-57/62
Redox reactions .... Np: TrU.D1-34/7, 75/80, 83/100
Separation by coprecipitation ...... Np: TrU.D1-106/12
Solubility in salt melts .... Np: TrU.D2-450, 452

– .... Americium alloys .... Np: TrU.B3-267/71
– .... Americium ions
Chemistry in organic solvents
see Am ... Americium: Extraction

Complex chemistry .... Np: TrU.D1-137/69
Coprecipitation .... Np: TrU.D1-106/12
Disproportionation .... Np: TrU.D1-34/6
Electrochemical behavior .... Np: TrU.D1-5/26
Hydration .... Np: TrU.D1-132/3
Hydrolysis .... Np: TrU.D1-133/7
Ion exchange reactions
Anion exchange reactions ...... Np: TrU.D2-409/17, 431/3
Cation exchange reactions ..... Np: TrU.D2-373/95
Ion exchanger reactions .... Np: TrU.D2-402/8
Reactions with chelating resins . . Np: TrU.D2-436/7
Redox exchange reactions ...... Np: TrU.D2-437

Isotopic exchange reactions .... Np: TrU.D1-36/7
Physical properties of solutions .... Np: TrU.D1-1/4
Preparation from aqueous solutions Np: TrU.D1-113/6, 120/6
Redox reactions .... Np: TrU.D1-75/80, 83/100

– .... Americium systems
Am-Al .... Np: TrU.B3-270
Am-Ba .... Np: TrU.B3-267

Am .... Americium systems

Am–Be .... Np: TrU.B3-270
Am–Bi .... Np: TrU.B3-271
Am–Hg .... Np: TrU.B3-270
Am–Ir .... Np: TrU.B3-269
Am–La .... Np: TrU.B3-267
Am–Li .... Np: TrU.B3-267
Am–Ni .... Np: TrU.B3-268
Am–Np .... Np: TrU.B2-7
Am–Os .... Np: TrU.B3-269
Am–Pd .... Np: TrU.B3-269
Am–Pt .... Np: TrU.B3-269
Am–Pu .... Np: TrU.B2-19, 21
Np: TrU.B3-268
Am–Rh .... Np: TrU.B3-269
Am–Sb .... Np: TrU.B3-270/1
Am–Th .... Np: TrU.B3-267

Am^{3+} .... Am^{3+}

Complex chemistry .... Np: TrU.D1-138/69
Coprecipitation .... Np: TrU.D1-106/12
Hydration .... Np: TrU.D1-132/3
Hydrolysis .... Np: TrU.D1-133/6
Ion exchange reactions .... Np: TrU.D2-373/95, 402/5, 409/17, 436
Isotopic exchange reactions .... Np: TrU.D1-36
Physical properties .... Np: TrU.D1-2/4
Precipitation .... Np: TrU.D1-113/6, 120/6
Redox potentials .... Np: TrU.D1-7/21
Redox reactions .... Np: TrU.D1-75/80, 83/100
Stability in salt melts .... Np: TrU.D2-452

Am^{4+} .... Am^{4+}

Complex chemistry .... Np: TrU.D1-148
Disproportionation .... Np: TrU.D1-34/5
Hydration .... Np: TrU.D1-132/3
Hydrolysis .... Np: TrU.D1-134
Precipitation .... Np: TrU.D1-114, 116
Redox potentials .... Np: TrU.D1-7/21
Redox reactions .... Np: TrU.D1-75/80, 83/100

$AmBe_{13}$ .... $AmBe_{13}$ .... Np: TrU.B3-270
AmBi .... AmBi .... Np: TrU.B3-271
$AmBr^{2+}$ .... $AmBr^{2+}$ .... Np: TrU.D1-144
$AmBr_4C_{18}H_9N_2O_3$ ... $Am(OH)(NC_9H_4Br_2O)_2$ .... Np: TrU.D2-317
$AmCKO_5$ .... $KAmO_2CO_3$ .... Np: TrU.D1-80, 122
$AmCNS^{2+}$ .... $AmSCN^{2+}$ .... Np: TrU.D1-145
$AmCNaO_5$ .... $NaAmO_2CO_3$ .... Np: TrU.D1-80
$AmC_2H_3O_2{}^{2+}$ .... $Am(CH_3COO)^{2+}$ .... Np: TrU.D1-152/3
$AmC_2H_3O_2S^{2+}$ ..... $Am(HSCH_2COO)^{2+}$ .... Np: TrU.D1-153
$AmC_2H_3O_3{}^{2+}$ .... $Am(HOCH_2COO)^{2+}$ .... Np: TrU.D1-153
$AmC_2H_3O_4$ .... $AmO_2(CH_3COO)$ .... Np: TrU.D1-153

$AmC_2H_4NO_2^{2+}$ ..... $Am(NH_2CH_2COO)^{2+}$ ..... Np: TrU.D1-155
$AmC_2N_2S_2^{+}$ ..... $Am(SCN)_2^{+}$ ..... Np: TrU.D1-145
$AmC_2O_4^{+}$ ..... $Am(C_2O_4)^{+}$ ..... Np: TrU.D1-158
$AmC_3HO_{10}^{4-}$ ..... $[Am(OH)(CO_3)_3]^{4-}$ ..... Np: TrU.D1-147
$AmC_3H_5O_3^{2+}$ ..... $Am(CH_3CH(OH)COO)^{2+}$ ..... Np: TrU.D1-154
$AmC_3K_5O_{11}$ ..... $K_5AmO_2(CO_3)_3$ ..... Np: TrU.D1-80
$AmC_3N_3S_3$ ..... $Am(SCN)_3$ ..... Np: TrU.D1-145
$AmC_4H_4O_6^{+}$ ..... $Am(C_4H_4O_6)^{+}$ ..... Np: TrU.D1-159
$AmC_4H_5NO_4^{+}$ ..... $AmNH(CH_2COO)_2^{+}$ ..... Np: TrU.D1-162
$AmC_4H_6O_4^{+}$ ..... $Am(CH_3COO)_2^{+}$ ..... Np: TrU.D1-152
$AmC_4H_6O_4S_2^{+}$ ..... $Am(HSCH_2COO)_2^{+}$ ..... Np: TrU.D1-153
$AmC_4H_6O_6^{+}$ ..... $Am(HOCH_2COO)_2^{+}$ ..... Np: TrU.D1-153
$AmC_4H_6O_6^{-}$ ..... $AmO_2(CH_3COO)_2^{-}$ ..... Np: TrU.D1-153
$AmC_4H_7O_3^{2+}$ ..... $Am((CH_3)_2C(OH)COO)^{2+}$ ..... Np: TrU.D1-154
$AmC_4N_4S_4^{-}$ ..... $Am(SCN)_4^{-}$ ..... Np: TrU.D1-145
$AmC_4O_8^{-}$ ..... $Am(C_2O_4)_2^{-}$ ..... Np: TrU.D1-158
$AmC_5H_7NO_4^{+}$ ..... $AmN(CH_3)(CH_2COO)_2^{+}$ ..... Np: TrU.D1-162
$AmC_5H_9O_4$ ..... $Am(OH)_2(CH_3COCHCOCH_3)$ ..... Np: TrU.D2-242
$AmC_6H_5O_7$ ..... $Am(C_6H_5O_7)$ ..... Np: TrU.D1-159
$AmC_6H_6NO_6$ ..... $Am(N(CH_2COO)_3)$ ..... Np: TrU.D1-163
$AmC_6H_8NO_7S$ ..... $Am(OSO_2(CH_2)_2N(CH_2COO)_2)$ ..... Np: TrU.D1-163
$AmC_6H_9NO_5^{+}$ ..... $AmN(C_2H_4OH)(CH_2COO)_2^{+}$ ..... Np: TrU.D1-162
$AmC_6H_9NO_7S^{+}$ ..... $Am(OSO_2(CH_2)_2N(CH_2COO)_2H)^{+}$ ..... Np: TrU.D1-163
$AmC_6H_9O_6$ ..... $Am(CH_3COO)_3$ ..... Np: TrU.D1-152
$AmC_6H_9O_9$ ..... $Am(HOCH_2COO)_3$ ..... Np: TrU.D1-153
$AmC_6H_{10}O_6^{+}$ ..... $Am(CH_3CH(OH)COO)_2^{+}$ ..... Np: TrU.D1-154
$AmC_6O_{12}^{3-}$ ..... $Am(C_2O_4)_3^{3-}$ ..... Np: TrU.D1-158
$AmC_7H_8NO_6$ ..... $Am(N(CH_2CH_2COO)(CH_2COO)_2)$ ..... Np: TrU.D1-164
$AmC_7H_9NO_6^{+}$ ..... $Am(N(CH_2CH_2COO)(CH_2COO)_2H)^{+}$ ..... Np: TrU.D1-164
$AmC_8F_3H_4O_2S^{2+}$ ..... $Am(SC_4H_3COCHCOCF_3)^{2+}$ ..... Np: TrU.D1-167/8
$AmC_8H_8O_{12}^{-}$ ..... $Am(C_4H_4O_6)_2^{-}$ ..... Np: TrU.D1-159
$AmC_8H_9O_{12}$ ..... $Am(C_4H_4O_6)(C_4H_5O_6)$ ..... Np: TrU.D1-159
$AmC_8H_{10}N_2O_8^{-}$ ..... $Am(NH(CH_2COO)_2)_2^{-}$ ..... Np: TrU.D1-162
$AmC_8H_{11}NO_6^{+}$ ..... $Am(N(CH_2CH_2CH_2COO)(CH_2COO)_2H)^{+}$ ..... Np: TrU.D1-164
$AmC_8H_{14}O_6^{+}$ ..... $Am((CH_3)_2C(OH)COO)_2^{+}$ ..... Np: TrU.D1-154
$AmC_8H_{18}N_2O_6P_2^{-}$ ..... $Am(C_2H_4(NHC_3H_6PO_3)_2)^{-}$ ..... Np: TrU.D1-168
$AmC_8H_{19}N_2O_6P_2$ ..... $Am(C_2H_4(NHC_3H_6PO_3)_2H)$ ..... Np: TrU.D1-168/9
$AmC_8H_{20}N_2O_6P_2^{+}$ ..... $Am(C_2H_4(NHC_3H_6PO_3)_2H_2)^{+}$ ..... Np: TrU.D1-168/9
$AmC_8H_{21}N_2O_6P_2^{2+}$ ..... $Am(C_2H_4(NHC_3H_6PO_3)_2H_3)^{2+}$ ..... Np: TrU.D1-168/9
$AmC_9H_{13}NO_6^{+}$ ..... $Am(N(C_4H_8COO)(CH_2COO)_2H)^{+}$ ..... Np: TrU.D1-164
$AmC_9H_{15}O_9$ ..... $Am(CH_3CH(OH)COO)_3$ ..... Np: TrU.D1-154
$AmC_{10}H_{10}N_2O_4^{+}$ ..... $Am(NC_5H_4CH_2N(CH_2COO)_2)^{+}$ ..... Np: TrU.D1-163
$AmC_{10}H_{12}N_2O_8^{-}$ ..... $AmC_2H_4N_2(CH_2COO)_4^{-}$ ..... Np: TrU.D1-164/5
$AmC_{10}H_{13}N_2O_8$ ..... $AmC_2H_4N_2(CH_2COOH)(CH_2COO)_3$ ..... Np: TrU.D1-165
$AmC_{10}H_{14}N_2O_8^{-}$ ..... $Am(N(CH_3)(CH_2COO)_2)_2^{-}$ ..... Np: TrU.D1-162
$AmC_{10}H_{15}N_2O_7$ ..... $AmC_2H_4N_2(C_2H_4OH)(CH_2COO)_3$ ..... Np: TrU.D1-164
$AmC_{11}H_8NO_6$ ..... $AmN(C_6H_4COO)(CH_2COO)_2$ ..... Np: TrU.D1-163
$AmC_{11}H_{12}N_2O_4^{+}$ ..... $Am(CH_3C_5H_3NCH_2N(CH_2COO)_2)^{+}$ ..... Np: TrU.D1-163
$AmC_{12}H_8N_2O_6$ ..... $AmO_2(NC_5H_4COO)_2$ ..... Np: TrU.D1-156
$AmC_{12}H_8N_2O_8$ ..... $AmO_2(ONC_5H_4COO)_2$ ..... Np: TrU.D1-156

$AmC_{12}H_{10}O_{14}^{3-}$ ..... $Am(C_6H_5O_7)_2^{3-}$ ..... Np: TrU.D1-159
$AmC_{12}H_{12}N_2O_{12}^{3-}$ ... $Am(N(CH_2COO)_3)_2^{3-}$ ..... Np: TrU.D1-163
$AmC_{12}H_{18}N_2O_{10}^{-}$ .... $Am(N(C_2H_4OH)(CH_2COO)_2)_2^{-}$ ..... Np: TrU.D1-162
$AmC_{12}H_{21}O_9$ .... $Am((CH_3)_2C(OH)COO)_3$ ..... Np: TrU.D1-154
$AmC_{14}H_{16}N_2O_{12}^{3-}$ ... $Am(N(CH_2CH_2COO)(CH_2COO)_2)_2^{3-}$ ..... Np: TrU.D1-164
$AmC_{14}H_{18}N_2O_8^{-}$ .... $AmC_6H_{10}N_2(CH_2COO)_4^{-}$ ..... Np: TrU.D1-165/6
$AmC_{14}H_{18}N_3O_{10}^{2-}$ ... $Am(CH_2COO)N(C_2H_4N(CH_2COO)_2)_2^{2-}$ ..... Np: TrU.D1-166
$AmC_{14}H_{19}N_2O_8$ ..... $AmC_6H_{10}N_2(CH_2COOH)(CH_2COO)_3$ ..... Np: TrU.D1-166
$AmC_{14}H_{19}N_3O_{10}^{-}$ .... $AmH(CH_2COO)N(C_2H_4N(CH_2COO)_2)_2^{-}$ ..... Np: TrU.D1-166
$AmC_{14}H_{20}N_2O_{10}^{-}$ .... $AmC_2H_4(OC_2H_4N(CH_2COO)_2)_2^{-}$ ..... Np: TrU.D1-166
$AmC_{16}F_6H_8O_4S_2^{+}$ ... $Am(SC_4H_3COCHCOCF_3)_2^{+}$ ..... Np: TrU.D1-168
$AmC_{18}ClH_{12}N_2O_6$ ... $Am(NC_9H_6O)_2ClO_4$ ..... Np: TrU.D2-316
$AmC_{18}H_{13}N_2O_3$ ..... $Am(NC_9H_6O)_2(OH)$ ..... Np: TrU.D1-167
$AmC_{18}H_{13}N_3O_8$ ..... $HAmO_2(NC_5H_4COO)_3$ ..... Np: TrU.D1-156
$AmC_{18}H_{21}O_{21}$ ...... $Am(C_6H_7O_7)_3$ ..... Np: TrU.D1-157
$AmC_{18}H_{24}N_4O_{12}^{3-}$ ... $AmC_2H_4(N(CH_2COO)C_2H_4N(CH_2COO)_2)_2^{3-}$ .. Np: TrU.D1-166
$AmC_{18}H_{25}N_4O_{12}^{2-}$ ... $AmH[C_2H_4(N(CH_2COO)C_2H_4N(CH_2COO)_2)_2]^{2-}$ Np: TrU.D1-166
$AmC_{18}H_{26}N_4O_{12}^{-}$ .... $AmH_2[C_2H_4(N(CH_2COO)C_2H_4N(CH_2COO)_2)_2]^{-}$ Np: TrU.D1-166
$AmC_{18}H_{35}N_2O_6$ ..... $Am(CH_3COO)(C_7H_{15}CONHO)_2$ ..... Np: TrU.D2-323
$AmC_{20}H_{20}N_4O_8^{-}$ .... $Am(NC_5H_4CH_2N(CH_2COO)_2)_2^{-}$ ..... Np: TrU.D1-163
$AmC_{20}H_{30}N_4O_{14}^{3-}$ ... $Am(C_2H_4N_2(C_2H_4OH)(CH_2COO)_3)_2^{3-}$ ..... Np: TrU.D1-164
$AmC_{22}H_{16}N_2O_{12}^{3-}$ ... $Am(N(C_6H_4COO)(CH_2COO)_2)_2^{3-}$ ..... Np: TrU.D1-163
$AmC_{24}F_9H_{12}O_6S_3$ ... $Am(SC_4H_3COCHCOCF_3)_3$ ..... Np: TrU.D2-364
$AmC_{27}Cl_6H_{12}N_3O_3$ ... $Am(NC_9H_4Cl_2O)_3$ ..... Np: TrU.D1-167/8
Np: TrU.D2-316/7
$AmC_{27}H_{18}N_3O_3$ ..... $Am(NC_9H_6O)_3$ ..... Np: TrU.D1-167
Np: TrU.D2-316
$AmC_{30}F_9H_{18}O_6$ ..... $Am(C_6H_5COCHCOCF_3)_3$ ..... Np: TrU.D2-242
$AmC_{30}F_9H_{24}O_7S_3$ ... $Am(SC_4H_3COCHCOCF_3)_3 \cdot C_4H_9COCH_3$ ... Np: TrU.D2-363
$AmC_{30}F_9H_{24}O_{10}$ ..... $Am(OC_4H_3COCHCOCF_3)_3 \cdot C_4H_9COCH_3$ ... Np: TrU.D2-363
$AmC_{32}F_9H_{30}O_7S_4$ ... $Am(SC_4H_3COCHCOCF_3)_3 \cdot (C_4H_9)_2SO$ .... Np: TrU.D2-363
$AmC_{32}F_{12}H_{17}O_8S_4$ ... $Am(SC_4H_3COCHCOCF_3)_3 \cdot SC_4H_3COCH_2COCF_3$ Np: TrU.D2-367
$AmC_{34}H_{76}N_7O_{15}$ .... $(CH_3(C_8H_{17})_2NH)_2Am(NO_3)_5$ ..... Np: TrU.D2-337
$AmC_{36}F_9H_{30}O_7$ ..... $Am(C_6H_5COCHCOCF_3)_3 \cdot C_4H_9COCH_3$ .... Np: TrU.D2-363
$AmC_{36}F_9H_{36}O_8S_3$ ... $Am(SC_4H_3COCHCOCF_3)_3 \cdot 2\ C_4H_9COCH_3$. Np: TrU.D2-363/4
$AmC_{36}F_9H_{36}O_{11}$ ..... $Am(OC_4H_3COCHCOCF_3)_3 \cdot 2\ C_4H_9COCH_3$. Np: TrU.D2-363
$AmC_{36}F_9H_{39}O_{10}PS_3$ .. $Am(SC_4H_3COCHCOCF_3)_3 \cdot (C_4H_9O)_3PO$ ... Np: TrU.D2-363
$AmC_{36}H_{45}O_7$ .... $Am((CH_3)_2CHC_7H_4O_2)_3 \cdot C_4H_9COCH_3$ ..... Np: TrU.D2-363
$AmC_{40}F_6H_{62}NO_9P_2S_2$. $Am(SC_4H_3COCHCOCF_3)_2(NO_3) \cdot 2\ (C_4H_9)_3PO$ Np: TrU.D2-364
$AmC_{40}F_9H_{48}O_8S_3$ ... $Am(SC_4H_3COCHCOCF_3)_3 \cdot 2\ C_4H_9CH(C_2H_5)CH_2OH$ Np: TrU.D2-363/4
$AmC_{40}F_9H_{48}O_8S_5$ ... $Am(SC_4H_3COCHCOCF_3)_3 \cdot 2\ (C_4H_9)_2SO$ ... Np: TrU.D2-363
$AmC_{42}F_9H_{24}O_6$ ..... $Am(C_{10}H_7COCHCOCF_3)_3$ ..... Np: TrU.D2-244
$AmC_{42}F_9H_{42}O_8$ ..... $Am(C_6H_5COCHCOCF_3)_3 \cdot 2\ C_4H_9COCH_3$ .. Np: TrU.D2-363
$AmC_{42}H_{57}O_8$ .... $Am((CH_3)_2CHC_7H_4O_2)_3 \cdot 2\ C_4H_9COCH_3$ ... Np: TrU.D2-363
$AmC_{48}F_9H_{66}O_8P_2S_3$ .. $Am(SC_4H_3COCHCOCF_3)_3 \cdot 2\ (C_4H_9)_3PO$ ... Np: TrU.D2-363/4

Formula	Compound	Reference
$AmC_{48}F_9H_{66}O_{12}P_2S_3$	$Am(SC_4H_3COCHCOCF_3)_3 \cdot 2\ (C_4H_9O)_2(C_4H_9)PO$	Np: TrU.D2-363/4
$AmC_{48}F_9H_{66}O_{14}P_2S_3$	$Am(SC_4H_3COCHCOCF_3)_3 \cdot 2\ (C_4H_9O)_3PO$	Np: TrU.D2-363/4
$AmC_{56}H_{116}N_7O_{14}$	$Am(C_7H_{15}CONHO)_3 \cdot 2\ (C_7H_{15}CONHOH)_2$	Np: TrU.D2-323
$AmC_{58}H_{77}N_5O_8P$	$Am(N(C_6H_5)NC(CH_3)C(COC_6H_5)CO)_2(NO_3) \cdot (C_8H_{17})_3PO$	Np: TrU.D2-365, 368
$AmC_{60}F_9H_{42}O_8P_2S_3$	$Am(SC_4H_3COCHCOCF_3)_3 \cdot 2\ (C_6H_5)_3PO$	Np: TrU.D2-363
$AmC_{60}F_9H_{42}O_{14}P_2S_3$	$Am(SC_4H_3COCHCOCF_3)_3 \cdot 2\ (C_6H_5O)_3PO$	Np: TrU.D2-363/4
$AmC_{72}F_9H_{114}O_8P_2S_3$	$Am(SC_4H_3COCHCOCF_3)_3 \cdot 2\ (C_8H_{17})_3PO$	Np: TrU.D2-363/4
$AmC_{75}H_{93}N_6O_{14}P_2$	$Am(N(C_6H_5)NC(CH_3)C(COC_6H_5)CO)_3 \cdot 2\ (C_4H_9O)_3PO$	Np: TrU.D2-365, 368
$AmC_{99}H_{141}N_6O_8P_2$	$Am(N(C_6H_5)NC(CH_3)C(COC_6H_5)CO)_3 \cdot 2\ (C_8H_{17})_3PO$	Np: TrU.D2-365, 368
$AmCl^{2+}$	$AmCl^{2+}$	Np: TrU.D1-3, 138/9, 143
$AmCl_2^+$	$AmCl_2^+$	Np: TrU.D1-3, 138/9, 143
$AmCl_3$	$AmCl_3$	Np: TrU.D1-3, 138
$AmCs_4F_8$	Cs_4AmF_8	Np: TrU.D1-116
AmF^{2+}	AmF^{2+}	Np: TrU.D1-142
AmF_3	AmF_3	Np: TrU.D1-141
		Np: TrU.D2-440
AmF_8K_4	K_4AmF_8	Np: TrU.D1-116
$AmGeO_4$	$AmGeO_4$	Pa: SVol.2-25
$AmHO^{2+}$	$Am(OH)^{2+}$	Np: TrU.D1-136
$AmHO_4S^{2+}$	$Am(HSO_4)^{2+}$	Np: TrU.D1-147, 149
$AmHO_9P_3^+$	$AmHP_3O_9^+$	Np: TrU.D1-147
$AmH_2O_4P^{2+}$	$AmH_2PO_4^{2+}$	Np: TrU.D1-147
$AmIr_2$	$AmIr_2$	Np: TrU.B3-269
$AmKMo_2O_8$	$KAm(MoO_4)_2$	Mo: SVol.B1-218
$AmK_5Mo_4O_{16}$	$K_5Am(MoO_4)_4$	Mo: SVol.B1-218
$AmLiMo_2O_8$	$LiAm(MoO_4)_2$	Mo: SVol.B1-199
$AmMo_2NaO_8$	$NaAm(MoO_4)_2$	Mo: SVol.B1-210
$AmMo_4Na_5O_{16}$	$Na_5Am(MoO_4)_4$	Mo: SVol.B1-210
$AmNO_3^{2+}$	$AmNO_3^{2+}$	Np: TrU.D1-139, 145
$AmN_2O_6^+$	$Am(NO_3)_2^+$	Np: TrU.D1-145
AmN_3O_9	$Am(NO_3)_3$	Np: TrU.D1-145
$AmNi_2$	$AmNi_2$	Np: TrU.B3-268
AmO_2^+	AmO_2^+	
	Complex chemistry	Np: TrU.D1-141, 148, 153
	Coprecipitation	Np: TrU.D1-106/12
	Disproportionation	Np: TrU.D1-35/6
	Ion exchange reactions	Np: TrU.D2-407
	Isotopic exchange reactions	Np: TrU.D1-36/7
	Precipitation	Np: TrU.D1-114, 116, 122, 126
	Redox potentials	Np: TrU.D1-7/21
	Redox reactions	Np: TrU.D1-75/80, 83/100
AmO_2^{2+}	AmO_2^{2+}	
	Complex chemistry	Np: TrU.D1-141
	Coprecipitation	Np: TrU.D1-106/12
	Disproportionation	Np: TrU.D1-36

AmO_2^{2+} .... AmO_2^{2+}
Ion exchange reactions .... Np: TrU.D2-431/3
Isotopic exchange reactions .... Np: TrU.D1-37
Precipitation .... Np: TrU.D1-114, 122/3
Redox potentials .... Np: TrU.D1-7/21
Redox reactions .... Np: TrU.D1-75/80, 83/100

AmO_4S^+ .... $Am(SO_4)^+$ .... Np: TrU.D1-147, 149
$AmO_8S_2^-$ .... $Am(SO_4)_2^-$ .... Np: TrU.D1-149
$AmOs_2$ .... $AmOs_2$ .... Np: TrU.B3-269
$AmPd_3$ .... $AmPd_3$ .... Np: TrU.B3-269
$AmPt_2$ .... $AmPt_2$ .... Np: TrU.B3-269
$AmPt_5$ .... $AmPt_5$ .... Np: TrU.B3-269
$AmRh_2$ .... $AmRh_2$ .... Np: TrU.B3-269
$AmRh_3$ .... $AmRh_3$ .... Np: TrU.B3-269
AmSb .... AmSb .... Np: TrU.B3-270/1
$Am_2C_6O_{12}$ .... $Am_2(C_2O_4)_3 \cdot 10H_2O$ .... Np: TrU.D1-120

Ar .... Argon
Sorption on tungsten .... W: SVol.B1-1/14

– .... Argon systems
Ar-W .... W: SVol.B1-1/42

ArB^+ .... ArB^+ .... B: B-Verb.10-270
$ArBF_3$ .... $Ar \cdot BF_3$ .... B: B-Verb.10-270
ArB_2F_6 .... $Ar \cdot 2\ BF_3$ .... B: B-Verb.10-270
ArB_3F_9 .... $Ar \cdot 3\ BF_3$ .... B: B-Verb.10-270
ArB_6F_{18} .... $Ar \cdot 6\ BF_3$ .... B: B-Verb.10-270
ArB_8F_{24} .... $Ar \cdot 8\ BF_3$ .... B: B-Verb.10-270
$ArB_{16}F_{48}$ .... $Ar \cdot 16\ BF_3$ .... B: B-Verb.10-270
ArCOS .... ArOCS .... C: MVol.D5-144/5

As .... Arsenic systems
As-Al-Ga-Sn .... Sn: MVol.C4-112
As-Cd-Sn .... Sn: MVol.C4-70
As-Cl-In-Sn .... Sn: MVol.C4-132
As-Ga-Ge-Sn .... Sn: MVol.C4-247
As-Ga-Sn .... Sn: MVol.C4-109/12
As-Ge-Se-Sn .... Sn: MVol.C4-247
As-Ir .... Ir: SVol.2-247
As-Sb-Sn .... Sn: MVol.C2-295
As-Se-Sn .... Sn: MVol.C2-281/4
As-Sn .... Sn: MVol.C2-271/3
As-Sn-Te .... Sn: MVol.C2-284
As-Sn-Zn .... Sn: MVol.C4-25/6
As-SnTe .... Sn: MVol.C2-284

$AsAuC_{30}FeH_{30}NO_3P$. $(CO)_2Fe(AuP(C_6H_5)_3)(NO)As(C_2H_5)_2C_6H_5$ ... Fe: Org.Verb.B1-133/5
AsB .... BAs .... B: B-Verb.3-144
$AsBBrC_8H_{20}N$ .... $(C_2H_5)_2NBBrAs(C_2H_5)_2$ .... B: B-Verb.3-145
$AsBBrC_9F_6H_{17}N_2$ ... $[((CH_3)_3N)(CH_3C_5H_4N)BHBr][AsF_6]$ .... B: B-Verb.10-107
$AsBBrC_{27}H_{29}N_2O_6$ ... $[((CH_3)_3N)(CH_3C_5H_4N)BHBr][As(C_6H_4O_2)_3]$. B: B-Verb.10-107, 111

Formula	Compound		Reference
$AsBBrH_3$	$(H_2AsBHBr)_n$	B:	B-Verb.19-199
$AsBBrH_7Si$	$H_3SiAsH_2 \cdot H_2BBr$	B:	B-Verb.9-53
$AsBBr_2C_{10}Cl_2F_6H_8N_2$	$[(ClC_5H_4N)_2BBr_2][AsF_6]$	B:	B-Verb.10-120
$AsBBr_2C_{24}H_{18}$	$(C_6H_5)_2AsB(C_6H_4Br)_2$	B:	B-Verb.3-145
$AsBBr_3C_3H_9$	$(CH_3)_3As \cdot BBr_3$	B:	B-Verb.3-145
		B:	B-Verb.19-305/7
$AsBC_2Cl_2F_6HN$	$(CF_3)_2As(BCl_2)NH$	F:	PerFHalOrg.3-200/3
$AsBC_2F_2H_9N$	$(CH_3)_2NAsF_2BH_3$	B:	B-Verb.10-259
$AsBC_2F_5H_6N$	$F_2AsN(CH_3)_2 \cdot BF_3$	B:	B-Verb.15-105
$AsBC_3Cl_3H_9$	$(CH_3)_3As \cdot BCl_3$	B:	B-Verb.3-145
		B:	B-Verb.19-305/7
$AsBC_3Cl_4H_{10}$	$[(CH_3)_3AsH]BCl_4$	B:	B-Verb.3-145
$AsBC_3F_3H_9$	$(CH_3)_3As \cdot BF_3$	B:	B-Verb.19-305/7
$AsBC_3F_3H_9O$	$(CH_3)_3AsO \cdot BF_3$	B:	B-Verb.19-106
$AsBC_3H_9I_3$	$(CH_3)_3As \cdot BI_3$	B:	B-Verb.3-145
$AsBC_3H_{12}$	$(CH_3)_3As \cdot BH_3$	B:	B-Verb.3-121, 145
		B:	B-Verb.18-59/62, 65/6
$AsBC_4H_{12}S$	$(CH_3)_2AsSB(CH_3)_2$	B:	B-Verb.3-36
$AsBC_6F_3H_{15}O$	$(C_2H_5)_3AsO \cdot BF_3$	B:	B-Verb.19-106
$AsBC_6F_6H_{20}NP$	$[((CH_3)_3N)((CH_3)_3As)BH_2][PF_6]$	B:	B-Verb.10-64, 72/3
$AsBC_6H_{16}N_2$	$(CH_2NCH_3)_2BAs(CH_3)_2$	B:	B-Verb.5-265
$AsBC_6H_{18}$	$(C_2H_5)_3As \cdot BH_3$	B:	B-Verb.3-145
$AsBC_6H_{20}IN$	$[((CH_3)_3N)((CH_3)_3As)BH_2]I$	B:	B-Verb.10-64, 72
$AsBC_6H_{20}N^+$	$[((CH_3)_3N)((CH_3)_3As)BH_2]^+$	B:	B-Verb.10-64, 72/3
$AsBC_8F_4H_{20}S$	$[(C_2H_5)_3AsSC_2H_5]BF_4$	B:	B-Verb.3-146
$AsBC_8F_6H_{16}N_2$	$[(C_5H_5N)(N(CH_3)_3)H_2B]AsF_6$	B:	B-Verb.3-147
		B:	B-Verb.10-58, 65, 67
$AsBC_8H_{22}N_2$	$[(CH_3)_2N]_2BAs(C_2H_5)_2$	B:	B-Verb.3-145
		B:	B-Verb.5-213/4
$AsBC_9F_3H_{21}O$	$(n\text{-}C_3H_7)_3AsO \cdot BF_3$	B:	B-Verb.19-106
$AsBC_9F_6H_{18}N_2$	$[((CH_3)_3N)(CH_3C_5H_4N)BH_2][AsF_6]$	B:	B-Verb.3-147
		B:	B-Verb.10-59, 65
$AsBC_9F_6H_{26}P_2$	$[((C_2H_5)_3P)((CH_3)_3As)BH_2][PF_6]$	B:	B-Verb.10-97/8, 100, 102/3
$AsBC_9H_{26}IP$	$[((C_2H_5)_3P)((CH_3)_3As)BH_2]I$	B:	B-Verb.3-146
		B:	B-Verb.10-97, 100
$AsBC_9H_{26}P^+$	$[((C_2H_5)_3P)((CH_3)_3As)BH_2]^+$	B:	B-Verb.10-97/8, 100, 102/3
$AsBC_{10}F_6H_{12}N_2$	$[(C_5H_5N)_2BH_2][AsF_6]$	B:	B-Verb.3-147
		B:	B-Verb.10-77, 84
$AsBC_{10}F_6H_{30}NPSi$	$[(CH_3)_3NBH_2(CH(Si(CH_3)_3)As(CH_3)_3)][PF_6]$	B:	B-Verb.10-57, 67, 70
$AsBC_{10}H_{30}NSi^+$	$[(CH_3)_3NBH_2(CH(Si(CH_3)_3)As(CH_3)_3)]^+$	B:	B-Verb.10-57, 67, 70
$AsBC_{11}F_4H_{26}S$	$[(C_3H_7)_3AsSC_2H_5]BF_4$	B:	B-Verb.3-146
$AsBC_{12}F_3H_{27}O$	$(n\text{-}C_4H_9)_3AsO \cdot BF_3$	B:	B-Verb.19-106
$AsBC_{12}F_6H_{16}N_2$	$[(CH_3C_5H_4N)_2BH_2][AsF_6]$	B:	B-Verb.10-79, 83
$AsBC_{16}H_{20}N_2$	$(CH_2NCH_3)_2BAs(C_6H_5)_2$	B:	B-Verb.5-265
$AsBC_{18}H_{18}$	$(C_6H_5)_3As \cdot BH_3$	B:	B-Verb.3-145
$AsBC_{20}F_4H_{20}S$	$[(C_6H_5)_3AsSC_2H_5]BF_4$	B:	B-Verb.3-146
$AsBC_{21}Cl_2CoH_{15}O_3$	$Cl_2BCo[(CO)_3As(C_6H_5)_3]$	B:	B-Verb.3-161
$AsBC_{22}F_4H_{24}S$	$[(C_6H_5)_3AsSC_4H_9]BF_4$	B:	B-Verb.3-146

$AsBC_{23}H_{30}$	$[(CH_3)_4As][B(CH_3)(C_6H_5)_3]$	B:	B-Verb.8-208
$AsBC_{24}Cl_3F_3GeH_{20}$	$[(C_6H_5)_4As][GeCl_3 \cdot BF_3]$	B:	B-Verb.3-162
$AsBC_{24}Cl_3F_3H_{20}Sn$	$[(C_6H_5)_4As][SnCl_3 \cdot BF_3]$	B:	B-Verb.3-162
$AsBC_{24}Cl_6GeH_{20}$	$[(C_6H_5)_4As][GeCl_3 \cdot BCl_3]$	B:	B-Verb.3-162
$AsBC_{24}Cl_6H_{20}Sn$	$[(C_6H_5)_4As][SnCl_3 \cdot BCl_3]$	B:	B-Verb.3-162
$AsBC_{24}F_4H_{20}$	$[(C_6H_5)_4As]BF_4$	B:	B-Verb.3-146
$AsBC_{24}H_{20}$	$(C_6H_5)_2AsB(C_6H_5)_2$	B:	B-Verb.3-145
$AsBC_{24}H_{24}$	$[(C_6H_5)_4As]BH_4$	B:	B-Verb.3-146
$AsBC_{26}H_{24}$	$(C_6H_5)_2AsB(C_6H_4CH_3)_2$	B:	B-Verb.3-145
$AsBC_{27}ClH_{29}N_2O_6$	$[((CH_3)_3N)(CH_3C_5H_4N)BHCl][As(C_6H_4O_2)_3]$	B:	B-Verb.3-147
		B:	B-Verb.10-107, 111
$AsBC_{27}CuH_{25}N_6$	$[HB(C_3H_3N_2)_3]CuAs(C_6H_5)_3$	B:	B-Verb.5-26, 28
$AsBC_{27}H_{30}$	$[(CH_3)(C_6H_5)_3As][B(CHCH_2)_4]$	B:	B-Verb.8-165
$AsBC_{33}CoH_{25}O_3$	$(C_6H_5)_2BCo(CO)_3As(C_6H_5)_3$	B:	B-Verb.3-179
$AsBC_{48}H_{32}$	$[(C_6H_4C_6H_4)_2As][B(C_6H_4C_6H_4)_2]$	B:	B-Verb.8-168/9
$AsBC_{48}H_{36}$	$[(C_6H_5)As(C_6H_4)_2(C_6H_4)_2B(C_6H_5)_3]$	B:	B-Verb.8-214
$AsBC_{48}H_{40}$	$[(C_6H_5)_4As][B(C_6H_5)_4]$	B:	B-Verb.8-174, 179, 188, 201
$AsBC_{49}Fe_2H_{41}IO_5S_2$	$[(CO)_5As(C_6H_5)_3Fe_2I(SCH_3)_2]B(C_6H_5)_4$	Fe:	Org.Verb.C1-12/4
$AsBC_{96}H_{77}IrP_4$	$[Ir(((C_6H_5)_2PC_6H_4)_3As)(P(C_6H_5)_3)]B(C_6H_5)_4$	Ir:	SVol.2-260
$AsBCa_2H_4O_8$	$Ca_2[B(OH)_4AsO_4]$		
	$= B_2O_3 \cdot As_2O_5 \cdot 4\ CaO \cdot 4\ H_2O$	B:	B-Verb.7-210
$AsBCl_3H_3$	$AsH_3 \cdot BCl_3$	B:	B-Verb.3-145
$AsBCl_3N$	$AsCl_3BN$	B:	B-Verb.3-146
$AsBH_3NO_4$	$BAsO_4 \cdot NH_3 \cdot 3\ H_2O$	B:	B-Verb.3-147
$AsBH_4$	$[H_2AsBH_2]_n$	B:	B-Verb.3-145
$AsBO_4$	$BAsO_4$	B:	B-Verb.3-146/7
–	$BAsO_4 \cdot 3\ H_2O$	B:	B-Verb.3-147
$AsB_2C_2H_{11}$	$(CH_3)_2AsB_2H_5$	B:	B-Verb.10-1
$AsB_2C_2H_{12}Na$	$Na[(CH_3)_2As(BH_3)_2]$	B:	B-Verb.3-144
$AsB_2C_5H_{20}N$	$(CH_3)_3NBH_2As(CH_3)_2BH_3$	B:	B-Verb.10-29
$AsB_2C_6H_{18}N_3$	$(CH_3)_2AsN(BCH_3NCH_3)_2$	B:	B-Verb.5-202/3
$AsB_3C_{24}H_{28}$	$[(C_6H_5)_4As][B_3H_8]$	B:	B-Verb.3-146
		B:	B-Verb.18-193/4
$AsB_3C_{62}CoH_{57}N_4O_4$	$(C_6H_5)_3As[(ONCCH_3)_2B(C_6H_5)_2]_2CoB(C_6H_5)_2$	B:	B-Verb.3-178/9
$AsB_4F_6H_3$	$AsH_3 \cdot B(BF_2)_3$	B:	B-Verb.3-145
AsB_5	B_5As	B:	B-Verb.3-144
AsB_6	B_6As	B:	B-Verb.3-144
AsB_7	B_7As	B:	B-Verb.3-144
$AsB_9BrC_2H_{11}$	$BrC_2AsB_9H_{11}$	B:	B-Verb.6-29, 132
$AsB_9BrC_6H_{14}N$	$(C_5H_4BrN)CAsB_9H_{10}$	B:	B-Verb.6-56
AsB_9CGeH_{10}	$CGeAsB_9H_{10}$	B:	B-Verb.6-58, 149/50
$AsB_9CH_{11}^-$	$[CAsB_9H_{11}]^-$	B:	B-Verb.6-50/2, 145
$AsB_9C_2H_{14}$	$CAsB_9H_{11}(CH_3)$	B:	B-Verb.6-52/4, 145
$AsB_9C_3H_{14}$	$CH_3C_2AsB_9H_{11}$	B:	B-Verb.6-29, 131
$AsB_9C_5H_{18}$	$(CH_3)_3C_2AsB_9H_9$	B:	B-Verb.6-29, 131
$AsB_9C_5H_{23}N$	$(CH_3)_4N[CAsB_9H_{11}]$	B:	B-Verb.6-51
$AsB_9C_6CoH_{15}$	$(C_5H_5)Co(CAsB_9H_{10})$	B:	B-Verb.6-66
$AsB_9C_6CrH_{11}O_5^-$	$[Cr(CO)_5CAsB_9H_{11}]^-$	B:	B-Verb.6-54/7, 146
$AsB_9C_6H_{11}MoO_5^-$	$[Mo(CO)_5CAsB_9H_{11}]^-$	B:	B-Verb.6-54/7, 146

Formula	Compound	Element	Reference
$AsB_9C_6H_{11}O_5W^-$	$[W(CO)_5CAsB_9H_{11}]^-$	B:	B-Verb.6-54/7, 146
$AsB_9C_6H_{20}$	$(n-C_4H_9)C_2AsB_9H_{11}$	B:	B-Verb.6-29, 131
$AsB_9C_6H_{22}O$	$(As(CH_3)_2)(C_2H_5O)C_2B_9H_{11}$	B:	B-Verb.6-36/7, 140
$AsB_9C_8H_{16}$	$C_6H_5C_2AsB_9H_{11}$	B:	B-Verb.6-29, 37, 131
$AsB_9C_{10}CrH_{23}NO_5$	$[Cr(CO)_5CAsB_9H_{11}]N(CH_3)_4$	B:	B-Verb.6-57
$AsB_9C_{10}H_{20}$	$C_6H_5AsC_2B_9H_9(CH_3)_2$	B:	B-Verb.6-37
$AsB_9C_{10}H_{23}MoNO_5$	$[Mo(CO)_5CAsB_9H_{11}]N(CH_3)_4$	B:	B-Verb.6-57
$AsB_9C_{10}H_{23}NO_5W$	$[W(CO)_5CAsB_9H_{11}]N(CH_3)_4$	B:	B-Verb.6-57
$AsB_{10}BrCH_{10}$	$CAsB_{10}H_{10}Br$	B:	B-Verb.6-48/50
$AsB_{10}CClH_{10}$	$CAsB_{10}H_{10}Cl$	B:	B-Verb.6-48/50
$AsB_{10}CH_{11}$	$CAsB_{10}H_{11}$	B:	B-Verb.6-39/42, 46/7, 143
–	$CAsB_{10}H_{10}D$	B:	B-Verb.6-39/42
$AsB_{10}C_2ClH_{10}O$	$CAsB_{10}H_{10}(COCl)$	B:	B-Verb.6-43/4
$AsB_{10}C_2H_{10}NaO_2$	$CAsB_{10}H_{10}(COONa)$	B:	B-Verb.6-45
$AsB_{10}C_2H_{11}O_2$	$CAsB_{10}H_{10}(COOH)$	B:	B-Verb.6-43/6
$AsB_{10}C_2H_{13}Hg$	$CAsB_{10}H_{10}(HgCH_3)$	B:	B-Verb.6-43/6
$AsB_{10}C_4H_{15}O_2$	$CAsB_{10}H_{10}(COOC_2H_5)$	B:	B-Verb.6-43/4
$AsB_{10}C_8H_{15}O$	$CAsB_{10}H_{10}(COC_6H_5)$	B:	B-Verb.6-43/4
$AsB_{10}C_{12}H_{23}$	$[(C_6H_5)_2As]B_{10}H_{13}$	B:	B-Verb.3-145
		B:	B-Verb.20-143
$AsB_{10}C_{24}H_{32}Na$	$[(C_6H_5)_4As]Na[B_{10}H_{12}]$	B:	B-Verb.20-178/9
$AsB_{10}C_{24}H_{35}$	$[(C_6H_5)_4As][B_{10}H_{15}]$	B:	B-Verb.20-167
$AsB_{10}C_{28}H_{37}$	$[(CH_3)_2C_2B_{10}H_{11}][(C_6H_5)_4As]$	B:	B-Verb.12-302
$AsB_{12}C_3H_{20}^-$	$[(CH_3)_3As \cdot B_{12}H_{11}]^-$	B:	B-Verb.20-237
$AsB_{12}C_3H_{23}O$	$H_3OB_{12}H_{11}As(CH_3)_3$	B:	B-Verb.3-146
$AsB_{12}C_6H_{29}S$	$[(CH_3)_3S][(CH_3)_3As \cdot B_{12}H_{11}]$	B:	B-Verb.20-237
$AsB_{12}C_7H_{32}N$	$[(CH_3)_4N][(CH_3)_3As \cdot B_{12}H_{11}]$	B:	B-Verb.3-146
		B:	B-Verb.20-237
$AsB_{18}C_{28}ClH_{42}Ni$	$As(C_6H_5)_4[(C_2B_9H_{11})_2NiCl]$	Ni:	Org.Verb.1-399
$AsB_{18}C_{29}H_{42}NNiS$	$As(C_6H_5)_4[(C_2B_9H_{11})_2NiNCS]$	Ni:	Org.Verb.1-399
$AsB_{20}C_3H_{20}OP$	$CO(CPB_{10}H_{10})CAsB_{10}H_{10}$	B:	B-Verb.6-43/5
$AsB_{20}C_5H_{23}O$	$CO(CB_{10}H_{10}CCH_3)CAsB_{10}H_{10}$	B:	B-Verb.6-43/5
$AsB_{20}C_7GeH_{29}$	$(CH_3)_2Ge(CB_{10}H_{10}C)_2AsCH_3$	B:	B-Verb.12-137, 152/3, 283
$AsB_{20}C_7H_{27}O$	$CH_3As(CB_{10}H_{10}C)_2CHOCH_3$	B:	B-Verb.12-138, 152, 284
$AsB_{20}C_7H_{29}Si$	$(CH_3)_2Si(CB_{10}H_{10}C)_2AsCH_3$	B:	B-Verb.12-136, 152/3, 283
$AsB_{30}C_{24}H_{46}$	$As(CB_{10}H_{10}CC_6H_5)_3$	B:	B-Verb.12-62
$AsBa_{0.5}O_4Th_{0.5}$	$Ba_{0.5}Th_{0.5}AsO_4$	Th:	SVol.C2-33
$AsBiBr_2C_{36}H_{30}$	$[(C_6H_5)_4As][(C_6H_5)_2BiBr_2]$	Bi:	Org.Verb.-141
$AsBiBr_3C_{30}H_{25}$	$[(C_6H_5)_4As][C_6H_5BiBr_3]$	Bi:	Org.Verb.-143
$AsBiC_{30}Cl_3H_{25}$	$[(C_6H_5)_4As][C_6H_5BiCl_3]$	Bi:	Org.Verb.-142
$AsBiC_{30}H_{25}I_3$	$[(C_6H_5)_4As][C_6H_5BiI_3]$	Bi:	Org.Verb.-143
$AsBiC_{36}Cl_2H_{30}$	$[(C_6H_5)_4As][(C_6H_5)_2BiCl_2]$	Bi:	Org.Verb.-140
$AsBiC_{36}H_{30}I_2$	$[(C_6H_5)_4As][C_6H_5)_2BiI_2]$	Bi:	Org.Verb.-141
$AsBiC_{38}H_{30}N_2S_2$	$[(C_6H_5)_4As][(C_6H_5)_2Bi(SCN)_2]$	Bi:	Org.Verb.-141
$AsBrC_2F_6$	$(CF_3)_2AsBr$	F:	PerFHalOrg.3-203, 205, 207
$AsBrC_4F_{10}$	$(C_2F_5)_2AsBr$	F:	PerFHalOrg.3-204, 206

Formula	Compound	Element	Reference
$AsBrC_{12}FeH_{15}HgNO_3$	$(CO)_2Fe(HgBr)(NO)As(C_2H_5)_2C_6H_5$	Fe:	Org.Verb.B1-133/5
$AsBrC_{21}H_{20}Ni$	$C_3H_5Ni(As(C_6H_5)_3)Br$	Ni:	Org.Verb.2-13, 15
$AsBrC_{23}H_{20}Ni$	$C_5H_5Ni(As(C_6H_5)_3)Br$	Ni:	Org.Verb.2-117, 119
$AsBrC_{30}F_{10}H_{15}OTl$	$(C_6F_5)_2TlBr \cdot (C_6H_5)_3AsO$	F:	PerFHalOrg.4-136/7, 140
$AsBrC_{30}F_{10}H_{15}Tl$	$(C_6F_5)_2TlBr \cdot (C_6H_5)_3As$	F:	PerFHalOrg.4-136/7, 141
$AsBrC_{32}H_{27}NTe$	$[(C_6H_5)_4As][C_6H_5CH_2Te(Br)CN]$	C:	MVol.D6-260
$AsBr_2CF_3$	CF_3AsBr_2	F:	PerFHalOrg.3-203, 205, 207
$AsBr_2C_2Cl_3N_2$	$AsCl_3 \cdot 2\ BrCN$	C:	MVol.D3-236
$AsBr_2C_6F_5$	$C_6F_5AsBr_2$	F:	PerFHalOrg.3-204
$AsBr_2C_{24}ClH_{20}Sn$	$[(C_6H_5)_4As]SnBr_2Cl$	Sn:	MVol.C3-113
$AsBr_2C_{26}H_{20}IrO_2$	$(C_6H_5)_4As[Ir(CO)_2Br_2]$	Ir:	SVol.2-147
$AsBr_2C_{28}ClFeH_{20}Hg_2O_4$	$[As(C_6H_5)_4][(CO)_4Fe(HgBr)_2Cl]$	Fe:	Org.Verb.B2-153, 157
$AsBr_3$	$AsBr_3$ systems		
	$AsBr_3$-$SnBr_4$	Sn:	MVol.C2-281
$AsBr_3C_{24}H_{20}Sn$	$[(C_6H_5)_4As]SnBr_3$	Sn:	MVol.C3-113
$AsBr_3C_{28}FeH_{20}O_4Sn$	$[As(C_6H_5)_4][(CO)_4FeSnBr_3]$	Fe:	Org.Verb.B2-143, 150
$AsBr_3F_6S$	$[SBr_3]\ [AsF_6]$	S:	SVol.2-305
$AsBr_4C_{10}H_{13}O_2Sn$	$SnBr_4 \cdot CH_3OCOC_6H_4As(CH_3)_2$	Sn:	MVol.C6-179
$AsBr_4C_{20}H_{17}O_2Sn$	$SnBr_4 \cdot CH_3OCOC_6H_4As(C_6H_5)_2$	Sn:	MVol.C6-179
$AsBr_4C_{21}FeH_{15}O_3Sn$	$(CO)_3Fe(As(C_6H_5)_3)(Br)SnBr_3$	Fe:	Org.Verb.B1-195/6
$AsBr_4C_{22}H_{21}O_2Sn$	$SnBr_4 \cdot CH_3OCOC_6H_4As(C_6H_4CH_3)_2$	Sn:	MVol.C6-179
$AsBr_4C_{24}H_{20}Ir$	$[(C_6H_5)_4As][IrBr_4] \cdot 3\ C_2H_5OH$	Ir:	SVol.2-147
$AsBr_4C_{25}H_{20}IrO$	$[As(C_6H_5)_4][IrBr_4(CO)]$	Ir:	SVol.2-147
$AsBr_4C_{26}H_{20}IrO_2$	$(C_6H_5)_4As[Ir(CO)_2Br_4]$	Ir:	SVol.2-147
$AsBr_4C_{30}H_{38}IrO_3$	$[(C_6H_5)_4As][IrBr_4] \cdot 3\ C_2H_5OH$	Ir:	SVol.2-147
$AsBr_4C_{60}H_{50}IrSb_2$	$[(C_6H_5)_4As][IrBr_4(Sb(C_6H_5)_3)_2]$	Ir:	SVol.2-263
$AsBr_5C_2N_2$	$AsBr_3 \cdot 2\ BrCN$	C:	MVol.D3-236
$AsBr_6C_{24}H_{20}U$	$As(C_6H_5)_4UBr_6$	U:	SVol.C9-139/41
$AsCCl_2F_3$	CF_3AsCl_2	F:	PerFHalOrg.3-203, 205, 207/10
$AsCF_3^+$	CF_3As^+	F:	PerFHalOrg.3-216
$AsCF_3H_2$	CF_3AsH_2	F:	PerFHalOrg.3-191/5
–	CF_3AsD_2	F:	PerFHalOrg.3-191/2, 194
$AsCF_3H_2O_3$	$CF_3As(O)(OH)_2$	F:	PerFHalOrg.3-197/200
$AsCF_3H_3P$	$CF_3As(H)PH_2$	F:	PerFHalOrg.3-224
$AsCF_3H_4P_2$	$CF_3As(PH_2)_2$	F:	PerFHalOrg.3-222/4
$AsCF_3I_2$	CF_3AsI_2	F:	PerFHalOrg.3-203, 205, 207/10
$AsCF_3O_2$	$[CF_3As(O)O]_n$	F:	PerFHalOrg.3-197, 198
$AsCF_4^+$	CF_3AsF^+	F:	PerFHalOrg.3-216
AsC_2ClF_6	$(CF_3)_2AsCl$	F:	PerFHalOrg.3-203, 205, 207/10
$AsC_2Cl_2F_3$	$CF_2CFAsCl_2$	F:	PerFHalOrg.3-204, 206
$AsC_2Cl_2F_5$	$C_2F_5AsCl_2$	F:	PerFHalOrg.3-204, 205, 207
$AsC_2Cl_2F_6N$	$(CF_3)_2NAsCl_2$	F:	PerFHalOrg.3-201
$AsC_2Cl_3F_6$	$(CF_3)_2AsCl_3$	F:	PerFHalOrg.3-203, 205, 207

Formula	Compound		Reference
AsC_2CsF_{10}	$Cs[(CF_3)_2AsF_4]$	F:	PerFHalOrg.3-213
$AsC_2F_3H_3I$	$CF_3(CH_3)AsI$	F:	PerFHalOrg.3-221
$AsC_2F_4^+$	CF_3AsCF^+	F:	PerFHalOrg.3-216
$AsC_2F_5^+$	$CF_3AsCF_2^+$	F:	PerFHalOrg.3-216
$AsC_2F_5H_2$	$C_2F_5AsH_2$	F:	PerFHalOrg.3-191, 194
$AsC_2F_5H_2O_3$	$C_2F_5As(O)(OH)_2$	F:	PerFHalOrg.3-197
AsC_2F_6H	$(CF_3)_2AsH$	F:	PerFHalOrg.3-191/5
–	$(CF_3)_2AsD$	F:	PerFHalOrg.3-191/4
AsC_2F_6HO	$(CF_3)_2AsOH$	F:	PerFHalOrg.3-199
$AsC_2F_6HO_2$	$(CF_3)_2As(O)OH$	F:	PerFHalOrg.3-197/200
$AsC_2F_6H_2N$	$(CF_3)_2AsNH_2$	F:	PerFHalOrg.3-200/1
$AsC_2F_6H_2P$	$(CF_3)_2AsPH_2$	F:	PerFHalOrg.3-222, 224/5
AsC_2F_6I	$(CF_3)_2AsI$	F:	PerFHalOrg.3-203, 205, 207/10
AsC_2F_7	$(CF_3)_2AsF$	F:	PerFHalOrg.3-203, 205, 207, 211
–	$C_2F_5AsF_2$	F:	PerFHalOrg.3-204, 205
AsC_2F_8P	$(CF_3)_2AsPF_2$	F:	PerFHalOrg.3-222, 224
$AsC_2F_{10}^-$	$[(CF_3)_2AsF_4]^-$	F:	PerFHalOrg.3-213
$AsC_2H_7O_4Sn$	$[HAsO_4Sn(CH_3)_2]_x$	Sn:	Org.Verb.6-28
$AsC_3Cl_2F_9$	$(CF_3)_3AsCl_2$	F:	PerFHalOrg.3-203, 205, 207
AsC_3CsF_{12}	$Cs[(CF_3)_3AsF_3]$	F:	PerFHalOrg.3-213
$AsC_3F_6H_3$	$(CF_3)_2AsCH_3$	F:	PerFHalOrg.3-221
$AsC_3F_6H_3O$	$(CF_3)_2AsOCH_3$	F:	PerFHalOrg.3-207, 210
–	$(CF_3)_2C(OH)AsH_2$	F:	PerFHalOrg.3-191, 194
$AsC_3F_6H_4N$	$(CF_3)_2AsNHCH_3$	F:	PerFHalOrg.3-208
AsC_3F_6N	$(CF_3)_2AsCN$	F:	PerFHalOrg.3-213/4, 219, 221
AsC_3F_6NS	$(CF_3)_2AsNCS$	F:	PerFHalOrg.3-200/3
$AsC_3F_8^+$	$(CF_3)_2AsCF_2^+$	F:	PerFHalOrg.3-216
AsC_3F_9	$(CF_3)_3As$	F:	PerFHalOrg.3-213/7, 220/1
$AsC_3F_9^+$	$(CF_3)_3As^+$	F:	PerFHalOrg.3-216
AsC_3F_9S	$(CF_3)_2AsSCF_3$	F:	PerFHalOrg.3-212
AsC_3F_{11}	$(CF_3)_3AsF_2$	F:	PerFHalOrg.3-203, 205, 207
$AsC_3F_{12}^-$	$[(CF_3)_3AsF_3]^-$	F:	PerFHalOrg.3-213
$AsC_3H_7N_2S_4$	$(CH_3)As[S_2CNH_2]_2$	C:	MVol.D6-140
$AsC_3H_9O_3Sn$	$[(CH_3)_2SnOAs(O)(CH_3)O]_x$	Sn:	Org.Verb.6-29
AsC_4ClF_{10}	$(C_2F_5)_2AsCl$	F:	PerFHalOrg.3-204, 205
$AsC_4ClF_{12}N_2O_2$	$[(CF_3)_2NO]_2AsCl$	F:	PerFHalOrg.7-101, 119
$AsC_4Cl_2F_9$	$(CF_3CFCl)_2AsF$	F:	PerFHalOrg.3-204, 205
$AsC_4Cl_2H_3NiO_3$	$(CO)_3NiAs(CH_3)Cl_2$	Ni:	Org.Verb.1-187, 189
$AsC_4Cl_8H_{12}NTe$	$[(CH_3)_4N][AsCl_2]TeCl_6$	Te:	SVol.B2-94
$AsC_4F_3GeH_{10}$	$CF_3As(H)Ge(CH_3)_3$	F:	PerFHalOrg.3-195
$AsC_4F_3H_{10}Sn$	$CF_3As(H)Sn(CH_3)_3$	F:	PerFHalOrg.3-195, 196
$AsC_4F_6H_6N$	$(CF_3)_2AsNHC_2H_5$	F:	PerFHalOrg.3-208
–	$(CH_3)_2NAs(CF_3)_2$	F:	PerFHalOrg.3-207/8
$AsC_4F_6H_6P$	$(CF_3)_2AsP(CH_3)_2$	F:	PerFHalOrg.3-195, 224/5

$AsC_4F_6H_6P$	$(CF_3)_2PAs(CH_3)_2$	F:	PerFHalOrg.3-182
$AsC_4F_6H_7N_2$	$(CF_3)_2AsN(H)N(CH_3)_2$	F:	PerFHalOrg.3-207, 209
$AsC_4F_9H_3NO$	$(CF_3)_2NOAs(CH_3)CF_3$	F:	PerFHalOrg.7-128
$AsC_4F_{10}I$	$(C_2F_5)_2AsI$	F:	PerFHalOrg.3-204, 206/7
AsC_4F_{11}	$(C_2F_5)_2AsF$	F:	PerFHalOrg.3-204, 205, 207
$AsC_4F_{12}HNP$	$(CF_3)_2AsN(H)P(CF_3)_2$	F:	PerFHalOrg.3-200/3
$AsC_4F_{12}N$	$(CF_3)_2AsN(CF_3)_2$	F:	PerFHalOrg.3-200/3
$AsC_4F_{12}NO$	$(CF_3)_2AsON(CF_3)_2$	F:	PerFHalOrg.3-197, 199
$AsC_4F_{12}OP$	$(CF_3)_2AsOP(CF_3)_2$	F:	PerFHalOrg.3-199
$AsC_4F_{12}P$	$(CF_3)_2AsP(CF_3)_2$	F:	PerFHalOrg.3-222/4
$AsC_4H_{12}I_3Sn$	$[(CH_3)_4As]SnI_3$	Sn:	MVol.C3-119/20
$AsC_5ClF_6FeO_5$	$[(CO)_5FeCl]AsF_6$	Fe:	Org.Verb.B3-253
$AsC_5F_3H_6$	$(CH_3)_2AsCCCF_3$	F:	PerFHalOrg.4-106
$AsC_5F_3H_{12}N_2$	$CF_3As[N(CH_3)_2]_2$	F:	PerFHalOrg.3-209
$AsC_5F_5FeO_5$	$(CO)_5FeAsF_5$	Fe:	Org.Verb.B3-250
$AsC_5F_6GeH_9$	$(CF_3)_2AsGe(CH_3)_3$	F:	PerFHalOrg.3-195, 196
$AsC_5F_6H_9Si$	$(CF_3)_2AsSi(CH_3)_3$	F:	PerFHalOrg.3-195
$AsC_5F_6H_9Sn$	$(CF_3)_2AsSn(CH_3)_3$	F:	PerFHalOrg.3-195, 196
$AsC_5F_9NO_5ReS$	$[Re(CO)_5NSF_3][AsF_6]$	S:	S-N-Verb.1-155
$AsC_5F_{12}H_3NP$	$(CF_3)_2AsN(CH_3)P(CF_3)_2$	F:	PerFHalOrg.3-207/8
$AsC_5F_{15}N_2$	$CF_3As[N(CF_3)_2]_2$	F:	PerFHalOrg.3-200/3
$AsC_5F_{15}N_2O_2$	$CF_3As[ON(CF_3)_2]_2$	F:	PerFHalOrg.3-197, 199
$AsC_5FeH_5O_4$	$(CO)_4FeAs(CH_3)H_2$	Fe:	Org.Verb.C2-50
$AsC_5H_6NO_6U$	$(C_5H_5NH)[UO_2AsO_4] \cdot H_2O$	U:	SVol.C3-177
$AsC_5H_{11}OS_2$	$(CH_3)_2As[S_2COC_2H_5]$	C:	MVol.D4-254
$AsC_5H_{15}Sn$	$(CH_3)_3SnAs(CH_3)_2$	Sn:	Org.Verb.5-81
$AsC_6ClFeH_6O_4$	$(CO)_4FeAs(CH_3)_2Cl$	Fe:	Org.Verb.B2-101, 121/2
$AsC_6Cl_2F_5$	$C_6F_5AsCl_2$	F:	PerFHalOrg.3-204, 206/7
$AsC_6F_4^+$	$C_6F_4As^+$	F:	PerFHalOrg.3-216
$AsC_6F_5^+$	$C_6F_5As^+$	F:	PerFHalOrg.3-216
AsC_6F_5O	$[C_6F_5AsO]_n$	F:	PerFHalOrg.3-197, 198
$AsC_6F_6^+$	$C_6F_5AsF^+$	F:	PerFHalOrg.3-216
$AsC_6F_6H_9O$	$(CF_3)_2AsOC(CH_3)_3$	F:	PerFHalOrg.3-207, 210
$AsC_6F_6H_{12}N_3O$	$(CF_3)_2NOAs[N(CH_3)_2]_2$	F:	PerFHalOrg.7-128
AsC_6F_9	$(CF_2CF)_3As$	F:	PerFHalOrg.3-213/4, 217, 221
$AsC_6F_{10}H_5O$	$(C_2F_5)_2AsOC_2H_5$	F:	PerFHalOrg.3-211
AsC_6F_{12}	$[C_6F_6][AsF_6]$	F:	PerFHalOrg.3-204, 206/7
AsC_6F_{15}	$(C_2F_5)_3As$	F:	PerFHalOrg.3-213/4, 217, 221
$AsC_6F_{18}NP_2$	$(CF_3)_2AsN[P(CF_3)_2]_2$	F:	PerFHalOrg.3-200/3
$AsC_6F_{18}N_3O_3$	$[(CF_3)_2NO]_3As$	F:	PerFHalOrg.7-101, 119, 126
$AsC_6FeH_7O_4$	$(CO)_4FeAs(CH_3)_2H$	Fe:	Org.Verb.C2-50
$AsC_6H_9NiO_3$	$(CO)_3NiAs(CH_3)_3$	Ni:	Org.Verb.1-187, 189
$AsC_6H_9NiO_6$	$(CO)_3NiAs(OCH_3)_3$	Ni:	Org.Verb.1-187, 189
$AsC_6H_9O_3S_6$	$As[S_2COCH_3]_3$	C:	MVol.D4-254
$AsC_6H_{13}OS_2$	$(CH_3)_2As[S_2COC_3H_7]$	C:	MVol.D4-254
$AsC_7F_3Ge_2H_{18}$	$CF_3As[Ge(CH_3)_3]_2$	F:	PerFHalOrg.3-195

$AsC_7F_3H_{18}Sn_2$	$CF_3As[Sn(CH_3)_3]_2$	F:	PerFHalOrg.3-195, 196
$AsC_7F_6H_3$	$CH_3As(CCCF_3)_2$	F:	PerFHalOrg.4-106
$AsC_7F_{12}H_9N_2O_2$	$[(CF_3)_2NO]_2As(CH_3)_3$	F:	PerFHalOrg.7-128
$AsC_7F_{12}H_{10}N_2P$	$(CF_3)_2AsN(H)P(CF_3)_2 \cdot N(CH_3)_3$	F:	PerFHalOrg.3-201
$AsC_7F_{15}H_6N_2O_2$	$[(CF_3)_2NO]_2As(CH_3)_2CF_3$	F:	PerFHalOrg.7-128
$AsC_7F_{18}H_3N_2O_2$	$[(CF_3)_2NO]_2As(CF_3)_2CH_3$	F:	PerFHalOrg.7-128
$AsC_7FeH_9O_4$	$(CO)_4FeAs(CH_3)_3$	Fe:	Org.Verb.B2-83, 85, 99, 103, 120
$AsC_8F_{10}H_{10}N$	$(C_2F_5)_2AsN(C_2H_5)_2$	F:	PerFHalOrg.3-211
$AsC_8FeH_{12}NO_4$	$(CO)_4FeAs(CH_3)_2N(CH_3)_2$	Fe:	Org.Verb.B2-100
$AsC_8Fe_2H_6NO_7$	$(CO)_4Fe(As(CH_3)_2)Fe(CO)_2NO$	Fe:	Org.Verb.C1-221
$AsC_8H_{11}O_3Sn$	$[(CH_3)_2SnOAs(O)(C_6H_5)O]_x$	Sn:	Org.Verb.6-29
$AsC_8H_{15}NiO_3Si$	$(CO)_3NiAs(CH_3)_2(Si(CH_3)_3)$	Ni:	Org.Verb.1-187, 189
$AsC_8H_{15}NiO_3Sn$	$(CO)_3NiAs(CH_3)_2(Sn(CH_3)_3)$	Ni:	Org.Verb.1-187, 189
$AsC_8H_{17}OS_2$	$(CH_3)_2As[S_2COC_5H_{11}]$	C:	MVol.D4-254
$AsC_9ClFe_2H_{12}O_5S_2$	$(CO)_3Fe(SCH_3)_2Fe(CO)_2As(CH_3)_2Cl$	Fe:	Org.Verb.C1-93
$AsC_9CoFeH_6O_7$	$(CO)_4Fe(As(CH_3)_2)Co(CO)_3$	Fe:	Org.Verb.B2-170/2
$AsC_9F_3H_{16}N_2$	$CF_3As[N(CH_2)_4]_2$	F:	PerFHalOrg.3-209
$AsC_9F_3H_{16}N_2O_2$	$CF_3As[N(CH_2CH_2)_2O]_2$	F:	PerFHalOrg.3-210
$AsC_9F_3H_{20}N_2$	$CF_3As[N(CH_2CH_3)_2]_2$	F:	PerFHalOrg.3-209
$AsC_9F_6FeH_5O_2$	$[C_5H_5FeAs(CF_3)_2(CO)_2]$	F:	PerFHalOrg.3-225
$AsC_9F_6H_5MoO_2$	$[C_5H_5MoAs(CF_3)_2(CO)_2]$	F:	PerFHalOrg.3-225
AsC_9F_9	$(CF_3CC)_3As$	F:	PerFHalOrg.3-213/8, 221
$AsC_9FeH_9O_7$	$(CO)_4FeAs(OCH_2)_3CCH_3$	Fe:	Org.Verb.B2-82
$AsC_9Fe_2H_6NO_8$	$(CO)_4FeAs(CH_3)_2Fe(CO)_3NO$	Fe:	Org.Verb.C1-67/9
$AsC_9H_8NO_{16}U_2$	$(C_5H_5NH)[(UO_2)_2(HC_2O_4)_2AsO_4] \cdot 4\ H_2O$	U:	SVol.C3-177
$AsC_9H_{15}NiO_3$	$(CO)_3NiAs(C_2H_5)_3$	Ni:	Org.Verb.1-187
$AsC_9H_{15}NiO_6$	$(CO)_3NiAs(OC_2H_5)_3$	Ni:	Org.Verb.1-187
$AsC_9H_{15}O_3S_6$	$As[S_2COC_2H_5]_3$	C:	MVol.D4-245, 254
$AsC_{10}Cl_2FeH_9$	$C_5H_5FeC_5H_4AsCl_2$	Fe:	Org.Verb.A6-203
$AsC_{10}Cl_4FeH_{10}$	$[Fe(C_5H_5)_2]AsCl_4$	Fe:	Org.Verb.A1-224
$AsC_{10}Cl_4H_{13}O_2Sn$	$SnCl_4 \cdot CH_3OCOC_6H_4As(CH_3)_2$	Sn:	MVol.C6-179
$AsC_{10}CoFeH_6O_8$	$(CO)_4FeAs(CH_3)_2Co(CO)_4$	Fe:	Org.Verb.B2-84/5, 102
$AsC_{10}F_6IMn_2O_8$	$(CF_3)_2AsIMn_2(CO)_8$	F:	PerFHalOrg.3-207
$AsC_{10}F_{10}H_5$	$(C_2F_5)_2AsC_6H_5$	F:	PerFHalOrg.3-211
$AsC_{10}FeH_6MnO_8$	$(CO)_4Fe(As(CH_3)_2)Mn(CO)_4$	Fe:	Org.Verb.B2-170/1
$AsC_{10}FeH_{15}O_4$	$(CO)_4FeAs(C_2H_5)_3$	Fe:	Org.Verb.B2-99
$AsC_{10}FeH_{18}N_3O_4$	$(CO)_4FeAs(N(CH_3)_2)_3$	Fe:	Org.Verb.B2-100, 121
$AsC_{11}F_3H_{20}N_2$	$CF_3As[N(CH_2)_5]_2$	F:	PerFHalOrg.3-210
$AsC_{11}FeH_6MnO_9$	$(CO)_4FeAs(CH_3)_2Mn(CO)_5$	Fe:	Org.Verb.B2-84/5, 101, 123
$AsC_{11}Fe_2H_{12}NO_7$	$(CO)_6Fe_2(OCN(CH_3)_2)(As(CH_3)_2)$	Fe:	Org.Verb.C1-237/8
$AsC_{11}Fe_2H_{15}NO_7P$	$(CO)_4FeAs(CH_3)_2Fe(CO)_2(NO)P(CH_3)_3$	Fe:	Org.Verb.C1-67/8
$AsC_{11}Fe_2H_{15}NO_{10}P$	$(CO)_6Fe_2As(CH_3)_2(NO)P(OCH_3)_3$	Fe:	Org.Verb.C1-67/8, 70
$AsC_{11}Fe_2H_{18}NO_5S_2$	$(CO)_5Fe_2(SCH_3)_2(As(CH_3)_2N(CH_3)_2)$	Fe:	Org.Verb.C1-93
$AsC_{12}ClF_{10}$	$(C_6F_5)_2AsCl$	F:	PerFHalOrg.3-204, 206, 207
$AsC_{12}CoFeH_{15}O_{10}P$	$(CO)_4FeAs(CH_3)_2Co(CO)_3P(OCH_3)_3$	Fe:	Org.Verb.B2-84/5, 102
$AsC_{12}F_{10}^+$	$(C_6F_5)_2As^+$	F:	PerFHalOrg.3-216
$AsC_{12}FeH_{15}HgINO_3$	$(CO)_2Fe(HgI)(NO)As(C_2H_5)_2C_6H_5$	Fe:	Org.Verb.B1-133/5

Formula	Compound		Reference
$AsC_{12}Fe_2H_{18}N_2O_7P$	$(CO)_6Fe_2As(CH_3)_2(NO)P(CH_3)_2N(CH_3)_2$	Fe:	Org.Verb.C1-67/8
$AsC_{12}H_{15}O_2S_4$	$C_6H_5As[S_2COC_2H_5]_2$	C:	MVol.D4-254
$AsC_{12}H_{21}O_3S_6$	$As[S_2COC_3H_7]_3$	C:	MVol.D4-254
$AsC_{12}H_{21}O_3Se_6$	$As[Se_2COC_3H_7]_3$	C:	MVol.D6-221
$AsC_{13}CoFe_2H_{12}O_9S_2$	$(CO)_5Fe_2(SCH_3)_2As(CH_3)_2Co(CO)_4$	Fe:	Org.Verb.C1-93
$AsC_{13}F_3H_{28}N_2$	$CF_3As[N(CH_2CH_2CH_3)_2]_2$	F:	PerFHalOrg.3-209
$AsC_{13}FeGe_3H_{27}O_4$	$(CO)_4FeAs(Ge(CH_3)_3)_3$	Fe:	Org.Verb.B2-100
$AsC_{13}FeH_{11}MoO_6$	$(CO)_4Fe(As(CH_3)_2)Mo(CO)_2C_5H_5$	Fe:	Org.Verb.B2-170
$AsC_{13}FeH_{11}O_6W$	$(CO)_4Fe(As(CH_3)_2)W(CO)_2C_5H_5$	Fe:	Org.Verb.B2-170
$AsC_{13}FeH_{13}O_4$	$(CO)_4FeAs(CH_3)_2CH_2C_6H_5$	Fe:	Org.Verb.B2-99
$AsC_{13}FeH_{15}MnO_8P$	$(CO)_4Mn(P(CH_3)_3)(As(CH_3)_2)Fe(CO)_4$	Fe:	Org.Verb.C2-54
$AsC_{13}FeH_{15}MnO_{11}P$	$(CO)_4FeAs(CH_3)_2Mn(CO)_4P(OCH_3)_3$	Fe:	Org.Verb.B2-84/5, 102
$AsC_{13}FeH_{17}O$	$C_5H_5FeC_5H_4CH_2OAs(CH_3)_2$	Fe:	Org.Verb.A2-20
$AsC_{13}FeH_{27}O_4Sn_3$	$(CO)_4FeAs(Sn(CH_3)_3)_3$	Fe:	Org.Verb.B2-100
$AsC_{13}Fe_2H_{11}O_6$	$C_5H_5Fe(CO)_2As(CH_3)_2Fe(CO)_4$	Fe:	Org.Verb.C2-49
$AsC_{14}F_{12}H_{13}N_2Pd$	$[C_6H_5(CH_3)_2As]Pd(C(CF_3)_2NH)_2$	F:	PerFHalOrg.7-74/5
$AsC_{14}FeH_{11}MoO_7$	$(CO)_4FeAs(CH_3)_2Mo(CO)_3C_5H_5$	Fe:	Org.Verb.B2-84/5, 101
$AsC_{14}FeH_{11}O_7W$	$(CO)_4FeAs(CH_3)_2W(CO)_3C_5H_5$	Fe:	Org.Verb.B2-84/5, 101
$AsC_{14}Fe_2H_{12}MnO_{10}S_2$	$(CO)_5Fe_2(SCH_3)_2As(CH_3)_2Mn(CO)_5$	Fe:	Org.Verb.C1-93
$AsC_{14}Fe_2H_{12}O_{10}ReS_2$	$(CO)_5Fe_2(SCH_3)_2As(CH_3)_2Re(CO)_5$	Fe:	Org.Verb.C1-93
$AsC_{14}H_{19}O_2S_4$	$C_6H_5As[S_2COC_3H_7]_2$	C:	MVol.D4-254
$AsC_{15}CoFeH_{24}N_3O_7P$	$(CO)_4FeAs(CH_3)_2Co(CO)_3P(N(CH_3)_2)_3$	Fe:	Org.Verb.B2-84/5, 102
$AsC_{15}H_{15}OS_2$	$(C_6H_5)_2As[S_2COC_2H_5]$	C:	MVol.D4-254
$AsC_{15}H_{27}O_3S_6$	$As[S_2COC_4H_9]_3$	C:	MVol.D4-254
$AsC_{15}H_{33}Ni$	$(CH_3)_2C_3H_3Ni(As(C_3H_7\text{-}i)_3)CH_3$	Ni:	Org.Verb.2-37, 40
$AsC_{16}FeH_{10}O_4^-$	$[(CO)_4FeAs(C_6H_5)_2]^-$	Fe:	Org.Verb.B2-52
$AsC_{16}FeH_{20}MoO_6P$	$C_5H_5Mo(CO)_2(P(CH_3)_3)As(CH_3)_2Fe(CO)_4$	Fe:	Org.Verb.C2-49
$AsC_{16}FeH_{24}MnN_3O_8P$	$(CO)_4FeAs(CH_3)_2Mn(CO)_4P(N(CH_3)_2)_3$	Fe:	Org.Verb.B2-84/5, 102
$AsC_{16}H_7N_2O_{11}PuS_2^{2-}$	$Pu(O_3AsC_6H_4NNC_{10}H_3(O)_2(SO_3)_2)^{2-}$	Np:	TrU.D1-168/9
$AsC_{16}H_8N_2O_{10}PuS_2^-$	$Pu(O_3AsC_6H_4NNC_{10}H_4(O)(SO_3)_2)^-$	Np:	TrU.D1-168
$AsC_{16}H_{10}N_2O_{11}PuS_2^+$	$PuH_3(O_3AsC_6H_4NNC_{10}H_3(O)_2(SO_3)_2)^+$	Np:	TrU.D1-168/9
$AsC_{16}H_{12}N_2O_{12}PuS_2^+$	$PuH_4(OH)(O_3AsC_6H_4NNC_{10}H_3(O)_2(SO_3)_2)^+$	Np:	TrU.D1-168/9
$AsC_{16}H_{17}OS_2$	$(C_6H_5)_2As[S_2COC_3H_7]$	C:	MVol.D4-254
$AsC_{17}CoFeH_{17}O_7P$	$(CO)_4FeAs(CH_3)_2Co(CO)_3P(CH_3)_2C_6H_5$	Fe:	Org.Verb.B2-84/5, 102
$AsC_{17}CrFe_2H_{17}O_8S_2$	$(CO)_5Fe_2(SCH_3)_2As(CH_3)_2Cr(CO)_3C_5H_5$	Fe:	Org.Verb.C1-93
$AsC_{17}F_3H_{36}N_2$	$CF_3As[N(CH_2CH_2CH_2CH_3)_2]_2$	F:	PerFHalOrg.3-209
$AsC_{17}FeH_{23}MoNO_6P$	$C_5H_5Mo(CO)_2(P(CH_3)_2N(CH_3)_2)As(CH_3)_2Fe(CO)_4$	Fe:	Org.Verb.C2-49
$AsC_{17}Fe_2H_{17}MoO_8S_2$	$(CO)_5Fe_2(SCH_3)_2As(CH_3)_2Mo(CO)_3C_5H_5$	Fe:	Org.Verb.C1-93
$AsC_{17}Fe_2H_{17}O_8S_2W$	$(CO)_5Fe_2(SCH_3)_2As(CH_3)_2W(CO)_3C_5H_5$	Fe:	Org.Verb.C1-93
$AsC_{17}H_{39}Sn$	$(C_4H_9)_3SnCH_2CH_2CH_2As(CH_3)_2$	Sn:	Org.Verb.2-289
$AsC_{18}Cl_2F_{15}$	$(C_6F_5)_3AsCl_2$	F:	PerFHalOrg.3-204, 206/7
$AsC_{18}Cl_2H_{15}OSn$	$SnCl_2 \cdot (C_6H_5)_3AsO$	Sn:	MVol.C5-62
$AsC_{18}Cl_{10}H_{15}U_2$	$2\ UCl_5 \cdot (C_6H_5)_3As$	U:	SVol.E1-52
$AsC_{18}F_2H_{15}$	$(C_6H_5)_3AsF_2$	F:	PerFHalOrg.5-144
$AsC_{18}F_2H_{17}N_2O_3S$	$(C_6H_5)_3AsO \cdot NH_2SO_2NF_2$	S:	S-N-Verb.1-176
$AsC_{18}F_{14}^+$	$(C_6F_5)_2AsC_6F_4^+$	F:	PerFHalOrg.3-216
$AsC_{18}F_{15}$	$(C_6F_5)_3As$	F:	PerFHalOrg.3-213/4, 216, 218/9, 221
$AsC_{18}F_{15}^+$	$(C_6F_5)_3As^+$	F:	PerFHalOrg.3-216

$AsC_{18}F_{15}O$	$(C_6F_5)_3AsO$	F:	PerFHalOrg.3-196/8
$AsC_{18}FeH_{15}N_2O_2$	$Fe(NO)_2(As(C_6H_5)_3)$	Fe:	Org.Verb.B1-43, 46
$AsC_{18}FeH_{17}MnO_8P$	$(CO)_4FeAs(CH_3)_2Mn(CO)_4P(CH_3)_2C_6H_5$	Fe:	Org.Verb.B2-84/5, 101
$AsC_{18}H_{27}O_2S_4$	$C_6H_5As[S_2COC_5H_{11}]_2$	C:	MVol.D4-254
$AsC_{18}H_{30}N_3NiO_3$	$(CO)_3NiAs(NC_5H_{10})_3$	Ni:	Org.Verb.1-187
$AsC_{18}H_{41}Sn$	$(C_4H_9)_3SnCH_2CH_2CH_2CH_2As(CH_3)_2$	Sn:	Org.Verb.2-291
$AsC_{19}CrFe_2H_5O_{13}$	$(CO)_8Fe_2As(C_6H_5)Cr(CO)_5$	Fe:	Org.Verb.C2-55
$AsC_{19}F_{10}FeH_5O_2$	$[(C_6F_5)_2AsFe(CO)_2C_5H_5]$	F:	PerFHalOrg.3-207, 225
$AsC_{19}F_{10}H_5MoO_2$	$[C_5H_5MoAs(C_6F_5)_2(CO)_2]$	F:	PerFHalOrg.3-225
$AsC_{19}FeH_{13}O_4$	$(CO)_4FeAs(C_6H_5)_2CH_2CCH$	Fe:	Org.Verb.B2-100
$AsC_{19}FeH_{15}N_2O_3$	$COFe(NO)_2As(C_6H_5)_3$	Fe:	Org.Verb.B1-42/3, 45/6, 48, 141/2
$AsC_{19}H_{18}I_6Pa$	$[As(C_6H_5)_3CH_3]PaI_6$	Pa:	SVol.2-73
$AsC_{19}H_{43}Sn$	$(C_4H_9)_3SnCH_2CH_2CH_2As(C_2H_5)_2$	Sn:	Org.Verb.2-293
$AsC_{20}Cl_2FeH_{15}NO_3P$	$(CO)_2Fe(AsCl_2)(NO)P(C_6H_5)_3$	Fe:	Org.Verb.B1-133/5
$AsC_{20}Cl_3H_{41}IrP_2$	$Ir(P(C_2H_5)_3)_2(As(CH_3)_2(C_6H_5))Cl_3$	Ir:	SVol.2-200/2
$AsC_{20}Cl_4H_{17}O_2Sn$	$SnCl_4 \cdot CH_3OCOC_6H_4As(C_6H_5)_2$	Sn:	MVol.C6-179
$AsC_{20}F_8H_{25}Ni$	$(CH_2C(CH_3)CH_2C_2F_4)_2Ni(As(CH_3)_2C_6H_5)$	Ni:	Org.Verb.2-85
$AsC_{20}F_{10}H_5MoO_3$	$[(C_6F_5)_2AsMo(CO)_3C_5H_5]$	F:	PerFHalOrg.3-207
$AsC_{20}FeH_{17}O_4$	$(CO)_4FeAs(C_6H_5)_2CH_2C(CH_3)CH_2$	Fe:	Org.Verb.B2-99
$AsC_{20}FeH_{17}O_6$	$(CO)_4FeAs(C_6H_5)_2CH_2COOC_2H_5$	Fe:	Org.Verb.B2-99
$AsC_{20}H_{20}IrOS_2$	$[(C_6H_5)_3As]IrH_2[S_2COCH_3]$	C:	MVol.D4-261
$AsC_{20}H_{44}IrP_2$	$Ir(P(C_2H_5)_3)_2(As(CH_3)_2(C_6H_5))H_3$	Ir:	SVol.2-195
$AsC_{21}Cl_2FeH_{15}O_3$	$(CO)_3Fe(As(C_6H_5)_3)Cl_2$	Fe:	Org.Verb.B1-180/3
$AsC_{21}Cl_3H_{12}NiO_3$	$(CO)_3NiAs(C_6H_4Cl)_3$	Ni:	Org.Verb.1-187, 189
$AsC_{21}Cl_4FeH_{15}O_3Sn$	$(CO)_3Fe(As(C_6H_5)_3)(Cl)SnCl_3$	Fe:	Org.Verb.B1-195/6
$AsC_{21}F_3H_{12}NiO_3$	$(CO)_3NiAs(C_6H_4F)_3$	Ni:	Org.Verb.1-187
$AsC_{21}F_4FeH_{16}O_3P$	$(CO)_3Fe(CH_3)_2AsC_2(CF_2)_2P(C_6H_5)_2$	Fe:	Org.Verb.B1-170/1, 175, 179
$AsC_{21}FeH_{15}I_2O_3$	$(CO)_3Fe(As(C_6H_5)_3)I_2$	Fe:	Org.Verb.B1-180/2
$AsC_{21}H_{15}NiO_3$	$(CO)_3NiAs(C_6H_5)_3$	Ni:	Org.Verb.1-187, 189
$AsC_{21}H_{20}INi$	$C_3H_5Ni(As(C_6H_5)_3)I$	Ni:	Org.Verb.2-13, 15
$AsC_{21}H_{22}IrOS_2$	$[(C_6H_5)_3As]IrH_2[S_2COC_2H_5]$	C:	MVol.D4-261
$AsC_{22}Cl_3H_{37}IrP_2$	$Ir(P(CH_3)_2C_6H_5)_2(As(C_2H_5)_3)Cl_3$	Ir:	SVol.2-212
$AsC_{22}Cl_4H_{21}O_2Sn$	$SnCl_4 \cdot CH_3OCOC_6H_4As(C_6H_4CH_3)_2$	Sn:	MVol.C6-179
$AsC_{22}F_4FeH_{16}O_4P$	$(CO)_4FeC_4F_4(P(C_6H_5)_2)As(CH_3)_2$	Fe:	Org.Verb.B2-83, 85, 98, 118/9
$AsC_{22}F_6FeH_{16}O_3P$	$(CO)_3Fe(CH_3)_2AsC_2(CF_2)_3P(C_6H_5)_2$	Fe:	Org.Verb.B1-170/1, 175, 179
$AsC_{22}FeH_{15}O_4$	$(CO)_4FeAs(C_6H_5)_3$	Fe:	Org.Verb.B2-99, 120/1
		Fe:	Org.Verb.B4-252
$AsC_{22}FeH_{17}O_4$	$H_2Fe(CO)_4 \cdot As(C_6H_5)_3$	Fe:	Org.Verb.B2-48
$AsC_{22}FeH_{20}NO_2$	$C_3H_5Fe(CO)(NO)As(C_6H_5)_3$	Fe:	Org.Verb.B5-7/9
$AsC_{22}FeH_{21}N_2O_3$	$COFe(NO)_2As(C_6H_4CH_3)_3$	Fe:	Org.Verb.B1-42/3, 45
$AsC_{22}FeH_{33}O_4$	$(CO)_4FeAs(C_6H_{11})_3$	Fe:	Org.Verb.B2-99
$AsC_{22}H_{21}O_3S_4U$	$UO_2(CH_3CSS)_2 \cdot (C_6H_5)_3AsO$	U:	SVol.E1-196, 199
$AsC_{22}H_{21}O_7U$	$[UO_2(CH_3COO)_2 \cdot (C_6H_5)_3AsO]_2$	U:	SVol.E1-196, 198/9
$AsC_{23}Cl_3GeH_{20}Ni$	$C_5H_5Ni(As(C_6H_5)_3)GeCl_3$	Ni:	Org.Verb.2-130
$AsC_{23}F_6FeH_{16}O_4P$	$(CO)_4FeC_5F_6(P(C_6H_5)_2)As(CH_3)_2$	Fe:	Org.Verb.B2-98, 120
$AsC_{23}FeH_{17}O_4$	$(CO)_4FeAs(C_6H_5)_2CH_2C_6H_5$	Fe:	Org.Verb.B2-100

Formula	Compound		Reference
$AsC_{23}Fe_2H_{25}O_2$	$(C_5H_5FeC_5H_4CH_2O)_2AsCH_3$	Fe:	Org.Verb.A6-169
$AsC_{23}H_{20}INi$	$C_5H_5Ni(As(C_6H_5)_3)I$	Ni:	Org.Verb.2-117, 120
$AsC_{23}H_{51}Sn$	$(C_4H_9)_3SnCH_2CH_2CH_2As(C_4H_9)_2$	Sn:	Org.Verb.2-297
$AsC_{24}Cl_3H_{20}Sn$	$[(C_6H_5)_4As]SnCl_3$	Sn:	MVol.C3-95/6
$AsC_{24}Cl_3H_{33}IrP_2$	$Ir(P(CH_3)_2C_6H_5)_2(As(CH_3)_2C_6H_5)Cl_3$	Ir:	SVol.2-213
$AsC_{24}Cl_6H_{20}Pa$	$[As(C_6H_5)_4]PaCl_6$	Pa:	SVol.2-58
$AsC_{24}Cl_6H_{20}U$	$[As(C_6H_5)_4]UCl_6$	U:	SVol.C9-60/3, 140
$AsC_{24}F_2H_{20}NOS$	$[(C_6H_5)_4As][NS(O)F_2]$	S:	S-N-Verb.1-61
$AsC_{24}F_2H_{20}NO_4S_2$	$[(C_6H_5)_4As]N(SO_2F)_2$	S:	S-N-Verb.1-50, 71
$AsC_{24}F_2H_{20}N_3O_3S_3$	$[(C_6H_5)_4As][NS(O)F]_2[NSO]$	S:	S-N-Verb.1-21
$AsC_{24}F_3H_{20}Sn$	$[(C_6H_5)_4As]SnF_3$	Sn:	MVol.C3-86
$AsC_{24}F_4Fe_2H_{16}O_6P$	$(C_4F_4(P(C_6H_5)_2)As(CH_3)_2)Fe_2(CO)_6$	Fe:	Org.Verb.C2-130/2, 134
$AsC_{24}F_6FeH_{15}O_2S_2$	$(CO)_2Fe(As(C_6H_5)_3)S_2C_2(CF_3)_2$	Fe:	Org.Verb.B1-101/2, 108
$AsC_{24}F_6H_{20}NO_2S_2$	$[(C_6H_5)_4As]SF_5NSO_2F$	S:	S-N-Verb.1-146
$AsC_{24}H_{20}I_6U$	$[As(C_6H_5)_4]UI_6$	U:	SVol.C9-176
$AsC_{24}H_{20}MnO_4$	$(C_6H_5)_4AsMnO_4$	Mn:	MVol.C2-34, 37, 41, 202
$AsC_{24}H_{21}NiO_3$	$(CO)_3NiAs(C_6H_4CH_3)_3$	Ni:	Org.Verb.1-187, 189
$AsC_{24}H_{23}Ni$	$C_5H_5Ni(As(C_6H_5)_3)CH_3$	Ni:	Org.Verb.2-144
$AsC_{24}H_{27}Ni$	$(CH_3)_2C_3H_3Ni(As(C_6H_5)_3)CH_3$	Ni:	Org.Verb.2-37
$AsC_{24}H_{53}Sn$	$(C_4H_9)_3Sn(CH_2)_4As(C_4H_9)_2$	Sn:	Org.Verb.2-297
$AsC_{25}Cl_4H_{23}Sn$	$[(C_6H_5)_4As][CH_3SnCl_4]$	Sn:	Org.Verb.6-222
$AsC_{25}CuH_{20}S_3$	$[(C_6H_5)_4As][Cu(CS_3)]$	C:	MVol.D4-224
$AsC_{25}F_2H_{21}N_2O_5S_2$	$[(C_6H_5)_4As][FSO_2NCONHSO_2F]$	S:	S-N-Verb.1-103
$AsC_{25}F_3H_{20}NO_4S$	$[(C_6H_5)_4As][CF_3N(O)SO_3]$	F:	PerFHalOrg.7-146
$AsC_{25}F_3H_{21}OPS_2$	$[(C_6H_5)_4As][CF_3PS_2OH]$	F:	PerFHalOrg.3-151, 156
–	$[(C_6H_5)_4As][CF_3P(S)(SH)O]$	F:	PerFHalOrg.3-151
$AsC_{25}F_3H_{21}O_2PS$	$[(C_6H_5)_4As][CF_3P(S)(OH)O]$	F:	PerFHalOrg.3-151, 157
$AsC_{25}F_4H_{20}NO_2S$	$[(C_6H_5)_4As][FSO_2NCF_3]$	F:	PerFHalOrg.7-57
		S:	S-N-Verb.1-100
$AsC_{25}F_5H_{18}I_3Tl$	$[(C_6H_5)_3AsCH_3][Tl(C_6F_5)I_3]$	F:	PerFHalOrg.4-137, 146
$AsC_{25}F_6Fe_2H_{16}O_6P$	$(C_5F_6P(C_6H_5)_2As(CH_3)_2)Fe_2(CO)_6$	Fe:	Org.Verb.C2-130/2, 134
$AsC_{25}F_8FeH_{15}O_3$	$(CF_2CF_2)_2Fe(CO)_3As(C_6H_5)_3$	Fe:	Org.Verb.B4-202, 207
$AsC_{25}FeH_{21}O_4$	$(CO)_4FeAs(C_6H_4CH_3)_3$	Fe:	Org.Verb.B2-99
$AsC_{25}Fe_2H_{21}IO_5S_2^+$	$[(CO)_5(As(C_6H_5)_3)Fe_2I(SCH_3)_2]^+$	Fe:	Org.Verb.C1-12/4
$AsC_{25}Fe_2H_{21}I_2O_5S_2$	$I(CO)_5Fe_2(SCH_3)_2As(C_6H_5)_3I$	Fe:	Org.Verb.C1-13
$AsC_{25}Fe_2H_{21}O_5S_2$	$(CO)_3Fe(SCH_3)_2Fe(CO)_2As(C_6H_5)_3$	Fe:	Org.Verb.C1-95
$AsC_{25}H_{20}NSe$	$(C_6H_5)_4AsSeCN$	C:	MVol.D6-223
$AsC_{25}H_{20}NTe$	$(C_6H_5)_4AsTeCN$	C:	MVol.D6-259/60
$AsC_{25}H_{22}NS_2$	$[(C_6H_5)_4As][S_2CNH_2]$	C:	MVol.D6-140
$AsC_{26}Cl_3H_{26}Sn$	$[(C_6H_5)_4As][(CH_3)_2SnCl_3]$	Sn:	Org.Verb.6-35
$AsC_{26}Cl_3H_{37}IrP_2$	$Ir(P(CH_3)_2C_6H_5)_2(As(C_2H_5)_2C_6H_5)Cl_3$	Ir:	SVol.2-212
$AsC_{26}Cl_4H_{25}Sn$	$[(C_6H_5)_4As][C_2H_5SnCl_4]$	Sn:	Org.Verb.6-232
$AsC_{26}F_4Fe_2H_{16}O_8P$	$(CO)_4Fe((C_6H_5)_2PC_4F_4As(CH_3)_2)Fe(CO)_4$	Fe:	Org.Verb.C1-36, 40/1
$AsC_{26}FeH_{24}N_2O_2P$	$Fe(NO)_2((C_6H_5)_2AsC_2H_4P(C_6H_5)_2)$	Fe:	Org.Verb.B1-143
$AsC_{26}Fe_2H_{21}NO_7P$	$(CO)_6Fe_2As(CH_3)_2(NO)P(C_6H_5)_3$	Fe:	Org.Verb.C1-67/8
$AsC_{26}H_{20}NSe_3$	$[(C_6H_5)_4As][NCC(Se)Se_2]$	C:	MVol.D6-200
$AsC_{27}CoFeH_{21}O_7P$	$(CO)_4FeAs(CH_3)_2Co(CO)_3 \cdot P(C_6H_5)_3$	Fe:	Org.Verb.B2-84/5, 102
$AsC_{27}FeH_{24}N_2O_3P$	$COFe(NO)_2(C_6H_5)_2PC_2H_4As(C_6H_5)_2$	Fe:	Org.Verb.B1-45, 47
$AsC_{27}Fe_2H_{15}O_9Pt$	$(CO)_8Fe_2Pt(CO)(As(C_6H_5)_3)$	Fe:	Org.Verb.C1-216/7, 219

$AsC_{27}Fe_2H_{25}O_5S_2$ ... $(CO)_3Fe(SC_2H_5)_2Fe(CO)_2As(C_6H_5)_3$ ... Fe: Org.Verb.C1-96

$AsC_{27}H_{23}O_2PtS_4$ ... $[(C_6H_5)_4As][Pt(S_2CO)(S_2COCH_3)]$ ... C: MVol.D4-208, 261

$AsC_{27}H_{24}Mo_{12}N_3O_{40}Th$ $(C_9H_7N)_3H_3AsTh(Mo_3O_{10})_4$ ... Th: SVol.C2-33

$AsC_{27}H_{43}Sn$ ... $(C_4H_9)_3SnCH_2CH_2CH_2As(C_6H_5)_2$ ... Sn: Org.Verb.2-298

$AsC_{28}Cl_3FeGeH_{20}O_4$. $[As(C_6H_5)_4][(CO)_4FeGeCl_3]$ ... Fe: Org.Verb.B2-142, 150

$AsC_{28}Cl_3FeH_{20}Hg_2O_4$ $[As(C_6H_5)_4][(CO)_4Fe(HgCl)_2Cl]$ ... Fe: Org.Verb.B2-153, 157, 160

$AsC_{28}Cl_3FeH_{20}O_4Sn$. $[As(C_6H_5)_4][(CO)_4FeSnCl_3]$ ... Fe: Org.Verb.B2-143, 150

$AsC_{28}Cl_3H_{30}Sn$ ... $[(C_6H_5)_4As][(C_2H_5)_2SnCl_3]$ ... Sn: Org.Verb.6-61

$AsC_{28}Cl_4H_{29}Sn$ ... $[(C_6H_5)_4As][C_4H_9SnCl_4]$ ... Sn: Org.Verb.6-247/8

$AsC_{28}Cl_4H_{30}O_2U$ ... $As(C_6H_5)_4[U(OC_2H_5)_2Cl_4]$ ... U: SVol.C9-90

$AsC_{28}FeH_{20}O_4P$ ... $(CO)_4FeAs(C_6H_5)_2P(C_6H_5)_2$ ... Fe: Org.Verb.B2-100

– ... $(CO)_4FeP(C_6H_5)_2As(C_6H_5)_2$ ... Fe: Org.Verb.B2-97

$AsC_{28}FeH_{21}MnO_8P$ .. $(CO)_4FeAs(CH_3)_2Mn(CO)_4P(C_6H_5)_3$ ... Fe: Org.Verb.B2-84/5, 101, 123

$AsC_{28}H_{25}O_2PdS_4$ ... $[(C_6H_5)_4As][Pd(S_2CO)(S_2COC_2H_5)]$ ... C: MVol.D4-208, 261

$AsC_{28}H_{25}O_2PtS_4$ ... $[(C_6H_5)_4As][Pt(S_2CO)(S_2COC_2H_5)]$ ... C: MVol.D4-208, 261

$AsC_{28}H_{35}N_2O_3S_4U$ ... $UO_2(S_2CN(C_2H_5)_2)_2 \cdot (C_6H_5)_3AsO$ ... U: SVol.E1-196, 198/9

$AsC_{28}H_{35}N_2O_3Se_4U$ .. $UO_2(Se_2CN(C_2H_5)_2)_2 \cdot (C_6H_5)_3AsO$ ... U: SVol.E1-196, 198/9

$AsC_{28}H_{44}IrP_2$ ... $Ir(P(C_2H_5)_2(C_6H_5))_2H_3As(CH_3)_2(C_6H_5)$ ... Ir: SVol.2-207

$AsC_{29}FeH_{23}O_3$ ... $(C_6H_5CHCHCHO)Fe(CO)_2As(C_6H_5)_3$ ... Fe: Org.Verb.B5-115

$AsC_{29}H_{25}Ni$ ... $C_5H_5Ni(As(C_6H_5)_3)C_6H_5$ ... Ni: Org.Verb.2-145

$AsC_{29}H_{27}O_2PtS_4$ ... $[(C_6H_5)_4As][Pt(S_2CO)(S_2COC_3H_7)]$ ... C: MVol.D4-208, 261

$AsC_{30}ClF_{10}H_{15}OTl$ ... $(C_6F_5)_2TlCl \cdot (C_6H_5)_3AsO$ ... F: PerFHalOrg.4-136/7, 139

$AsC_{30}ClF_{10}H_{15}Tl$ ... $(C_6F_5)_2TlCl \cdot (C_6H_5)_3As$ ... F: PerFHalOrg.4-136/7, 141

$AsC_{30}ClH_{26}N_2S_2Sn$ .. $[(C_6H_5)_4As][(CH_3)_2ClSn(SC(CN))_2]$ ... Sn: Org.Verb.6-30

$AsC_{30}Cl_3H_{34}Sn$ ... $[(C_6H_5)_4As][(C_3H_7)_2SnCl_3]$ ... Sn: Org.Verb.6-68

$AsC_{30}Cl_4H_{25}Sn$ ... $[(C_6H_5)_4As][C_6H_5SnCl_4]$ ... Sn: Org.Verb.6-279

$AsC_{30}F_{10}H_{15}NO_4Tl$ .. $(C_6F_5)_2TlNO_3 \cdot (C_6H_5)_3AsO$ ... F: PerFHalOrg.4-136/7, 140

$AsC_{30}FeH_{23}O_4$ ... $C_6H_5C_2H_2CH(O)Fe(CO)_3As(C_6H_5)_3$ ... Fe: Org.Verb.B4-199, 205/6

$AsC_{30}FeH_{25}O_3$ ... $(C_6H_5CHCHCOCH_3)Fe(CO)_2As(C_6H_5)_3$ ... Fe: Org.Verb.B5-116

$AsC_{30}FeH_{30}NO_3Sn$ .. $(CO)_2Fe(Sn(C_6H_5)_3)(NO)As(C_2H_5)_2C_6H_5$ ... Fe: Org.Verb.B1-133/5

$AsC_{30}Fe_2H_{21}O_5S_2$ ... $(CO)_5As(C_6H_5)_3Fe_2(SC_6H_3(CH_3)S)$ ... Fe: Org.Verb.C1-123/4

$AsC_{30}Fe_3H_{27}$ ... $(C_5H_5FeC_5H_4)_3As$ ... Fe: Org.Verb.A6-265

$AsC_{30}H_{36}N_3Ni$ ... $(i\text{-}C_3H_7NC)_3NiAs(C_6H_5)_3 \cdot NH_3$ ... Ni: Org.Verb.1-323, 324

$AsC_{30}H_{39}N_4Ni$ ... $(i\text{-}C_3H_7NC)_3NiAs(C_6H_5)_3 \cdot NH_3$ ... Ni: Org.Verb.1-323, 324

$AsC_{31}FeH_{25}O_4$ ... $C_6H_5C_2H_2C(O)CH_3Fe(CO)_3As(C_6H_5)_3$ ... Fe: Org.Verb.B4-200, 205/6

$AsC_{32}Cl_3H_{38}Sn$ ... $[(C_6H_5)_4As][(C_4H_9)_2SnCl_3]$ ... Sn: Org.Verb.6-99

$AsC_{32}Cl_3H_{44}IrO_4P_3$ .. $[Ir(P(CH_3)_2C_6H_5)_3(As(CH_3)_2C_6H_5)Cl_2]ClO_4$ .. Ir: SVol.2-217/8

$AsC_{32}F_5H_{20}O_3W$ ... $C_5H_5W(CO)_2[(C_6H_5)_3As]C(O)C_6F_5$ ... F: PerFHalOrg.4-36

$AsC_{32}F_{13}H_{15}O_3Tl$ ... $(C_6F_5)_2TlOC(O)CF_3 \cdot (C_6H_5)_3AsO$ ... F: PerFHalOrg.4-136, 140

$AsC_{32}H_{23}K_2O_9U$ ... $K_2UO_2(OC_6H_4COO)_2 \cdot (C_6H_5)_3AsO$ ... U: SVol.E1-196, 199

$AsC_{32}H_{25}O_3S_4U$ ... $UO_2(C_6H_5CSS)_2 \cdot (C_6H_5)_3AsO$ ... U: SVol.E1-196, 199

$AsC_{32}H_{25}O_7U$ ... $UO_2(C_7H_5O_2)_2 \cdot (C_6H_5)_3AsO$ ... U: SVol.E1-197, 201

$AsC_{33}Fe_3H_{33}O_3$ ... $(C_5H_5FeC_5H_4CH_2O)_3As$ ... Fe: Org.Verb.A6-278

$AsC_{33}H_{25}NiO_3P_2$ ... $(CO)_3NiAs(C_6H_5)(P(C_6H_5)_2)_2$ ... Ni: Org.Verb.1-187, 189

$AsC_{33}H_{35}O_3PtS_6$ ... $[(C_6H_5)_4As][Pt(S_2COC_2H_5)_3]$ ... C: MVol.D4-261

$AsC_{34}FeGeH_{25}O_4$ ... $(CO)_4FeAs(C_6H_5)_2Ge(C_6H_5)_3$ ... Fe: Org.Verb.B2-100

$AsC_{34}FeH_{27}O_3$ ...... $[As(C_6H_5)_4][C_7H_7Fe(CO)_3]$ .... Fe: Org.Verb.B5-78
$AsC_{34}H_{37}N_2O_7U$ .... $UO_2(C_2H_4(NHCCH_3CHCOCHC(CH_3)O)_2)$ · $(C_6H_5)_3AsO$ .... U: SVol.E1-196, 201
$AsC_{35}FeH_{27}O_3$ ...... $(C_6H_5C_2H_2COC_6H_5)Fe(CO)_2As(C_6H_5)_3$ ..... Fe: Org.Verb.B5-117
$AsC_{35}FeH_{28}NO_2$ .... $(C_6H_5(CH)_3NC_6H_5)Fe(CO)_2As(C_6H_5)_3$ ...... Fe: Org.Verb.B5-114, 118
$AsC_{35}Fe_2H_{23}N_2O_5$ ... $(CO)_3Fe(C_{12}H_8N_2)Fe(CO)_2As(C_6H_5)_3$ ...... Fe: Org.Verb.C1-158/60
$AsC_{35}Fe_2H_{25}O_5S_2$ ... $(CO)_3Fe(SC_6H_5)_2Fe(CO)_2As(C_6H_5)_3$ .... Fe: Org.Verb.C1-98
$AsC_{36}Cl_3H_{30}Sn$ ..... $[(C_6H_5)_4As][(C_6H_5)_2SnCl_3]$ .... Sn: Org.Verb.6-174/5
$AsC_{36}Cl_4H_{28}IrN_2$ .... $[As(C_6H_5)_4][Ir(C_{12}H_8N_2)Cl_4]$ .... Ir: SVol.2-91
$AsC_{36}Cl_4H_{50}IrP_2$ .... $[(C_6H_5)_4As][Ir(P(C_2H_5)_3)_2Cl_4]$ .... Ir: SVol.2-203
$AsC_{36}F_{15}H_{15}In$ ...... $(C_6F_5)_3In$ · $As(C_6H_5)_3$ .... F: PerFHalOrg.4-129/30
$AsC_{36}F_{15}H_{15}InO$ ..... $(C_6F_5)_3In$ · $OAs(C_6H_5)_3$ .... F: PerFHalOrg.4-129/30
$AsC_{36}FeH_{27}O_4$ ...... $C_6H_5C_2H_2C(O)C_6H_5Fe(CO)_3As(C_6H_5)_3$ ..... Fe: Org.Verb.B4-201, 205/6
$AsC_{36}H_{41}O_3PtS_6$ .... $[(C_6H_5)_4As][Pt(S_2COC_3H_7)_3]$ .... C: MVol.D4-261
$AsC_{41}F_4Fe_2H_{31}O_5PSb$ $(C_4F_4P(C_6H_5)_2As(CH_3)_2)Fe_2(CO)_5Sb(C_6H_5)_3$ Fe: Org.Verb.C2-135/7, 139
$AsC_{41}F_4Fe_2H_{31}O_5P_2$. $(C_4F_4P(C_6H_5)_2As(CH_3)_2)Fe_2(CO)_5P(C_6H_5)_3$ .. Fe: Org.Verb.C2-135/8
$AsC_{41}F_4Fe_2H_{31}O_8P_2$. $(C_4F_4P(C_6H_5)_2As(CH_3)_2)Fe_2(CO)_5P(OC_6H_5)_3$ Fe: Org.Verb.C2-135/8
$AsC_{42}H_{35}O_3Sn_2$ ..... $[(C_6H_5)_3SnO]_2As(O)C_6H_5$ .... Sn: Org.Verb.5-174
$AsC_{42}H_{55}N_6O_3S_4U$ ... $[(C_2H_5)_4N]_2[UO_2(C_4N_2S_2)_2$ · $(C_6H_5)_3AsO]$ ... U: SVol.E1-197, 201
$AsC_{44}Cl_4H_{50}IrP_2$ .... $[(C_6H_5)_4As][Ir(P(C_2H_5)_2C_6H_5)_2Cl_4]$ .... Ir: SVol.2-215
$AsC_{44}Cr_2H_{33}O_{13}U$ ... $UO_2(CH_3COCHCOC_6H_5Cr(CO)_3)_2$ · $(C_6H_5)_3AsO$ U: SVol.E1-197, 201
$AsC_{46}FeH_{35}O_4Sn$ ... $[As(C_6H_5)_4][(CO)_4FeSn(C_6H_5)_3]$ .... Fe: Org.Verb.B2-143
$AsC_{46}FeH_{35}O_7U$ .... $UO_2Fe(C_5H_4COCHCOC_6H_5)_2$ · $(C_6H_5)_3AsO$ U: SVol.E1-196, 201
$AsC_{47}F_4Fe_2H_{38}O_4P_3$. $(C_4F_4P(C_6H_5)_2As(CH_3)_2)Fe_2(CO)_4(CH_2(P(C_6H_5)_2)_2)$ Fe: Org.Verb.C2-135/7, 141
$AsC_{48}F_4Fe_2H_{38}O_5P_3$. $(C_4F_4P(C_6H_5)_2As(CH_3)_2)Fe_2(CO)_5(CH_2(P(C_6H_5)_2)_2)$ Fe: Org.Verb.C2-135/7, 139
$AsC_{48}F_4Fe_2H_{40}O_4P_3$. $(C_4F_4P(C_6H_5)_2As(CH_3)_2)Fe_2(CO)_4(C_2H_4(P(C_6H_5)_2)_2)$ Fe: Org.Verb.C2-135/7, 141
$AsC_{48}H_{37}O_5S_2U$ ..... $UO_2(C_6H_5COCHCSC_6H_5)_2$ · $(C_6H_5)_3AsO$ ... U: SVol.E1-197, 201
$AsC_{49}F_4Fe_2H_{40}O_5P_3$. $(C_4F_4P(C_6H_5)_2As(CH_3)_2)Fe_2(CO)_5(C_2H_4(P(C_6H_5)_2)_2)$ Fe: Org.Verb.C2-135/7, 139
$AsC_{49}F_4Fe_2H_{42}O_4P_3$ .. $(C_4F_4P(C_6H_5)_2As(CH_3)_2)Fe_2(CO)_4(C_3H_6(P(C_6H_5)_2)_2)$ Fe: Org.Verb.C2-135/7, 141, 143
$AsC_{50}F_4Fe_2H_{42}O_5P_3$. $(C_4F_4P(C_6H_5)_2As(CH_3)_2)Fe_2(CO)_5(C_3H_6(P(C_6H_5)_2)_2)$ Fe: Org.Verb.C2-135/7, 139
$AsC_{50}F_8Fe_2H_{36}O_4P_3$. $(C_4F_4P(C_6H_5)_2As(CH_3)_2)Fe_2(CO)_4(C_4F_4(P(C_6H_5)_2)_2)$ Fe: Org.Verb.C2-135/7, 141
– .... $(C_4F_4(P(C_6H_5)_2)_2)Fe_2(CO)_4(C_4F_4P(C_6H_5)_2As(CH_3)_2)$ Fe: Org.Verb.C2-135/7, 143
$AsC_{50}H_{71}N_6O_3S_4U$ ... $[(C_3H_7)_4N]_2[UO_2(C_4N_2S_2)_2$ · $(C_6H_5)_3AsO]$ ... U: SVol.E1-197, 201
$AsC_{51}F_{10}Fe_2H_{36}O_4P_3$. $(C_4F_4P(C_6H_5)_2As(CH_3)_2)Fe_2(CO)_4(C_5F_6(P(C_6H_5)_2)_2)$ Fe: Org.Verb.C2-135/7, 141
$AsC_{52}F_{12}Fe_2H_{36}O_4P_3$. $(C_4F_4P(C_6H_5)_2As(CH_3)_2)Fe_2(CO)_4(C_6F_8(P(C_6H_5)_2)_2)$ Fe: Org.Verb.C2-135/7, 141
$AsC_{52}FeH_{40}NO_4P_2$ ... $[N(P(C_6H_5)_3)_2][(CO)_4FeAs(C_6H_5)_2]$ .... Fe: Org.Verb.B2-52
$AsC_{52}FeH_{40}N_5O$ .... $[As(C_6H_5)_4][(C_6H_5NC)_4FeNO]$ .... Fe: Org.Verb.B4-71/2
$AsC_{54}ClH_{45}IrO_2P_2$ ... $Ir(O_2)(P(C_6H_5)_3)_2(As(C_6H_5)_3)Cl$ .... Ir: SVol.2-22

$AsC_{54}Cl_2H_{46}IrP_2$ .... $Ir(P(C_6H_5)_3)_2HCl_2(As(C_6H_5)_3)$ .... Ir: SVol.2-229
$AsC_{54}FH_{45}Ir$ .... $Ir(As(C_6H_5)_3)_3F$ .... Ir: SVol.2-254
$AsC_{54}H_{45}O_4Sn_3$ .... $[(C_6H_5)_3Sn]_3AsO_4$ .... Sn: Org.Verb.5-174
$AsC_{56}Fe_2H_{45}O_7U$ .... $UO_2(C_5H_5FeC_5H_4COCHCOC_6H_5)_2 \cdot (C_6H_5)_3AsO$
Fe: Org.Verb.A3-39
U: SVol.E1-197, 201
$AsC_{57}Cl_2H_{52}IrP_2$ .... $Ir(P(C_6H_5)_3)_2HCl_2(As(C_6H_4CH_3)_3)$ .... Ir: SVol.2-229
$AsC_{58}Cl_2FeH_{50}N_5O_8$ $[(CH_3C_6H_4NC)_5Fe(As(C_6H_5)_3)][ClO_4]_2$ .... Fe: Org.Verb.B4-95
$AsC_{58}F_4Fe_2H_{46}O_4PSb_2$
$(C_4F_4P(C_6H_5)_2As(CH_3)_2)Fe_2(CO)_4(Sb(C_6H_5)_3)_2$
Fe: Org.Verb.C2-135/7, 139
$AsC_{58}F_4Fe_2H_{46}O_{10}P_3$. $(C_4F_4P(C_6H_5)_2As(CH_3)_2)Fe_2(CO)_4(P(OC_6H_5)_3)_2$
Fe: Org.Verb.C2-135/7, 139, 144
$AsC_{58}FeH_{50}N_5^{2+}$ .... $[(CH_3C_6H_4NC)_5Fe(As(C_6H_5)_3)]^{2+}$ .... Fe: Org.Verb.B4-95
$AsC_{60}ClF_6H_{50}IrN_2P_3$ .. $[Ir(N_2C_6H_5)(As(C_6H_5)_3)Cl(P(C_6H_5)_3)_2]PF_6$ ... Ir: SVol.2-65
$AsC_{60}Cl_4H_{50}IrSb_2$ ... $(C_6H_5)_4As[IrCl_4(Sb(C_6H_5)_3)_2]$ .... Ir: SVol.2-262
$AsC_{75}F_4Fe_2H_{61}O_{12}P_4$. $(C_4F_4P(C_6H_5)_2As(CH_3)_2)Fe_2(CO)_3(P(OC_6H_5)_3)_3$
Fe: Org.Verb.C2-135/7, 140
$AsC_{75}H_{60}P_3S_9$ .... $[(C_6H_5)_4P]_3[As(CS_3)_3]$ .... C: MVol.D4-223
$AsCa_{0.5}O_4Th_{0.5}$ .... $Ca_{0.5}Th_{0.5}AsO_4$ .... Th: SVol.C2-33
$AsCd_{0.5}O_4Th_{0.5}$ .... $Cd_{0.5}Th_{0.5}AsO_4$ .... Th: SVol.C2-33
$AsCeH_2O_3^{3+}$ .... $CeH_2AsO_3^{3+}$ .... Sc: MVol.B7-80
$AsCeH_3O_3^{4+}$ .... $CeH_3AsO_3^{4+}$ .... Sc: MVol.B7-80
$AsCeO_3$ .... $CeAsO_3 \cdot 3\ H_2O$ .... Sc: MVol.B7-29
$AsCl_2O_3Sn$ .... $SnAsCl_2O_3$ .... Sn: MVol.C2-281
$AsCl_3$ .... $AsCl_3$ systems
$AsCl_3$-$SnCl_4$ .... Sn: MVol.C2-280/1
$AsCl_3F_6S$ .... $[SCl_3][AsF_6]$ .... S: SVol.2-273, 290/1
$AsCl_3F_6Te$ .... $[TeCl_3][AsF_6]$ .... Te: SVol.B2-94, 108, 112
$AsCl_5OS$ .... $[SOCl][AsCl_4]$ .... S: SVol.1-56
$AsCl_{11}SnTe$ .... $[TeCl_3][AsCl_2][SnCl_6]$ .... Sn: MVol.C2-284
$AsCl_{16}SbTe_2$ .... $[TeCl_3]_2[AsCl_4]SbCl_6$ .... Te: SVol.B2-94
$AsCsO_6U$ .... $CsUO_2AsO_4 \cdot n\ H_2O$ .... U: SVol.C3-174/5
$AsDyO_3$ .... $DyAsO_3 \cdot 3\ H_2O$ .... Sc: MVol.B7-29
$AsErO_3$ .... $ErAsO_3 \cdot 3\ H_2O$ .... Sc: MVol.B7-29
AsF^+ .... AsF^+ .... F: PerFHalOrg.3-216
$AsFHKO_6U$ .... $K[UO_2AsO_4 \cdot HF] \cdot H_2O$ .... U: SVol.C3 177
AsF_2^+ .... AsF_2^+ .... F: PerFHalOrg.3-216
AsF_2NOS^+ .... $AsNSOF_2^+$ .... S: S-N-Verb.1-89
AsF_3NOS^+ .... $FAsNSOF_2^+$ .... S: S-N-Verb.1-89
$AsF_3N_2O_3S_2^+$ .... $OAsNSOF(NSOF_2)^+$ .... S: S-N-Verb.1-89
AsF_4NOS^+ .... $AsF_2NSOF_2^+$ .... S: S-N-Verb.1-89
$AsF_4N_2O_2S_2^+$ .... $As(NSOF_2)_2^+$ .... S: S-N-Verb.1-89
$AsF_5N_2O_2S_2^+$ .... $FAs(NSOF_2)_2^+$ .... S: S-N-Verb.1-89
$AsF_6N_3O_3S_3$ .... $As[NS(O)F_2]_3$ .... S: S-N-Verb.1-88/9
$AsF_6N_3O_3S_3^+$ .... $As(NSOF_2)_3^+$ .... S: S-N-Verb.1-89
AsF_7OS .... $SOF_2 \cdot AsF_5$ .... S: SVol.1-17
AsF_7S_2 .... $[S_2F][AsF_6]$ .... S: SVol.2-22, 31
AsF_7Sn .... $[SnF][AsF_6] = SnF_2 \cdot AsF_5$ .... Sn: MVol.C2-280

Formula	Compound	Reference
AsF_8NS	$NSF_3 \cdot AsF_5$	S: S-N-Verb.1-155
AsF_9S	$[SF_3][AsF_6]$	S: SVol.2-86, 210
$AsF_{11}OTeXe$	$[XeTeF_5O][AsF_6]$	Te: SVol.B2-58
$AsF_{15}O_3Te_3$	$As(TeF_5O)_3$	Te: SVol.B2-35
$AsF_{17}O_3Te_3$	$[AsF_2(TeF_5O)_3]$	Te: SVol.B2-35
AsGa	GaAs, doped with Sn	Sn: MVol.C4-111/2
–	GaAs systems	
	GaAs-Sn	Sn: MVol.C4-109/11
$AsGdO_3$	$GdAsO_3 \cdot 3\ H_2O$	Sc: MVol.B7-29
$AsHO_4Sn$	$SnHAsO_4$	Sn: MVol.C2-277/8
$AsHO_6U$	$UO_2HAsO_4 \cdot n\ H_2O$	U: SVol.C3-173/4, 175
AsH_3O_7U	$(H_3O)UO_2AsO_4 \cdot 3\ H_2O$	U: SVol.C3-173, 175/6
AsH_4NO_6U	$NH_4UO_2AsO_4 \cdot 3\ H_2O$	U: SVol.C3-174/6
AsH_6KO_9U	$[K(H_2O)_3][UO_2AsO_4]$	U: SVol.C3-176
$AsH_9O_{10}U$	$H(H_2O)_4[UO_2AsO_4]$	
	$= [(H_3O)(H_2O)_3][UO_2AsO_4]$	U: SVol.C3-173/4, 176
$AsH_{10}NO_9U$	$[NH_4(H_2O)_3][UO_2AsO_4]$	U: SVol.C3-176
AsI_3	AsI_3 solid solutions	
	AsI_3-TeI_4	Te: SVol.B3-62
–	AsI_3 systems	
	AsI_3-SnI_4	Sn: MVol.C2-281
AsIn	InAs solid solutions	
	InAs-$CdSnAs_2$	Sn: MVol.C4-127/32
	InAs-$SnIn_2Te$	Sn: MVol.C4-133/4
–	InAs systems	
	InAs-$CdSnAs_2$	Sn: MVol.C4-127
	InAs-$CdSnAs_2$-$ZnSnAs_2$	Sn: MVol.C4-132
	InAs-Sn	Sn: MVol.C4-124
	InAs-$ZnSnAs_2$	Sn: MVol.C4-125/7
AsInSeSn	SnAs · InSe	Sn: MVol.C4-132/3
AsIrS	IrAsS	Ir: SVol.2-248
AsIrSe	IrAsSe	Ir: SVol.2-248
AsIrTe	IrAsTe	Ir: SVol.2-248
$AsKO_6U$	$K[UO_2AsO_4] \cdot H_2O$	U: SVol.C3-177
–	$K[UO_2AsO_4] \cdot 3\ H_2O$	U: SVol.C3-174/5, 178
$AsLaO_3$	$LaAsO_3 \cdot 3\ H_2O$	Sc: MVol.B7-29
$AsLiMoO_6$	$LiMoO_2AsO_4$	Mo: SVol.B1-125
$AsLiO_6U$	$LiUO_2AsO_4$	U: SVol.C3-175
$AsLuO_3$	$LuAsO_3 \cdot 3\ H_2O$	Sc: MVol.B7-29
$AsMn_3N$	Mn_3AsN solid solutions	
	Mn_3AsN-Mn_3SbN	Mn: MVol.C3-260
$AsNdO_3$	$NdAsO_3 \cdot 3\ H_2O$	Sc: MVol.B7-29
AsO_3Pr	$PrAsO_3 \cdot 3\ H_2O$	Sc: MVol.B7-29
AsO_3Sm	$SmAsO_3 \cdot 3\ H_2O$	Sc: MVol.B7-29
AsO_3Y	$YAsO_3 \cdot 3\ H_2O$	Sc: MVol.B7-29
AsO_3Yb	$YbAsO_3 \cdot 3\ H_2O$	Sc: MVol.B7-29
AsO_4Pa^{2+}	$Pa(AsO_4)^{2+}$	Pa: SVol.2-158
$AsO_4Pb_{0.5}Th_{0.5}$	$Pb_{0.5}Th_{0.5}AsO_4$	Th: SVol.C2-33

Formula	Compound		Reference
$AsO_4Sr_{0.5}Th_{0.5}$	$Sr_{0.5}Th_{0.5}AsO_4$	Th:	SVol.C2-33
AsO_6RbU	$RbUO_2AsO_4 \cdot n\ H_2O$	U:	SVol.C3-174/5
AsO_6U^-	$[UO_2AsO_4]^-$	U:	SVol.C3-173/7
AsPu	PuAs	Np:	TrU.B2-41
AsSn	SnAs	Sn:	MVol.C2-271/7
–	SnAs systems		
	SnAs-InSe	Sn:	MVol.C4-132/3
	SnAs-SnSb	Sn:	MVol.C2-295
	SnAs-SnTe	Sn:	MVol.C2-284
	$SnAs\text{-}SnTe\text{-}Sn_4As_3$	Sn:	MVol.C2-284
$As_{1.5}NaO_{12}Th_2V_{1.5}$	$NaTh_2(V_{0.5}As_{0.5}O_4)_3$	Th:	SVol.C2-33, 125
$As_2B_{0.4}Mg_6O_{11.6}$	$0.2\ B_2O_3 \cdot As_2O_5 \cdot 6\ MgO$	B:	B-Verb.7-210
$As_2BBrC_{15}F_{12}H_{15}N_3$	$[(C_5H_5N)_3BBr][AsF_6]_2$	B:	B-Verb.3-147
		B:	B-Verb.10-163
$As_2BBrC_{21}F_{12}H_{27}N_3$	$[((CH_3)_2C_5H_3N)_3BBr][AsF_6]_2$	B:	B-Verb.3-147
		B:	B-Verb.10-163
$As_2BBr_3C_{37}ClH_{30}ORh$	$[(C_6H_5)_3As]_2Rh(CO)Cl \cdot BBr_3$	B:	B-Verb.3-164
$As_2BC_6ClH_{20}$	$[((CH_3)_3As)_2BH_2]Cl$	B:	B-Verb.10-97, 100, 103
$As_2BC_6F_6H_{20}P$	$[(As(CH_3)_3)_2BH_2][PF_6]$	B:	B-Verb.3-146
		B:	B-Verb.10-97/8, 100/1, 103
$As_2BC_6H_{20}^+$	$[((CH_3)_3As)_2BH_2]^+$	B:	B-Verb.3-146
		B:	B-Verb.10-97/101, 103
$As_2BC_6H_{20}I$	$[((CH_3)_3As)_2BH_2]I$	B:	B-Verb.10-97, 99, 103
$As_2BC_{12}H_{30}N$	$[(C_2H_5)_2As]_2BN(C_2H_5)_2$	B:	B-Verb.3-145
		B:	B-Verb.5-224/5
$As_2BC_{36}F_4H_{30}IrN_2O_2$	$[Ir(NO)_2(As(C_6H_5)_3)_2]BF_4$	Ir:	SVol.2-59
$As_2BC_{36}F_4H_{31}IrNO_2$	$[Ir(NO)(OH)(As(C_6H_5)_3)_2]BF_4$	Ir:	SVol.2-60
$As_2BC_{41}F_4H_{35}Ni$	$[C_5H_5Ni(As(C_6H_5)_3)_2][BF_4]$	Ni:	Org.Verb.2-137, 139
$As_2BC_{44}F_4FeH_{35}N_2O_2$	$[(CO)_2Fe(As(C_6H_5)_3)_2N_2C_6H_5]BF_4$	Fe:	Org.Verb.B1-117/9
$As_2B_{1.6}Mg_6O_{13.4}$	$0.8\ B_2O_3 \cdot As_2O_5 \cdot 6\ MgO$	B:	B-Verb.7-210
$As_2B_2BiC_{54}F_8H_{45}O_2$	$[(C_6H_5)_3Bi((C_6H_5)_3AsO)_2](BF_4)_2$	Bi:	Org.Verb.-138/9
$As_2B_2BiC_{57}F_8H_{51}O_3$	$[(C_6H_5)_3Bi((C_6H_5)_3AsO)_2](BF_4)_2 \cdot (CH_3)_2CO$	Bi:	Org.Verb.-138/9
$As_2B_2Ca_4O_{12}$	$B_2O_3 \cdot As_2O_5 \cdot 4\ CaO \cdot 4\ H_2O = Ca_2[B(OH)_4AsO_4]$	B:	B-Verb.7-210
$As_2B_3C_{36}CuH_{38}$	$[(C_6H_5)_3As]_2CuB_3H_8$	B:	B-Verb.3-188
$As_2B_8C_{48}H_{48}$	$[(C_6H_5)_4As]_2B_8H_8$	B:	B-Verb.3-146
$As_2B_8H_8S$	$B_8H_8As_2S$	B:	B-Verb.3-146
$As_2B_9C_6H_{23}$	$(As(CH_3)_2)_2C_2B_9H_{11}$	B:	B-Verb.6-36/7, 140
$As_2B_9C_8H_{27}$	$(As(CH_3)_2)_2(CH_3)_2C_2B_9H_9$	B:	B-Verb.6-36/7, 140
$As_2B_{10}C_6H_{22}$	$B_{10}H_{10}C_2(As(CH_3)_2)_2$	B:	B-Verb.12-62, 264, 271
$As_2B_{10}C_8H_{22}NiO_2$	$(CO)_2NiB_{10}H_{10}C_2(As(CH_3)_2)_2$	B:	B-Verb.12-129, 148, 151
		Ni:	Org.Verb.1-155/6
$As_2B_{10}C_9FeH_{22}O_3$	$(CO)_3Fe(CH_3)_2AsCB_{10}H_{10}CAs(CH_3)_2$	B:	B-Verb.12-129, 148, 151
		Fe:	Org.Verb.B1-170/1, 175
$As_2B_{10}C_{10}H_{22}MoO_4$	$B_{10}H_{10}C_2(As(CH_3)_2)_2Mo(CO)_4$	B:	B-Verb.12-129, 148
$As_2B_{10}C_{12}H_{42}$	$[(C_2H_5)_3As]_2B_{10}H_{12}$	B:	B-Verb.3-146
		B:	B-Verb.20-153/4, 161
$As_2B_{10}C_{26}H_{30}$	$B_{10}H_{10}C_2(As(C_6H_5)_2)_2$	B:	B-Verb.12-62

Formula	Compound		Reference
$As_2B_{10}C_{36}H_{42}$	$[(C_6H_5)_3As]_2B_{10}H_{12}$	B:	B-Verb.3-146
		B:	B-Verb.20-153
$As_2B_{10}C_{48}Cl_2H_{52}Sn$	$[(C_6H_5)_4As]_2[(B_{10}H_{12})SnCl_2]$	B:	B-Verb.3-198
$As_2B_{10}C_{48}H_{50}$	$[(C_6H_5)_4As]_2[B_{10}H_{10}]$	B:	B-Verb.20-182, 184, 187
$As_2B_{10}C_{48}H_{52}$	$[(C_6H_5)_4As]_2[B_{10}H_{12}]$	B:	B-Verb.20-178/80
$As_2B_{10}C_{50}Cl_2H_{58}Sn$	$[(C_6H_5)_4As]_2[B_{10}H_{12}Sn(CH_3)_2Cl_2]$	Sn:	Org.Verb.6-37
$As_2B_{10}C_{52}Cl_2H_{62}Sn$	$[(C_6H_5)_4As]_2[B_{10}H_{12}(C_2H_5)_2SnCl_2]$	Sn:	Org.Verb.6-62
$As_2B_{11}Br_4C_{48}H_{47}$	$[(C_6H_5)_4As]_2[Br_4B_{11}H_7]$	B:	B-Verb.3-146
		B:	B-Verb.14-296
$As_2B_{11}C_{48}H_{51}$	$[(C_6H_5)_4As]_2[B_{11}H_{11}]$	B:	B-Verb.3-146
		B:	B-Verb.20-214
$As_2B_{12}C_6H_{28}$	$[(CH_3)_3As]_2 \cdot B_{12}H_{10}$	B:	B-Verb.20-237
$As_2B_{12}C_{48}H_{52}$	$[(C_6H_5)_4As]_2[B_{12}H_{12}]$	B:	B-Verb.3-146
		B:	B-Verb.20-221
$As_2B_{12}Ca_4H_{12}MgO_{34}$	$Ca_4Mg[AsO_4(B_6O_7(OH)_6)]_2 \cdot 14\ H_2O$ $= 6\ B_2O_3 \cdot As_2O_5 \cdot 4\ CaO \cdot MgO \cdot 20\ H_2O$	B:	B-Verb.7-210/11
$As_2B_{12}Ca_4MgO_{28}$	$6\ B_2O_3 \cdot As_2O_5 \cdot 4\ CaO \cdot MgO \cdot 20\ H_2O$ $= Ca_4Mg[AsO_4(B_6O_7(OH)_6)]_2 \cdot 14\ H_2O$	B:	B-Verb.7-210/1
As_2B_{13}	$B_{13}As_2$	B:	B-Verb.3-144
$As_2B_{18}C_{44}Cr_2FeH_{92}N_2O_{10}$	$[N(C_4H_9)_4]_2[(Cr(CO)_5CAsB_9H_{10})_2Fe]$	Fe:	Org.Verb.B1-19, 21
$As_2B_{20}C_6H_{26}$	$CH_3As(CB_{10}H_{10}C)_2AsCH_3$	B:	B-Verb.12-135, 152/3, 282
$As_2B_{20}C_{50}H_{64}N_2Ni$	$[As(C_6H_5)_4]_2[(H_2NCB_{10}H_{10})_2Ni]$	Ni:	Org.Verb.1-100
$As_2BaO_{12}U_2$	$Ba(UO_2AsO_4)_2 \cdot n\ H_2O$	U:	SVol.C3-137, 178/9
$As_2Ba_3O_8$	$Ba_3(AsO_4)_2$ solid solutions $Ba_3(AsO_4)_2$-$Ba_3(MnO_4)_2$	Mn:	MVol.C2-9, 258/9
$As_2BiC_{54}Cl_2H_{45}O_{10}$	$[(C_6H_5)_3Bi((C_6H_5)_3AsO)_2](ClO_4)_2$	Bi:	Org.Verb.-138/9
$As_2BiC_{54}F_{12}H_{45}O_2P_2$	$[(C_6H_5)_3Bi((C_6H_5)_3AsO)_2](PF_6)_2$	Bi:	Org.Verb.-138/9
$As_2BiC_{54}H_{45}N_2O_8$	$[(C_6H_5)_3Bi((C_6H_5)_3AsO)_2](NO_3)_2$	Bi:	Org.Verb.-140
$As_2BiC_{54}H_{45}O_2^{2+}$	$[(C_6H_5)_3Bi((C_6H_5)_3AsO)_2]^{2+}$	Bi:	Org.Verb.-138/9
$As_2BiC_{56.5}F_{12}H_{49}O_3P_2$	$[(C_6H_5)_3Bi((C_6H_5)_3AsO)_2](PF_6)_2 \cdot 0.5\ CH_2(COCH_3)_2$	Bi:	Org.Verb.-138/9
$As_2BiC_{57}F_{12}H_{51}O_3P_2$	$[(C_6H_5)_3Bi((C_6H_5)_3AsO)_2](PF_6)_2 \cdot (CH_3)_2CO$	Bi:	Org.Verb.-138/9
$As_2BiC_{59}F_{12}H_{55}O_3P_2$	$[(C_6H_5)_3Bi((C_6H_5)_3AsO)_2](PF_6)_2 \cdot (C_2H_5)_2CO$	Bi:	Org.Verb.-138/9
$As_2Bi_2C_{72}Cl_2H_{60}O_{11}$	$[((C_6H_5)_3Bi(C_6H_5)_3AsO)_2O](ClO_4)_2$	Bi:	Org.Verb.-145
$As_2Bi_4O_{14}U$	$(BiO)_4(UO_2)(AsO_4)_2 \cdot 3\ H_2O$	U:	SVol.C3-179, 183
$As_2BrC_{18}F_3H_{22}Ni$	$CF_2CFNi(As(CH_3)_2C_6H_5)_2Br$	Ni:	Org.Verb.1-27, 29
$As_2BrC_{22}F_{10}H_{16}Tl$	$(C_6F_5)_2TlBr \cdot C_6H_4[As(CH_3)_2]_2$	F:	PerFHalOrg.4-137, 144
$As_2Br_2C_{10}Fe_2H_{12}O_6$	$Br(CO)_3Fe(As(CH_3)_2)_2Fe(CO)_3Br$	Fe:	Org.Verb.C1-8, 11/2
$As_2Br_2C_{12}FeH_{16}O_2$	$(CO)_2Fe[(CH_3)_2AsC_6H_4As(CH_3)_2]Br_2$	Fe:	Org.Verb.B1-101/2, 107, 110
$As_2Br_2C_{26}H_{24}O_4U$	$UO_2Br_2 \cdot C_2H_4(AsO(C_6H_5)_2)_2$	U:	SVol.E1-200/1
$As_2Br_2C_{28}H_{28}O_4U$	$UO_2Br_2 \cdot C_2H_4(CH_2AsO(C_6H_5)_2)_2$	U:	SVol.E1-200/1
$As_2Br_2C_{36}ClH_{30}NOsS$	$[OsClBr_2(NS)(As(C_6H_5)_3)_2]$	S:	SVol.2-261
$As_2Br_2C_{36}H_{30}IrNO$	$Ir(NO)Br_2(As(C_6H_5)_3)_2$	Ir:	SVol.2-59/60

Formula	Compound	Reference
$As_2Br_2C_{38}FeH_{30}O_2$	$(CO)_2Fe(As(C_6H_5)_3)_2Br_2$	Fe: Org.Verb.B1-101/2, 106, 110
$As_2Br_2C_{39}FeH_{30}HgO_3$	$(CO)_3Fe(As(C_6H_5)_3)_2 \cdot HgBr_2$	Fe: Org.Verb.B1-192/3
$As_2Br_2C_{48}H_{40}I_2Mn$	$[(C_6H_5)_4As]_2MnI_2Br_2$	Mn: MVol.C5-343
$As_2Br_3C_{18}H_{26}IrS_2$	$Ir((CH_3)_2AsC_6H_4SCH_3)_2Br_3 \cdot C_2H_5OH$	Ir: SVol.2-259
$As_2Br_3C_{20}H_{32}IrOS_2$	$Ir((CH_3)_2AsC_6H_4SCH_3)_2Br_3 \cdot C_2H_5OH$	Ir: SVol.2-259
$As_2Br_3C_{36}H_{30}Ir$	$Ir(As(C_6H_5)_3)_2Br_3$	Ir: SVol.2-253
$As_2Br_4C_6H_{18}O_2U$	$UBr_4 \cdot 2\ (CH_3)_3AsO$	U: SVol.E1-194/5
$As_2Br_4C_{12}H_{30}O_2U$	$UBr_4 \cdot 2\ (C_2H_5)_3AsO$	U: SVol.E1-194/5
$As_2Br_4C_{12}H_{30}Sn$	$SnBr_4 \cdot 2\ (C_2H_5)_3As$	Sn: MVol.C6-177/8
$As_2Br_4C_{14}H_{24}Sn$	$SnBr_4 \cdot (C_2H_5)_2AsC_6H_4As(C_2H_5)_2$	Sn: MVol.C6-180
$As_2Br_4C_{16}H_{23}Ir$	$H[Ir(As(CH_3)_2C_6H_5)_2Br_4]$	Ir: SVol.2-255
$As_2Br_4C_{16}H_{26}IrN$	$NH_4[Ir(As(CH_3)_2C_6H_5)_2Br_4]$	Ir: SVol.2-255
$As_2Br_4C_{21}H_{28}IrN$	$C_5H_5NH[Ir(As(CH_3)_2C_6H_5)_2Br_4]$	Ir: SVol.2-255
$As_2Br_4C_{26}H_{24}O_2Sn$	$SnBr_4 \cdot C_2H_4(AsO(C_6H_5)_2)_2$	Sn: MVol.C6-181
$As_2Br_4C_{26}H_{27}Ir$	$H[Ir(AsCH_3(C_6H_5)_2)_2Br_4]$	Ir: SVol.2-255
$As_2Br_4C_{26}H_{30}IrN$	$NH_4[Ir(AsCH_3(C_6H_5)_2)_2Br_4]$	Ir: SVol.2-255
$As_2Br_4C_{28}H_{28}O_2Sn$	$SnBr_4 \cdot C_4H_8(AsO(C_6H_5)_2)_2$	Sn: MVol.C6-181
$As_2Br_4C_{36}H_{30}O_2Sn$	$SnBr_4 \cdot 2\ (C_6H_5)_3AsO$	Sn: MVol.C6-180/1
$As_2Br_4C_{36}H_{30}O_2U$	$UBr_4 \cdot 2\ (C_6H_5)_3AsO$	U: SVol.E1-194/5
$As_2Br_4C_{36}H_{30}Sn$	$SnBr_4 \cdot 2\ (C_6H_5)_3As$	Sn: MVol.C6-178
$As_2Br_4C_{38}H_{36}Mn$	$[(CH_3)(C_6H_5)_3As]_2MnBr_4$	Mn: MVol.C5-289, 305
$As_2Br_4C_{48}H_{40}Mn$	$[(C_6H_5)_4As]_2MnBr_4$	Mn: MVol.C5-305
$As_4Br_4C_{72}H_{60}O_4U$	$UBr_4 \cdot 4\ (C_6H_5)_3AsO$	U: SVol.E1-194, 196
$As_2Br_5C_{48}H_{40}Ir$	$[(C_6H_5)_4As]_2[IrBr_5]$	Ir: SVol.2-147
$As_2Br_5C_{48}H_{40}OU$	$[As(C_6H_5)_4]_2UOBr_5$	U: SVol.C9-149/50
$As_2Br_5C_{49}H_{40}IrO$	$[(C_6H_5)_4As]_2[Ir(CO)Br_5]$	Ir: SVol.2-147
$As_2Br_6C_{48}H_{40}Ir$	$[(C_6H_5)_4As]_2[IrBr_6]$	Ir: SVol.2-147
$As_2Br_8C_{40}H_{34}O_4Sn_2$	$[SnBr_3((C_6H_5)_2AsC_6H_4COOCH_3)_2][SnBr_5]$	Sn: MVol.C6-179
$As_2Br_8C_{44}H_{42}O_4Sn_2$	$[SnBr_3((CH_3C_6H_4)_2AsC_6H_4COOCH_3)_2][SnBr_5]$	Sn: MVol.C6-179
$As_2Br_{10}Te$	$[AsBr_2]_2TeBr_6$	Te: SVol.B3-12
$As_2C_2F_6H_2O_5$	$CF_3(OH)As(O)O(O)As(OH)CF_3$	F: PerFHalOrg.3-197, 198
$As_2C_2H_2K_2O_{16}U_2$	$K_2[(UO_2)_2(HAsO_4)_2(C_2O_4)] \cdot 4\ H_2O$	U: SVol.C3-177
$As_2C_4F_6H_6$	$(CF_3)_2AsAs(CH_3)_2$	F: PerFHalOrg.3-224/5
$As_2C_4F_{12}$	$(CF_3)_2AsAs(CF_3)_2$	F: PerFHalOrg.3-222/5
$As_2C_4F_{12}Fe_2N_4O_4$	$(NO)_2Fe(As(CF_3)_2)_2Fe(NO)_2$	Fe: Org.Verb.B1-143
$As_2C_4F_{12}HN$	$[(CF_3)_2As]_2NH$	F: PerFHalOrg.3-200/1
$As_2C_4F_{12}Hg$	$[(CF_3)_2As]_2Hg$	F: PerFHalOrg.3-225
$As_2C_4F_{12}O$	$(CF_3)_2AsOAs(CF_3)_2$	F: PerFHalOrg.3-196/9
$As_2C_4F_{12}S$	$(CF_3)_2AsSAs(CF_3)_2$	F: PerFHalOrg.3-212
$As_2C_4Fe_2H_{12}N_4O_4$	$(NO)_2Fe(As(CH_3)_2)_2Fe(NO)_2$	Fe: Org.Verb.B1-143
$As_2C_5F_3H_{12}P$	$[(CH_3)_2As]_2PCF_3$	F: PerFHalOrg.3-7
$As_2C_5F_{12}FeHgO_5$	$[(CO)_5FeHg][AsF_6]_2$	Fe: Org.Verb.B3-251
$As_2C_6Cl_4H_{18}O_2U$	$UCl_4 \cdot 2\ (CH_3)_3AsO$	U: SVol.E1-194/5
$As_2C_6Fe_2H_{12}N_4O_6$	$[(NO)_2(CO)Fe]_2As(CH_3)_2As(CH_3)_2$	Fe: Org.Verb.C1-67, 69/70
$As_2C_6H_{18}O_4Sn$	$(CH_3)_2Sn(OAs(O)(CH_3)_2)_2$	Sn: Org.Verb.6-29
$As_2C_6H_{18}O_8Sn_3$	$[(OAs(O)OSn(CH_3)_2O)_2Sn(CH_3)_2]_x$	Sn: Org.Verb.6-29
$As_2C_8F_6FeH_{12}N_2O_2$	$Fe(NO)_2(C_2(CF_3)_2(As(CH_3)_2)_2)$	Fe: Org.Verb.B1-143

Formula	Compound		Reference
$As_2C_8F_{20}O$	$(C_2F_5)_2AsOAs(C_2F_5)_2$	F:	PerFHalOrg.3-197, 198
$As_2C_8FeH_{12}O_4$	$(CO)_4FeAs(CH_3)_2As(CH_3)_2$	Fe:	Org.Verb.B2-28
$As_2C_8H_{14}NiO_2$	$(CO)_2Ni(CH_3)_2AsCHCHAs(CH_3)_2$	Ni:	Org.Verb.1-155/6
$As_2C_8H_{18}NiO_2$	$(CO)_2Ni(As(CH_3)_3)_2$	Ni:	Org.Verb.1-144/5
$As_2C_8H_{18}NiO_2Si$	$(CO)_2Ni(CH_3)_2Si(As(CH_3)_2)_2$	Ni:	Org.Verb.1-155/6
$As_2C_8H_{18}NiO_2Sn$	$(CO)_2Ni(CH_3)_2Sn(As(CH_3)_2)_2$	Ni:	Org.Verb.1-155/6
$As_2C_8H_{18}NiO_8$	$(CO)_2Ni(As(OCH_3)_3)_2$	Ni:	Org.Verb.1-144
$As_2C_8H_{24}I_6Sn$	$[(CH_3)_4As]_2[SnI_6]$	Sn:	MVol.C3-120
$As_2C_9FeH_{18}O_3$	$(CO)_3Fe(As(CH_3)_3)_2$	Fe:	Org.Verb.B1-148/50, 155, 164
$As_2C_{10}CdF_{12}Fe_2O_{10}$	$[Cd(Fe(CO)_5)_2][AsF_6]_2$	Fe:	Org.Verb.B3-130
		Fe:	Org.Verb.C1-62
$As_2C_{10}Cl_4H_{16}Pa$	$PaCl_4 \cdot (CH_3)_2AsC_6H_4As(CH_3)_2$	Pa:	SVol.2-94/5
$As_2C_{10}Cl_4H_{16}Sn$	$SnCl_4 \cdot (CH_3)_2AsC_6H_4As(CH_3)_2$	Sn:	MVol.C6-180
$As_2C_{10}Cl_4H_{16}U$	$UCl_4[C_6H_4(As(CH_3)_2)_2]$	U:	SVol.E1-51/2
$As_2C_{10}F_8H_6$	$CH_3As(C_4F_4)_2AsCH_3$	F:	PerFHalOrg.4-31
$As_2C_{10}F_{12}Fe_2I_2O_6$	$I(CO)_3Fe(As(CF_3)_2)_2Fe(CO)_3I$	Fe:	Org.Verb.C1-9, 12
$As_2C_{10}F_{12}Fe_2O_6$	$(CO)_3Fe(As(CF_3)_2)_2Fe(CO)_3$	Fe:	Org.Verb.C1-164, 169, 179
$As_2C_{10}FeH_{16}N_2O_2$	$Fe(NO)_2((CH_3)_2AsC_6H_4As(CH_3)_2)$	Fe:	Org.Verb.B1-143
$As_2C_{10}Fe_2H_{12}I_2O_6$	$I(CO)_3Fe(As(CH_3)_2)_2Fe(CO)_3I$	Fe:	Org.Verb.C1-8, 11/2
$As_2C_{10}Fe_2H_{12}O_6$	$(CO)_3Fe(As(CH_3)_2)_2Fe(CO)_3$	Fe:	Org.Verb.C1-164/5, 177/9
$As_2C_{11}Cl_3H_{19}Ti$	$CH_3TiCl_3 \cdot C_6H_4(As(CH_3)_2)_2$	Ti:	Org.Verb.1-34/5
$As_2C_{11}F_6FeH_{12}O_3$	$(CO)_3Fe(CH_3)_2As(CCF_3)_2As(CH_3)_2$	Fe:	Org.Verb.B1-170/1, 174
$As_2C_{11}Fe_3O_{11}$	$As_2Fe_3(CO)_{11}$	Fe:	Org.Verb.B2-27
$As_2C_{12}Cl_4H_{30}O_2U$	$UCl_4 \cdot 2\ (C_2H_5)_3AsO$	U:	SVol.E1-3/4, 194/5, 198
$As_2C_{12}Cl_4H_{30}Sn$	$SnCl_4 \cdot 2\ (C_2H_5)_3As$	Sn:	MVol.C6-177/8
$As_2C_{12}F_4FeH_{12}O_4$	$(CO)_4FeC_4F_4(As(CH_3)_2)_2$	Fe:	Org.Verb.B2-100, 121
$As_2C_{12}F_4H_{12}NiO_2$	$(CO)_2NiC_6F_4(As(CH_3)_2)_2$	Ni:	Org.Verb.1-155
$As_2C_{12}F_{12}H_6$	$CH_3As(C_5F_6)_2AsCH_3$	F:	PerFHalOrg.4-31
$As_2C_{12}F_{12}Mn_2O_8$	$[MnAs(CF_3)_2(CO)_4]_2$	F:	PerFHalOrg.3-225
$As_2C_{12}F_{18}$	$(CF_3)_6C_6As_2$	F:	PerFHalOrg.3-187, 189/91
$As_2C_{12}FeH_{16}IO_2$	$(CO)_2Fe[(CH_3)_2AsC_6H_4As(CH_3)_2]I$	Fe:	Org.Verb.B1-101
$As_2C_{12}FeH_{16}I_2O_2$	$(CO)_2Fe[(CH_3)_2AsC_6H_4As(CH_3)_2]I_2$	Fe:	Org.Verb.B1-101/2, 107, 110
$As_2C_{12}Fe_2H_{12}O_8$	$(CO)_4Fe((CH_3)_2AsAs(CH_3)_2)Fe(CO)_4$	Fe:	Org.Verb.C1-36, 41
$As_2C_{12}Fe_2H_{12}O_9$	$(CO)_4Fe((CH_3)_2AsOAs(CH_3)_2)Fe(CO)_4$	Fe:	Org.Verb.C1-36, 41
$As_2C_{12}H_{13}O_8Pa$	$H_3PaO_2(C_6H_5AsO_3)_2$	Pa:	SVol.2-84/5, 182
$As_2C_{12}H_{16}NiO_2$	$(CO)_2NiC_6H_4(As(CH_3)_2)_2$	Ni:	Org.Verb.1-155/6
$As_2C_{12}H_{30}NiO_2Si_2$	$(As(CH_3)_2Si(CH_3)_3)_2Ni(CO)_2$	Ni:	Org.Verb.1-247, 248
$As_2C_{12}H_{30}NiO_2Sn_2$	$(As(CH_3)_2Sn(CH_3)_3)_2Ni(CO)_2$	Ni:	Org.Verb.1-247, 248
$As_2C_{13}F_6H_{16}NiO$	$(CF_3)_2CONi(C_6H_4(As(CH_3)_2)_2)$	Ni:	Org.Verb.1-60, 64
$As_2C_{13}FeH_{16}O_3$	$(CO)_3Fe(CH_3)_2AsC_6H_4As(CH_3)_2$	Fe:	Org.Verb.B1-170/1, 174, 178/9
$As_2C_{13}FeH_{18}O_9$	$(CO)_3Fe(As(OCH_2)_3CCH_3)_2$	Fe:	Org.Verb.B1-149
$As_2C_{14}Cl_4H_{24}Sn$	$SnCl_4 \cdot (C_2H_5)_2AsC_6H_4As(C_2H_5)_2$	Sn:	MVol.C6-180
$As_2C_{14}F_4Fe_2H_{12}O_6$	$(CO)_3Fe(C_4F_4(As(CH_3)_2)_2)Fe(CO)_3$	Fe:	Org.Verb.C2-130/1, 133/4
$As_2C_{14}F_4Fe_2H_{18}O_4S_2$	$(CO)_4(C_4F_4(As(CH_3)_2)_2)Fe_2(SCH_3)_2$	Fe:	Org.Verb.C1-107

Formula	Compound		Reference
$As_2C_{14}F_6FeH_{16}S_2^+$	$[Fe(C_6H_4(As(CH_3)_2)_2)C_2S_2(CF_3)_2]^+$	Fe:	Org.Verb.B1-34/5
$As_2C_{14}F_6Fe_2H_{12}O_6$	$[(CF_3)_2C_2(As(CH_3)_2)_2]Fe_2(CO)_6$	Fe:	Org.Verb.C2-130/2, 134
$As_2C_{14}F_8H_{16}Ni$	$C_4F_8Ni(CH_3)_2AsC_6H_4As(CH_3)_2$	Ni:	Org.Verb.1-345, 346
$As_2C_{14}H_{18}O_6Sn$	$(CH_3)_2Sn[OAsO(C_6H_5)OH]_2$	Sn:	Org.Verb.6-25
$As_2C_{14}H_{22}Ni_2$	$(C_5H_5NiAs(CH_3)_2)_2$	Ni:	Org.Verb.2-339, 341
$As_2C_{14}H_{30}NiO_2$	$(CO)_2Ni(As(C_2H_5)_3)_2$	Ni:	Org.Verb.1-144
$As_2C_{14}H_{30}NiO_8$	$(CO)_2Ni(As(OC_2H_5)_3)_2$	Ni:	Org.Verb.1-144
$As_2C_{15}F_6FeH_{16}OS_2$	$COFe[C_6H_4(As(CH_3)_2)_2]S_2C_2(CF_3)_2$	Fe:	Org.Verb.B1-34/6
$As_2C_{15}F_6FeH_{16}OS_2^+$	$[COFe(C_6H_4(As(CH_3)_2)_2)S_2C_2(CF_3)_2]^+$	Fe:	Org.Verb.B1-34/5
$As_2C_{15}F_6Fe_2H_{12}O_6$	$(CO)_3Fe(C_5F_6(As(CH_3)_2)_2)Fe(CO)_3$	Fe:	Org.Verb.C2-130/1
$As_2C_{15}FeH_{20}I_2NiO$	$CONi[(As(CH_3)_2C_5H_4)_2Fe]I_2$	Ni:	Org.Verb.1-121/2
$As_2C_{15}FeH_{36}N_6O_3$	$(CO)_3Fe(As(N(CH_3)_2)_3)_2$	Fe:	Org.Verb.B1-149
$As_2C_{16}F_4Fe_2H_{12}O_8$	$(CO)_4Fe(C_4F_4(As(CH_3)_2)_2)Fe(CO)_4$	Fe:	Org.Verb.C1-36, 41
$As_2C_{16}F_8Fe_2H_{12}O_6$	$(CO)_3Fe(C_6F_8(As(CH_3)_2)_2)Fe(CO)_3$	Fe:	Org.Verb.C2-130/2
$As_2C_{16}F_{10}FeO_4$	$(CO)_4Fe(AsC_6F_5)_2$	Fe:	Org.Verb.B2-152
$As_2C_{16}F_{12}H_{16}NiO_2$	$(CF_3)_4C_2O_2Ni[(CH_3)_2AsC_6H_4As(CH_3)_2]$	Ni:	Org.Verb.1-66
$As_2C_{16}F_{12}H_{17}NNiO$	$(CF_3)_4C_2HNONi[(CH_3)_2AsC_6H_4As(CH_3)_2]$	Ni:	Org.Verb.1-66
$As_2C_{16}F_{16}$	$C_6F_5AsC(CF_3)C(CF_3)AsC_6F_5$	F:	PerFHalOrg.3-187/9
$As_2C_{16}FeH_{20}NiO_2$	$(CO)_2Ni[Fe(C_5H_4As(CH_3)_2)_2]$	Ni:	Org.Verb.1-156
$As_2C_{16}Fe_2H_{30}HgN_2O_6$	$[NO(CO)_2FeAs(C_2H_5)_3]_2Hg$	Fe:	Org.Verb.C1-64/5
$As_2C_{16}H_{22}I_4Sn$	$SnI_4 \cdot 2\ C_6H_5As(CH_3)_2$	Sn:	MVol.C6-179
$As_2C_{17}Fe_2H_{12}O_8$	$(CO)_4Fe(C_6H_4(AsCH_3)_2CH_2)Fe(CO)_4$	Fe:	Org.Verb.C1-37
$As_2C_{18}ClF_3H_{22}Ni$	$CF_2CFNi(As(CH_3)_2C_6H_5)_2Cl$	Ni:	Org.Verb.1-27, 29
$As_2C_{18}Cl_2F_2H_{22}Ni$	$CFClCFNi(As(CH_3)_2C_6H_5)_2Cl$	Ni:	Org.Verb.1-28, 29
–	$CF_2CClNi(As(CH_3)_2C_6H_5)_2Cl$	Ni:	Org.Verb.1-28, 29
$As_2C_{18}Cl_3H_{26}IrS_2$	$Ir((CH_3)_2AsC_6H_4SCH_3)_2Cl_3$	Ir:	SVol.2-259
$As_2C_{18}Cl_4H_{42}Ir$	$Ir(As(n\text{-}C_3H_7)_3)_2Cl_4$	Ir:	SVol.2-256
$As_2C_{18}Fe_2H_{16}O_8$	$(CO)_4Fe(C_6H_4(As(CH_3)_2)_2)Fe(CO)_4$	Fe:	Org.Verb.C1-36
$As_2C_{18}Fe_2H_{16}O_8Pt$	$(CO)_8Fe_2Pt[C_6H_4(As(CH_3)_2)_2]$	Fe:	Org.Verb.C1-216/7
$As_2C_{18}H_{26}I_3IrS_2$	$Ir((CH_3)_2AsC_6H_4SCH_3)_2I_3$	Ir:	SVol.2-259
$As_2C_{20}Cl_2Fe_2H_{18}O$	$(C_5H_5FeC_5H_4AsCl)_2O$	Fe:	Org.Verb.A6-203/4
$As_2C_{20}Cl_2H_{42}IrP$	$Ir(P(CH_3)_2C_6H_5)(As(C_2H_5)_3)_2HCl_2$	Ir:	SVol.2-209/10
$As_2C_{20}F_6H_{22}Ni$	$(CFCF_2)_2Ni(As(CH_3)_2C_6H_5)_2$	Ni:	Org.Verb.1-343, 344
–	$(CF_3CCCF_3)Ni(As(CH_3)_2C_6H_5)_2$	Ni:	Org.Verb.1-374, 375
$As_2C_{20}F_6H_{24}Ni$	$(CHFCF_2)_2Ni(As(CH_3)_2C_6H_5)_2$	Ni:	Org.Verb.1-343, 344
$As_2C_{20}F_8H_{22}Ni$	$C_4F_8Ni(As(CH_3)_2C_6H_5)_2$	Ni:	Org.Verb.1-345, 346
$As_2C_{20}FeH_{20}Ni_2O_6$	$(CO)_3Ni[Fe(C_5H_4As(CH_3)_2)_2]Ni(CO)_3$	Ni:	Org.Verb.2-264
$As_2C_{20}Fe_2H_{16}O_6$	$(CO)_3Fe(As(C_6H_5)CH_3)_2Fe(CO)_3$	Fe:	Org.Verb.C1-164/5, 169, 179
$As_2C_{20}Fe_2H_{18}O_2$	$(C_5H_5FeC_5H_4As)_2O_2$	Fe:	Org.Verb.A6-203/4
$As_2C_{20}Fe_2H_{20}O_3$	$(C_5H_5FeC_5H_4AsOH)_2O$	Fe:	Org.Verb.A6-204
$As_2C_{21}Cl_4H_{28}IrN$	$(C_5H_5NH)[Ir(As(CH_3)_2C_6H_5)_2Cl_4]$	Ir:	SVol.2-255
$As_2C_{21}F_6FeH_{16}O_3$	$(CO)_3Fe(C_6H_5(CH_3)As)_2C_2(CF_3)_2$	Fe:	Org.Verb.B1-170/1, 174, 178
$As_2C_{22}F_{10}H_{16}Ni$	$(C_6F_5)_2Ni((CH_3)_2AsC_6H_4As(CH_3)_2)$	Ni:	Org.Verb.1-93
$As_2C_{24}F_6Fe_2H_{16}O_6$	$((CF_3)_2C_2(As(CH_3)C_6H_5)_2)Fe_2(CO)_6$	Fe:	Org.Verb.C2-130/2, 134/5
$As_2C_{24}F_{20}$	$(C_6F_5)_2AsAs(C_6F_5)_2$	F:	PerFHalOrg.3-222, 223, 225
$As_2C_{24}F_{20}O$	$(C_6F_5)_2AsOAs(C_6F_5)_2$	F:	PerFHalOrg.3-197, 198

$As_2C_{24}F_{20}S$	$(C_6F_5)_2AsSAs(C_6F_5)_2$	F:	PerFHalOrg.3-212/3
$As_2C_{24}Fe_2H_{30}HgN_2O_6$	$[NO(CO)_2Fe(As(C_2H_5)_2C_6H_5)]_2Hg$	Fe:	Org.Verb.C1-64/5
$As_2C_{26}Cl_2H_{24}O_4U$	$UO_2Cl_2 \cdot C_2H_4(AsO(C_6H_5)_2)_2$	U:	SVol.E1-200/1
$As_2C_{26}Cl_4H_{24}O_2Sn$	$SnCl_4 \cdot C_2H_4(AsO(C_6H_5)_2)_2$	Sn:	MVol.C6-181
$As_2C_{26}Cl_4H_{24}O_2U$	$UCl_4 \cdot C_2H_4(AsO(C_6H_5)_2)_2$	U:	SVol.E1-200/1
$As_2C_{26}Cl_4H_{27}Ir$	$H[Ir(AsCH_3(C_6H_5)_2)_2Cl_4]$	Ir:	SVol.2-255
$As_2C_{26}Cl_4H_{30}IrN$	$NH_4[Ir(AsCH_3(C_6H_5)_2)_2Cl_4]$	Ir:	SVol.2-255
$As_2C_{26}FeH_{24}N_2O_2$	$Fe(NO)_2(C_2H_4(As(C_6H_5)_2)_2)$	Fe:	Org.Verb.B1-143
$As_2C_{26}H_{24}I_2O_4U$	$UO_2I_2 \cdot C_2H_4(AsO(C_6H_5)_2)_2$	U:	SVol.E1-200/1
$As_2C_{28}Cl_2H_{28}O_4U$	$UO_2Cl_2 \cdot C_2H_4(CH_2AsO(C_6H_5)_2)_2$	U:	SVol.E1-200/1
$As_2C_{28}Cl_4H_{28}O_2Sn$	$SnCl_4 \cdot C_4H_8(AsO(C_6H_5)_2)_2$	Sn:	MVol.C6-181
$As_2C_{28}Cl_4H_{28}O_2U$	$UCl_4 \cdot C_2H_4(CH_2AsO(C_6H_5)_2)_2$	U:	SVol.E1-200/1
$As_2C_{28}F_{18}H_{22}Ni$	$C_6(CF_3)_6Ni(As(CH_3)_2C_6H_5)_2$	Ni:	Org.Verb.2-80, 82, 222, 224
$As_2C_{28}FeH_{20}O_4$	$(CO)_4FeAs(C_6H_5)_2As(C_6H_5)_2$	Fe:	Org.Verb.B2-100
$As_2C_{28}H_{24}NiO_2$	$(CO)_2Ni(C_6H_5)_2As(CH_2)_2As(C_6H_5)_2$	Ni:	Org.Verb.1-155
$As_2C_{28}H_{28}I_2O_4U$	$UO_2I_2 \cdot C_2H_4(CH_2AsO(C_6H_5)_2)_2$	U:	SVol.E1-200/1
$As_2C_{29}FeH_{24}O_3$	$(CO)_3Fe(C_6H_5)_2AsCH_2CH_2As(C_6H_5)_2$	Fe:	Org.Verb.B1-170/1, 174
$As_2C_{29}FeH_{26}O_3$	$(CO)_3Fe(As(C_6H_5)_2CH_3)_2$	Fe:	Org.Verb.B1-148/50, 155
$As_2C_{30}F_{20}Fe_2O_6$	$(CO)_3Fe(As(C_6F_5)_2)_2Fe(CO)_3$	Fe:	Org.Verb.C1-164, 169, 179/80
$As_2C_{30}F_{20}O_6Ru_2$	$[(C_6F_5)_2AsRu(CO)_3]_2$	F:	PerFHalOrg.3-225
$As_2C_{30}Fe_2H_{20}O_6$	$(CO)_3Fe(As(C_6H_5)_2)_2Fe(CO)_3$	Fe:	Org.Verb.C1-164, 169, 179
$As_2C_{30}Fe_2H_{26}HgN_2O_6$	$[NO(CO)_2Fe(As(C_6H_5)_2CH_3)]_2Hg$	Fe:	Org.Verb.C1-64/5
$As_2C_{31}F_4Fe_2H_{27}O_8P$	$(C_4F_4(As(CH_3)_2)_2)Fe_2(CO)_5P(OC_6H_5)_3$	Fe:	Org.Verb.C2-135/8, 144
$As_2C_{31}Fe_2H_{28}O_4S_2$	$(CO)_4(CH_2(As(C_6H_5)_2)_2)Fe_2(SCH_3)_2$	Fe:	Org.Verb.C1-107
$As_2C_{32}Fe_2H_{28}O_5S_2$	$(CO)_5Fe_2(SCH_3)_2(CH_2(As(C_6H_5)_2)_2)$	Fe:	Org.Verb.C1-95
$As_2C_{32}Fe_2H_{30}O_4S_2$	$(CO)_4(C_2H_4(As(C_6H_5)_2)_2)Fe_2(SCH_3)_2$	Fe:	Org.Verb.C1-107
$As_2C_{32}H_{24}NiO_2$	$(CO)_2NiC_6H_4(As(C_6H_5)_2)_2$	Ni:	Org.Verb.1-155
$As_2C_{33}F_8Fe_2H_{27}O_8P$	$(C_6F_8(As(CH_3)_2)_2)Fe_2(CO)_5P(OC_6H_5)_3$	Fe:	Org.Verb.C2-135/8
$As_2C_{33}Fe_2H_{30}O_5S_2$	$(CO)_5Fe_2(SCH_3)_2(C_2H_4(As(C_6H_5)_2)_2)$	Fe:	Org.Verb.C1-95
$As_2C_{33}Fe_2H_{32}O_4S_2$	$(CO)_4(CH_2(As(C_6H_5)_2)_2)Fe_2(SC_2H_5)_2$	Fe:	Org.Verb.C1-107
$As_2C_{34}Cl_3H_{35}IrP$	$Ir[((C_6H_5)_2As)_2C_2H_4][(CH_3)_2C_6H_5P]Cl_3$	Ir:	SVol.2-219, 258
$As_2C_{34}FeH_{28}I_2Ni$	$Ni(Fe(C_5H_4As(C_6H_5)_2)_2)I_2$	Ni:	Org.Verb.1-156
$As_2C_{34}Fe_2H_{32}O_5S_2$	$(CO)_5Fe_2(SC_2H_5)_2(CH_2(As(C_6H_5)_2)_2)$	Fe:	Org.Verb.C1-96
$As_2C_{34}Fe_2H_{34}O_4S_2$	$(CO)_4(C_2H_4(As(C_6H_5)_2)_2)Fe_2(SC_2H_5)_2$	Fe:	Org.Verb.C1-107
$As_2C_{35}FeH_{28}I_2NiO$	$CONi[(As(C_6H_5)_2C_5H_4)_2Fe]I_2$	Ni:	Org.Verb.1-122
$As_2C_{35}Fe_2H_{34}O_5S_2$	$(CO)_5Fe_2(SC_2H_5)_2(C_2H_4(As(C_6H_5)_2)_2)$	Fe:	Org.Verb.C1-96
$As_2C_{36}ClH_{30}IIrNO_5$	$[Ir(NO)I(As(C_6H_5)_3)_2]ClO_4$	Ir:	SVol.2-59
$As_2C_{36}ClH_{30}IrN_2O_6$	$[Ir(NO)_2(As(C_6H_5)_3)_2]ClO_4$	Ir:	SVol.2-59
$As_2C_{36}Cl_2H_{30}IrNO$	$Ir(NO)Cl_2(As(C_6H_5)_3)_2$	Ir:	SVol.2-59
$As_2C_{36}Cl_2H_{30}IrNO_5$	$[Ir(NO)Cl(As(C_6H_5)_3)_2]ClO_4$	Ir:	SVol.2-59
$As_2C_{36}Cl_2H_{30}O_4U$	$UO_2Cl_2 \cdot 2\ (C_6H_5)_3AsO$	U:	SVol.E1-196, 199
$As_2C_{36}Cl_3H_{30}Ir$	$Ir(As(C_6H_5)_3)_2Cl_3$	Ir:	SVol.2-253
$As_2C_{36}Cl_3H_{30}NOsS$	$[OsCl_3(NS)(As(C_6H_5)_3)_2]$	S:	SVol.2-261
$As_2C_{36}Cl_4H_{30}Ir$	$Ir(As(C_6H_5)_3)_2Cl_4$	Ir:	SVol.2-256
$As_2C_{36}Cl_4H_{30}O_2Sn$	$SnCl_4 \cdot 2\ (C_6H_5)_3AsO$	Sn:	MVol.C6-180/1
$As_2C_{36}Cl_4H_{30}O_2U$	$UCl_4 \cdot 2\ (C_6H_5)_3AsO$	U:	SVol.E1-194/5

Formula	Compound	Reference
$As_2C_{36}Cl_4H_{30}Sn$	$SnCl_4 \cdot 2\,(C_6H_5)_3As$	Sn: MVol.C6-178
$As_2C_{36}F_4H_{30}O_2Sn$	$SnF_4 \cdot 2\,(C_6H_5)_3AsO$	Sn: MVol.C6-180/1
$As_2C_{36}FeH_{28}NiO_2$	$(CO)_2Ni(Fe(C_5H_4As(C_6H_5)_2)_2)$	Ni: Org.Verb.1-156
$As_2C_{36}FeH_{30}N_2O_2$	$Fe(NO)_2(As(C_6H_5)_3)_2$	Fe: Org.Verb.B1-43, 48, 142, 187
$As_2C_{36}Fe_2H_{40}N_2O_6Pb$	$(C_6H_5)_2Pb(Fe(CO)_2(NO)As(C_2H_5)_2C_6H_5)_2$	Fe: Org.Verb.C1-67/9
$As_2C_{36}Fe_2H_{40}N_2O_6Sn$	$(C_6H_5)_2Sn(Fe(CO)_2(NO)As(C_2H_5)_2C_6H_5)_2$	Fe: Org.Verb.C1-67/9
$As_2C_{36}H_{30}I_2IrNO$	$Ir(NO)I_2(As(C_6H_5)_3)_2$	Ir: SVol.2-60
$As_2C_{36}H_{30}I_3Ir$	$Ir(As(C_6H_5)_3)_2I_3$	Ir: SVol.2-253
$As_2C_{36}H_{30}I_4O_2Sn$	$SnI_4 \cdot 2\,(C_6H_5)_3AsO$	Sn: MVol.C6-180/1
$As_2C_{36}H_{30}N_2O_{10}U$	$UO_2(NO_3)_2 \cdot 2\,(C_6H_5)_3AsO$	U: SVol.E1-196, 198/9
$As_2C_{36}H_{33}Ir$	$Ir(As(C_6H_5)_3)_2H_3$	Ir: SVol.2-249
$As_2C_{37}FeH_{30}N_2O_3$	$COFe(NO)_2(As(C_6H_5)_3)_2$	Fe: Org.Verb.B1-43, 46
$As_2C_{37}H_{31}IrO$	$Ir(As(C_6H_5)_3)_2H(CO)$	Ir: SVol.2-249
$As_2C_{37}H_{33}IrO$	$Ir(As(C_6H_5)_3)_2H_3(CO)$	Ir: SVol.2-249
$As_2C_{38}ClH_{35}IrNO_6$	$[Ir(NO)(OC_2H_5)(As(C_6H_5)_3)_2]ClO_4$	Ir: SVol.2-60
$As_2C_{38}Cl_2FeH_{30}O_2$	$(CO)_2Fe(As(C_6H_5)_3)_2Cl_2$	Fe: Org.Verb.B1-101/2, 105, 110
$As_2C_{38}Cl_3H_{35}IrNO_2$	$Ir(As(C_6H_5)_3)_2Cl_3(C_2H_5ONO)$	Ir: SVol.2-254
$As_2C_{38}Cl_4H_{36}Mn$	$[CH_3(C_6H_5)_3As]_2MnCl_4$	Mn: MVol.C5-84/7, 209/10
$As_2C_{38}F_4Fe_2H_{36}O_4P_2$	$(C_4F_4P(C_6H_5)_2As(CH_3)_2)Fe_2(CO)_4((C_6H_5)_2PC_2H_4As(CH_3)_2)$	Fe: Org.Verb.C2-135/7, 143
$As_2C_{38}F_6FeH_{30}NO_3P$	$[(CO)_2Fe(As(C_6H_5)_3)_2NO]PF_6$	Fe: Org.Verb.B1-130/1
$As_2C_{38}FeH_{30}NO_3{}^+$	$[(CO)_2Fe(As(C_6H_5)_3)_2NO]^+$	Fe: Org.Verb.B1-130/1
$As_2C_{38}H_{30}N_2O_4S_2U$	$UO_2(NCS)_2 \cdot 2\,(C_6H_5)_3AsO$	U: SVol.E1-196, 199
$As_2C_{38}H_{30}NiO_2$	$(CO)_2Ni(As(C_6H_5)_3)_2$	Ni: Org.Verb.1-144/5
$As_2C_{38}H_{30}NiO_8$	$(CO)_2Ni(As(OC_6H_5)_3)_2$	Ni: Org.Verb.1-144/5
$As_2C_{38}H_{36}I_4Mn$	$[CH_3(C_6H_5)_3As]_2MnI_4$	Mn: MVol.C5-322/3, 326
$As_2C_{38}H_{36}I_6Pa$	$[As(C_6H_5)_3CH_3]_2PaI_6$	Pa: SVol.2-72/3
$As_2C_{38}H_{36}MnN_4O_{12}$	$[CH_3(C_6H_5)_3As]_2[Mn(NO_3)_4]$	Mn: MVol.C3-295/7
$As_2C_{39}Cl_2FeH_{30}HgO_3$	$(CO)_3Fe(As(C_6H_5)_3)_2 \cdot HgCl_2$	Fe: Org.Verb.B1-192/3
$As_2C_{39}F_6H_{31}NPd$	$[(C_6H_5)_3As]_2Pd[C(CF_3)_2NH]$	F: PerFHalOrg.7-74/5
$As_2C_{39}F_8FeH_{32}O_3P_2$	$(CO)_3Fe(C_4F_4(As(CH_3)_2)P(C_6H_5)_2)_2$	Fe: Org.Verb.B1-148/50, 153, 160/1
$As_2C_{39}FeH_{30}O_3$	$(CO)_3Fe(As(C_6H_5)_3)_2$	Fe: Org.Verb.B1-148/50, 155, 164
$As_2C_{39}FeH_{31}O_3{}^+$	$[(CO)_3Fe(As(C_6H_5)_3)_2H]^+$	Fe: Org.Verb.B1-164
$As_2C_{39}FeH_{66}O_3$	$(CO)_3Fe(As(C_6H_{11})_3)_2$	Fe: Org.Verb.B1-148/50, 155
$As_2C_{39}H_{37}IrOS_2$	$IrH_2(As(C_6H_5)_3)_2(S_2COC_2H_5)$	Ir: SVol.2-250
$As_2C_{40}CdFe_2H_{30}N_2O_6$	$[NO(CO)_2FeAs(C_6H_5)_3]_2Cd$	Fe: Org.Verb.C1-63
$As_2C_{40}Cl_8H_{34}O_4Sn_2$	$[SnCl_3((C_6H_5)_2AsC_6H_4COOCH_3)_2][SnCl_5]$	Sn: MVol.C6-179
$As_2C_{40}F_8Fe_2H_{32}O_4P_2$	$(C_4F_4P(C_6H_5)_2As(CH_3)_2)_2Fe_2(CO)_4$	Fe: Org.Verb.C2-135/7, 142, 145
$As_2C_{40}FeH_{28}Ni_2O_6$	$Fe(C_5H_4As(C_6H_5)_2)_2(Ni(CO)_3)_2$	Ni: Org.Verb.1-248
$As_2C_{40}Fe_2H_{30}HgN_2O_6$	$[NO(CO)_2FeAs(C_6H_5)_3]_2Hg$	Fe: Org.Verb.C1-64/5
$As_2C_{40}H_{36}O_8U$	$UO_2(CH_3COO)_2 \cdot 2\,(C_6H_5)_3AsO$	U: SVol.E1-196, 198/9
$As_2C_{40}H_{42}IrPS_2$	$Ir(As(C_6H_5)_3)_2H_2((C_2H_5)_2PS_2)$	Ir: SVol.2-250
$As_2C_{41}F_3H_{36}IrO_2$	$Ir(As(C_6H_5)_3)_2H_2(CF_3COCHCOCH_3)$	Ir: SVol.2-250
$As_2C_{41}F_4Fe_2H_{31}O_5P$	$(C_4F_4P(C_6H_5)_2As(CH_3)_2)Fe_2(CO)_5As(C_6H_5)_3$	Fe: Org.Verb.C2-135/7, 139

$As_2C_{41}F_6H_{33}IrO_2$ .... $Ir(As(C_6H_5)_3)_2H_2(CF_3COCHCOCF_3)$ .... Ir: SVol.2-250
$As_2C_{41}F_{10}Fe_2H_{32}O_4P_2$ $(C_5F_6P(C_6H_5)_2)As(CH_3)_2Fe_2(CO)_4$ $(C_4F_4P(C_6H_5)_2As(CH_3)_2)$ .... Fe: Org.Verb.C2-135/7, 142, 143
$As_2C_{41}Fe_2H_{32}O_4S_2$ .. $(CO)_4(CH_2(As(C_6H_5)_2)_2)Fe_2(SC_6H_5)_2$ .... Fe: Org.Verb.C1-108
$As_2C_{41}H_{38}IrN$ .... $Ir(As(C_6H_5)_3)_2H_3(C_5H_5N)$ .... Ir: SVol.2-249
$As_2C_{41}H_{39}IrO_2$ ...... $Ir(As(C_6H_5)_3)_2H_2(CH_3COCHCOCH_3)$ .... Ir: SVol.2-250
$As_2C_{41}H_{42}IrNS_2$ ..... $IrH_2(As(C_6H_5)_3)_2(S_2CN(C_2H_5)_2)$ .... Ir: SVol.2-250
$As_2C_{42}F_4Fe_2H_{36}O_4P_2$. $[C_4F_4(P(C_6H_5)_2)_2]Fe_2(CO)_4[C_6H_4(As(CH_3)_2)_2]$ Fe: Org.Verb.C2-135/7, 140
$As_2C_{42}F_{12}Fe_2H_{32}O_4P_2$ $(C_5F_6P(C_6H_5)_2As(CH_3)_2)_2Fe_2(CO)_4$ .... Fe: Org.Verb.C2-135/7, 142
$As_2C_{42}Fe_2H_{34}O_4S_2$ .. $(CO)_4(C_2H_4(As(C_6H_5)_2)_2)Fe_2(SC_6H_5)_2$ ...... Fe: Org.Verb.C1-108
$As_2C_{42}Fe_2H_{36}O_4S_2$ .. $(CO)_4(As(C_6H_5)_3)_2Fe_2(SCH_3)_2$ .... Fe: Org.Verb.C1-103
$As_2C_{44}Cl_8H_{42}O_4Sn_2$ .. $[SnCl_3((CH_3C_6H_4)_2AsC_6H_4COOCH_3)_2][SnCl_5]$ Sn: MVol.C6-179
$As_2C_{44}H_{42}O_{14}U_2$ .... $[UO_2(CH_3COO)_2 \cdot (C_6H_5)_3AsO]_2$ .... U: SVol.E1-196, 198/9
$As_2C_{45}F_6FeH_{36}NO_2P$. $[(C_6H_5)_3AsC_8H_6Fe(CO)(NO)As(C_6H_5)_3][PF_6]$. Fe: Org.Verb.B5-11/3
$As_2C_{45}FeH_{36}NO_2^+$ ... $[(C_6H_5)_3AsC_8H_6Fe(CO)(NO)As(C_6H_5)_3]^+$ ... Fe: Org.Verb.B5-11/3
$As_2C_{48}Cl_2H_{40}I_2Mn$ ... $[(C_6H_5)_4As]_2MnI_2Cl_2$ .... Mn: MVol.C5-342
$As_2C_{48}Cl_2H_{102}O_4U$ ... $UO_2Cl_2 \cdot 2\ (C_8H_{17})_3AsO$ .... U: SVol.E1-195, 199
$As_2C_{48}Cl_4H_{40}Mn$ .... $[(C_6H_5)_4As]_2MnCl_4$ .... Mn: MVol.C5-209/10
$As_2C_{48}Cl_5H_{40}OU$ .... $[As(C_6H_5)_4]_2UOCl_5$ .... U: SVol.C9-90/2
$As_2C_{48}Cl_6H_{40}Ir$ ..... $[(C_6H_5)_4As]_2[IrCl_6]$ .... Ir: SVol.2-128
$As_2C_{48}Cl_6H_{40}Sn$ .... $[(C_6H_5)_4As]_2[SnCl_6]$ .... Sn: MVol.C3-110
$As_2C_{48}Cl_6H_{40}Te$ .... $[(C_6H_5)_4As]_2TeCl_6$ .... Te: SVol.B2-118, 119
$As_2C_{48}Cl_6H_{40}U$ ..... $[As(C_6H_5)_4]_2UCl_6$ .... U: SVol.C9-52, 55, 58/9
$As_2C_{48}Cl_{10}H_{40}Te_2$ ... $[(C_6H_5)_4As]_2Te_2Cl_{10}$ .... Te: SVol.B2-143/4
$As_2C_{48}F_4Fe_2H_{40}O_4P_2$. $(C_4F_4P(C_6H_5)_2As(CH_3)_2)Fe_2(CO)_4$ $((C_6H_5)_2PC_2H_4As(C_6H_5)_2)$ .... Fe: Org.Verb.C2-135/7, 142, 143
$As_2C_{48}F_{10}H_{30}NO_5Tl$ .. $(C_6F_5)_2TlNO_3 \cdot 2\ (C_6H_5)_3AsO$ .... F: PerFHalOrg.4-138, 144
$As_2C_{48}H_{40}I_4Mn$ ..... $[(C_6H_5)_4As]_2MnI_4$ .... Mn: MVol.C5-326
$As_2C_{48}H_{40}I_6U$ .... $[As(C_6H_5)_4]_2UI_6$ .... U: SVol.C9-177
$As_2C_{48}H_{40}MnN_4O_{12}$ .. $[(C_6H_5)_4As]_2[Mn(NO_3)_4]$ .... Mn: MVol.C3-297
$As_2C_{48}H_{40}NO_7S_2$ .... $[(C_6H_5)_4As]_2[NO(SO_3)_2]$ .... S: S-N-Verb.1-57
$As_2C_{48}H_{40}N_{18}Sn$ .... $[(C_6H_5)_4As]_2[Sn(N_3)_6]$ .... Sn: MVol.C3-84
$As_2C_{49}H_{40}S_3$ .... $[(C_6H_5)_4As]_2CS_3$ .... C: MVol.D4-223
$As_2C_{49}H_{40}S_4$ .... $[(C_6H_5)_4As]_2CS_4$ .... C: MVol.D4-231
$As_2C_{50}CdH_{40}S_6$ ..... $[(C_6H_5)_4As]_2[Cd(CS_3)_2]$ .... C: MVol.D4-224
$As_2C_{50}Cl_2H_{46}Sn$ .... $[(C_6H_5)_4As]_2[(CH_3)_2SnCl_2]$ .... Sn: Org.Verb.6-37
$As_2C_{50}Cl_4H_{46}Sn$ .... $[(C_6H_5)_4As]_2[(CH_3)_2SnCl_4]$ .... Sn: Org.Verb.6-35
$As_2C_{50}F_8Fe_2H_{42}O_{10}P_2$ $(C_6F_8(As(CH_3)_2)_2)Fe_2(CO)_4(P(OC_6H_5)_3)_2$ .... Fe: Org.Verb.C2-135/7, 139
$As_2C_{50}H_{40}NiS_6$ ..... $[(C_6H_5)_4As]_2[Ni(CS_3)_2]$ .... C: MVol.D4-218, 225
$As_2C_{50}H_{40}NiS_7$ ..... $[(C_6H_5)_4As]_2[Ni(CS_3)(CS_4)]$ .... C: MVol.D4-225, 231
$As_2C_{50}H_{40}NiS_8$ ..... $[(C_6H_5)_4As]_2[Ni(CS_4)_2]$ .... C: MVol.D4-231
$As_2C_{50}H_{40}O_2PtS_4$ ... $[(C_6H_5)_4As]_2[Pt(S_2CO)_2]$ .... C: MVol.D4-208
$As_2C_{50}H_{40}PdS_6$ ..... $[(C_6H_5)_4As]_2[Pd(CS_3)_2]$ .... C: MVol.D4-226
$As_2C_{50}H_{40}PtS_6$ ..... $[(C_6H_5)_4As]_2[Pt(CS_3)_2]$ .... C: MVol.D4-226
$As_2C_{50}H_{40}PtS_8$ ..... $[(C_6H_5)_4As]_2[Pt(CS_4)_2]$ .... C: MVol.D4-231
$As_2C_{50}H_{40}S_6Sn$ ..... $[(C_6H_5)_4As]_2[Sn(CS_3)_2]$ .... C: MVol.D4-223
$As_2C_{50}H_{40}S_6Zn$ ..... $[(C_6H_5)_4As]_2[Zn(CS_3)_2]$ .... C: MVol.D4-224
$As_2C_{50}H_{46}N_{12}Sn$ .... $[(C_6H_5)_4As]_2[(CH_3)_2Sn(N_3)_4]$ .... Sn: Org.Verb.6-36

Formula	Compound		Reference
$As_2C_{52}ClH_{44}IrO_2P_2$	$[Ir((C_6H_5)_2P(CH)_2As(C_6H_5)_2)_2O_2]Cl$	Ir:	SVol.2-260
$As_2C_{52}ClH_{44}IrP_2$	$[Ir((C_6H_5)_2P(CH)_2As(C_6H_5)_2)_2]Cl$	Ir:	SVol.2-260
$As_2C_{58}H_{46}N_4S_4Sn$	$[(C_6H_5)_4As]_2[(CH_3)_2Sn(S(CCN)_2S)_2]$	Sn:	Org.Verb.6-30
$As_2C_{60}H_{40}N_6S_6Sn$	$[(C_6H_5)_4As]_2[Sn(NC(S)CC(S)CN)_3]$	Sn:	MVol.C6-88
$As_2C_{66}H_{50}N_6O_2S_4U$	$((C_6H_5)_4As)_2[UO_2(C_2S_2(CN)_2)_2(C_5H_5N)_2]$	U:	SVol.E1-33, 37/8
$As_2CaH_4O_{20}U_4$	$Ca(UO_2)_4(AsO_4)_2(OH)_4 \cdot 6\ H_2O$	U:	SVol.C3-178
$As_2CaO_{12}U_2$	$Ca(UO_2AsO_4)_2 \cdot 8\ H_2O$	U:	SVol.C3-176, 179
–	$Ca(UO_2)_2(AsO_4)_2 \cdot 10\ H_2O$	U:	SVol.C3-179
As_2CdGe	$CdGeAs_2$ solid solutions		
	$CdGeAs_2$-$CdSnAs_2$	Sn:	MVol.C4-245/7
As_2CdSn	$CdSnAs_2$		
	Bond	Sn:	MVol.C4-74/5
	Chemical reactions	Sn:	MVol.C4-70
	Crystallographic properties	Sn:	MVol.C4-72/4
	Electrical properties	Sn:	MVol.C4-78/87
	Formation	Sn:	MVol.C4-76
	Magnetic properties	Sn:	MVol.C4-81, 85/7
	Mechanical properties	Sn:	MVol.C4-75/6
	Mössbauer effect	Sn:	MVol.C4-74/5
	Optical properties	Sn:	MVol.C4-88/92
	Preparation	Sn:	MVol.C4-71/2
	Review	Sn:	MVol.C4-70/1
	Thermal properties	Sn:	MVol.C4-75/7
–	$CdSnAs_2$, doped with Ga, GaAs, GaSb	Sn:	MVol.C4-112, 113
–	$CdSnAs_2$ solid solutions		
	$CdSnAs_2$-$CdGeAs_2$	Sn:	MVol.C4-245/7
	$CdSn(As_xP_{1-x})_2$	Sn:	MVol.C4-95
	$(CdSnAs_2)_{1-x}(2\ CdSe)_x$	Sn:	MVol.C4-93/4
	$(CdSnAs_2)_{1-x}(2\ CdTe)_x$	Sn:	MVol.C4-94/5
	$CdSnAs_2$-InAs	Sn:	MVol.C4-127/32
	$CdSnAs_2$-InSb	Sn:	MVol.C4-137
–	$CdSnAs_2$ systems		
	$CdSnAs_2$-InAs	Sn:	MVol.C4-127
	$CdSnAs_2$-InAs-$ZnSnAs_2$	Sn:	MVol.C4-132
	$CdSnAs_2$-$ZnSnAs_2$	Sn:	MVol.C4-92/3
As_2Cd_3	Cd_3As_2 systems:		
	Cd_3As_2-$ZnSnAs_2$	Sn:	MVol.C4-93
$As_2Cl_4F_6S$	$2\ AsF_3 \cdot SCl_4$	S:	SVol.2-235
$As_2Cl_4O_3Sn$	$SnCl_4 \cdot As_2O_3$	Sn:	MVol.C2-281
$As_2Cl_4O_5Sn_2$	$2\ SnOCl_2 \cdot As_2O_3$	Sn:	MVol.C2-281
As_2Cl_8OS	$(SO)(AsCl_4)_2$	S:	SVol.1-56
$As_2Cl_{10}Sn$	$[AsCl_2]_2[SnCl_6]$	Sn:	MVol.C2-281, 284
$As_2Cl_{12}O_3S_3$	$(SO)_3(AsCl_6)_2$	S:	SVol.1-56/7
$As_2CoO_{12}U_2$	$Co(UO_2AsO_4)_2 \cdot 8\ H_2O$	U:	SVol.C3-177/8
–	$Co(UO_2AsO_4)_2 \cdot 12\ H_2O$	U:	SVol.C3-177
$As_2CuO_{12}U_2$	$Cu(UO_2AsO_4)_2 \cdot 8\ H_2O$	U:	SVol.C3-178, 188
–	$Cu(UO_2AsO_4)_2 \cdot 12\ H_2O$	U:	SVol.C3-177/8
–	$Cu(UO_2AsO_4)_2 \cdot n\ H_2O$	U:	SVol.C3-188

$As_2F_{12}O_2SSe_8Te_2$ . . . $Te_2Se_8(AsF_6)_2 \cdot SO_2$ . . . Te: SVol.B3-131
$As_2FeO_{12}U_2$ . . . $Fe(UO_2AsO_4)_2 \cdot 8\ H_2O$ . . . U: SVol.C3-177/9
– . . . $Fe(UO_2AsO_4)_2 \cdot 12\ H_2O$ . . . U: SVol.C3-177/9
$As_2H_{0.12}Li_{1.88}O_8Th$ . . . $Th(Li_{0.94}H_{0.06}AsO_4)_2 \cdot H_2O$ . . . Th: SVol.C2-33
$As_2H_{0.4}Li_{1.6}O_8Th$ . . . $Th(Li_{0.8}H_{0.2}AsO_4)_2 \cdot 0.2\ H_2O$ . . . Th: SVol.C2-33
$As_2H_2O_8Sn$ . . . $Sn(HAsO_4)_2$ . . . Sn: MVol.C2-278/80
– . . . $Sn(HAsO_4)_2 \cdot H_2O$ . . . Sn: MVol.C2-278
$As_2H_2O_8Th$ . . . $Th(HAsO_4)_2$ . . . Th: SVol.C2-32
– . . . $Th(HAsO_4)_2 \cdot n\ H_2O$ . . . Th: SVol.C2-31/2
$As_2H_2O_{12}U_2$ . . . $H_2(UO_2AsO_4)_2 \cdot 8\ H_2O$ . . . U: SVol.C3-173, 179
$As_2H_3KO_{13}U_2$ . . . $K(H_3O)(UO_2AsO_4)_2 \cdot 6\ H_2O$ . . . U: SVol.C3-175/6
$As_2H_4O_9Sn$ . . . $SnO(H_2AsO_4)_2$ . . . Sn: MVol.C2-278
$As_2H_6N_6O_6S_3$ . . . $As(NHSO_2NH)_3As$ . . . S: S-N-Verb.1-192
$As_2InSeSn_2$ . . . 2 SnAs · InSe . . . Sn: MVol.C4-132/3
As_2Ir . . . $IrAs_2$ . . . Ir: SVol.2-247
$As_2Li_2O_8Th$ . . . $Li_2Th(AsO_4)_2$ . . . Th: SVol.C2-33
– . . . $Th(LiAsO_4)_2 \cdot H_2O$ . . . Th: SVol.C2-32
$As_2MgO_{12}U_2$ . . . $Mg(UO_2AsO_4)_2 \cdot 4\ H_2O$ . . . U: SVol.C3-85
– . . . $Mg(UO_2AsO_4)_2 \cdot 8\ H_2O$ . . . U: SVol.C3-177/8
– . . . $Mg(UO_2AsO_4)_2 \cdot 10\ H_2O$ . . . U: SVol.C3-177/8
– . . . $Mg(UO_2AsO_4)_2 \cdot 12\ H_2O$ . . . U: SVol.C3-177/8
– . . . $Mg(UO_2AsO_4)_2 \cdot n\ H_2O$ . . . U: SVol.C3-85, 177
$As_2Mg_7O_{16}Sn_2$ . . . $7\ MgO \cdot 2\ SnO_2 \cdot As_2O_5$ . . . Sn: MVol.C3-168
$As_2Na_2O_8Th$ . . . $Na_2Th(AsO_4)_2$ . . . Th: SVol.C2-33
$As_2Na_2O_{12}U_2$ . . . $Na_2(UO_2AsO_4)_2 \cdot n\ H_2O$ . . . U: SVol.C3-174/6
As_2O_5 . . . As_2O_5 systems
As_2O_5-MgO-SnO_2 . . . Sn: MVol.C3-168
As_2O_6U . . . $UO_2O \cdot As_2O_3$ . . . U: SVol.C3-173
As_2O_7Sn . . . $SnO_2 \cdot As_2O_5$ . . . Sn: MVol.C2-278
As_2O_7Th . . . $ThAs_2O_7$ . . . Th: SVol.C2-31/2
– . . . $ThO_2 \cdot As_2O_5 \cdot n\ H_2O$ . . . Th: SVol.C2-32
$As_2O_8Pa^-$ . . . $Pa(AsO_4)_2^-$ . . . Pa: SVol.2-158
$As_2O_8Sn_3$ . . . $Sn_3(AsO_4)_2$ . . . Sn: MVol.C2-280
$As_2O_9U_2$ . . . $2\ UO_2O \cdot As_2O_3$ . . . U: SVol.C3-173
$As_2O_{12}U_3$ . . . $3\ UO_2O \cdot As_2O_3$ . . . U: SVol.C3-173
$As_2O_{14}U_3$ . . . $(UO_2)_3(AsO_4)_2 \cdot 12\ H_2O$ . . . U: SVol.C3-179
As_2S_4Sn . . . $SnAs_2S_4$ . . . Sn: MVol.C2-281
As_2S_8Sn . . . $SnAs_2S_8$ . . . Sn: MVol.C2-281
As_2Se_3 . . . As_2Se_3 systems
As_2Se_3-SnSe . . . Sn: MVol.C2-281/4
As_2Se_4Sn . . . $SnAs_2Se_4$ . . . Sn: MVol.C2-281/2
As_2SnZn . . . $ZnSnAs_2$ . . . Sn: MVol.C4-25/39
– . . . $ZnSnAs_2$ systems
$ZnSnAs_2$-$CdSnAs_2$ . . . Sn: MVol.C4-92/3
$ZnSnAs_2$-$CdSnAs_2$-InAs . . . Sn: MVol.C4-132
$ZnSnAs_2$-Cd_3As_2 . . . Sn: MVol.C4-93
$ZnSnAs_2$-InAs . . . Sn: MVol.C4-125/7
As_2Sn_3 . . . Sn_3As_2 . . . Sn: MVol.C2-271/7

As_2Sn_3 .... Sn_3As_2 systems
Sn_3As_2-$ZnAs_2$ .... Sn: MVol.C4-25/6
Sn_3As_2-Zn_3As_2 .... Sn: MVol.C4-25/6

As_2Te_3 .... As_2Te_3 systems
As_2Te_3-SnTe .... Sn: MVol.C2-284
As_2Te_3-SnTe-Te .... Sn: MVol.C2-284

As_2Zn .... $ZnAs_2$ systems
$ZnAs_2$-Sn_3As_2 .... Sn: MVol.C4-25/6

As_2Zn_3 .... Zn_3As_2 systems
Zn_3As_2-Sn_3As_2 .... Sn: MVol.C4-25/6

$As_{2.25}LiO_{12}Th_2V_{0.75}$ .. $LiTh_2(V_{0.25}As_{0.75}O_4)_3$ .... Th: SVol.C2-33, 125
$As_3BBiBrC_{48}CoH_{50}$ .. $[CoBrBi(C_6H_4As(CH_3)_2)_3][B(C_6H_5)_4]$ .... Bi: Org.Verb.-153
$As_3BBiBrC_{48}H_{50}Ni$ .. $[NiBrBi(C_6H_4As(CH_3)_2)_3][B(C_6H_5)_4]$ .... Bi: Org.Verb.-153
$As_3BBiC_{48}ClH_{50}Ni$ .. $[NiClBi(C_6H_4As(CH_3)_2)_3][B(C_6H_5)_4]$ .... Bi: Org.Verb.-153
$As_3BBiC_{48}FH_{50}Ni$ ... $[NiFBi(C_6H_4As(CH_3)_2)_3][B(C_6H_5)_4]$ .... Bi: Org.Verb.-153
$As_3BBiC_{48}H_{50}INi$ .... $[NiIBi(C_6H_4As(CH_3)_2)_3][B(C_6H_5)_4]$ .... Bi: Org.Verb.-153
$As_3BBr_2C_{17}CuH_{23}$ ... $[CH_3As(C_6H_4As(CH_3)_2)_2]CuBBr_2$ .... B: B-Verb.3-161
$As_3BC_{20}F_4H_{20}Ni_4$ .... $[(C_5H_5)_4Ni_4As_3][BF_4]\cdot CH_2Cl_2$ .... Ni: Org.Verb.2-388/9
$As_3BC_{20}F_{18}H_{16}N_4$ ... $[(NC_5H_4C_5H_4N)_2B][AsF_6]_3$ .... B: B-Verb.10-169
$As_3BC_{21}Cl_2F_4H_{22}Ni_4$. $[(C_5H_5)_4Ni_4As_3][BF_4]\cdot CH_2Cl_2$ .... Ni: Org.Verb.2-388/9
$As_3BC_{28}F_{18}H_{36}N_4$ ... $[(C_2H_5C_5H_4N)_4B][AsF_6]_3$ .... B: B-Verb.10-168
$As_3BC_{55}CuH_{48}N$ .... $Cu[H_3BCN][(C_6H_5)_3As]_3$ .... B: B-Verb.8-48/9
$As_3BC_{72}H_{67}NNi$ ..... $[C_6H_5Ni((As(C_6H_5)_2C_2H_4)_3N)]B(C_6H_5)_4$ ..... Ni: Org.Verb.1-38, 44/5
$As_3BC_{96}H_{77}IrP_2$ ..... $[Ir(((C_6H_5)_2AsC_6H_4)_3P)(P(C_6H_5)_3)]B(C_6H_5)_4$ .. Ir: SVol.2-260
$As_3B_3C_6F_{18}H_6$ ...... $[(CF_3)_2AsBH_2]_3$ .... F: PerFHalOrg.3-187, 189/91
$As_3B_3C_6H_{24}$ .... $[(CH_3)_2AsBH_2]_3$ .... B: B-Verb.3-145
B: B-Verb.10-29
As_3B_{13} .... $B_{13}As_3$ .... B: B-Verb.3-144
$As_3B_{13}C_9H_{40}$ .... $[((CH_3)_3As)_2BH_2][(CH_3)_3As\cdot B_{12}H_{11}]$ ...... B: B-Verb.3-146
B: B-Verb.10-97, 99, 103
B: B-Verb.20-237
$As_3Ba_5HO_{13}$ .... $Ba_5(AsO_4)_3OH$ solid solutions
$Ba_5(AsO_4,MnO_4)_3OH$ .... Mn: MVol.C2-9, 261
$As_3BiC_{24}Cl_6Fe_2H_{30}$ .. $[FeCl_2Bi(C_6H_4As(CH_3)_2)_3][FeCl_4]$ .... Bi: Org.Verb.-153
$As_3BiC_{24}H_{30}$ .... $Bi[C_6H_4As(CH_3)_2]_3$ .... Bi: Org.Verb.-92
$As_3BiC_{75}H_{60}S_9$ ..... $[(C_6H_5)_4As]_3[Bi(CS_3)_3]$ .... C: MVol.D4-223
$As_3BrC_{14}H_{31}Ni^+$ .... $[((CH_3)_2AsC_3H_6)_2As(C_4H_7)NiBr]^+$ .... Ni: Org.Verb.1-341
$As_3BrC_{24}Cl_2H_{33}Ir$ .... $Ir(As(CH_3)_2C_6H_5)_3Cl_2Br$ .... Ir: SVol.2-253
$As_3BrC_{54}H_{47}Ir$ ...... $Ir(As(C_6H_5)_3)_3H_2Br$ .... Ir: SVol.2-250
$As_3Br_2C_{39}H_{40}Ir$ ..... $Ir(AsCH_3(C_6H_5)_2)_3HBr_2$ .... Ir: SVol.2-251/2
$As_3Br_2C_{42}H_{46}Ir$ ..... $Ir(AsC_2H_5(C_6H_5)_2)_3HBr_2$ .... Ir: SVol.2-251/2
$As_3Br_2C_{45}H_{52}Ir$ ..... $Ir(As(n-C_3H_7)(C_6H_5)_2)_3HBr_2$ .... Ir: SVol.2-251/2
$As_3Br_2C_{54}H_{46}Ir$ ..... $Ir(As(C_6H_5)_3)_3HBr_2$ .... Ir: SVol.2-251/2
$As_3Br_3C_{24}H_{33}Ir$ ..... $Ir(As(CH_3)_2C_6H_5)_3Br_3$ .... Ir: SVol.2-253
$As_3Br_3C_{36}H_{39}Ir$ ..... $Ir(As(CH_3)_2C_{10}H_7)_3Br_3$ .... Ir: SVol.2-253
$As_3Br_3C_{39}H_{39}Ir$ ..... $Ir(AsCH_3(C_6H_5)_2)_3Br_3$ .... Ir: SVol.2-253
$As_3Br_3C_{42}H_{33}Ir$ ..... $Ir(((C_6H_5)_2AsC_6H_4)_2AsC_6H_5)Br_3$ .... Ir: SVol.2-258

Formula	Compound		Reference
$As_3Br_3C_{42}H_{45}Ir$	$Ir(AsC_2H_5(C_6H_5)_2)_3Br_3$	Ir:	SVol.2-253
$As_3Br_3C_{54}H_{45}Ir$	$Ir(As(C_6H_5)_3)_3Br_3$	Ir:	SVol.2-142
$As_3Br_5C_{24}H_{33}HgIr$	$Ir(As(CH_3)_2C_6H_5)_3Br_3(HgBr_2)$	Ir:	SVol.2-264/5
$As_3C_{14}ClH_{31}Ni^+$	$[((CH_3)_2AsC_3H_6)_2As(C_4H_7)NiCl]^+$	Ni:	Org.Verb.1-341
$As_3C_{14}H_{31}I_2Ni$	$[((CH_3)_2AsC_3H_6)_2As(C_4H_7)NiI_2]$	Ni:	Org.Verb.1-341
$As_3C_{18}Cl_2H_{46}Ir$	$Ir(As(C_2H_5)_3)_3HCl_2$	Ir:	SVol.2-251/2
$As_3C_{18}Cl_3H_{45}Ir$	$Ir(As(C_2H_5)_3)_3Cl_3$	Ir:	SVol.2-253
$As_3C_{18}Cl_5H_{45}HgIr$	$Ir(As(C_2H_5)_3)_3Cl_3(HgCl_2)$	Ir:	SVol.2-264/5
$As_3C_{18}H_{18}O_{10}Pa$	$H_3PaO(C_6H_5AsO_3)_3$	Pa:	SVol.2-232
$As_3C_{24}Cl_2H_{33}IIr$	$Ir(As(CH_3)_2C_6H_5)_3Cl_2I$	Ir:	SVol.2-253
$As_3C_{24}Cl_2H_{33}IrNO_2$	$Ir(As(CH_3)_2C_6H_5)_3Cl_2NO_2$	Ir:	SVol.2-254
$As_3C_{24}Cl_2H_{33}IrNO_3$	$Ir(As(CH_3)_2C_6H_5)_3Cl_2NO_3$	Ir:	SVol.2-254
$As_3C_{24}Cl_3H_{33}Ir$	$Ir(As(CH_3)_2C_6H_5)_3Cl_3$	Ir:	SVol.2-253
$As_3C_{24}Cl_4H_{34}Ir$	$[As(CH_3)_2(C_6H_5)H][Ir(As(CH_3)_2C_6H_5)_2Cl_4]$	Ir:	SVol.2-256
$As_3C_{24}Cl_5H_{33}HgIr$	$Ir(As(CH_3)_2C_6H_5)_3Cl_3(HgCl_2)$	Ir:	SVol.2-264/5
$As_3C_{25}Cl_2H_{36}IrO$	$Ir(As(CH_3)_2C_6H_5)_3Cl_2OCH_3$	Ir:	SVol.2-254
$As_3C_{26}Cl_2H_{36}IrO_2$	$Ir(As(CH_3)_2C_6H_5)_3Cl_2(COOCH_3)$	Ir:	SVol.2-254
$As_3C_{27}Cl_3H_{39}IrO_3$	$Ir(As(CH_3)_2(o\text{-}CH_3OC_6H_4))_3Cl_3$	Ir:	SVol.2-256
$As_3C_{27}Cl_4H_{64}Ir$	$[As(n\text{-}C_3H_7)_3H][Ir(As(n\text{-}C_3H_7)_3)_2Cl_4]$	Ir:	SVol.2-256
$As_3C_{29}Cl_3H_{38}IrNO_4$	$[Ir(As(CH_3)_2C_6H_5)_3(C_5H_5N)Cl_2]ClO_4$	Ir:	SVol.2-254
$As_3C_{30}ClH_{47}Ir$	$Ir(As(C_2H_5)_2C_6H_5)_3H_2Cl$	Ir:	SVol.2-250
$As_3C_{30}Cl_2H_{46}Ir$	$Ir(As(C_2H_5)_2C_6H_5)_3HCl_2$	Ir:	SVol.2-251/2
$As_3C_{30}Cl_3H_{45}Ir$	$Ir(As(C_2H_5)_2C_6H_5)_3Cl_3$	Ir:	SVol.2-253
$As_3C_{30}Cl_3H_{48}IrN_3$	$Ir(As(CH_3)_2(o\text{-}(CH_3)_2NC_6H_4))_3Cl_3$	Ir:	SVol.2-256
$As_3C_{30}Cl_5H_{45}HgIr$	$Ir(As(C_2H_5)_2C_6H_5)_3Cl_3(HgCl_2)$	Ir:	SVol.2-264/5
$As_3C_{30}F_8Fe_2H_{28}O_4P$	$(C_4F_4P(C_6H_5)_2As(CH_3)_2)Fe_2(CO)_4(C_4F_4(As(CH_3)_2)_2)$	Fe:	Org.Verb.C2-135/7, 141, 143
$As_3C_{30}H_{48}Ir$	$Ir(As(C_2H_5)_2C_6H_5)_3H_3$	Ir:	SVol.2-249
$As_3C_{31}F_8Fe_2H_{28}O_5P$	$(C_4F_4(As(CH_3)_2)_2)Fe_2(CO)_5(C_4F_4(P(C_6H_5)_2)As(CH_3)_2)$	Fe:	Org.Verb.C2-135/8, 139, 144
$As_3C_{31}F_{10}Fe_2H_{28}O_4P$	$(C_5F_6(As(CH_3)_2)_2Fe_2(CO)_4C_4F_4P(C_6H_5)_2As(CH_3)_2)$	Fe:	Org.Verb.C2-135/7, 140, 142
$As_3C_{32}F_4Fe_2H_{32}O_4P$	$(C_4F_4P(C_6H_5)_2As(CH_3)_2)Fe_2(CO)_4(C_6H_4(As(CH_3)_2)_2)$	Fe:	Org.Verb.C2-135/7, 141, 143, 145
$As_3C_{32}F_{12}Fe_2H_{28}O_4P$	$(C_4F_4P(C_6H_5)_2As(CH_3)_2)Fe_2(CO)_4(C_6F_8(As(CH_3)_2)_2)$	Fe:	Org.Verb.C2-135/7, 141, 142
$As_3C_{33}F_{12}Fe_2H_{28}O_5P$	$(C_6F_8(As(CH_3)_2)_2)Fe_2(CO)_5(C_4F_4(P(C_6H_5)_2)As(CH_3)_2)$	Fe:	Org.Verb.C2-135/8
$As_3C_{34}Cl_3H_{35}Ir$	$Ir(((C_6H_5)_2As)_2C_2H_4)((CH_3)_2C_6H_5As)Cl_3$	Ir:	SVol.2-257/8
$As_3C_{36}Cl_3H_{39}Ir$	$Ir(As(CH_3)_2C_{10}H_7)_3Cl_3$	Ir:	SVol.2-253
$As_3C_{39}Cl_2H_{40}Ir$	$Ir(AsCH_3(C_6H_5)_2)_3HCl_2$	Ir:	SVol.2-251/2
$As_3C_{39}Cl_3H_{39}Ir$	$Ir(AsCH_3(C_6H_5)_2)_3Cl_3$	Ir:	SVol.2-253
$As_3C_{39}H_{39}I_3Ir$	$Ir(AsCH_3(C_6H_5)_2)_3I_3$	Ir:	SVol.2-253
$As_3C_{39}H_{40}I_2Ir$	$Ir(AsCH_3(C_6H_5)_2)_3HI_2$	Ir:	SVol.2-251/2

$As_3C_{42}Cl_2H_{46}Ir$ ..... $Ir(AsC_2H_5(C_6H_5)_2)_3HCl_2$ ..... Ir: SVol.2-251/2
$As_3C_{42}Cl_3H_{33}Ir$ ..... $Ir(((C_6H_5)_2AsC_6H_4)_2AsC_6H_5)Cl_3$ ..... Ir: SVol.2-258
$As_3C_{42}Cl_3H_{45}Ir$ ..... $Ir(AsC_2H_5(C_6H_5)_2)_3Cl_3$ ..... Ir: SVol.2-253
$As_3C_{42}H_{45}I_3Ir$ ..... $Ir(AsC_2H_5(C_6H_5)_2)_3I_3$ ..... Ir: SVol.2-253
$As_3C_{42}H_{46}I_2Ir$ ..... $Ir(AsC_2H_5(C_6H_5)_2)_3HI_2$ ..... Ir: SVol.2-251/2
$As_3C_{43}F_3H_{40}Ni$ ..... $(CF_2CHF)Ni(As(C_6H_5)_2CH_2)_3CCH_3$ ..... Ni: Org.Verb.1-378/9
$As_3C_{43}F_4H_{39}Ni$ ..... $(CF_2CF_2)Ni(As(C_6H_5)_2CH_2)_3CCH_3$ ..... Ni: Org.Verb.1-378/9
$As_3C_{44}F_6H_{39}Ni$ ..... $(CF_3CFCF_2)Ni(As(C_6H_5)_2CH_2)_3CCH_3$ ..... Ni: Org.Verb.1-378, 380
$As_3C_{45}Cl_2H_{52}Ir$ ..... $Ir(As(n\text{-}C_3H_7)(C_6H_5)_2)_3HCl_2$ ..... Ir: SVol.2-251/2
$As_3C_{45}F_8H_{39}Ni$ ..... $C_4F_8Ni(As(C_6H_5)_2CH_2)_3CCH_3$ ..... Ni: Org.Verb.1-378, 380
$As_3C_{45}H_{52}I_2Ir$ ..... $Ir(As(n\text{-}C_3H_7)(C_6H_5)_2)_3HI_2$ ..... Ir: SVol.2-251/2
$As_3C_{47}F_4Fe_2H_{38}O_4P$. $(C_4F_4P(C_6H_5)_2As(CH_3)_2)Fe_2(CO)_4(CH_2(As(C_6H_5)_2)_2)$ Fe: Org.Verb.C2-135/7, 143
$As_3C_{48}F_4Fe_2H_{40}O_4P$. $(C_4F_4P(C_6H_5)_2As(CH_3)_2)Fe_2(CO)_4(C_2H_4(As(C_6H_5)_2)_2)$ Fe: Org.Verb.C2-135/7, 141, 143
$As_3C_{48}H_{47}NNi^+$ ..... $[C_6H_5Ni((As(C_6H_5)_2CH_2CH_2)_3N)]^+$ ..... Ni: Org.Verb.1-44/5
$As_3C_{49}F_4Fe_2H_{40}O_5P$. $(C_4F_4P(C_6H_5)_2As(CH_3)_2)Fe_2(CO)_5(C_2H_4(As(C_6H_5)_2)_2)$ Fe: Org.Verb.C2-135/7, 139
$As_3C_{54}ClH_{45}Ir$ ..... $Ir(As(C_6H_5)_3)_3Cl$ ..... Ir: SVol.2-254
$As_3C_{54}ClH_{46}IrNO_5$ ... $[Ir(NO)H(As(C_6H_5)_3)_3]ClO_4$ ..... Ir: SVol.2-59
$As_3C_{54}ClH_{47}Ir$ ..... $Ir(As(C_6H_5)_3)_3H_2Cl$ ..... Ir: SVol.2-250
$As_3C_{54}ClH_{47}IrO_4$ .... $[Ir(As(C_6H_5)_3)_3H_2]ClO_4$ ..... Ir: SVol.2-250
$As_3C_{54}Cl_2H_{46}Ir$ ..... $Ir(As(C_6H_5)_3)_3HCl_2$ ..... Ir: SVol.2-251/2
$As_3C_{54}H_{45}IrNO$ ..... $Ir(NO)(As(C_6H_5)_3)_3$ ..... Ir: SVol.2-59
$As_3C_{54}H_{46}I_2Ir$ ..... $Ir(As(C_6H_5)_3)_3HI_2$ ..... Ir: SVol.2-251/2
$As_3C_{54}H_{47}IIr$ ..... $Ir(As(C_6H_5)_3)_3H_2I$ ..... Ir: SVol.2-250
$As_3C_{54}H_{48}Ir$ ..... $Ir(As(C_6H_5)_3)_3H_3$ ..... Ir: SVol.2-249
$As_3C_{58}F_4Fe_2H_{46}O_4P$. $(C_4F_4P(C_6H_5)_2As(CH_3)_2)Fe_2(CO)_4(As(C_6H_5)_3)_2$ Fe: Org.Verb.C2-135/7, 139
$As_3C_{62}Fe_3H_{45}O_{11}$ .... $[Fe(As(C_6H_5)_3O)_3][Fe_2(CO)_8]$ ..... Fe: Org.Verb.C2-9
$As_3C_{75}CoH_{60}S_9$ ..... $[(C_6H_5)_4As]_3[Co(CS_3)_3]$ ..... C: MVol.D4-225
$As_3C_{75}CrH_{60}S_9$ ..... $[(C_6H_5)_4As]_3[Cr(CS_3)_3]$ ..... C: MVol.D4-224
$As_3C_{75}H_{60}S_9Sb$ ..... $[(C_6H_5)_4As]_3[Sb(CS_3)_3]$ ..... C: MVol.D4-223
$As_3Cl_{21}Sn_2Te$ ..... $[TeCl_3][AsCl_2]_3[SnCl_6]_2$ ..... Sn: MVol.C2-284
As_3Ir ..... $IrAs_3$ ..... Ir: SVol.2-247
$As_3LiO_{12}Th_2$ ..... $LiTh_2(AsO_4)_3$ ..... Th: SVol.C2-33
– ..... $LiTh_2(AsO_4)_3$ solid solutions
$LiTh_2(V_{1-x}As_xO_4)_3$ ..... Th: SVol.C2-33, 125
$As_3NaO_{12}Th_2$ ..... $NaTh_2(AsO_4)_3$ ..... Th: SVol.C2-33
– ..... $NaTh_2(AsO_4)_3$ solid solutions
$NaTh_2(V_{1-x}As_xO_4)_3$ ..... Th: SVol.C2-33, 125
As_3Sn_4 ..... Sn_4As_3 ..... Sn: MVol.C2-271/7
– ..... Sn_4As_3 systems
Sn_4As_3-Bi-Sn ..... Sn: MVol.C2-300
Sn_4As_3-Sn-SnTe ..... Sn: MVol.C2-284
Sn_4As_3-SnAs-SnTe ..... Sn: MVol.C2-284
Sn_4As_3-SnTe ..... Sn: MVol.C2-284

Formula	Compound		Reference
$As_4BBr_2C_{44}H_{52}Ir$	$[Ir(((CH_3)_2As)_2C_6H_4)_2Br_2]B(C_6H_5)_4$	Ir:	SVol.2-257
$As_4BC_{44}Cl_2H_{52}Ir$	$[Ir(((CH_3)_2As)_2C_6H_4)_2Cl_2]B(C_6H_5)_4$	Ir:	SVol.2-257
$As_4BC_{45}ClFeH_{52}O$	$COFeCl[C_6H_4(As(CH_3)_2)_2]_2B(C_6H_5)_4$	Fe:	Org.Verb.B1-31
$As_4BC_{56}H_{64}IrO_2$	$[Ir(O_2)(As(CH_3)_2C_6H_5)_4][B(C_6H_5)_4]$	Ir:	SVol.2-24
$As_4BC_{78}ClH_{63}Ir$	$[Ir(((C_6H_5)_2AsC_6H_4)_3As)HCl]B(C_6H_5)_4$	Ir:	SVol.2-259
$As_4BC_{96}H_{77}IrP$	$[Ir(((C_6H_5)_2AsC_6H_4)_3As)(P(C_6H_5)_3)]B(C_6H_5)_4$	Ir:	SVol.2-259
$As_4B_4C_8F_{24}H_8$	$[(CF_3)_2AsBH_2]_4$	F:	PerFHalOrg.3-187, 190
$As_4B_4C_8H_{32}$	$[(CH_3)_2AsBH_2]_4$	B:	B-Verb.3-145
$As_4BrC_{20}H_{32}IrO_2$	$[Ir(O_2)(((CH_3)_2As)_2C_6H_4)_2]Br$	Ir:	SVol.2-257
$As_4BrC_{20}H_{34}Ir$	$[Ir(((CH_3)_2As)_2C_6H_4)_2H_2]Br$	Ir:	SVol.2-257
$As_4BrC_{72}H_{58}IrP^+$	$[Ir(((C_6H_5)_2AsC_6H_4)_3As)(P(C_6H_5)_3)HBr]^+$	Ir:	SVol.2-259
$As_4Br_2C_{20}H_{32}IrNO_3$	$[Ir(((CH_3)_2As)_2C_6H_4)_2Br_2]NO_3$	Ir:	SVol.2-257
$As_4Br_2C_{32}H_{44}Ir$	$Ir(As(CH_3)_2C_6H_5)_4Br_2$	Ir:	SVol.2-251
$As_4Br_2C_{36}H_{52}IrO_4$	$Ir(As(CH_3)_2(p-CH_3OC_6H_4))_4Br_2$	Ir:	SVol.2-251
$As_4Br_2C_{84}Cl_2H_{70}Ir_2S_2$	$[Ir(SC_6H_4Br)HCl(As(C_6H_5)_3)_2]_2$	Ir:	SVol.2-163
$As_4Br_3C_{20}H_{32}Ir$	$[Ir(((CH_3)_2As)_2C_6H_4)_2Br_2]Br$	Ir:	SVol.2-257
$As_4C_4F_{12}$	$(CF_3As)_4$	F:	PerFHalOrg.3-187/91
$As_4C_9H_{18}N_6NiO_3$	$As_4(NCH_3)_6Ni(CO)_3$	Ni:	Org.Verb.1-248, 249
$As_4C_{10}Fe_2H_{12}O_6$	$(CO)_3Fe(AsCH_3)_4Fe(CO)_3$	Fe:	Org.Verb.C1-164/5, 169, 180
$As_4C_{12}H_{18}N_6Ni_2O_6$	$(CO)_3NiAs_4(NCH_3)_6Ni(CO)_3$	Ni:	Org.Verb.2-264
$As_4C_{14}Fe_2H_{20}O_6$	$(CO)_3Fe(AsC_2H_5)_4Fe(CO)_3$	Fe:	Org.Verb.C1-164/5, 169, 180
$As_4C_{15}H_{18}N_6Ni_3O_9$	$As_4(NCH_3)_6(Ni(CO)_3)_3$	Ni:	Org.Verb.2-374
$As_4C_{18}H_{18}N_6Ni_4O_{12}$	$As_4(NCH_3)_6(Ni(CO)_3)_4$	Ni:	Org.Verb.2-386
$As_4C_{20}Cl_2H_{32}Ir^+$	$[Ir(((CH_3)_2As)_2C_6H_4)_2Cl_2]^+$	Ir:	SVol.2-125, 257
$As_4C_{20}Cl_2H_{32}IrNO_3$	$[Ir(((CH_3)_2As)_2C_6H_4)_2Cl_2]NO_3$	Ir:	SVol.2-257
$As_4C_{20}Cl_3H_{32}Ir$	$[Ir(((CH_3)_2As)_2C_6H_4)_2Cl_2]Cl$	Ir:	SVol.2-256
$As_4C_{20}Cl_3H_{32}IrO_4$	$[Ir(((CH_3)_2As)_2C_6H_4)_2Cl_2]ClO_4$	Ir:	SVol.2-257
$As_4C_{20}Cl_5H_{32}Pa$	$PaCl_5 \cdot 2\ (CH_3)_2AsC_6H_4As(CH_3)_2$	Pa:	SVol.2-94/5
$As_4C_{20}FeH_{32}I_2$	$Fe[(CH_3)_2AsC_6H_4As(CH_3)_2]_2I_2$	Fe:	Org.Verb.B1-29
$As_4C_{20}H_{32}Ni$	$[(CH_3)_2AsC_6H_4As(CH_3)_2]_2Ni$	Ni:	Org.Verb.1-248, 249
$As_4C_{21}FeH_{32}O$	$COFe[(CH_3)_2AsC_6H_4As(CH_3)_2]_2$	Fe:	Org.Verb.B1-29
$As_4C_{22}F_{12}Fe_2H_{24}O_4$	$(C_5F_6(As(CH_3)_2)_2)_2Fe_2(CO)_4$	Fe:	Org.Verb.C2-135/7, 140
$As_4C_{24}Cl_4H_{60}O_4U$	$UCl_4 \cdot 4\ (C_2H_5)_3AsO$	U:	SVol.E1-194/5
$As_4C_{24}F_6H_{34}Ni$	$(CHFCF_2)_2Ni(C_6H_4(As(CH_3)_2)_2)_2$	Ni:	Org.Verb.1-344
$As_4C_{24}F_{16}Fe_2H_{24}O_4$	$(C_6F_8(As(CH_3)_2)_2)_2Fe_2(CO)_4$	Fe:	Org.Verb.C2-135/7, 140
$As_4C_{24}F_{20}$	$(C_6F_5As)_4$	F:	PerFHalOrg.3-187/91
$As_4C_{24}F_{20}S_4$	$(C_6F_5AsS)_4$	F:	PerFHalOrg.3-187, 190
$As_4C_{30}Fe_2H_{20}O_6$	$(CO)_3Fe(AsC_6H_5)_4Fe(CO)_3$	Fe:	Org.Verb.C1-164, 169, 180
$As_4C_{52}Cl_4H_{48}O_4U$	$UCl_4 \cdot 2\ C_2H_4(AsO(C_6H_5)_2)_2$	U:	SVol.E1-200/1
$As_4C_{53}FeH_{48}N_2O_2$	$Fe(NO)_2(C(CH_2As(C_6H_5)_2)_4)$	Fe:	Org.Verb.B1-143
$As_4C_{54}ClH_{42}Ir$	$Ir(((C_6H_5)_2AsC_6H_4)_3As)Cl$	Ir:	SVol.2-259
$As_4C_{56}Fe_2H_{50}O_4S_2$	$(CO)_4(CH_2(As(C_6H_5)_2)_2)_2Fe_2(SCH_3)_2$	Fe:	Org.Verb.C1-103
$As_4C_{57}H_{48}Ni_2O_4$	$(CO)_2Ni[((C_6H_5)_2AsCH_2)_4C]Ni(CO)_2$	Ni:	Org.Verb.2-261/2
$As_4C_{58}Fe_2H_{54}O_4S_2$	$(CO)_4(CH_2(As(C_6H_5)_2)_2)_2Fe_2(SC_2H_5)_2$	Fe:	Org.Verb.C1-104
–	$(CO)_4(C_2H_4(As(C_6H_5)_2)_2)_2Fe_2(SCH_3)_2$	Fe:	Org.Verb.C1-103
$As_4C_{60}Fe_2H_{58}O_4S_2$	$(CO)_4(C_2H_4(As(C_6H_5)_2)_2)_2Fe_2(SC_2H_5)_2$	Fe:	Org.Verb.C1-104

Formula	Compound	Reference
$As_4C_{72}ClH_{58}IrP^+$	$[Ir(((C_6H_5)_2AsC_6H_4)_3As)(P(C_6H_5)_3)HCl]^+$	Ir: SVol.2-259
$As_4C_{72}ClH_{62}IrO_4$	$[Ir(As(C_6H_5)_3)_4H_2]ClO_4$	Ir: SVol.2-250
$As_4C_{72}Cl_2H_{60}O_{14}U$	$UO_2(ClO_4)_2 \cdot 4\ (C_6H_5)_3AsO$	U: SVol.E1-196, 199
$As_4C_{72}Cl_4H_{60}O_4U$	$UCl_4 \cdot 4\ (C_6H_5)_3AsO$	U: SVol.E1-194
$As_4C_{72}H_{58}IIrP^+$	$[Ir(((C_6H_5)_2AsC_6H_4)_3As)(P(C_6H_5)_3)HI]^+$	Ir: SVol.2-259
$As_4C_{72}H_{60}I_2O_4Sn^{2+}$	$[SnI_2((C_6H_5)_3AsO)_4]^{2+}$	Sn: MVol.C6-181
$As_4C_{72}H_{60}I_4O_4Sn$	$SnI_4 \cdot 4\ (C_6H_5)_3AsO$	Sn: MVol.C6-181
$As_4C_{73}H_{58}IrNPS^+$	$[Ir(((C_6H_5)_2AsC_6H_4)_3As)(P(C_6H_5)_3)HSCN]^+$	Ir: SVol.2-259
$As_4C_{75}F_4Fe_2H_{61}O_3P$	$(C_4F_4P(C_6H_5)_2As(CH_3)_2)Fe_2(CO)_3(As(C_6H_5)_3)_3$	Fe: Org.Verb.C2-135/7, 140
$As_4C_{75}H_{60}S_9$	$[(C_6H_5)_4As]_3[As(CS_3)_3]$	C: MVol.D4-223
$As_4C_{84}Cl_2F_2H_{70}Ir_2S_2$	$[Ir(SC_6H_4F)HCl(As(C_6H_5)_3)_2]_2$	Ir: SVol.2-163
$As_4C_{84}Cl_2H_{72}Ir_2S_2$	$[Ir(SC_6H_5)HCl(As(C_6H_5)_3)_2]_2$	Ir: SVol.2-163
$As_4C_{84}H_{68}Ir_2N_2O_4S_2$	$[Ir(SC_6H_4NO_2)(As(C_6H_5)_3)_2]_2$	Ir: SVol.2-164
$As_4C_{86}Cl_2H_{76}Ir_2O_2S_2$	$[Ir(SC_6H_4OCH_3)HCl(As(C_6H_5)_3)_2]_2$	Ir: SVol.2-163
$As_4C_{92}Cl_2H_{82}Ir_2S_2$	$[Ir(SC_6H_4CH_3)HCl(As(C_6H_5)_3)_2]_2 \cdot C_6H_6$	Ir: SVol.2-163
$As_4C_{110}Cl_6H_{102}O_4Sn_3$	$3\ (CH_3)_2SnCl_2 \cdot 4\ (C_6H_5)_3AsCHC(O)C_6H_5$	Sn: Org.Verb.6-37
$As_4N_{14}O_{19}Sn$	$Na_2SnO_3 \cdot 4\ Na_3AsO_4 \cdot 48\ H_2O$	Sn: MVol.C3-36
$As_4O_{12}Th$	$Th(AsO_3)_4$	Th: SVol.C2-31
–	$ThO_2 \cdot 2\ As_2O_5 \cdot n\ H_2O$	Th: SVol.C2-32
$As_4O_{16}Th_3$	$Th_3(AsO_4)_4$	Th: SVol.C2-31/2
–	$3\ ThO_2 \cdot 2\ As_2O_5 \cdot n\ H_2O$	Th: SVol.C2-32
$As_4S_{36}Sn_3$	$Sn_3As_4S_{36}$	Sn: MVol.C2-281
$As_5C_5F_{15}$	$(CF_3As)_5$	F: PerFHalOrg.3-187
$As_5Na_{17}O_{23}Sn$	$Na_2SnO_3 \cdot 5\ Na_3AsO_4 \cdot 60\ H_2O$	Sn: MVol.C3-36
$As_6Br_4C_{18}H_{54}O_6U$	$UBr_4 \cdot 6\ (CH_3)_3AsO$	U: SVol.E1-194/5, 199
$As_6Br_4C_{36}H_{90}O_6U$	$UBr_4 \cdot 6\ (C_2H_5)_3AsO$	U: SVol.E1-194/5
$As_6Br_4C_{54}H_{78}Ir_2$	$Ir_2(As(CH_3)_2(p\text{-}CH_3C_6H_4))_6Br_4$	Ir: SVol.2-255
$As_6Br_4C_{78}H_{78}Ir_2$	$Ir_2(AsCH_3(C_6H_5)_2)_6Br_4$	Ir: SVol.2-255
$As_6C_{18}Cl_4H_{54}O_6U$	$UCl_4 \cdot 6\ (CH_3)_3AsO$	U: SVol.E1-194/5, 199
$As_6C_{18}H_{54}I_4O_6U$	$UI_4 \cdot 6\ (CH_3)_3AsO$	U: SVol.E1-194/5, 199
$As_6C_{36}H_{90}I_4O_6U$	$UI_4 \cdot 6\ (C_2H_5)_3AsO$	U: SVol.E1-194/5
$As_6C_{38}Cu_2FeH_{46}O_4$	$(CO)_4FeCu_2(CH_3As(C_6H_4As(CH_3)_2)_2)_2$	Fe: Org.Verb.B2-167
$As_6C_{54}Cl_4H_{78}Ir_2$	$Ir_2(As(CH_3)_2(p\text{-}CH_3C_6H_4))_6Cl_4$	Ir: SVol.2-255
$As_6C_{54}H_{78}I_4Ir_2$	$Ir_2(As(CH_3)_2(p\text{-}CH_3C_6H_4))_6I_4$	Ir: SVol.2-255
$As_6C_{60}Fe_6H_{54}$	$(C_5H_5FeC_5H_4As)_6$	Fe: Org.Verb.A6-292
$As_6C_{78}Cl_4H_{78}Ir_2$	$Ir_2(AsCH_3(C_6H_5)_2)_6Cl_4$	Ir: SVol.2-255
$As_6C_{78}Cl_6H_{66}Ir_2$	$Ir_2(((C_6H_5)_2As)_2C_2H_2)_3Cl_6$	Ir: SVol.2-258
$As_6C_{78}Cl_6H_{72}Ir_2$	$Ir_2(((C_6H_5)_2As)_2C_2H_4)_3Cl_6$	Ir: SVol.2-258
$As_6C_{78}Cl_6H_{78}Ir_2$	$[Ir(AsCH_3(C_6H_5)_2)_4Cl_2][Ir(AsCH_3(C_6H_5)_2)_2Cl_4]$	Ir: SVol.2-251
$As_6C_{78}H_{78}I_4Ir_2$	$Ir_2(AsCH_3(C_6H_5)_2)_6I_4$	Ir: SVol.2-255
$As_6O_{23}Th_4$	$Th_4As_6O_{23}$	Th: SVol.C2-31/2
$As_8CdIn_2Sn_3Zn_2$	$CdZn_2In_2Sn_3As_8$	Sn: MVol.C4-132
$As_8CdIn_4Sn_2Zn$	$CdZnIn_4Sn_2As_8$	Sn: MVol.C4-132
$As_8Cd_2In_2Sn_3Zn$	$Cd_2ZnIn_2Sn_3As_8$	Sn: MVol.C4-132
$As_8I_2MnO_{12}$	$MnI_2 \cdot 4\ As_2O_3 \cdot 12\ H_2O$	Mn: MVol.C5-321
$As_9Bi_3C_{72}Cl_4H_{90}Ni_2O_{16}$	$Ni_2[Bi(C_6H_4As(CH_3)_2)_3]_3(ClO_4)_4$	Bi: Org.Verb.-153
As_xO_3W	As_xWO_3 (tungsten oxide bronzes)	W: SVol.B3-39/41

Au . . . Gold systems
Au-Ir . . . Ir: SVol.1-129
Au-Pu . . . Np: TrU.B2-19, 22
Np: TrU.B3-105/11

$AuBBrC_6Cl_4H_{19}N_2$ . . . $[((CH_3)_3N)_2BHBr][AuCl_4]$ . . . B: B-Verb.10-106
$AuBC_6Cl_4H_{20}N_2$ . . . $[((CH_3)_3N)_2BH_2][AuCl_4]$ . . . B: B-Verb.10-55, 65
$AuBC_6Cl_4H_{20}P_2$ . . . $[((CH_3)_3P)_2BH_2][AuCl_4]$ . . . B: B-Verb.10-97, 99, 103
$AuBC_6Cl_5H_{19}N_2$ . . . $[((CH_3)_3N)_2BHCl][AuCl_4]$ . . . B: B-Verb.10-106
$AuBC_8Cl_4H_{22}N_2$ . . . $[CH_3C_3H_5(N(CH_3)_2)_2BH_2][AuCl_4]$ . . . B: B-Verb.10-91
$AuBC_{10}Cl_4H_{14}O_4$ . . . $[(CH(COCH_3)_2)_2B][AuCl_4]$ . . . B: B-Verb.10-144, 159
$AuBC_{30}H_{25}P$ . . . $(C_6H_5)_3PAuB(C_6H_5)_2$ . . . B: B-Verb.3-178
AuB_3H_{12} . . . $Au[BH_4]_3$ . . . B: B-Verb.8-35/6
$AuB_9C_2H_{11}^{2-}$ . . . $[(C_2B_9H_{11})Au]^{2-}$ . . . B: B-Verb.6-62
$AuB_9C_{54}H_{59}P_3$ . . . $[(C_6H_5)_3P]_3AuB_9H_{14}$ . . . B: B-Verb.3-196
$AuB_{10}C_{84}H_{99}P_4$ . . . $[(CH_3C_6H_4)_3P]_4Au[B_{10}H_{15}]$ . . . B: B-Verb.20-166
$AuB_{11}C_{36}H_{44}P_2$ . . . $[(C_6H_5)_3P]_2AuB_{11}H_{14}$ . . . B: B-Verb.3-201
$AuB_{18}C_4H_{22}^-$ . . . $[(C_2B_9H_{11})_2Au]^-$ . . . B: B-Verb.6-62
$AuBrC_{10}Cl_4FeH_9$ . . . $[C_5H_5FeC_5H_4Br][AuCl_4]$ . . . Fe: Org.Verb.A1-380
$AuCCl_4N$ . . . $AuCl_3 \cdot ClCN$ . . . C: MVol.D3-215
$AuC_2H_8N_4S_2^+$ . . . $[Au((NH_2)_2CS)_2]^+$ . . . C: MVol.D6-108
$AuC_3H_5OS_2$ . . . $Au[S_2COC_2H_5]$ . . . C: MVol.D4-257
$AuC_4H_7OS_2$ . . . $Au[S_2COC_3H_7]$ . . . C: MVol.D4-257
$AuC_5H_{11}OS_2$ . . . $(CH_3)_2Au[S_2COC_2H_5]$ . . . C: MVol.D4-257
$AuC_5H_{11}O_2S$ . . . $(CH_3)_2Au[SOCOC_2H_5]$ . . . C: MVol.D4-237
$AuC_6FeH_9NO_4P$ . . . $(CO)_3Fe(NO)AuP(CH_3)_3$ . . . Fe: Org.Verb.B1-196/7
$AuC_6FeH_9NO_7P$ . . . $(CO)_3Fe(NO)AuP(OCH_3)_3$ . . . Fe: Org.Verb.B1-196/7
$AuC_8H_{18}OPS_2$ . . . $[(C_2H_5)_3P]Au[S_2COCH_3]$ . . . C: MVol.D4-257
$AuC_{10}Cl_4FeH_9I$ . . . $[C_5H_5FeC_5H_4I][AuCl_4]$ . . . Fe: Org.Verb.A1-380
$AuC_{10}Cl_4FeH_{10}$ . . . $[Fe(C_5H_5)_2]AuCl_4$ . . . Fe: Org.Verb.A1-226
$AuC_{10}Cl_5FeH_9$ . . . $[C_5H_5FeC_5H_4Cl][AuCl_4]$ . . . Fe: Org.Verb.A1-380
$AuC_{10}H_{22}OPS_2$ . . . $[(C_2H_5)_3P]Au[S_2COC_3H_7]$ . . . C: MVol.D4-257
$AuC_{11}H_{24}OPS_2$ . . . $[(C_2H_5)_3P]Au[S_2COC_4H_9]$ . . . C: MVol.D4-257
$AuC_{12}Cl_4FeH_{12}O$ . . . $[C_5H_5FeC_5H_4COCH_3]AuCl_4$ . . . Fe: Org.Verb.A2-301
$AuC_{12}Cl_4FeH_{12}O_2$ . . . $[C_5H_5FeC_5H_4COOCH_3]AuCl_4$ . . . Fe: Org.Verb.A3-111
$AuC_{12}Cl_4FeH_{14}$ . . . $[C_5H_5FeC_5H_4C_2H_5][AuCl_4]$ . . . Fe: Org.Verb.A1-278
$AuC_{12}H_{26}OPS_2$ . . . $[(C_2H_5)_3P]Au[S_2COC_5H_{11})$ . . . C: MVol.D4-257
$AuC_{16}Cl_4FeH_{14}$ . . . $[C_5H_5FeC_5H_4C_6H_5][AuCl_4]$ . . . Fe: Org.Verb.A1-332
$AuC_{17}Cl_4FeH_{14}O$ . . . $[C_5H_5FeC_5H_4COC_6H_5]AuCl_4$ . . . Fe: Org.Verb.A2-301
$AuC_{18}F_{12}H_{15}IrP_5$ . . . $Ir(Au(P(C_6H_5)_3))(PF_3)_4$ . . . Ir: SVol.2-194
$AuC_{18}F_{15}$ . . . $Au(C_6F_5)_3$ . . . F: PerFHalOrg.4-111
$AuC_{20}H_{18}OPS_2$ . . . $[(C_6H_5)_3P]Au[S_2COCH_3]$ . . . C: MVol.D4-257
$AuC_{21}Cl_3FeH_{12}NO_4P$. $(CO)_3Fe(NO)AuP(C_6H_4Cl)_3$ . . . Fe: Org.Verb.B1-196/7
$AuC_{21}FeH_{15}NO_4P$ . . . $(CO)_3Fe(NO)AuP(C_6H_5)_3$ . . . Fe: Org.Verb.B1-196/7
$AuC_{21}FeH_{21}NO_4P$ . . . $(CO)_3Fe(NO)AuP(C_6H_5)_2C_6H_{11}$ . . . Fe: Org.Verb.B1-196/7
$AuC_{21}FeH_{27}NO_4P$ . . . $(CO)_3Fe(NO)AuP(C_6H_{11})_2C_6H_5$ . . . Fe: Org.Verb.B1-196/7
$AuC_{21}H_{20}OPS_2$ . . . $[(C_6H_5)_3P]Au[S_2COC_2H_5]$ . . . C: MVol.D4-257
$AuC_{22}H_{22}OPS_2$ . . . $[(C_6H_5)_3P]Au[S_2COC_3H_7]$ . . . C: MVol.D4-257
$AuC_{24}F_5H_{15}P$ . . . $(C_6H_5)_3PAuC_6F_5$ . . . F: PerFHalOrg.4-111
$AuC_{24}FeH_{21}NO_4P$ . . . $(CO)_3Fe(NO)AuP(C_6H_4CH_3)_3$ . . . Fe: Org.Verb.B1-196/7

Formula	Compound	Ref.
$AuC_{30}ClF_{10}H_{15}P$	$(C_6F_5)_2AuClP(C_6H_5)_3$	F: PerFHalOrg.4-136
$AuC_{30}FeH_{26}OP$	$C_5H_5FeC_5H_4COCH_2AuP(C_6H_5)_3$	Fe: Org.Verb.A2-228
$AuC_{38}ClFe_2H_{33}P$	$(C_5H_5FeC_5H_4)_2Au(Cl)P(C_6H_5)_3$	Fe: Org.Verb.A6-215
$AuC_{38}FeH_{30}NO_3P_2$	$(CO)_2Fe(AuP(C_6H_5)_3)(NO)P(C_6H_5)_3$	Fe: Org.Verb.B1-133/5
$AuC_{38}FeH_{30}NO_6P_2$	$(CO)_2Fe(AuP(C_6H_5)_3)(NO)P(OC_6H_5)_3$	Fe: Org.Verb.B1-133/5
$AuC_{39}Cl_5FeH_{30}O_9P_2$	$[(CO)_3Fe(P(OC_6H_5)_3)_2Cl]AuCl_4$	Fe: Org.Verb.B1-184/5
$AuMn_3N$	Mn_3AuN	Mn: MVol.C3-253/4
$AuPu$	$PuAu$	Np: TrU.B2-40
		Np: TrU.B3-105/10
$Au_2BC_{46}F_4FeH_{39}P_2$	$[C_5H_5FeC_5H_4(AuP(C_6H_5)_3)_2]BF_4$	Fe: Org.Verb.A1-91
		Fe: Org.Verb.A6-206
$Au_2C_{40}FeH_{28}O_4P_2$	$(CO)_4FeAu_2(C_6H_5)_2PC_{12}H_8P(C_6H_5)_2$	Fe: Org.Verb.B2-167/8
$Au_2C_{40}FeH_{30}O_4P_2$	$(CO)_4Fe(AuP(C_6H_5)_3)_2$	Fe: Org.Verb.B2-167
$Au_2C_{46}FeH_{39}P_2$	$C_5H_5FeC_5H_4Au_2(P(C_6H_5)_3)_2$	Fe: Org.Verb.A1-219
Au_2Pu	$PuAu_2$	Np: TrU.B2-40
		Np: TrU.B3-107/10
Au_3Pu	$PuAu_3$	Np: TrU.B2-40
		Np: TrU.B3-107/10
$Au_4Br_3C_{30}Cl_{12}Fe_3H_{27}$	$3\,[(C_5H_5FeC_5H_4Br)(AuCl_4)] \cdot Au$	Fe: Org.Verb.A1-380
$Au_4C_{30}Cl_{12}Fe_3H_{27}I_3$	$3\,[(C_5H_5FeC_5H_4I)(AuCl_4)] \cdot Au$	Fe: Org.Verb.A1-380
$Au_4C_{30}Cl_{12}Fe_3H_{30}$	$3\,[(Fe(C_5H_5)_2)AuCl_4] \cdot Au$	Fe: Org.Verb.A1-226
$Au_4C_{30}Cl_{15}Fe_3H_{27}$	$3\,[(C_5H_5FeC_5H_4Cl)(AuCl_4)] \cdot Au$	Fe: Org.Verb.A1-380
$Au_4C_{36}Cl_{12}Fe_3H_{36}O_3$	$3\,[C_5H_5FeC_5H_4COCH_3]AuCl_4 \cdot Au$	Fe: Org.Verb.A2-301
$Au_4C_{36}Cl_{12}Fe_3H_{36}O_6$	$3\,[C_5H_5FeC_5H_4COOCH_3]AuCl_4 \cdot Au$	Fe: Org.Verb.A3-111
$Au_4C_{36}Cl_{12}Fe_3H_{42}$	$3\,[(C_5H_5FeC_5H_4C_2H_5)(AuCl_4)] \cdot Au$	Fe: Org.Verb.A1-278
$Au_4C_{48}Cl_{12}Fe_3H_{42}$	$3\,[(C_5H_5FeC_5H_4C_6H_5)(AuCl_4)] \cdot Au$	Fe: Org.Verb.A1-332
$Au_4C_{51}Cl_{12}Fe_3H_{42}O_3$	$3\,[C_5H_5FeC_5H_4COC_6H_5]AuCl_4 \cdot Au$	Fe: Org.Verb.A2-301
Au_4Pu	$PuAu_4$	Np: TrU.B2-40
		Np: TrU.B3-107/10
Au_5Pu	$PuAu_5$	Np: TrU.B2-40
		Np: TrU.B3-107/10
Au_5Pu_4	Pu_4Au_5	Np: TrU.B3-109/10
$Au_{21}Pu_9$	Pu_9Au_{21}	Np: TrU.B3-109/10